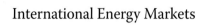
International Energy Markets

INTERNATIONAL ENERGY MARKETS

UNDERSTANDING PRICING, POLICIES, AND PROFITS

2ND EDITION

CAROL A. DAHL

Copyright© 2015 by
PennWell Corporation
1421 South Sheridan Road
Tulsa, Oklahoma 74112-6600 USA

800.752.9764
+1.918.831.9421
sales@pennwell.com
www.pennwellbooks.com
www.pennwell.com

Marketing Manager: Sarah De Vos
National Account Executive: Barbara McGee

Director: Mary McGee
Managing Editor: Stephen Hill
Production Manager: Sheila Brock
Production Editor: Tony Quinn
Book Designer: Susan E. Ormston
Cover Designer: Elizabeth Wollmershauser

Library of Congress Cataloging-in-Publication Data

Dahl, Carol A. (Carol Ann), 1947-
  International energy markets : understanding pricing, policies, and profits / Carol A. Dahl. -- 2nd edition.
      pages cm
  Includes bibliographical references and index.
  ISBN 978-1-59370-291-5
  1. Energy industries. 2. International economic relations. I. Title.
  HD9502.A2D335 2014
  333.79--dc23
                                                                        2014029321

Printed in the United States of America

3 4 5   19 18

With love to Jim for his patience, forbearance,
and unfailing love and support.

Contents

Chapter 4

Chapter 5

Chapter 6

Chapter 7

Chapter 15

Chapter 16

Chapter 17

Chapter 18

Chapter 19

Chapter 20

Chapter 21

Note: An online glossary for this text can be found at http://dahl.mines.edu/glossary.pdf.

Acknowledgments

The technical, political, and economic challenges of global energy industries and markets are fascinating! My work has given me the privilege and joy of traveling all over the globe, while observing, applying economic reasoning, and teaching about these industries. This book has evolved out of these experiences and a desire to share the fascination that these industries offer. It has benefited immeasurably from the many gifted students, coauthors, and colleagues who have provided me with intellectual challenge through their questions, suggestions, answers, and comments on my thoughts and work over the years. I have learned much from them, and for this I am grateful. Although too numerous to mention fully, I particularly want to remember and thank the following professional colleagues, mentors, students, and assistants.

In Memory of Professional Colleagues and Mentors No Longer with Us

Professor Emeritus Morris Adelman (1917–2014), Massachusetts Institute of Technology

Professor Emeritus Richard Gordon (1934–2014), Penn State University

Professor Lennart Hjalmarsson (1944–2012), Gothenberg University

Professor Emeritus Alan Manne (1925–2005), Stanford University

Professor Robert McCrae (1948–2006) University of Calgary

Professor Dennis O'Brien (1936–2005), University of Oklahoma

Professor Lee Schipper (1947–2011), Stanford University/University of California, Berkeley

Professor Campbell Watkins (1939–2005), University of Aberdeen

Professional Colleagues and Mentors

Associate Professor Edward Balistreri, Colorado School of Mines

Professor Jean Thomas Bernard, University of Ottawa

Professor Roy Boyd, University of Ohio, Athens

Professor Janie Chermak, University of New Mexico

Professor Graham Davis, Colorado School of Mines

Dr. Dorothea El Mallakh, International Center for Energy and Economic Development

James Jensen, Jensen and Associates

Professor Fred Joutz, George Washington University

Professor G. S. Laumas, Southern Illinois State University, Carbondale

Professor Ronald Ripple, University of Tulsa

Professor Christopher Sims, Princeton University
Professor Helen Tauchen, University of North Carolina
Professor John Tilton, Colorado School of Mines
Professor Zhen Wang, Chinese University of Petroleum, Beijing
Professor Myrna Wooders, Vanderbilt University
Professor Adonis Yatchew, University of Toronto
Dr. Mine Yücel, Federal Reserve Bank of Dallas
Associate Professor Lei Zhang, Chinese University of Mining and Technology

Students

Dana M. Abdulbaqi
Matt Adkins
Muhammad Akimaya*
Basil Al-Ajmi
Abdulaziz Al-Aqeel
Nader Al-Arfaj
Abdulrahman Albelushi
Ali A. Albinali
Fernando Albornoz
Javier Alcaraz
Nasser Al-Dossary
Alex Alexanian
Mansoor Al-Harthy
Eyas Al-Homouz
Bodour Al-Humaid*
Faisal Al-Jalahmah
Jared Allen
Patrick Allen
Abdulmalik Al-Mushary
Ayed Al-Qahtani
Nayef Al-Sadhan
Mohamed Al-Shami
Serhat Altun
Ramon Alvarado
Patricio Amuchastegui
Claire Anderson
ZiKai Ang
Benjamin Anglen

Rezki Anindhito
Yahya Anouti*
Luis Aragon-Castro
Anibal Araya Dominguez
Michael Asheim
Wisam J Assiri
Zauresh Atakhanova*
Robin Auer
Ayawo Awanyo*
Hamit Aydin
Felipe Azocar*
Rafael Bacigalupo
Yang Bai*
Gregory S. Baker
Paul Bakke
Babak Banaei
Olivier Bardet
Emmanuel Barrois
Matt Bates
Marshall Bates
Celia Bauluz
Guo Beiyao
Bayee Besong
Bradley Blair
Nathan Bock
Zachery S. Bogle
Alfred Bograh
Thitisak Boonpramote
Gregory Bosunga-Sileki

Mehdi Bouguetaia
Ghislain Boukoubi
John Boxall
Diego Bravo
Steve Brochu
Laura Burke
Jeffrey Campbell
Jili Cao
Irene Castillo
Daniel Celta
Juan Pablo Centeno
Diego Chaves Mesa
Joseph G. Chavez
Lingling Chen
Jean Christophe Cheylus
Nikko Collida
John S. Collins
Cory T. Cook
Leila Dagher
Lauren Davis
Guillaume de Bonnieres
Jason Deardorff
William John Decooman
Kevin DeGeorge
Francisco J. Del Rio
Mitchell Derubis
Sarmad Diab
Elizabeth Dinan
Nicholas Dodge

Clayton Doke
David Domagala
Matthew Doyle
Remi Duchateau
Thomas Duggan
James Easton*
Carlos Echazu
Nicholas B. Edgar
Babatunde Egunjobi
Meftun Erdogan*
Britton Escoe
Carl Evans
Tsepho Falatsa
Yiwei Fan
Pan Feng
Matthew Ford
Cory Forgrave
Danielle Fox
Gerardo Franco
Benjamin Freestone
Alfonso Garcia
Leonardo Garcia
  Manriquez
Anant Kumar Garg
Chad S. Garrett
Rachel Gelman
Benoit Gervais
Andrew Gilmore
Marie Ginies*
Martha Jeanne Gitt
Joseph Godell
Judith Gomez
Ignacio Gonzalez
  Gonzalez
William A. Good
Greta Goto
Abram D. Grae
Marc A. Guerra
Amar Gujral

Amar M. Gujral*
Ernesto Guzman
Sarah Riyadh Hajm*
Detlef Hallerman
David Hammond
Dalila Hanifi
Zachary Havens*
Adam Hayes
Jonathan Hebert
John Tyler Hodge*
Jacob Hohn
Peter Howie
Thomas Hunt
Jihun Hyun
Felipe Illanes Treswalt
Daniel Ingelido
Chad Isaacs
Jeremiah J. Jantzen
Pirat Jaroonpattanapo
Javier Jativa
Nils Jenson
Daniel Jerrett
Hankyung Jhung
Daniel Johnson
Christopher Johnson
Nicolas Jonard
Bo Jonsson
Eric Juan
Sirichai Kaewseekhao
Mohamed Sami Kamel
Dmitriy Kamenetskiy
Miharu Kanai
Shin Won Kang
Durga Kar
John M. Kelly
Mohammad Kemal
Catherine Kesge
Sattor Khakimov
Hyotae Kim

Derek Klingeman
James Knight-Dominick
Jinwen Ko
Hassan Kouao-Bile
Erik Kuhn
Kurtubi Kurtubi
Ricardo Labo
William Lambert
Rong Lao
Adrian LaraGarcia
Quinn Larwood
Changkeun Lee
Shan Lei
Ella Lein
Mathieu LeRenard
Changwei Li
Guijun Li
Klemens Lichtenberger
Fengliang Liu
Brice Lodugnon
Hermann Logsend
Alejandro Lombardia
Ivan London
Brice Ludognon
Pongkit Luksamepicheat
Marc Maestracci
Anthony Makowski
ShaneF. Mathers
Thomas Matson
Joe Mazumdar
John I. Mbibi
Richard McAllister
Maureen McDaniel
Lisa McDonald
Jeannette McGill
Nicole McMurrer
Samantha Mead
Remco Meeuwis
Augusto Mendonça

Alban Meteyer
Ariel Miara
Jeremy Miller
Seong Ki Min
Sofia Pilar Morales*
Gail Vallance Mosey
Damdinjav Munkbold
Shinsuke Murakami
Toru Muta
Precious Myeni
Ian Myslivec
Lauren Nagy
Balazs Nagy
Lawrence Nemetz
John Nowoslawski
Grant Nulle
Benjamin Nyarko
Steve Nzegaing
Patrick O'Brien
Kelsey O'Connor
Clifton Oertli
Kehinde Ogunsekam
Takashi Ohoka
Yris Olaya
Javier OliveroAlvares
Onikepo Omotade
Reginald Onyirioha
Donal O'Sullivan
Cheikh Ould El Moctar
  Salem
Sarp Ozkan*
Ahmet Ozturk
Anton Padin Deben
Nicholas J. Paduano
Berengere Papin
Fanyu Pei
Vianney Pellegrin
William Penfold
Nanfa Pennap

Jan Pfeifer
Steven Piper
Benjamin Pope
Gian Porro
Kaibin Qiu
Pri Rakhmanto
Jennifer Rano
Irfan Rau
Alban Reboul-Salze
Gabriel Regus
John Reinsma
Michael Rice
David Richardson
Robert Rindlaub
Marc Rock
Patrick Rogers
Carlos Roman
Joseph Romani*
Philippe Ronvaux
Marcela Rosas*
Lindsay Rothfelder
Theodore Royer
Sara Russell
Luciano Feirreira Sa
Dedi Sadagori
Francisco Sanchez
Maria Sanchez
Jeff Sanders
Martha Sandia
Rafael Santamaria
  Bolivar
Wisnu Santoso
Stephanie Savage
William Sconce
Stephen Sednek
Eric Segal
Ayoub Semaan
Aggei Semonov*
Naphorn Sermsakulwat

Omar Sharaf
John Sisa
Timothy Smith
Braeton J. Smith
Jacob N. Smith
Tayo Soyemi
Ronald Stanis
Kaleigh Starr
Kenneth Steirer
Laura Stevenson
Didier Strebelle*
Baruno Subroto
Donald Sult
Yohan Sumaiku*
Agus Supriano
Harianto Tarigan
James Clay Terry
Shiju Thomas
Andrew Tillotson
Tanner Trimble
Robert E. Tucker
Daniel Vagasky
Claudio Valencia
Nicholas L. Van Gundy
Matthew Vargo
Arturo L. Vasquez
  Cordano*
Jennifer Velasco
Erick Vila
Laura Vimmerstedt
Christopher R. Waldeck
Brant Wiedel
Piotr Wilczek
Michael S. Wilson
Brent Wilson
Rachel Wimbish
Christopher Worley
Robert Wright
Christopher Wyatt

Ross Wyeno
Yiping Xia
Jing Xie
Xhao Xu*
Yuji Yamamoto
Peifang Yang
Woo-Jin Yoon
Luky Yusgiantoro*
Nahl H. Zahran
Abdel Moughit Zellou*
Carlos Florencio Zerpa
Xiaolin Zhang
Xiaosong Zhang
Rong Zhao
Guo Zhao

Professor Carol Dahl, author, lecturer,
adjunct professor, professor emeritus,
Mineral and Energy Economics Program,
Colorado School of Mines, Golden, CO 80401 USA

* Indicates the student has also served as my research assistant.

Figures

Tables

1

Introduction to Our Journey

Energy economists want to get the price right. Politicians can't define obscene energy prices but know them when they see them. Energy traders believe that everything has a price and they know it, but if you outlaw price, only outlaws will know it.

—Modified from Unknown Author

Whether you are an energy economist, a politician, an energy trader, or an energy consumer, energy and its price will be of interest to you. Energy in all its forms can help you live an easier and more comfortable life. In the 1950s, it was touted that nuclear power would introduce an era when energy would be a nonscarce resource and we would have power too cheap to meter. Thus, we would be in a bear market with perpetually decreasing prices. Unfortunately, this prediction has not yet come to pass. Useful energy is still scarce, since we need capital, labor, and technical know-how to convert abundant but heterogeneous energy resources into the forms we all use. Sometimes useful energy is scarcer than at other times, with prices rising in a bull market or falling in a bear market. These shifts happen as consumers and producers react to changes in the market, including income, expectations, depletion, costs, and technology. But whether the market is running with the bulls or hibernating with bears, we want to use our energy resources wisely.

Energy is just one of four basic factors of production, the others being nonenergy natural resources, capital, and labor. Nor is energy the largest building block by value. Labor generally claims that prize. For example, labor's share of the gross domestic product (GDP) is somewhat greater than 60% in the United States, while energy expenditures have not exceeded 10% of US GDP since 1985 (Mutikani 2012). Nevertheless, energy is just as crucial as the other factors. It is the ability to do work, and without work, none of the other factors could be made into the products that we enjoy every day.

My goal in this text is to consider how to use this precious factor wisely. However, the principles here apply to choices for all factors and facets of our lives where scarcity is present. Life is full of choices. Economics is about making good choices in the presence of scarcity. Since we still feel the pinch of scarcity in the energy realm, I will present the economic fundamentals, along with technical and institutional knowledge needed to implement sound economic, business, and government policy decisions relating to energy industries.

Some Scientific Principles

Although useful energy is scarce and, hence, is not free, it is hard to imagine truly running out of energy (E) any time soon, as it is all about us. Energy, as Einstein's famous equation ($E = mc^2$) points out, is strongly tied to another fundamental concept in the universe, mass (m). Historically, this interchangeability of mass and energy and other scientific principles has led to major technological energy breakthroughs in the generation of electricity, transistors, nuclear fission and fusion, microwaves, lasers, iPads, and more. Science and technology are likely our best hope for even more spectacular breakthroughs in the future, as we transition out of fossil fuels and into renewable, and perhaps even to an as-yet-undiscovered source of clean, abundant energy. In the process, we need the underlying science to know if the source is possible. We also need the engineering skills to produce and deliver it, the socio-cultural sensitivity to disseminate it to a global audience, and the economic savvy to do it all at a profit. We will touch on all these attributes, with a focus on the economic aspect in this text, as well as in related material that will be posted and updated at http://dahl.mines.edu.

Let's begin with some scientific principles related to energy. Scientists refer to four fundamental forces that govern all fundamental interactions between objects: gravity, electromagnetic, weak nuclear, and strong nuclear. These forces are responsible for all the familiar forms of energy we use. Conventional notions of force can be thought of as some sort of pressure on matter that can cause matter to move. Newton discovered the first of the four fundamental forces of physics—gravity. Although this is the weakest of the forces, it has the greatest reach and is also applicable to bodies at great distances from each other, such as galaxies.

The reach of a fundamental force is referred to as its *field*. The force exerted by gravity between two bodies is directly related to their masses and inversely related to the distance or space between them. Thus, the bigger the body, the more force it exerts. Bodies that are further apart exert less force.

So how does gravitation relate to energy? *Energy* can be thought of as the potential to do work, where *work* can be thought of as force acting over a distance. Thus, the force of gravity has the potential to cause water to flow from a higher to a lower elevation. If the force is exercised and water flows, it is called *kinetic energy*. If the water is constrained and not allowed to flow, the energy is stored and is called *potential kinetic energy*. However, the energy will become kinetic energy once the operator of the dam opens the sluice gates and allows the water to flow. Kinetic energy is measured in joules (J) according to the international system of units (SI) and the metric system. (Energy is also measured in calories, with 1 calorie equal to 4.1855 joules, or in nonmetric British thermal units [Btu], with 1 Btu = 1.055 thousand joules or 1 kilojoule [kJ].) Unfortunately, despite its many advantages, the metric system is not yet in popular use in the United States. Some advantages of the metric system include the following:

- The metric system is used worldwide, with the exception of the United States, Myanmar, and Burma.
- Units in multiples of tens make computations easier to learn, easier to use, and less error prone.
- Standardized prefixes make it easier to grasp related dimensions.
- Not having to convert across systems or to own redundant tools and equipment saves time and money for a seamless fit into the global economy.

Since the system was created by scientists to be logical, consistent, and easy to use, it is the worldwide standard in science. Metric equivalents and their abbreviations will be provided throughout. However, we still live in an imperfect world and have to navigate between the systems. Thus, while we are waiting and agitating for a more perfect metric world, popular US energy equivalents will be given as well.

The next unifying principle came from Maxwell, a couple of hundred years later. He discovered that electrostatics, magnetism, and light could all be explained under one unifying force and theory—the electromagnetic force and electromagnetism.

This second fundamental force is the attraction between oppositely charged particles and the repulsion between like-charged particles. Static charges create an electric field and are responsible for electricity. Moving charges create a magnetic field, and accelerating charges create electromagnetic radiation. This radiation, from the longest to the shortest wavelength, lowest to highest frequency, and lowest to higher energy, with some familiar uses, includes the following:

- *Radio waves.* Transmit signals for radio and television, and some radar uses.
- *Microwaves.* Cook food, cellular phone communication, radar, and monitor precipitation.
- *Infrared waves.* Medical imaging, remote controls, and night vision.
- *Visible light.* Normal vision, communication through fiber optic cables, and lasers.
- *Ultraviolet rays.* Sterilize bacteria and viruses, including those on clothes hung out to dry.
- *X-rays.* Medical imaging, tumor destruction, and security imaging.
- *Gamma rays.* Treatment of disease and checking pipeline welds.

In the period around 1900, Thomson, Rutherford, and others gave us a more consistent view of the atom, with a positive charge in the center and negatively charged electrons circling the nucleus. The electromagnetic force between positive and negative charges holds atoms together, and the residual force between electrons in one atom and protons in another holds molecules together. When molecules are formed or break apart, electromagnetic energy may be emitted or absorbed. Thus, electromagnetism is a unifying principle for all of chemistry. It is responsible for the heat and light when we burn fossil fuels.

Electromagnetic radiation travels at the speed of light (c) and can behave like a wave, with a crest and a trough. The *wavelength* is the distance from one crest through a trough to the next crest. The *frequency* (f) of the cycle is how many wavelengths the energy travels in a second and is called a *hertz*. The velocity (v) of the radiation in meters per second (m/s) equals the wavelength (λ) in meters times the frequency (v = λf). Electromagnetic radiation can also be thought of as photons—similar to little energy packets. In a photovoltaic cell, when light photons hit the semiconductors, this energy causes electrons to be emitted, forming a direct current.

The weak nuclear force was formulated by Fermi. It is another of the four fundamental forces that govern radioactivity. It allows neutrons in an atom to break into a proton, an electron (beta particle), and an antineutrino. It also allows larger alpha particles (two protons and two neutrons—the equivalent of a helium nucleus) to be emitted.

At this point, we still do not know what holds these positively charged protons together in the nucleus. The electromagnetic force suggests they should repel each other. The fourth fundamental force comes to the rescue, and it keeps everything from flying apart. It is the strongest force but has a very short range. It holds all the particles in the nucleus together. It too was hypothesized to exist in the 1930s, with the discovery that protons and neutrons make up the nucleus. When the strong force is broken by breaking apart elements heavier than iron, *fission energy* is liberated. When this force is exploited to fuse together elements lighter than iron, *fusion energy* is also liberated. However, fusing heavier elements than iron and breaking apart lighter elements than iron require a net input, rather than a release, of energy.

Although the current scientific knowledge is more complicated than the discussion above, this simplified discussion gives us some intuition about the basic energy forms we use. (For more information on newer ideas relating to the physics of energy, see http:\dahl.mines.edu\b0101.pdf.)

Energy is generated from the four fundamental forces, with commercial energy coming in six familiar forms:

1. *Mechanical energy* is associated with motion. Falling water resulting from gravity can turn a grinder, wind resulting from temperature differentials through electromagnetism can turn a wind turbine, and human and animal power can be used to move objects fueled by the chemical reaction of food.

2. *Chemical energy* is released when molecular bonds are broken or changed, as in the combustion of fossil fuels, such as coal, oil, and natural gas, or with biomatter, such as dung, wood, and crop residues. Such chemical energy from the electromagnetic force may be turned into mechanical energy, as in the internal combustion engine.

3. *Thermal energy* is a measure of the heat in the vibrations of molecules. It may result from friction. It may also be a product of the chemical energy of combustion. Geothermal energy, which is heat from within the earth, may be heat stored from the formation of the earth, supplemented with heating from pressure and radioactive decay (Cornell Center for Materials Research 1999).

4. *Radiant energy* is all forms of electromagnetic radiation. Solar energy is a critical source of radiant energy, with about 40% in the infrared and longer wavelength range, about 50% in the visible range, and about 10% in the ultraviolet or shorter wavelength range (University of California Museum of Paleontology, n.d.).

5. *Nuclear energy* from fusion and fission results from the strong nuclear force. It is changed to mechanical and other forms of energy in nuclear submarines, the explosions of nuclear weapons, and in nuclear power plants.

6. *Electrical energy* is the movement of electrons caused by the electromagnetic force. If the electrons travel one way through a wire, we have direct current. If the electrons continually reverse directions flowing back and forth, we have the more common alternating current.

In any system, we can transform energy from one form into another; for example, the mechanical energy of a stream can be turned into electricity by a hydro unit. The resulting electricity can be turned into heat and light in a home or can run a machine in a factory. With these changes, the first law of thermodynamics requires that the total amount of energy in an isolated system will always remain constant. Why then is energy scarcity a problem? The reason lies in the second law of thermodynamics, which requires that when energy is converted, it is reduced in quality and in its ability to do work. Thus, with each energy conversion, we have the same total amount of energy, but we have less available energy to do work. For example, the generation of electricity using a conventional thermal plant produces both heat and electricity. Although the heat generated may be used to warm oyster beds or might even provide district heat, it is often at too low a temperature or too far from a market to be otherwise usefully captured for work (Georgescu-Roegen 1979; Hinrichs 1996).

An understanding of the economical use of energy is interdisciplinary. Hence, in this book, we will combine knowledge of economics and mathematical analysis with institutional and technical information to better understand various energy markets. A discussion of the topics covered follows.

Outline of the Book

Since the advent of the big bang, theorized to have occurred some 13 billion years ago, energy has remained a fundamental component of the universe. Humans, who arrived only a few million years ago, have consumed only a small portion of the vast supply of energy on just one small planet. Part of the ascent of humans has been the process of learning how to use ever more of this supply of energy to help satisfy basic needs, along with space conditioning, transportation, and entertainment.

In chapter 2, we set the stage for the book by considering energy's geological past and the evolution of human energy use and technology. We also address methodologies for forecasting its use in the future and analyzing energy economy and environmental interactions.

In chapter 3, we consider our first market model, perfect competition. Markets consist of buyers and sellers getting together and exchanging goods or services. We can refer to a market for a particular good, such as coal, or a class of goods, such as energy. Economists often loosely refer to the market as the accumulation of all the consumers and producers buying and selling all goods and services.

Economists often favor competitive markets in a capitalist economy for allocating scarce resources. They feel that the discipline of the market helps to create efficiencies and minimize costs. The lure of profits helps attract capital away from shrinking markets to growing ones, spurs innovation, and promotes new products. With competition and decentralized decision making, capitalist economies are more flexible and personal freedom is enhanced.

Our discussion of competitive markets in a static framework is applied to the coal industry. Principles of demand and supply help us to understand how market prices are influenced and how energy industries evolve. Coal, once the linchpin of industrial economies, has been slowly surpassed, as markets have attracted resources toward oil and gas and away from coal. However, such trends can change, as we see with recent large increases in coal use in China. Demand and supply *elasticities*, which capture responsiveness to price and income, are developed and used to analyze such market changes. In turn, elasticities can also be used to recreate demand and supply curves.

Energy resources are often publicly owned and considered basic wealth to a society. As such, they are usually taxed, sometimes quite heavily. In chapter 4, we consider energy taxes in the context of a static model. Criteria for tax collection such as equity and fairness will be considered. Who pays, or the incidence of the tax, depends on how responsive demanders and suppliers are to market price. Measures of this price responsiveness (price elasticities), developed in chapter 3, will be used to show tax and subsidy incidence. Price controls, another way that governments interfere with markets, will also be considered.

Although economists often favor markets and private ownership for the allocation of goods and services, there are a number of cases where economists generally agree that markets fail and that room exists for the government to step in. One such case is a decreasing cost industry, in which the greater the production, the lower the unit costs. Such industries are considered natural monopolies.

For many years, the electricity industry's huge capital costs and economies of scale had been considered a natural monopoly. In such an industry, we prefer one producer on the grounds of greater efficiency, since the biggest producer has the lowest average cost. However, one private producer would be able to monopolize the industry and make monopoly profits.

In chapter 5, we consider the electricity industry, summarize the various technologies for generating electricity, and discuss how government ownership and price regulation have been used to try to control monopoly profits.

Problems with both government ownership and regulation, along with technical change in electricity generation, have led to deregulation, privatization, and restructuring of electricity generation in numerous markets, which is discussed in chapter 6. Classic deregulation examples in New Zealand, the United Kingdom, and Scandinavia will be considered, along with the horrific problems accompanying the restructuring of regulated markets in California.

If large producers have market power and are able to set prices, they can make monopoly profits. A classic example of this market failure is the Organization of Petroleum Exporting Countries (OPEC), which we consider in chapter 7. Some history of OPEC and models to explain OPEC's behavior are also given. Since OPEC cannot control non-OPEC production, it will be treated as a dominant firm, rather than a monopoly. Since OPEC is not a monolith but is comprised of 12 different countries, some of their differences will be noted as well.

With deregulation, the institutional arrangements or governance structures in markets are likely to evolve. Such structures include spot purchases, long-term contracts, or vertical integration. Transaction cost economics suggests that the market structure that survives is the one that minimizes transaction costs. Specificity of assets in the industry will influence market governance. For example, a pipeline is a very specific asset, transporting a specific good from one specific place to another, whereas a semi-truck is much less specific and can transport a variety of goods to and from a variety of places. Market governance is also influenced by the amount of uncertainty and the frequency of transactions, all of which influence transaction costs. In chapter 8, we consider transaction cost economics and apply it to changes in the US natural gas markets.

Market power for either buyers or sellers leads to an inefficient allocation of resources. If there is only one buyer in a market, we refer to this market structure as *monopsony*. One buyer is able to depress the buying price and reap monopsony profits. A multinational company with exclusive rights to buy energy resources in a small developing country with a weak government would be an example of market power on the part of the buyer. With the famous Red Line agreement in 1928, the multinational oil companies of the time carved up the Middle East and agreed not to compete with each other over resources, preserving their monopsony power. We consider the monopsony model in chapter 9 and apply it to Japan's purchases of liquefied natural gas (LNG) in the Asia-Pacific market.

A single multinational oil company dealing with a strong government in an energy-rich developing country would be an example of a *bilateral monopoly*, which is a *monopsonist* (one buyer in a market) buying from a *monopolist* (one seller in a market). In this case, the outcome is ambiguous and depends on the negotiation skills of the two players in the market. Chapter 9 concludes with pointers on negotiation.

A few buyers or a few sellers in a market constitute *oligopsony* and *oligopoly*, respectively. These models get more complicated, as their outcome depends on the strategies of all the players in the market. We consider these market structures in the context of game theory, with an application to the European natural gas market in chapter 10.

Energy production, transport, and consumption produce a variety of pollutants, which are summarized in chapter 11. Power plants have often polluted the air we breathe, and coal mine runoff has fouled our waters. Since private decision makers do not take into account these costs, which are external to them, private markets will not allocate energy efficiently. Policies that have been undertaken in response to externalities such as pollution are also presented and evaluated in chapter 11.

Another externality comes from public goods. A pure public good is one from which people cannot be excluded (*nonexcludability*), and where one person's consumption does not reduce another person's consumption (*nonrivalrous*). The classic example is a lighthouse. Anyone in the vicinity of a lighthouse can look at it, and one person looking at it does not generally restrict the ability of another to look at it. In making a private decision to produce such a good, individuals typically only take their own satisfaction or utility into account and too little of the good will likely be produced. Further, if one cannot be excluded from consumption, each consumer will want someone else to pay for the good—the *free rider* problem. Both effects cause a public good to be underprovided by the private market.

In poorer countries, a significant amount of biomass is consumed to provide energy. This consumption, along with the associated land clearing and timber harvest, is reducing the biodiversity on the planet, which might be considered a public good. In addition, the reduction in forest is reducing the capacity of flora to absorb CO_2. At the same time, the burning of fossil fuels, historically largely from industrial countries but increasingly from rapidly emerging markets, is increasing the amount of CO_2 in the atmosphere. It is generally agreed that this buildup will cause global climate change. The National Academies of Science for the G8 countries plus five other countries (Brazil, China, India, Mexico, and South Africa) in a joint statement have noted that "the need for urgent action to address climate change is now indisputable." They call for governments to set and implement a goal to reduce CO_2 emissions 50% below 1990 levels by 2050 (US National Academies 2009).

We do not precisely know the timing, extent, or the effects of this buildup, but expectations are that the climate will generally become warmer. With the melting of the polar caps, coastal areas will flood and weather patterns will change. Since many people enjoy the benefits of biological diversity and lower levels of CO_2, but the benefits are nonexcludable and nonrivalrous, they have the characteristics of public goods. In chapter 12, we consider an analysis of the provision of such public goods, as well as current policies toward global climate change.

Energy accidents cause death and destruction on a regular basis. Sovacool (2008) documents 279 major energy accidents from 1907 to 2007, which cost thousands and thousands of lives and billions of dollars of property damage. The more recent Fukushima nuclear accident in Japan and the Macondo oil spill in the Gulf of Mexico are also high-profile examples of what can go wrong. In chapter 13, we consider some of these more high-profile accidents.

Humans and technologies can always fail, especially in today's large, complex systems that find, produce, transform, transport, and distribute energy. Thus, such accidents, though unfortunate, to some extent are probably inevitable, and they typically produce negative externalities or damages to those nearby. With proper liability laws in place, those negative externalities can, in theory, be internalized through a country's legal system, and those who suffer the negative damages will be reimbursed. If managers can estimate the probability of an accident and the amount of damage they will have to pay, they can also spend resources on safety systems to reduce the likelihood of accidents.

However, economists think that liability laws and markets alone may not be able to get us to the optimal level of precautionary spending for a number of reasons. Managers may get the probabilities wrong and optimistically think that low probability events cannot happen, or may even be unable to imagine some events. Did anyone expect the tidal wave that led to the Fukushima power plant accident? Managers may lack information and not be familiar with safety technology. The size of damages paid after an accident by the company may also be truncated. If damages are large enough, the company may be unable or unwilling to pay and instead will go out of business.

The situation may also suffer from *principal agent* problems. Managers, as agents of shareholders, may share in profits when times are good, but may not suffer proportionate losses from an accident. The losses may be spread over shareholders and others in the company, or the manager may resign and go to greener, less-soiled pastures. For these reasons, governments may step in with regulations that require safety procedures and investments to reduce the likelihood of accidents. They may require funds to be put into escrow to cover damages in the event of an accident, or they may impose fines and even jail managers for safety violations. Such sanctions can be imposed before (*ex ante*) as well as after (*ex post*) an accident occurs. In chapter 13, we also consider such policies and how they should be designed to get the optimal amount of precautionary spending on safety procedures.

Chapters 2 through 13 contain static economic analysis of the allocation of energy resources. However, many energy sources, such as fossil fuels and uranium, are nonrenewable, depletable resources. For such fuels, if we produce the resource today, it will not be available for tomorrow, and dynamic analysis, in which we maximize net present value of all future production, is more appropriate. In chapter 14, we will look at a basic two-period model, with applications to oil production and leasing.

Dynamic analysis also has applications in allocating capital costs over time. In a very capital-intensive industry such as energy, it is important to be able to allocate such costs across units of production or consumption, even if the resource is renewable. Capital cost allocation procedures are developed in chapter 15. These are applied to the production of depletable resources with declining production, as well as capital investments where services do not decline until the equipment wears out, such as energy transport, renewable energy production, and services from household appliances.

Such costs are important inputs to the many market models considered in this text and have implications for energy supply. For example, Shell's blueprint scenario (2008) pegs renewables at 30% of total global energy consumption in 2050. However, which markets renewable sources penetrate and how fast they do so will be strongly influenced by not only their characteristics, but also their costs.

Economists believe that economic actors are rational and optimize given their preferences, and demanders are no exception. Demand functions representing consumer preferences are an important part of understanding energy markets and forecasting future consumption. Along with supply or costs, they are a basic building block of many energy models, both simple and complex. Chapter 16 offers a global look at energy demand by major sectors. We look behind demand curves at optimization decisions for energy use by consumers and producers and consider statistical problems encountered when estimating demand on real-world data.

The costs and demands discussed in chapter 15 and 16 can be used as components in other larger models. If such a model is composed of linear equations, it is usually easy to solve, even if the model is quite large. In linear programming, we maximize or minimize a linear objective function subject to linear constraints. We apply this technique to oil refining and energy transportation in chapter 17, as well as mention other more complicated nonlinear static and dynamic optimizing techniques.

In chapters 1 through 12 and 14 through 17, most of the analyses assume certainty. However, we face large uncertainties in most aspects of our lives, and with uncertainty comes risk, or the possibility of loss or gain. Energy, being no exception, is a risky business. Chapters 18 and 19 of the book deal with ways to manage financial risk. Government policies, the economy, and competition influence energy prices and costs. All three can provide unpleasant surprises, threatening not only profits, but, in some instances, a company's very survival. Should we want to hedge—for example, reduce the risk of price change—we have various choices, including organized futures markets, with standardized contracts where parties do not know who is on the other side of the trade. With futures markets, discussed in chapter 18, we can lock in future prices for energy products that we want to buy or sell in order to reduce and manage risk. Speculators who want to take on risk in hopes of a profit can also operate in futures markets. With futures contracts, a player locks in a price and has an obligation to buy or sell at this price, or what is more likely, to close out the position at the locked-in price before the contract comes due. Because delivery is rarely taken on futures contracts, such contracts on crude oil and products are sometimes referred to as *paper barrels*, and the market as a *paper refinery*.

Sometimes a player would rather provide a ceiling or a floor for the price of energy. A refinery might want to lock in a minimum price for its product and a maximum price for the crude oil it buys. To do so, it can buy or sell an option on a futures contract for these products. These standardized contracts, discussed in chapter 19, give the buyer the right, but not the obligation, to buy or sell a futures contract depending on whether a *call* or *put* option has been purchased. If it is not profitable, the option is allowed to expire. However, if the option is *in the money*, usually the buyer closes out an option for a cash settlement rather than taking delivery, as with futures contracts.

Modern industrial economies have grown rich and powerful with help from copious quantities of carbon fuels. As numerous emerging markets are now quickening their pace of development, they understandably seek to follow in the same path to prosperity. However, if these carbon paths are followed, we may find ruin rather than wealth at the end of the trail, for fossil fuels will not last forever. Nor does it now seem likely that the globe can absorb all the related carbon emissions and still sustain a global population that is more than 7 billion strong and growing. So how do we sustain this endless flow of energy services in the face of resource and absorptive capacity constraints?

Energy consumption is often considered to be an important input into growth and development. Both the quality and quantity of fuels can influence output and are indicators of economic well-being. For example, the *World Development Indicators* (World Bank, n.d.) includes consumption of commercial energy per capita, such as electricity, gasoline, and diesel fuel, as well as more traditional biocombustible fuels and waste.

Prior to the industrial revolution, households relied on renewable noncommercial energy to heat and light their homes. Wood, straw, and crop residue were prominent sources. However, as we have progressed and become richer, households have moved away from these less-efficient sources, transitioning to charcoal and kerosene, and eventually to cleaner and more convenient liquefied petroleum gas (LPG) and electricity. This movement to cleaner, more convenient, and more efficient fuels has been called a *movement up the energy ladder*. Worldwide, more than 1 billion people do not yet have electricity, and more than 2 billion still use traditional fuels to heat their homes and water and cook their foods. In chapter 20, we consider countries that get more than 50% of their fuel from traditional biomass, as well as issues they face, including poverty, income inequality, corruption, and problems of common ownership of bioresources. We also consider policy-induced moves to modern bioenergy in richer countries, as well as models to wisely manage commercial forests.

Energy not only supplies services to consumers, it provides revenues and incomes for producers. A number of fossil-rich countries, including Venezuela, Saudi Arabia, and Russia, receive large riches from these finite resources. Sustainability, from their point of view, is a continuance of income after the last economically viable fossil fuel has been extracted from their soil. In chapter 21, we consider issues of fossil-rich countries with more than 14% of their GDP from fossil fuels. Since such resource-rich countries often perform more poorly than non-resource-rich countries, some have considered fossil and other nonrenewable resources a curse. We will consider the hypothesized causes of the *resource curse*, as well as the performance of fossil-rich countries, including income per capita, health, inequality, corruption, and violence. Sustainability for them will require the investment of some of the resource rents into capital. This investment may be into international financial capital in the form of sovereign wealth funds or into other forms of capital including produced capital, human capital, technology, and institutional capital. Such investment is also briefly considered in the chapter.

Energy is a global business, with many large national, multinational, and transnational companies involved in its production and distribution. The models we consider in this book are powerful tools to help us better understand and manage our energy resources in such a global environment, and they are summed up in the concluding chapter. To succeed in this highly charged atmosphere also requires companies to understand the technologies, players, market structures, and policies that we discuss in this book. Companies should also understand the culture of both their employees and customers. It is important to develop a corporate culture that is compatible with both the corporate mission and vision statements, as well as with the national cultures with which the company does business. Thus, in addition to summing up what we have learned, we will also consider aspects of both national and corporate culture. Topics include how power is earned and distributed, people's view of themselves relative to others and to nature, and views on uncertainty and time.

2

Energy Lessons from the Past and Modeling the Future

Those who cannot remember the past are condemned to repeat it.

—George Santayana

Introduction

Energy markets continually evolve. How they evolve in the future will be influenced by many of the factors that we will discuss in this book, including energy resources, technology, population growth, demographics, climate change, costs, preferences, government policy, regulation, and risk. In this chapter, we will consider energy in the great historical panorama, which sets the stage for coming chapters. We will also consider models that help us forecast coming events, analyze policies, make business decisions, and simulate interactions between energy and other sectors.

Energy Geological History

Science suggests that the most cataclysmic energy event for the universe was at its beginning, with the big bang and subsequent inflation of the universe some 14 billion years ago (NASA 2013). These and other geological energy milestones are shown in table 2–1.

Although there is not total agreement or understanding of how the universe began, physicists generally believe it to have proceeded as follows. Before the big bang, time, space, matter, and energy did not exist. Then an anomaly caused negative gravity and positive energy to form from nothing. The net energy was zero, but the universe had zero size and infinite temperature. The negative gravity caused an expansion, and the high temperatures caused the formation of a small amount of matter in the form of subatomic particles from energy according to Einstein's $E = mc^2$. With the expansion, temperatures started to drop.

Table 2–1. Cosmological and geologic milestones in energy

Date	Event or Time Period	Comments
13 bya	Big bang	
5.5 bya	Sun formed	
4.6 bya	Earth formed	
4.5 to 0.544 bya	Precambrian	
4.5 to 3.8 bya	Hadean (Early)	Earth crust solidifies
3.8 to 2.5 bya	Archaeozoic (Middle)	First life forms release oxygen to atmosphere
2.5 to 0.544 bya	Proterozoic (Late)	First multicelled animals, one continent called Rodinia, oxygen buildup, mass extinction
544 mya to today	Phanerozoic	
544 to 245 mya	Paleozoic Era	Invertebrates, primitive amphibians
544 to 505 mya	Cambrian	Age of trilobites, explosion of life, all phyla develop, extinction of 50% of animals, continents begin to break up
505 to 440 mya	Ordovician	Primitive land plants and fish appear, North America covered by shallow seas, glaciation kills many species
440 to 410 mya	Silurian	First fish with jaws, insects, vascular land plants
410 to 360 mya	Devonian	Age of fishes, first amphibians, new insects, many extinctions
360 to 286 mya	Carboniferous	Huge forests and many ferns reduce carbon dioxide—global temperature cools, atmospheric moisture increases, first winged insects and reptiles
325 to 360 mya	Mississippian	
325 to 286 mya	Pennsylvanian	
286 to 245 mya	Permian	Age of amphibians, supercontinent Pangea, largest extinctions, earth's atmosphere approaches modern composition
245 to 65 mya	Mesozoic Era	Age of dinosaurs
245 to 208 mya	Triassic	First dinosaurs, true flies, and mammals, many reptiles, minor extinctions allow dinosaurs to flourish
208 to 146 mya	Jurassic	Many dinosaurs, first birds, first flowering plants, minor extinctions
146 to 65 mya	Cretaceous	Tectonic and volcanic activity high, first marsupials, butterflies, bees, and ants; many dinosaurs, continents as today, large extinction from comet collision
65 mya to today	Cenozoic Era	Age of mammals
65 to 1.8 mya	Tertiary	Modern plants and invertebrates
65 to 54 mya	Paleocene	First large mammals and primates
54 to 38 mya	Eocene	Lots of mammals, first rodents, and whales
38 to 23 mya	Oligocene	Many new mammals, grasses common
23 to 5 mya	Miocene	More mammals (horses, dogs, bears), modern birds, and monkeys
5 to 1.8 mya	Pliocene	First hominids, modern whales
1.8 mya to today	Quaternary	Age of humans
1.8 mya to 11,000 ya	Pleistocene	Appearance of humans, first mastodons, saber-toothed tigers, giant sloths, mass extinction at 10,000 years ago from glaciations

Date	Event or Time Period	Comments
11,000 ya to today	Holocene	Human civilization—domestication of plants and animals
1.8 mya to 4,000 BCE	Stone Age	Dates vary from region to region; hunters and gatherers use stones that are chipped and flaked to form tools for an increasing variety of uses—arrows, needles, axes, etc.; agriculture appears; pottery develops; humans harness fire to cook, keep warm, and scare off animals
4,000 BCE to 1,200 BCE	Bronze Age	Dates vary from region to region; bronze formed by heating tin and copper used for tools, ornaments, and weapons
1,200 BCE to 500 CE	Iron Age	Dates vary from region to region; more abundant iron replaces bronze for many applications

Sources: Scotese (2002); Zoom Dinosaurs (2010); GSA (2012).
Notes: mya, bya, ya = million years ago, billion years ago, and years ago, respectively.
BCE = before the current age or before year 1; CE = the current age or after year 1.

However, the cosmic inflation caused temperatures to shoot up again. The universe blew up and inflated by trillions of times its former size. As it got larger, it cooled down to almost absolute zero. The strong, weak, and electromagnetic forces separated, which pushed the temperature back up. During this time, electrons, quarks, and other basic particles constituting the basic mass of the universe formed. They combined into protons and neutrons. As the universe continued to expand and cool, these basic particles combined into primarily hydrogen, with some helium and small amounts of lithium.

Later, gases formed clouds that gravity and fusion turned into galaxies and stars. Fusion within stars created heavier elements up to iron, giving off energy. When a large star ran out of fuel, it exploded into a supernova, which created elements heavier than iron, including uranium. These heavier elements also formed into stars and planets. Around 5.5 billion years ago, our sun formed, which is still directly or indirectly the source of most of our usable supply of energy. Somewhat later, the earth formed, with a core of iron. After a million years or so, the crust solidified, although the interior still remains molten and is the source of our geothermal energy. After the earth's formation, water is thought to have accumulated from comets hitting the earth's surface and melting.

Life formed in the oceans during the Precambrian period more than half a billion years ago. Bacteria and blue-green algae used sunlight to photosynthesize carbon dioxide and water into glucose ($C_6H_{12}O_6$), oxygen (O_2), and water (H_2O) through the following chemical reaction:

$$6CO_2 + 12H_2O + sunlight \rightarrow C_6H_{12}O_6 + 6O_2 + 6H_2O$$

Glucose, in turn, changed into polysaccharides, such as starch and cellulose. These reactions released oxygen, paving the way for oxygen-using animals by the Proterozoic period.

Petroleum is known to have formed as early as the Precambrian, more than half a billion years ago. From the Precambrian up through the Devonian, marine organisms (mostly plants such as algae, phytoplankton, and bacteria) were deposited in the absence of oxygen, which prevented their decay. Such anaerobic conditions occurred if deposits were buried quite rapidly or if oxygen was absent for other reasons, such as deep water. Shallow marine areas had abundant plant life that got buried deeper and deeper. As sediment piled up, the material was subject to bacterial action, forming *kerogen*.

Heat and pressure eventually formed oil and gas from kerogen. Oil forms under pressure at temperatures of about 60°–120° Celsius (°C), or 140°–248° Fahrenheit (°F), while gas forms at temperatures of about 120°C–255°C (248°F–491°F). Such gas and oil may migrate in interconnected porous rock and accumulate in pools, when stopped by impermeable cap rock.

Supergiant oil fields are those with more than 5 billion barrels of oil initially, while *giant oil fields* have 1 to 5 billion barrels. However, these definitions can be a bit cloudy. Sometimes, the oil referred to is *oil in place*—a measure of how much oil is thought to be in the field; at other times, it is economically recoverable reserves, usually a much smaller number. Sandrea and Sandrea (2007) put a conservative measure of current average global oil recovery rate (eventually produced reserves divided by oil in place) at 22%. Better reservoir management and enhanced recovery techniques can clearly raise this average.

Although there are no universal definitions categorizing reserves, a common approach is to define *economically recoverable reserves* as those that have a 90% chance of being produced with current prices, technologies, and conditions. These reserves are variously called *proven*, *P90*, and *1P*. Similarly, reserves with a 50% or 10% chance are often designated as *probable* (also *P50*, *2P*) and *possible* (also *P10*, *3P*), respectively. For companies listed on public exchanges, these definitions and reporting requirements are sometimes prescribed by law. (For more discussion of reserve definition for various countries, see Society of Petroleum Engineers [SPE 2007] and Etherington, Pollen, and Zuccolo [2005].) For national oil companies, such transparency is usually not the case, and how they define *reserves* is fuzzier. Thus, the huge proven reserves additions for OPEC countries in the late 1980s, after production quotas were introduced in 1984, are more likely of political than geological origin.

The *Oil & Gas Journal* (*OGJ*) publishes proven reserve estimates for oil and gas by country in its annual December worldwide issue. They find that global proven reserves of oil, including condensate from natural gas wells, and reserves for natural gas for January 1, 2014, are 1,645 billion (10^9) barrels of oil and 6,886 trillion (10^{12}) cubic feet of natural gas. (One metric tonne of oil is about 7.5 barrels of oil and varies depending on the weight of the crude, and 1 cubic meter of natural gas is about 35.3 cubic feet.) OPEC is estimated to have about 75% and 50% of these reserves, respectively. Global production was fairly flat while reserves of crude oil and lease condensate increased in 2013, according to Xu and Bell (2013) in the annual "Worldwide Production" report of the *Oil & Gas Journal*. However, recalling the lack of audits for national oil company reserve estimates, the *OGJ* numbers likely provide an upper bound on proven reserves.

In addition to condensate, other liquids are separated from natural gas at natural gas plants. These natural gas liquids (NGLs), which include propane (C_3H_8), butane (C_4H_{10}), pentane (C_5H_{12}), and some heavier hydrocarbon chains, also contribute to our stock of hydrocarbons. The first two of these products (often referred to as C3 and C4) are gases at normal temperature and pressure, and pentane (C5) has a very low boiling point (36°C or 96.8°F) (Ophardt 2003). However, C3 and C4 become liquids under a moderate amount of pressure. An estimated additional 48.9 billion barrels of NGL proven reserves may also be extracted from natural gas wells. This estimate of world NGL proven reserves for the beginning of 2013 is arrived at by taking proven oil reserves from British Petroleum (2014), which include NGLs, and subtracting the proven reserves from Xu and Bell (2013).

Some of the world's largest oil fields, and the geological time of their formation, are shown in table 2–2. The reserves shown are the amount estimated from primary recovery or oil recovered from the natural pressure of the well. The amount of reserves that can be recovered can be increased by secondary recovery, which increases well pressure by injecting water, gas, steam, or other materials.

If the proven numbers are accurate, does this then imply that when these reserves are used up, all our oil and gas will be gone? Certainly not, because reserves are an inventory that we believe can be produced economically. However, inventories are expensive. A firm producing automobiles does not want to hold extra inventories. Nor does an oil and gas company want to invest in finding and developing extra hydrocarbon inventories. A company is likely to feel comfortable with an inventory of 10 years or so. As time proceeds, companies refine their estimates of reserves in known fields with *reserve additions*. They search and find new fields as needed and learn better technology to produce reserves more cheaply and increase recovery rates.

The above discussion of reserves largely applies to conventional reserves of oil and gas. Such reserves, when tapped into, have low enough viscosity to move on their own accord through permeable rock to the well bore. According to Tissot and Welte (1984), the world's conventional oil and gas reserves were formed during three geological time periods in the following proportions:

- Paleozoic: 14% of oil and 29% of natural gas
- Cretaceous: 54% of oil and 44% of natural gas
- Tertiary: 32% of oil and 27% of natural gas

During this time, global temperatures, except for occasional ice ages, tended to be much hotter than today (for an example, see Scotese [2002]).

Oil and gas reserves in source rock that is not very permeable or porous are typically referred to as *unconventional reserves*, which we will consider in the following sections.

Table 2–2. The world's largest oil fields

Field	Year Discovered	Country	Age of Reservoir	Primary Reserves* (billion barrels)
1 Ghawar	1948	Saudi Arabia	Jurassic	83.0
2 Burgan	1938	Kuwait	Cretaceous	72.0
3 West Qurna	1973	Iraq	Jurassic	43.0
4 Bolivar Coastal	1917	Venezuela	Mio.[a]–Eoc.[b]	32.0
5 Safaniya-Khafji	1951	Saudi Arabia/Neutral Zone	Cretaceous	30.0
6 Rumaila	1953	Iraq	Cretaceous	20.0
7 Cantarell	1976	Mexico	Paleocene	18.0
8 Ahwaz	1958	Iran	Oligo.[c]–Mio., Cret.[d]	17.5
9 Kirkuk	1927	Iraq	Oligo.–Eoc., Cret.	16.0
10 Daqing	1959	China	Cretaceous	16.0
11 Marun	1963	Iran	Oligo.–Mio.	16.0
12 Kashagan	2000	Kazakhstan	Devon.–Carb.[e]	16.0
13 Samotlor	1965	Russia	Cretaceous	16.0
14 Gachsaran	1927	Iran	Oligo.–Mio., Cret.	15.5
15 Shaybah	1998	Saudi Arabia	Cretaceous	15.0
16 Aghajari	1937	Iran	Oligo.–Mio., Cret.	14.0
17 Romashkino	1948	USSR	Carb.–Devon.[f]	14.3
18 Prudhoe Bay	1969	United States	Cret.–Trias.[g], Miss.[h]	13.0
19 Majnoon	1975	Iraq	Jurassic	12.6
20 Abqaiq	1941	Saudi Arabia	Jurassic	12.5

Sources: Tiratsoo (1986, 23); OGJ (2004).
Note: *Primary reserves include estimated ultimately recoverable reserves (EUR) for primary recovery only.
[a]Miocene, [b]Eocene, [c]Oligocene, [d]Cretaceous, [e]Carboniferous, [f]Devonian, [g]Triassic, and [h]Mississippian

Natural Gas

Three categories of gas that fall under this rubric are tight sands gas, coalbed methane (CBM), and shale gas. In each case, the free movement of gas is inhibited by the source rock (tight sandstones, unmineable coal seams, and shale, respectively). Thus, the rock must be fractured with the use of high-pressure water and a *proppant*, such as sand or chemicals, or both. The fracing creates a space in the source rock, and the proppant holds it open so the gas can flow. Such gas was once uneconomic, but with changes in price and technology, such as fracing and horizontal drilling, many of these resources have become commercial (Piccolo 2008).

Global production of these unconventional gas sources provided about 13% of total global natural gas production (3,169 billion cubic meters) in 2010. North Americans have been pioneers in the development of these unconventional gas resources. The United States and Canada, the number one and number three global natural gas producers, are estimated to have produced almost 85% of the unconventional gas in 2010. They get an estimated 50% and 25% of their production from unconventional gas, respectively, while Australia gets 10% of its gas production from its CBM. Fledgling CBM industries are being developed in China, India, and Indonesia (IEA 2012b).

Production for the easiest source (tight sands) started in the United States in the 1970s. Indeed, it has been produced long enough that the US and Canadian governments do not tend to report them any longer in their unconventional categories. The more difficult CBM did not get a sustainable start in North America until the 1990s.

Shale gas has become a particularly hot play in the United States, increasing from 4% of domestic production in 2005 to 43% by 2012. (EIA n.d.b.) This has affected LNG projects worldwide that had been gearing up to supply the huge US market, where declining conventional gas was not thought to be able to satisfy coming demand (*Economist* 2012b).

Shale gas production first took off in the Barnett Shale in Texas, which still dominates with about 50% of US production. However, production is rapidly growing elsewhere, particularly in the Marcellus Shale. The Marcellus, which follows a long, wide belt along the Appalachian Mountains from northern Tennessee to mid–New York State, is thought to hold between 40% and 50% of the US reserves (National Energy Technology Laboratory 2013).

Although shale gas is largely a US phenomenon, the International Energy Agency (IEA) (2011b) reckons that unconventional reserves are now as large as conventional ones, with their relative importance by country shown in figure 2–1. China has the largest reserves, followed by the United States, Argentina, Mexico, Australia, and Canada.

Although the potential looks huge, there are some clouds on the horizon that could prevent the dramatic US speed of development from being duplicated elsewhere. In the United States, the shale gas plays are near conventional plays, so infrastructure is already well developed. Deregulation has made US pipelines *open access*, or accessible to new producers. Drilling rigs have been readily available. With privately held mineral property rights in many of the US shale plays, NIMBY (not in my backyard) becomes IMBY as mineral owners laugh all the way to the bank to deposit their royalty checks. Few other places in the world allow private individuals to own subsurface rights (*Economist* 2012b). Environmental implications for shale gas are nontrivial and could even slow the euphoric frenzy in the US production. Shale gas is estimated to emit 3.5% more methane than conventional gas wells, and more if gas is vented. There have been concerns about chemicals and methane leaking into drinking water supplies.

Fracing has been around since the late 1940s, and the leakage problem has been solved with proper cement jobs on wells to prevent pollution to drinking water supplies. But it does require vigilance to ensure that proper techniques are implemented. Fracing also requires huge amounts of water that must be acquired and disposed of. China's lack of water in the west, where the bulk of reserves are located, may slow developments of its copious reserves. The fracing process and deep underground water disposal have been associated with an increased incidence of earthquakes. Although Ellsworth (2013) finds the earthquakes from the fracing process to be inconsequential, those from underground water disposal may on occasion be more damaging. Thus, more densely populated areas of the world, with no private mineral rights, may put up more resistance to shale gas. Already, France and Bulgaria have banned fracing (*Economist* 2012b), whereas the United Kingdom lifted a moratorium on fracing at the end of 2012 (Bakhsh 2013).

An additional unconventional gas source is methane hydrates. These combinations of methane and water in crystalline form are found in permafrost areas and in ocean sediments. Although resources are thought to be huge (by some estimates, more than twice conventional reserves), difficulty in producing them makes their costs prohibitive relative to the other unconventional sources (Pooladi-Darvish 2004).

Although these resources are too expensive to be considered reserves, they still may merit some attention. Some have expressed concern that increasing global temperatures might release some of the methane, a potent greenhouse gas (GHG), causing further climate change (Mascarelli 2009).

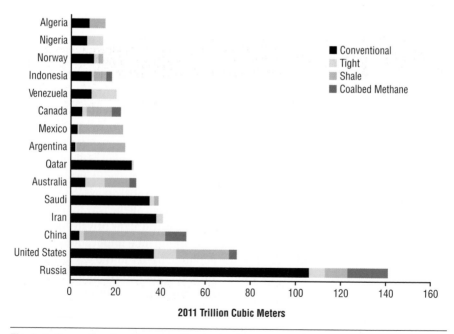

Fig. 2–1. Conventional and unconventional natural gas reserves by major country
Source: Economist (2012c).

Unconventional Oil Resources

Oil also comes from various sources. If the source rock is relatively impermeable, or the oil is too dense or viscous to flow through the source rock, or both, we have unconventional oil.

Two unconventional sources of oil, extra heavy oil and bitumen, are both dense and viscous.

Heavy crudes and bitumen

Heavy and extra heavy crudes have typically been degraded by the loss of lighter components and usually contain more sulfur and heavy metals than lighter crudes. Thus, they are both dirtier and much more difficult to extract. Extra heavy oils with the highest viscosity are called *bitumen*. The triennial report of the World Energy Council (2010) contains estimates for resources/reserves for these two unconventional oils, as shown in table 2–3. Most of the known extra heavy oil deposits (not including bitumen) are found in Venezuela. Natural bitumen that is also extra heavy but with an even higher viscosity than extra heavy oils is found in *oil sands*, which are a mixture of about 90% sand, clay, and water, with the remainder being bitumen. This higher viscosity oil is more expensive to produce than heavy oil, as it requires heat to be produced.

Table 2–3. Largest accumulations of estimated unconventional oil reserves

Location	Billion Barrels	Type of Resource	Definition	Reservoir Age[†]
United States	24.0	Tight Light*	Technically recoverable	U. Devonian–L. Miss.
Canada	0.5	Tight Light*	Proved and probable	U. Devonian–L. Miss.
Argentina	22.0	Tight Light*	Prospective resources	L. Cretaceous–U. Tertiary
Rest of World	—	Tight Light*	—	—
Venezuela	220	Heavy/X Oil	Technically recoverable	Oligocene–Miocene
Rest of World	1,400	Heavy/X Oil	Technically recoverable	—
Canada	170.4	Oil Sands	Remaining reserves	Lower Cretaceous
Kazakhstan	42.4	Oil Sands	Remaining reserves	—
Russia	28.4	Oil Sands	Remaining reserves	Miocene
Rest of World	2.0	Oil Sands	Remaining reserves	—
United States	3,706.8	Oil Shale	In place resources	Paleogene & Cretaceous
China	354.0	Oil Shale	In place resources	Upper Permian
Russia	247.9	Oil Shale	In place resources	Ordovician & Jurassic
Rest of World	477.4	Oil Shale	In place resources	—

Sources: WEC (2010); Petzet (2012); Selley (1985); and Selley (1997).
Note: *Tight light oil is sometimes referred to as shale oil, but care must be taken to distinguish it from the heavy oils recovered from oil shale. †Indicates age of majority of deposits; U. Devonian is Upper Devonian; L. Miss. is Lower Mississippian; L. Cretaceous is Lower Cretaceous; U. Tertiary is Upper Tertiary; — indicates unknown to author. Heavy/X is heavy and extra heavy.

American Petroleum Institute (API) gravity is the conventional way of measuring how heavy or dense a given stream of crude is. It is related to *specific gravity*, which is the weight of a given volume of a compound divided by the weight of an equal volume of water, as follows:

$$\text{API gravity} = [(141.5/\text{specific gravity}) - 131.5]$$

The specific gravity of water is 1, with API gravity of 10 degrees (°). Higher values for the API gravity are associated with lighter crude oils. For example, West Texas Intermediate (WTI) is a light oil with an API gravity around 40°, but Saudi Light is a bit heavier crude, with API gravity around 34°. Bitumen, which is extracted from oil sands, typically has an API of less than 12°. Denser oils are typically harder to move through source rocks. The actual characteristic that determines how difficult it is to move a liquid through a given source rock is its viscosity. The higher the viscosity, the harder it is to move the oil through the source rock. A standard international unit to measure viscosity is the millipascal-second (mPa-s) at 15°C. The viscosity of water is about 1 mPa-s at 20°C (Elert 2013). WTI has a viscosity of less than 9 mPa-s (Wang et al. 2003).

Just as conventional crude is heterogeneous, so too, is nonconventional oil. Cupcic (2003) gives a helpful classification for these heavy unconventional oils (B, C, and D classes) as shown in table 2–4.

Table 2–4. Categories of heavy unconventional oils

Class	API Density	Viscosity (mPa-s)	Fluid in Reservoir
A Class: Medium Heavy Oil	18°–25°	10–100	Mobile
B Class: Extra Heavy Oil	7°–20°	100–10,000	Mobile
C Class: Tar Sands and Bitumen	7°–12°	>10,000	Not mobile
D Class: Oil Shales			Not mobile

Source: Cupcic (2003).
Notes: mPa-s = a millipascal-second, which is also equal to a centipoise (cP). Large numbers indicate higher viscosity.

Oil shale

Some source rocks rich in kerogen, a precursor to oil, but which have not been subject to enough heat and pressure, are known as *oil shale* (D class in table 2–4). Vast amounts of hydrocarbons are locked up in oil shale, with major deposits shown in table 2–3. The US Rocky Mountain region has the largest deposits worldwide. Although there have been shale oil industries in various countries, beginning in France in 1838, the high cost of extracting and processing oil from the shale has caused much of this extraction to be phased out. Releasing the kerogen from the shale requires even more energy than recovering bitumen from oil sands.

Tight light oil

Just as for gas, some oil has gotten trapped in shales, and fracing of the shale can allow this oil to move to the well bore and be produced. This tight oil produced from shale is sometimes referred to as *shale oil*. However, such terminology is confusing, as the term shale oil has also been used to refer to oil produced from oil shale. Tight oil is not viscous, it is just trapped. It might reasonably be called *tight light oil*. For example, Bakken crude oil has an API gravity from 40° to 42°, with a viscosity around 2 mPa-s (Wang et al. 2011). As with shale gas, releasing this trapped oil has been a hot play in North America, and current technically recovered resources are estimated to be around 24 billion barrels (US EIA 2011b).

Coal Resources

Coal was formed from higher plants in swampy areas beginning in the Devonian. It usually started out as peat. The world still has large reserves of peat, according to the World Energy Council (WEC 2013). With increasing heat and pressure, the peat changed to brown coal, then hard coal, and then eventually to anthracite, corresponding to the increase in the ratio of carbon to oxygen and to hydrogen. The major deposits of coal worldwide range from the Carboniferous to the Pliocene (table 2–5).

Table 2–5. Major eras of coal formation

Time Period	N. America	Europe	Far East	Southern Hemisphere
Pliocene	+(Alaska)	+	−	−
Miocene	−	++	−	+(Australia)
Eocene	++	+	++	+(Australia)
Paleocene	++US Powder River Basin			
Cretaceous	++	−	++(Japan, China)	++
Jurassic	−	+	++	++(Australia)
Triassic	−	+	−	+(Australia)
Permian	−	−	++(China)	++(All Gondwanaland)
Carboniferous	++(US Appalachia & Midwest)	++	−	−

Sources: Tissot and Welte (1984, 232); Carroll (2011).
Note: ++ Coals very abundant, + abundant, − absent.

Energy's Human History

When humans appeared after the formation of hydrocarbons, they were mostly unaware of the energy riches that had accumulated under their feet. The sun warmed them and wood, rather than fossil fuels, was their fuel source. Human muscle was the primary means of transportation. Humans learned much about their plant and animal food sources. They domesticated plants and animals to have a more stable food supply and more sources of energy. These domestications led to two different lifestyles, herding and sedentary agriculture, around 11,000 years ago. With sedentary agriculture, slavery also developed as a source of energy, which was carried on in many areas until the 1800s (Ray 1979). For more information on human energy use up to the fossil fuel era, see http://dahl.mines.edu/b0201.pdf.

We know that about 1,000 years ago, the Chinese burned coal, because Marco Polo brought back the knowledge of these burnable black stones in 1275 CE. The Dutch subsequently discovered coal and exported it to England. However, the English soon overtook the Dutch in productive capacity, and by 1660, the English were mining and exporting much of the world's supply. Coal was also responsible for an early but ineffective environmental regulation forbidding the burning of coal in London toward the end of the 14th century.

In the early 1700s, Thomas Newcomen developed the steam engine to pump water from mines. This greatly increased the supply of coal. Abraham Darby replaced charcoal with coking coal for iron production, which greatly increased coal demand. A further improvement by James Watt in 1765 allowed the steam engine to run machines, which further promoted the Industrial Revolution. Steam engines allowed flexibility in siting plants, since they did not require running water. In the early 1800s, coal, along with the steam locomotive and steamboat, changed the face of transportation. Almost a century later, internal combustion engines were put in automobiles and aircraft. These innovations boosted the use of petroleum and had a profound impact on the 20th century.

The first oil well in North America was drilled in Oil Springs, Ontario, Canada, in about 1854. The modern oil era in the United States began with Colonel Drake in Titusville, Pennsylvania. He had been skimming oil off a creek to get kerosene but went bankrupt. In 1859, he decided to drill and struck oil at 69 feet. The discovery spawned the first US oil boomtown. This and a few other important events in recent oil and natural gas history are summarized in table 2–6. For a more complete listing of energy and related milestones, many of which will be discussed later in this book, see http:\dahl.mines.edu\t0206.pdf.

Table 2–6. A few oil and natural gas milestones in recent human history

Dates	Era or Event
Early 1800s	Gas began to be used in London to light street lamps; Murdock lit a factory using coal gas lamps; and kerosene, called coal oil, was extracted from coal.
1858	Lenoir created a two-stroke internal combustion engine.
1859	Drake discovered oil in Pennsylvania while looking for coal lamp oil substitutes.
1870	Rockefeller began refining oil into kerosene in Cleveland.
1879	Edison invented the electric light bulb.
1880s	Rothschild and Nobel produced oil in Russia; Royal Dutch produced oil in Indonesia.
1885	Daimler and Benz made the first automobile in Germany.
1896	Ford built his first automobile.

Dates	Era or Event
1901	Using the first rotary drill, Spindletop was discovered in East Texas.
1903	Wright brothers completed the first airplane flight powered by gasoline engine.
1908	Anglo Persian struck first oil in Persia.
1911	US government broke up Standard Oil.
1928	"Red Line" and "As Is" agreements limited international oil company competition.
1938	Mexico nationalized its oil industry; oil was discovered in Saudi Arabia and Kuwait.
1948	Jersey Standard (Exxon), Socony Vacuum (Mobil), California Standard (Chevron), and Texaco formed the Arabian American Oil Company (Aramco).
1951	Iran nationalized Anglo Persion Oil into the National Iranian Oil Company (NIOC).
1954	Western companies took over NIOC in Iran after the previous year's coup.
1959	Groningen gas field found in the Netherlands.
1960	OPEC was formed in response to cut in posted prices that reduced their tax revenue.
1968	Prudhoe Bay oil field was discovered in Alaska.
1969	Oil was discovered in the North Sea.
1970	US oil production peaked.
1971–79	Increased government participation and/or nationalizations occurred in OPEC countries (e.g., Iraq, Kuwait, Libya, Nigeria, Qatar, Saudi Arabia, the United Arab Emirates, Iran, and Iraq).
1973	Arab oil embargo of United States and the Netherlands began as result of Yom Kippur War.
	United States dismantled price controls implemented in 1971 as part of a price stabilization policy, but crude oil, oil product, and natural gas price controls were not lifted.
1974	International Energy Agency (IEA) was formed.
1975	Brazil implemented an ethanol program designed to eliminate fossil fuels in vehicles.
	United States authorized the Strategic Petroleum Reserve (SPR) and removed price controls on old oil.
1978	US Natural Gas Policy Act started natural gas price decontrol.
1979	Energy crisis resulted from the Iranian Revolution and oil production cuts. Iran renationalized NIOC.
1980	United States enacted windfall profit tax on crude oil.
1981	US domestic oil price controls were lifted.
1984	OPEC first established production quotas.
1985–86	Oil price plummeted more than 50% with Saudi Arabian netback pricing and increased production.
1991	Gulf War ousted Iraq from Kuwait.
1997	Qatar started exporting from world's largest LNG facility.
1997–98	Asian financial crisis caused oil prices to plummet.
1998	Landlocked Caspian Sea area became exploration hot spot, with export pipelines built in the following decade.
1999–2006	Oil company acquisitions and mergers: BP/Amoco; Exxon/Mobil; TotalFina/Elf; Repsol/YPF; Norsk Hydro/Saga; Chevron/Texaco; Conoco/Phillips; and Rosneft/Yukos.
2004	World Bank agreed to new lending rules intended to prevent the funding of corrupt regimes with revenues from oil and natural gas projects.
2005–08	Shale gas production took off in the United States.
2006	Russia temporarily cut natural gas supplies to Ukraine over a pricing dispute, which caused continuing security concerns in Western Europe.

Dates	Era or Event
2008	With booming world economy, spot oil prices peaked at $146 per barrel but fell with ensuing financial crisis in industrial countries.
2010	An explosion and fire on the Deepwater Horizon drilling rig caused 11 deaths and the largest ocean oil spill in history.
	Iran removed energy and other subsidies in exchange for cash transfers.
2011	Revolution in Libya and other Middle Eastern and African turmoil kept oil markets jittery.
2012	Prompted by the United States, the European Union embargoed Iranian oil over concerns about the Iranian nuclear program; Iran threatened to block off the Straits of Hormuz.
2013	Interim agreement and partial lifting of embargo on Iran. China passed the United States as the largest importer of petroleum.

Sources: Jenkins 1989; Gustafson 2012; Kalt 1981; PennWell 1997; Ratnikas 2013; Smil 1994; US EIA 1995; Yergin 1991, 2011. For more entries, updates, and sources, see http:\dahl.mines.edu\t0206.pdf.

One of the noteworthy things about the history of energy is the consistency of fuel and energy use, even from the earliest times, to provide heat, light, lubrication, transportation, mechanical power, and materials for war. Thus, while many of the basic needs remain unchanged, the means of providing them have changed, as humans have sought better ways to satisfy these needs with increasingly sophisticated technologies. Most fuels, except for electricity and nuclear power, have been known for centuries. Which fuels have claimed dominance has evolved.

In figure 2–2, I approximate and update a diagram from Nakićenović (1984). He summarized how fuel use has changed since 1860 and speculated how it will continue to evolve during the coming century. We can think of this as a simple forecasting or simulation model, where F is the share of energy supplied by a particular fuel. Nakićenović graphs $F/(1 - F)$ for five major energy sources. Thus, when coal supplied one-half of the world market, coal's $F/(1 - F) = 0.5/0.5 = 1$. The dotted white lines are his suggestions on how these shares have evolved and how they would devolve after 1940. Notice that the model fits well for wood through about 1900, and for coal, oil, and gas through about 1975, but never fits well for nuclear. He has solar-fusion reaching the 1% share with $F/(1 - F) = 0.01$ at 2025.

Currently fusion does not look so promising. Solar has been growing rapidly, but from such a small base that it does not yet register in the figure. This figure demonstrates the perils of simple extrapolations and shows us the need to look at underlying driver and structural changes in these markets, which we will do throughout the coming chapters.

Note: The black line is $F/(1 - F)$ for actual consumption and the dotted white line is its smoothed and projected consumption. F is the fuel share. The diagram was produced four decades ago and shows how rapidly our perceptions about the future change. He forecasted that oil would never reach the ascendancy of either wood or coal in terms of market share in the coming century, but that gas would. This forecast is still rather believable. However, the prominence of nuclear power he envisioned now seems doubtful. One would guess that solar and other renewables, such as wind, will probably be more prominent in the coming decade. We have seen changing energy patterns in the past and can expect the changes to continue in the future. In the next section we will review some models that can help us predict or shape this future.

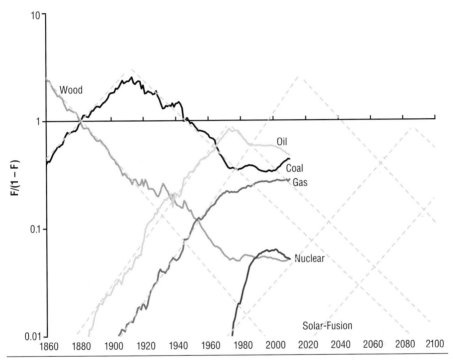

Fig. 2–2. World primary energy substitution

Sources: 1850–1994 from online data developed by Nakićenović (1984) and updated for use in Grübler (1998). These data are no longer available online but are said to be available from Publications Department, International Institute for Applied Systems Analysis, Laxenburg, Austria. The 1995–2010 data has been updated from US EIA (n.d.d) and World Bank (n.d.).
Note: The black line is $F/(1 - F)$ for actual consumption and the dotted white line is its smoothed and projected consumption. F is the fuel share.

Energy Modeling

An understanding of what influences the evolution of energy markets is valuable knowledge. It allows sellers to plan and develop the appropriate capacity and buyers to pick the appropriate mix of equipment and appliances to minimize the cost of producing energy services. It permits the government to make policy plans during peace and contingency plans for war, and enables financial institutions to pick the projects to back and the projects to reject. Because of this value, considerable resources have been spent in energy modeling, particularly since the late 1970s and early 1980s, when increases in prices and shortages focused policy attention very heavily on energy availability and cost. A variety of techniques have been used for energy modeling and forecasting. We will briefly consider the following techniques:

- historical trends
- simulation
- statistical analysis
- judgment
- surveys
- scenarios
- energy balances
- end-use models
- engineering or process models
- optimization models
- game theory
- economic experiments

Historical trends and system simulation

Historical trends or growth rates have long been used as simple tools for forecasting. In such a technique, if oil demand has been growing by 2.5% per annum, we assume that oil will continue to grow by this same rate. Then oil consumption at time t (O_t) is the following function of oil consumption now (O_0) and the exponential function (e):

$$O_t = e^{0.025t}O_0 \qquad\qquad (2\text{--}1)$$

For a general growth rate of r and small changes in t, we can represent ΔO by the derivative dO/dt and the growth rate by $(dO_t/dt)/O_t$. For the general exponential function $e^{rt}O_0$, $dO/dt = re^{rt}O_{t-1}$ and $(dO_t/dt)/O_t = re^{rt}O_{t-1}/O_t = re^{rt}O_{t-1}/e^{rt}O_{t-1} = r$. Using the above growth rate, if current oil consumption is O_0, then next year's forecast, O_1, would be $O_1 = e^{0.025}O_0$, and the forecast 10 years from now would be $O_{10} = e^{0.025\times10}O_0$. With this assumed constant growth rate, oil consumption would increase exponentially.

This technique works well when growth is constant and may be quite effective in the short run with a business-as-usual scenario. For example, in the United States during the 1960s and early 1970s, electricity consumption grew fairly consistently at 7% per year, and electric utilities found it quite easy to plan for capacity expansion. This growth continued until the 1973 energy price increases. Utility regulators then began to allow "fuel adjustment clauses" that allowed fuel price increases to be passed on to rate payers without the need for rate hearings. With the ensuing electricity price increases, electricity consumption growth fell considerably. The result was that electric utilities both in the United States and Europe found that they had built considerable excess capacity. We saw similar high electricity growth rates averaging between 7% and 8% a year during the decades of the 1980s and 1990s for China. But instead of decreasing, the rate jumped, averaging more than 10% from 2000 to 2008, leaving China with the need to scramble for enough new capacity to prevent shortages (BP 2013).

An early proponent of forecasting using an exponential growth rate was William Stanley Jevons, who forecasted England's coal consumption. Writing in 1865, when British coal consumption was around 84 million long tons (1 long ton = 1.016 metric tonne) per year, and using an exponential growth rate of 0.75%, he forecasted coal consumption would exceed 2,600 million tons by 1961 (Jevons 1965, XII.24). However, actual coal consumption for the entire United Kingdom was just under 200 million tons by the year 1960, falling to less than 60 million tons by 1999 (Gordon 1987; BP 2014). Thus, historical extrapolation of trends does not predict turning points very well, and it can also be very misleading in the long run, as consumption seldom increases exponentially over the long haul.

System simulation studies include multiple equations to represent the behavior of economic entities and interactions between entities to simulate outcomes. They can be particularly useful in systems with a lot of interaction and feedback that cannot easily be otherwise seen. However, as with any forecasting tool, they can also be quite misleading when simulating many decades into the future, especially with a weak understanding of the underlying inputs.

For example, the widely cited book by Meadows et al. (1972), used a system of variables related by mostly differential equations to investigate whether the earth's resources could provide sustainable growth rates into the 21st century. Their theme was the limits to growth, and they included exponential growth rates that painted a bleak picture for future sustainability, with depleting resources, eventual decreasing food and industrial output per capita, and increasing pollution. Their updates, Meadows et al. (1992) and Meadows et al. (2004), paint somewhat similar pictures. Prominent economists have been rather critical of these models for the ad hoc nature of their model inputs, the high degree of aggregation, and a failure to adequately take into account substitution across resources, policy response, and technical changes, particularly for simulations going out for a century and more (Nordhaus 1973; Nordhaus et al. 1992; Smil 2005).

Statistical models

More sophisticated statistical techniques, including univariate time series, multivariate time series, econometrics, and Bayesian econometrics estimation, will be reviewed in the context of demand modeling in chapter 16. With simple statistical models, a variable of interest such as oil price (P_i) is influenced or related to X_n variables. For example:

$$P_i = \sum_{j=1}^{n} \alpha_j X_j + \varepsilon_i \qquad (2\text{--}2)$$

The *X*s could be lagged prices (univariate time series) or lagged prices and other variables (multivariate time series), or other variables in the same time period (econometrics). These techniques can also be extended to multiple equation models. Such models may be inputs into simulation and other larger models.

Judgment

Time series and econometrics are explicitly based on historical information. However, sometimes we believe the future may not be like the past, and we want to include more personal judgment and rationality in the model. Such a forecast can be purely judgmental, with no formal model, and the experts base their forecast on what they *think* will happen. This judgment is likely based on the analysts' historical experience in the market, informal heuristics, and intuition, along with speculation on what will happen in the future.

Surveys

A more forward-looking tool for forecasting is to conduct surveys that ask firms about their plans. Trade journals regularly conduct and publish surveys about ongoing and future investments. For example, PennWell's *Oil & Gas Journal* publishes information from a number of surveys that they collect annually or biannually. They include the following:

- *Worldwide Pipeline Construction* (usually early February)
- *Worldwide Construction Update* (usually early April and late October)
- *EOR Survey* (biannually, usually mid-April)

Another unique biannual survey was conducted by the International Energy Workshop (IEW), a joint effort between Stanford University's Energy Modeling Forum (EMF, n.d.) and the International Institute for Applied System Analysis (n.d.). From 1981 to 1997, IEW collected energy forecasts and analyzed why the forecasts differed (IEW 2005). Figure 2–3 shows successive median forecasts compared to the actual oil price shown in the darker solid line. All forecasts indicated rising prices from 1990 to 2000, whereas the actual general price trend was down and considerably lower than earlier expectations. It is easy to see how the forecasts are anchored to the actual price at the time of the forecast. All forecasts from the 1990s for 2000 to 2010 were for an increase, although none forecasted how steeply prices would actually rise or where they would peak. Overall, only the forecast for 1987 actually touches any of the actual prices from 2005–2012.

This particular example for oil price forecasts is not unique, and many other examples can be found in the post-1970 period. For example, US EIA (2013b) compared their *Annual Energy Outlook* forecasts to actual outcomes for 1994 to 2012 for 25 key variables. They found an average absolute percent difference of 35.2% for real refinery acquisition cost of crude oil, 30.7% for real natural gas wellhead prices, and 18.6% for real coal prices to electricity generators. These forecasting errors will carry through to other forecasts within their model that rely on these prices. For example, natural gas and coal prices are inputs in electricity demand forecasting. (See US EIA [n.d.e] for documentation of the model used to produce their *Annual Energy Outlook*.)

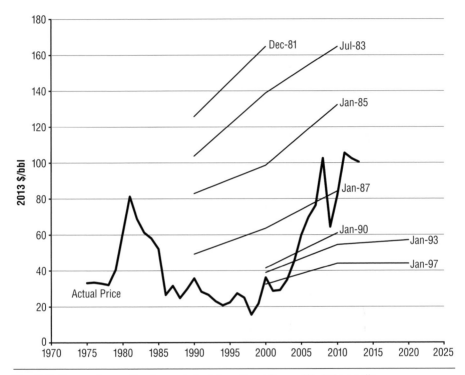

Fig. 2–3. Successive median forecasts by International Energy Workshop polls
Sources: Reproduced from Schrattenholzer (1998), updated from US EIA (n.d.d) and US BLS (n.d.).
Note: Oil prices have been updated from the original and converted to 2013 dollars.

Consensus Inc. provides a more recent survey example. They conduct surveys of economic forecasts, including the prices of crude oil, gasoline, gas oil, natural gas, coking coal, steaming coal, and uranium (Consensus Economics 2013).

These examples do not necessarily suggest that we should not forecast, but they do suggest we should have a healthy skepticism of forecasts. Too many unknown and unknowable events and features can intervene to make our forecasts go awry. Technologies and tastes change; policies are implemented.

Another such feature of forecasting is forecast feedback. Robinson (1992) noted that forecast results may be passively accepted, which could lead to one outcome, or they may be vigorously opposed, which could lead to another outcome. Forecasts cannot necessarily be compared to reality to ascertain their validity.

Hence, self-fulfilling prophecies would look good, whereas self-defeating ones would look bad. Forecasts of recession might lead to recession, whereas forecasts of high energy prices and shortages have been known to lead to surpluses and falling prices. Forecasts may also be used to influence an outcome, which Robinson calls the *theory of deliberate feedback*. An environmental group may forecast low electricity demand growth, while opposing a new nuclear power plant. The electric utility planning the new plant may forecast much higher electricity demand growth. These phenomena, as well as the types of models used, should be considered when interpreting forecasts.

Scenario planning

Due to the unreliability of forecasts decades or more into the future, another technique called scenario planning can be used. Scenarios are also useful when you think the future is unlikely to be similar to the past. This technique relies more on imagination and less on history. To build scenarios, you imagine future possible events and circumstances, use logic to determine what the events and circumstances imply, and then plan for possible contingencies. It is desirable to have scenarios that are logically consistent and span the range of possibilities. Such a range of scenarios allows planners to recognize both opportunities and potential hazards.

Shell Oil has used this technique fairly extensively (Skov 1995). In the early 1980s, Shell was considering whether to develop the Giant Troll field in Norway. At the time, Soviet gas was politically limited to around 35% of the European market. Some of the scenario games included removing this limit, which was thought to be a possibility at the end of the Cold War and an invitation for foreign investment into the Soviet Union. At the time of these scenarios, *glasnost*, the breakup of the Soviet Union, and the end of the Cold War were far off, but they were considered, and possible responses were entertained. More recently, Pacific Gas & Electric used scenarios to help prepare strategies after problems of cost run-ups with its Diablo Canyon nuclear power plant. You can follow Shell scenarios at Shell (n.d.).

The Art of the Long View (Schwartz 1991) is a seminal book on scenario planning that is still used in college courses today. Schwartz argues that scenario building is best done in teams that include some high-level management, a broad range of functions and divisions within the company, and imaginative people. (See http://dahl.mines.edu/b0202.pdf for his eight steps in scenario building.)

So, what sorts of scenarios might we envision for the coming decades? Possibilities to consider include the major driving force of population growth. Demographic shifts and elongation of human life span are likely to change work and leisure patterns, and, consequently, fuel use. A second major consideration is income growth. Will the industrial countries be able to continue their high lifestyles? Will developing countries continue to catch up? If so, will energy and natural resources or environmental carrying capacity become a constraint? If income continues to grow, how will the new income be spent? Will people want more goods, more services, or more leisure?

Environmental concerns, both local and global, will also influence fuel use. If we learn to economically sequester carbon so that fossil fuels pose less of a threat, then coal will remain a feasible long-term fuel. We could continue to use fossil fuels and move on to gas-to-liquids, gas hydrates, tar sands, and shale oils for an extended fossil-fuel age. If fossil fuels remain important, how will depletion and technological improvement change their relative prices?

Alternatively, environmental constraints or technology or both may move us to cleaner, more sustainable energy sources. Buildings and homes could be the main energy generators in a distributive utility approach. Power could be derived from solar, wind, fuel cells, or some hybrid and be stored in new and better batteries. Biomass waste could become a fuel instead of a nuisance. Efficiency could improve with appliances and other energy-using equipment employing intelligent sensors for power control. Fusion could become our mainstay in a centralized electricity grid. Indeed, we could even see a renaissance of fission, if issues such as waste storage and proliferation can be mastered and smaller, safer technologies evolve.

Energy market structure could influence the shape of our energy future. If OPEC keeps oil prices high, it could hasten the transition out of fossil fuels, while decreasing regulation could mean a movement toward cheaper fuels and higher fuel use. Government intervention to internalize the negative effects of fossil fuels could promote a cleaner world sooner.

Around one-fourth of total energy consumption in the Organization for Economic Cooperation and Development (OECD) countries goes to transportation; almost 20% of energy consumption in developing countries does so (IEA, n.d.f). Travel includes moving freight, commuting, recreation and tourism, socializing, shopping, other services, and industry travel. However, information technology might affect fuel use by streamlining traffic patterns, and telecommuting, teleconferencing, e-commerce, and changing geographical settlement patterns could change how many miles people and freight travel.

Technology will clearly influence all these uses, and the amount of investment capital available will affect its growth. Technological change will also influence the price and availability of various fuels. As technical change helps to conserve energy, less energy will be consumed per unit of energy service output. However, this conservation will make the energy service cheaper and possibly increase the demand for the energy service, causing a rebound effect, which cancels some of the decrease in energy use. For example, increasing miles per gallon would decrease gasoline consumption. However, with greater fuel efficiency, it is cheaper to drive a mile, and people may drive more. (For a good discussion of this effect, see Sorrell and Dimitropoulos [2008].) Sorrell (2007) concludes from a survey of the literature that rebound effects for household heating, cooling, and transport are likely to be between 10% and 30%.

Input-output

Models can be developed from components called *bottom up*, or they can be built by looking down from an aggregate level, called *top down*. Bottom-up models start with disaggregate data trying to look at an economic decision maker or particular technology. For example, you may look at demand for coal for various electricity generators. The bottom-up model lacks comprehensiveness but contains more detail. To find out total demand by all generators, you would need to aggregate over all generators. Top-down models typically look comprehensively at some energy system or subsystem but lack the intimate detail of the components of that system. (For good discussions of bottom-up and top-down models, see Evans and Hunt [2009].) You might estimate total end-use demand for energy and nonenergy goods but disaggregate a top-down model demand into total production for energy and nonenergy goods through input-output analysis.

The easiest way to demonstrate an input-output model is by a simple example. Suppose we have an economy with two composite goods, energy (E) and another good (NE). Both goods are used as intermediate inputs and as final goods for end uses. Suppose it takes 0.1 unit of energy to produce a unit of E and 0.2 units of energy to produce a unit of NE. It takes 0.3 units of NE to produce a unit of E and 0.4 units of NE to produce a unit of NE. Suppose that we have a good forecast of end-use demand for E and NE from a top-down model of $D_E = 80$ and $D_{NE} = 1,200$. We want to disaggregate to determine how much total production of E and NE are needed to satisfy this demand. Since both goods are used as end-use demand and intermediate goods, we know that production of E and NE will have to be larger than the end-use demands for E and NE.

Let's start with total requirements for E. Since each unit of E requires 0.1 unit of E as an input, the demand for E in the energy sector will be $0.1E$. Since each unit of NE requires 0.2 units of energy, the demand for E in the other sector will be $0.2NE$. The total requirement for E will be these intermediate requirements plus the end-use demand, as follows:

$$E = 0.1E + 0.2NE + 80 \qquad (2\text{–}3)$$

Similarly, the total requirement for NE would be the following:

$$NE = 0.3E + 0.4NE + 1,200 \qquad (2\text{–}4)$$

The solution to this system is $E = 600$ and $NE = 2,300$. You can see the computations at http://dahl.mines.edu/st02/st02.pdf in the answers to question 13. Alternatively, if you know matrix algebra, the solution is as follows:

$$x = (I - A)^{-1d} \qquad (2\text{–}5)$$

where

x is the vector of total production,

I is the identity matrix, $I = \begin{bmatrix} 1 & 0 \\ 0 & 1 \end{bmatrix}$,

A contains the input-output coefficients, $A = \begin{bmatrix} 0.1 & 0.2 \\ 0.3 & 0.4 \end{bmatrix}$, and

d is the vector of final demands $d = \begin{bmatrix} D_E \\ D_{NE} \end{bmatrix} = \begin{bmatrix} 80 \\ 1200 \end{bmatrix}$.

The superscript for $(I - A)$ indicates that we take the inverse of $(I - A)$. The beauty of the matrix approach is it can represent any number of sectors (n). Then I and A would be dimensions $n \times n$, and the corresponding x and d would be $n \times 1$. Matrix algebra and computer algorithms solve input-output models for hundreds of goods, which can be measured in either monetary or physical units. Such large actual input-output tables for dozens of countries are located at OECD (n.d.).

Such models have a variety of interesting applications whereby a change in one part of the economy can be traced through to see the implications for all the other sectors of the economy. For example, if environmental regulations, such as the European Union's light duty vehicle emission standards, increase the need for inputs, the coefficients in the model can be increased to reflect the change. (The latest standards with their timetable of implementation can be viewed at EU [2013].) In a more complicated setting, the standards could also be modeled with a pollution abatement sector added to the input-output model as in questions 52 and 53 in http://dahl.mines.edu/st02/st02.pdf.

The latest of these European vehicle standards is Euro6, effective in 2013. Various stages of these standards are being gradually adopted by a number of other countries, including China, India, and Indonesia. India has adopted Euro4. If India wanted to know the effect of adopting Euro5, they could modify and apply their input-output tables to find out.

Another important application is to compute the cradle-to-grave effect of a change in a given end demand. For example, suppose we wanted to know the total amount of crude oil embedded in 1 liter (0.264 gallons) of gasoline. First, we can quantify the amount of oil refined to yield gasoline. But that is not the end of the story. Oil products ran the tanker that delivered the oil to the refinery and the tank wagons that delivered the gasoline to the pump. Oil is likely embedded in every piece of equipment along the supply chain: the drilling rig, the tanker, the pipeline, and the filling station. Although this sounds like a very complicated problem, in reality, input-output models make the computation quite easy. To see how, return to our simple model above. But now write the model, equations (2–2) and (2–3), with general input-output coefficients a_{ij}, with i and j equal to 1 and 2:

$$E = a_{11}E + a_{12}NE + D_E$$

$$NE = a_{21}E + a_{22}NE + D_{NE}$$

Solve the above model for E and NE to get the solutions to this system as follows:

$$E = \frac{(1 - a_{22})D_E}{(1 - a_{11} - a_{22} + a_{11}a_{22} - a_{12}a_{21})} + \frac{a_{12}D_{NE}}{(1 - a_{11} - a_{22} + a_{11}a_{22} - a_{12}a_{21})}$$

$$NE = \frac{(1 - a_{11})D_{NE}}{(1 - a_{22} - a_{11} + a_{11}a_{22} - a_{12}a_{21})} - \frac{a_{21}D_E}{(1 - a_{22} - a_{11} + a_{11}a_{22} - a_{12}a_{21})}$$

(You can see the computations at http://dahl.mines.edu/st02/st02.pdf, question 46 answers.)

Now it is trivial to compute the cradle-to-grave change in production from a one unit increase of either product. The change in NE production from a one unit increase in end-use demand for energy (D_E) is the following derivative:

$$\frac{dNE}{dD_E} = \frac{a_{21}}{1 - a_{22} - a_{11} + a_{11}a_{22} - a_{12}a_{21}}$$

Similarly, we can get all the rest of the cradle-to-grave uses for an increase in end-use demand of each good as follows:

$$\frac{dE}{dD_E} = \frac{1 - a_{22}}{1 - a_{22} - a_{11} + a_{11}a_{22} - a_{12}a_{21}}$$

$$\frac{dE}{dD_{NE}} = \frac{a_{12}}{1 - a_{22} - a_{11} + a_{11}a_{22} - a_{12}a_{21}}$$

$$\frac{dNE}{dD_E} = \frac{1 - a_{11}}{1 - a_{22} - a_{11} + a_{11}a_{22} - a_{12}a_{21}}$$

We can easily compute these for the n good case using matrix algebra. Remember the solution in the matrix case: $x = (I - A)^{-1}d$, with I and A being $n \times n$, and x and d being $n \times 1$. Let τ_{ij} be the element in the ith row and the jth column of $(I - A)^{-1}$. Then the cradle-to-grave use of the good produced by sector i for a one-unit increase in end-use demand for good j is $\frac{dx_i}{dd_j} = \tau_{ij}$. A good source for sample cradle-to-grave modeling is the Economic Input-Output Life Cycle Assessment (EIO-LCA, n.d.).

Input-output modules may be components in other larger models. They are the starting point for computable general equilibrium models that essentially allow the input-output coefficients to change as prices change. They may tie model modules together or allow aggregation and disaggregation within models.

Energy balances

Input-output models are accounting models, since they keep track of how inputs and outputs are related to each other. Another accounting-oriented model is the material or energy balance approach (see chapter 16.) In a system-wide model, independent estimates are compiled for each major energy end use and its growth. Econometric models, tempered by expert judgment and institutional considerations, are often used to craft estimates. Independent estimates are also compiled of major energy supply sources. These supply components may come from engineering, econometric, geological, or combination models coupled with expert judgment.

Supplies and demands are then compiled and compared. Scenarios are built and assumptions challenged until supplies and demands are consistent. Often, the model is brought into balance by assuming that a backup energy type will be available to fill the gap. For example, the National Petroleum Council (1972) assumed that imported oil was the backup source for the United States. Although large integrated energy models today are typically much more complicated, all must ensure that energy balances are maintained for the model to be internally consistent.

Bottom-up end-use models

An end-use model considers the demand for an end-use service separately from energy required to produce that service. In the simplest end-use model, the consumer uses an energy service (S), such as miles travelled or space heated. The consumer combines energy with capital to produce this service. The capital requires a certain amount of energy to produce a unit of service (E/S). We call this E/S *fuel intensity*; the inverse of fuel intensity (S/E) is often called *energy efficiency*. We can relate energy consumption to energy intensity and energy efficiency as follows:

$$E = S\frac{E}{S} = \frac{S}{S/E} \tag{2-6}$$

Right away you can see the increased data requirements, as now we need to know both energy service demand and fuel intensity. If such data exists, energy service demand and energy intensity or energy efficiency might be estimated using statistical techniques on historical or survey data. For example, Ajanovic et al. (2012) survey the literature on such estimates for gasoline and diesel demand, where the service is distance travelled (kilometers or miles), and fuel efficiency is distance travelled per unit of energy (miles/gallon or kilometers/liter). These are typically modeled as some function of fuel prices and income. Alternatively, S could come from an extrapolation, while E/S could come from engineering estimates of various technologies. Or we could implement an efficiency standard to change E/S and try to model the energy implications of our policy.

Disaggregating equation (2–6) even further, we might want to model how much capital (K) to own, how intensely to use the capital (S/K), and the fuel intensity of our chosen capital. If our model is disaggregated and we are looking at service use for each unit of capital (S/K), we could aggregate energy use over all units of capital as follows:

$$E = K \frac{S}{K} \frac{E}{S}$$

(2–7)

We could get even more complicated, where we have capital of different ages or vintages, with old capital being more energy intensive. Or we could choose across capital of different intensity categories (big versus small cars) or capital with different fuel uses (gas versus electric clothes dryers). Such end-use models may be modules in larger integrated energy models. For example, the US National Energy Modeling System (NEMS) has end-use demand models in its residential demand module, its commercial demand module, and its transportation demand module (US EIA, n.d.e). (See Ajanovic et al. [2012] for more discussion of simple end-use transportation models, and Swan and Ugursal [2009] for residential end-use model methodologies.)

Process models

Engineering process models consider the process of converting an energy product to an energy service. It could be as simple as a coefficient that represents the amount of gas required per kilowatt-hour of electricity. Or it might be a production function, with output Q related to some set of inputs, such as capital, labor, and energy, as follows: $Q = f(K,L,E)$. Alternately, it might be as complex as a large multiequation energy optimization model that typically maximizes profits or minimizes costs. Such optimization models have often been used to model oil refining, energy transportation, and electricity systems. Such optimization techniques include mathematical programming (linear, nonlinear, integer, and mixed integer), Lagrangian techniques, calculus of variation, and optimal control theory, all of which will be touched upon in coming chapters or supplemental materials online.

Models could be stochastic, with inputs uncertain rather than deterministic. For example, we do not know if the wind will blow for the next hour, but past observation may indicate there is a 50% probability it will blow. We do not know if our drilling efforts will result in a dry hole or a well, but geological information might suggest there is a 10% chance it will be a dry hole. Both simulation and optimization models could have stochastic inputs.

Game theory models

Simulation models often follow equilibrium or optimization paths. However, game theory models, which are another form of simulation model, show interactions by a small number of players with a variety of options that affect the outcome of interest. A player's strategies are chosen to take into consideration the uncertainty of other player's actions, possible payoffs under different strategies, and the players' risk aversion. For example, the Norwegian government built game theory models of the European natural gas market to help decide when to add capacity when they were one of three external suppliers—Norway, Russia, and Algeria—to the continental European natural gas market. See chapter 10 for some game theory models in the context of the European natural gas market. For a straightforward introduction to game theory, see Rosenthal (2011). For more advanced modeling, see Tirole (1993), which is still a popular graduate textbook in industrial organization, with game theory models applied to industry. For a good introduction to game theory, see Watson (2013).

Experimental economics

In coming chapters, we will see that many economic models are predicated on assumptions about how economic players and institutions behave. Consumers are considered to be rational, with known preferences that do not change. Given their preferences and constraints, they are assumed to maximize their satisfaction, which is referred to as *utility*. Likewise, producers are thought to optimize by maximizing profits and minimizing costs. Under certain conditions, competitive markets are thought to maximize societies' economic welfare, but at other times, they are thought to fail. Experimental economists do experiments in controlled settings on real people to determine whether people's actual behavior conforms to assumptions of economic theory. Two pioneers in this field are Nobel laureates Vernon Smith and Daniel Kahneman. Dr. Smith's experiments in competitive markets have been instrumental in the design of new wholesale electricity markets, while Dr. Kahneman and coauthors have established numerous biases from rational behavior, especially regarding decisions and perceptions under uncertainty and across time. Much of this work is discussed in the fascinating book by Kahneman, *Thinking, Fast and Slow* (2011). Kagel and Roth (1995) review many of the issues that have been considered by experimental economists.

Summary

From the big bang until the Quaternary (the age of humans), energy has had a pervasive influence on the evolution of the natural universe. As humans have struggled and evolved, energy has been a key ingredient of economic progress as well. For most of human history, energy needs were met by renewable sources.

Between 500 million years ago and 2 million years ago, vast amounts of carbon became sequestered, with heat and pressure in the absence of oxygen turning them into fossil fuels. In recent history, especially since the Industrial Revolution, technology has helped to unlock and allow use of this vast cache.

First coal, then oil and natural gas, and finally nuclear energy have been harnessed to largely replace muscles, sunshine, wind, and trees. Although periodic crises have appeared and there has been concern about running out of fossil fuels, technologies so far have helped us dig deeper. Technology allows us to use seismic imaging and better software to see formerly hidden resources, force more oil and gas from underground, and squeeze more energy out of ever-poorer resources as we move from conventional to nonconventional oil and natural gas. Energy efficiency has improved in vehicles, power plants, equipment, appliances, and buildings, with more improvements yet to come.

However, inevitably the fossil age will come to an end. In the future, we expect energy will provide for the same needs as in the past—heating, lighting, cooling, transportation, communication, electricity, and mechanical power. How much will be needed and how it will be provided will vary. Predicting and shaping these needs will be aided by the application of economic modeling, expert judgment, and economic analysis. Accurate knowledge of prices and energy needs will allow producers to have capacity available, consumers to invest in the appropriate capital stock, banks to finance the best technologies, and governments to design optimal policies.

A number of modeling and analytical techniques have been briefly introduced in this chapter. Such models and techniques include extrapolating historical trends and simulation, statistical techniques, judgment, surveys, scenarios, input-output, energy balances, end-use models, engineering or process models, optimization models, game theory models, and experimental models. You may use these techniques as stand-alone simpler models, or in the future, you may combine and use them as building blocks in large integrated models of energy systems.

Models, such as input-ouput, may help us keep track of things and keep our models internally consistent. Models may be used to simulate policies, outcomes, and interactions. However, these simulations are only as good as the underlying assumptions about human behavior.

Some models are used for forecasting. These models can range from the simple (judgment and trends) to the complex (statistical), but as demonstrated, forecasting is fraught with difficulty. The future is unknown and probably unknowable, as a variety of economic, political, cultural, and technological issues influence the evolution of energy markets. All these issues are shrouded in uncertainty, making the forecaster's job difficult.

In addition, the forecasts themselves may influence the outcome. A good forecast may be believed, and actions may be taken to cause it to come true—a self-fulfilling prophecy. A bad forecast may be resisted, and actions may be taken to cause it to not come true—a self-defeating prophecy. Nevertheless, forecasting and imagining how the future might unfold is still a useful exercise. We will come back to many of the models mentioned above, and I urge you to use the models to make your own forecasts as well as to evaluate your own and others' forecasts.

The farther out the forecast, the more difficult the task becomes. We may need to look back less and forward more, turning to our imagination and scenarios. Once the sun winks out, humans, if we are still around, will probably have moved on, or we will have to find an alternative source of energy. But in the meantime, we have an abundance of energy, although some is hard to unlock. In the coming chapters, we discuss the analytical tools to help us use our energy bounty wisely.

3

Perfect Competition and the Coal Industry

All of us form models of the world in our heads, which help us to understand what happens around us.

—Professor Marji Lines,
Department of Statistical Science, University of Udine, Italy

Introduction

Coal, the world's most abundant and widely distributed fossil fuel, has a long and venerable history. It was used in households and industry in China 2,000 years ago, with the Chinese industry becoming well developed by the 12th century CE. While oil was not yet a twinkle in some wildcatter's eye, coal was fueling the Industrial Revolution in England. As industry spread to the European continent and later to North America and eventually Japan, coal was still king.

Water encroachment initially limited mine depth and thus coal supply. Then in 1712, Newcomen's coal-driven steam engine first pumped water from mines. The more efficient steam engine by Watt pumped water by 1765, and in the coming century, coal was the fuel source for paddle boats, trains, and other machinery. Coal was often the fuel of choice because of its higher energy content and relatively easily accessible reserves compared to a declining wood supply.

In the middle 1800s, Great Britain produced slightly more than 60% of the world's coal from around 3,000 mines or collieries. It exported around 15% of its production. The United States and Russia together accounted for another 20% of world production (fig. 3–1). By 1900, the United States had surpassed the United Kingdom in coal production, and in the 1920s, the United States consumed about one-half of the world's coal production.

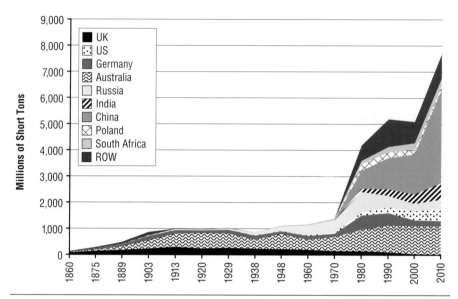

Fig. 3–1. World historical coal production by major country
Sources: Jevons (1965); Gordon (1987); and BP (2012).
Note: UK = Great Britain, later the United Kingdom; Germany = Prussia, then Saxony, then Germany; Russia = the Russian Empire, then USSR, then Russia; ROW = rest of the world.

Coal consumption has always been strongly tied to industrial production, and we see its sensitivity to war and the economic cycle. Thus, coal consumption fell from 1913 to 1920, largely as a result of falling consumption in the economic downturn in Germany after World War I and in Russia after the revolution. During the Great Depression, consumption fell once again, with the United States hit the hardest. Germany by then was rearming and showed modest increases in consumption, while Russia's economy, rather isolated from the rest of the world under Stalin, was rapidly industrializing. During this time, Russian coal production increased an average of 12% per year.

At the end of World War II, coal was still king. It was more than two-thirds of Japanese and former Soviet Union (FSU) energy consumption and production and more than three-fourths of energy consumption and production in Western Europe. It had an even higher share in Eastern Europe. Oil did not pass coal in the United States until 1951.

Postwar fears of coal shortages in Europe and Japan were assuaged as cheap Middle Eastern oil continued to make inroads, and Soviet oil and natural gas exports continued to increase. Oil consumption had passed coal consumption in all of these areas but Eastern Europe by the mid-1960s. Coal is now less than one-fourth of energy consumption in Europe, the former Soviet Union, and Japan. Production has fallen in Europe and has nearly been phased out in Japan. Coal was also dominant in China and India, as it is even today. Although coal's share of the total energy mix has fallen, production, consumption, and imports have increased to fuel the ongoing industrial revolution in these two behemoths (UN, n.d.a; US EIA, n.d.d).

Almost 70% of the world coal reduction in the 1990s was in the former Soviet Union, where coal consumption fell by one-half as their economies crashed. Germany's consumption fell by somewhat less than one-half as the country phased out high-cost mines as well as dirty lignite production for a cleaner former East Germany.

China passed the United States as the world's largest coal producer in 1987. Further, note the dramatic run up in production from 2000 to 2010. China's coal consumption increased on average by 8.5% a year, and India's increased 6.5%, reflecting their domestic coal reserves and the relative strength of their economies. China's real GDP increased an annual average of 10.0%, while India's GDP increased at an annual average of 7.5% during this period. More than 70% of the increase in production can be attributed to China, while almost 20% of the remainder is divided nearly equally between India and Indonesia. Indonesia is exporting much of its new capacity to a booming Asian market (BP 2012) (IMF, n.d.).

The current production shares of the world's more prominent coal producers are shown in figure 3–2. You can see the dominance of China, followed by the United States and India. All three produce largely for their domestic market, while China and India are net importers. Coal still dominates consumption in the domestic market in a number of these countries, including India, China, Poland, South Africa, and Australia.

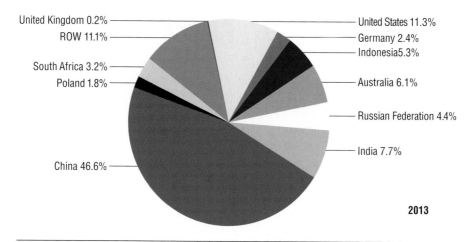

Fig. 3–2. Percent of world coal production by major producer in 2013
Source: BP (2014).
Notes: ROW = rest of the world. Total world production is 7,864.5 million tonnes. Multiply by 1.1 to get short tons.

Growth rates of coal production have varied over time, as well. They averaged more than 4% per year on a global basis from 1860 to 1913. More than 45% of this output came from a rapidly industrializing United States, with annual average increases of 6.8% during this period. Then with a world war and a world recession, global coal production in 1938 had not regained its prewar level from 1913. World production increased more than fourfold from 1938 to 1990. However, during this time period, it never regained its pre-1913 growth rates except in the 1970s, when the oil crisis and surging oil prices caused industry and electric utilities to switch toward cheaper and politically safer coal. Falling consumption in the 1990s was followed by growth rates after 2000 of 4.7%, rivaling those of the heady years before 1913, when coal was king.

Fossil fuel markets are rarely dull. We can see this from coal prices, which have fluctuated rather dramatically, as shown by US coal prices since 1800 (fig. 3–3).

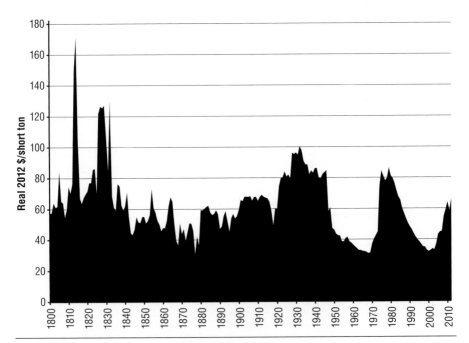

Fig. 3–3. US historical coal prices adjusted for inflation

Sources: The price variable is bituminous price deflated by the consumer price index. Price from 1949 to 2012 is from US EIA (n.d.b), which is extrapolated back to 1800 using the price of anthracite coal. Price of anthracite coal from 1800 to 1949 and consumer price index from 1800 to 1913 are from US DOC (1975). Consumer price index from 1913 to 2012 is from the US BLS (n.d.).

The price has been adjusted for inflation to real 2012 dollars. (See http://dahl.mines.edu/st03/st03.pdf, question 34, if you want to see how to use a price index to change nominal to real dollars.) It has averaged about $62 per short ton over the past 200 years. (Multiply by 1.1 to convert price to dollars per metric tonne.) It has fluctuated considerably, which is quite normal for fossil fuels. One way to measure this fluctuation is to use its deviation around the mean, called *standard deviation* (σ), as follows:

$$\sigma = \sqrt{\sum_{i=1}^{T} (P_i - \bar{P})^2 / (T - 1)} = \$21.97$$

where

P_i is the price in year i,

T is the sample size (201), and

$\bar{P}$ is the average price $= \sum_{i}^{T} (P_i)/T = \$62.31$.

We expect the majority of observations to be within one standard deviation of the mean and find this to be the case, since more than 70% of the prices are between $40.24 and $84.28 ($62.31 − $21.97 and $62.31 + $21.97). The maximum price was $170.65 in 1814, the last year of the War of 1812. It then made a choppy descent to its minimum price of $30.69 in 1877. Prices generally trended up from 1870 to 1913, when coal growth was strong. After an end-of-the-war dip in 1918 and 1919, it resumed an upward trajectory through the Roaring Twenties. This was followed by a sharp downward trend a couple of years into the Great Depression in the 1930s. Thereafter, as oil and gas and other sources made inroads, the trend was generally down until the late 1960s, when it again rose with the commodity boom of the early 1970s and the two oil shocks in 1973 and 1979. With the oil price collapse in late 1985, coal price also fell, but long-term contracts made its descent more leisurely. By 2000, after the Asian financial crisis, price was once again near its historical low. The lull was not long. With the recent commodity boom starting in 2004, price rebounded to its historical average.

These prices are the result of all the economic forces that have impinged on the coal market, including the strength of the economy, the prices of factors of production, and the price of other goods related to coal and technology. To help us better understand how production and coal prices have evolved, we turn to our first economic model.

Perfect Competition

The fundamental challenge in economics is to allocate scarce resources across competing uses. A consumer with a restricted budget may have to choose between buying gasoline or blue jeans. A wealthy consumer may have to choose between driving to a movie and ordering pay-per-view, where time rather than cost is the constraint. An underground coal mine owner may choose between longwall and room-and-pillar mining. A government may decide to allocate more research and development (R&D) funds to clean coal technologies or to renewable resources. In each case, the decision maker will want to make an optimal allocation.

Such allocations are typically done either by the private sector through markets or by the government. Economists tend to favor the private sector for such allocations under ideal market conditions. An *ideal market* requires that an industry is perfectly competitive, property rights are well-defined, necessary market information is easily accessible, externalities are few, and the industry does not have decreasing average costs as production increases. We will learn more about why this is an ideal case in this and subsequent chapters. With well-defined property rights, coal producers have exclusive rights to the coal reserves they own or lease; there are no external costs or benefits that accrue to others as the result of coal production or consumption. Again, we know this is a simplification. Both coal production and consumption produce pollution (a negative externality), as will be discussed in chapter 11. If an industry has decreasing costs, unit costs fall as the size of the firm increases. In such an industry, monopolies have a tendency to develop, as will be discussed in chapter 5.

To begin, let us build a perfectly competitive model for the coal industry. We start with this powerful model because economists hold it up as an ideal, and its components are building blocks of more complicated models to come. Even though competition is never perfect, it still offers insights into reasonably competitive markets.

If the model is competitive, we have free entry into and exit from the industry, and sufficient buyers and sellers so that no one can set the price. Thus, each buyer and seller must take the market price as given. Free entry and exit assures that no buyer or seller can develop and maintain market power. In a perfectly competitive market, market participants have perfect information. They know prices and other information relating to their market decision. For instance, coal sellers have information on coal mining technologies, factor prices, and output prices.

Although successive reforms in the coal industry have striven to make the Chinese coal market more competitive and productive, this model may not yet be truly representative of China, the world's biggest coal producer and consumer (Lin and Zhu 2001).

Historically, most of the coal in the People's Republic of China has been owned and produced by state-owned mines managed by a central government authority, with a fringe of small town and village enterprises (TVEs). The TVEs accounted for around 5% of production in 1949. The TVEs increased during the Cultural Revolution (1966–76), and by the time Deng Xiaoping opened up the Chinese economy in 1978, their share of coal production had increased to about 15%. From 1949 to 1985, China had a single coal price administered by the central government. Since the price was not increased in line with costs, the coal mining industry suffered from chronic deficits (Tu 2011).

Liberalization of the coal industry started when coal shortages caused the Chinese government to urge collectives and individuals to invest in the coal industry. The TVEs would be allowed to control and freely market this production.

This caused an amazing spurt of production among the TVEs. The number of small mines shot up, and their total production began encroaching on large state-owned coal company's markets until the TVEs overtook them in 1994. Then other factors, including a coal surplus and pressure from the big state-owned mines with excess capacity and high debt, led to a reversal of government policy. In addition, policy changes were influenced by problems with environmental degradation, tax evasion, and high accident rates in the TVE mines. The government started a campaign to close TVE mines and leave about 100 large state coal companies. This policy received considerable resistance from local authorities, who did not want to lose local sources of revenue from the TVEs (Su 2004). In 2009, the TVEs accounted for about 38% of Chinese coal production, with local and state governments accounting for another 12% (Tu 2011).

In the meantime, the large state coal mines were caught up in a series of economic reforms across China. Pilot reforms to increase manager authority and institute performance rewards in select industries from 1979 to 1983 were followed by increases in enterprise autonomy and manager incentives from 1984 to 1992 in all industries. Ownership restructuring began in 1993. Instead of being sole proprietorships, large state-owned firms were to be corporatized into limited liability companies. The government would still remain the controlling shareholder, but stock could be issued to other owners. The largest of these companies could have stock listed on international exchanges, with shares owned by foreigners (Lin and Zhu 2001). As of 2011, four large Chinese coal companies, Shenhua, China Coal, Yanzhou, and Hidili, were listed on the Hong Kong Stock Exchange, and Yanzhou and Yitai were listed on the New York Stock Exchange. Another 23 companies were only listed on local Chinese exchanges (*China Coal Report* 2011, 2).

Coal prices have also been increasingly liberalized over time. From 1985 to 2002, TVE mine output and new state-owned coal mine output above capacity could be sold at market prices. However, from 1994 to 2001, the government still imposed prices for coal sold to power plants. In 2002, the government took another step toward market pricing. By 2006, full market pricing was instituted, allowing coal producers and utilities to negotiate contract prices without interference from the central government. In 2008, the coal industry was decentralized and control was devolved to the state governments (Tu 2011).

In the latest five-year plan, the government announced that it would further consolidate the coal industry by mergers and acquisitions. The goal is to have 10 large coal mining companies, with a minimum of 100 million tonnes of annual capacity each, and 10 medium-sized companies, with a minimum of 50 million tonnes of annual capacity. Together, these 20 companies are to produce 60% of China's coal in 2015, which is targeted for a cap at 3,900 million tonnes. Further, the 10,000 coal companies existing in 2012 were to be reduced to 4,000 or less (*China Bystander* 2013).

Production for the 10 largest Chinese coal companies in 2010 is shown in table 3–1, which accounted for about 30% of China's coal production.

Table 3–1. Ten largest coal companies in China in 2010

Company	Production (10^6 tonnes)
Shenhua	327.0
China Coal Energy Group	125.0
Shanxi Coking	80.8
Shanxi Datong	74.5
Shaanxi Coal and Chemical	71.0
Anhui Huanan Mining	67.1
Henan Coal and Chemical Group	57.0
Shanzi Lu'an Group	55.1
Heilongjiang Longmei Group	54.9
Shanxi Yanzhou Coal Mining Group	49.7
Other Companies	2,277.9
China Total	**3,240.0**

Source: China Coal Report (2011, 2).

The United States, which is the second largest coal producer, has had a longer tradition of competition and private ownership. Peabody, the largest US coal company, controls less than 20% of the coal production, and the top four companies control less than 50% (see table 3–2). Australia, at fourth place, also has private ownership, with the largest four producers accounting for 41% of production.

Table 3–2. Ten largest US coal producers, 2010

Rank	Controlling Company Name	Production (1,000 short tons)	Percent of Total Production
1	Peabody Energy Corp.	202,237	18.5
2	Arch Coal Inc.	160,279	14.6
3	Alpha Natural Resources LLC	116,394	10.6
4	Cloud Peak Energy	95,596	8.7
5	CONSOL Energy Inc.	62,089	5.7
6	Cerrejon Coal Co.	33,300	3.0
7	Alliance Resource Operating Partners LP	32,949	3.0
8	Energy Future Holdings Corp.	32,610	3.0
9	Peter Kiewit Sons Inc.	29,998	2.7
10	NACCO Industries Inc.	27,904	2.5
	Other	302,272	27.7
	US Total	**1,095,628**	**100.0**

Source: US EIA (2012e).
Note: Multiply by 0.907 to convert to metric tonnes.

The third largest coal producer is India. Since its coal is essentially produced by large state-owned enterprises, it is not likely to be very competitive. Coal India, with seven subsidiaries, owned by the Indian government, is the largest coal company in the world. In 2010, it produced 569.9 million tonnes, about 80% of India's coal, with most of the remainder produced by Singareni Colliers, Ltd, jointly owned by the state of Andhra Pradesh and the Indian government (IEA 2011b).

Although not every coal consumer is able to buy from every coal producer because of transportation costs, excessive profits in one mining area are likely to bring new entrants in the form of other coal producers or other energy sources. (See IEA [n.d.c] for coal trade statistics.) This threat of entry by other producers is referred to as *market contestability*. As the coal industry is a global industry and coal is the second largest product by weight to be traded internationally after oil, this contestability can come from large foreign producers as well as other domestic companies. Examples of other large coal companies worldwide not already mentioned are given in table 3–3. Notice that the 10 largest companies in the table amount to less than 10% of global production.

Table 3–3. Ten additional large world coal producers

Company	Country	Production (10^6 tonnes)
BHP Billiton	UK-Australia	104
RWE	Germany	99
Anglo American	UK-South Africa	97
SUEX	Russia	89
Xstrata	UK-Switzerland	80
Rio Tinto	UK-Australia	73
Pt Bumi Resources	Indonesia	59
Kuzbassrazrezugol	Russia	50
Banpu	Thailand	43
Sasol	South Africa	43
Total of 10 Companies		633
Total World Production		**7,273**

Sources: IEA (2011b) and BP (2012).
Note: To convert from metric tonnes to short tons, multiply by 1.1.

An additional requirement for a competitive market is that the product be *homogeneous*, or each unit of the product is just like every other. For coal, we know this is not quite true. Coal can be broadly categorized by two uses. Coking or metallurgical coal is used in the production of iron and steel, and thermal or steam coal is burned to produce heat for electricity generation and other industrial processes. (For more on coking coal, see http://dahl.mines.edu/st03/st03.pdf, answer to question 33.)

Coal burned for process heat and to raise steam is not as valuable as coking coal but is more plentiful and widely used. Average import price *cif* (including customs, insurance, and freight) for coking coal into OECD countries in 2010 was $169.38 per tonne, while steam coal price averaged $99.23 (IEA, n.d.g). Global coking and steam coal consumption in that same year were 0.879 and 5.437 billion tonnes, respectively (IEA, n.d.b). Steam coal varies by energy content and impurities. For example, in the United States, eastern coal has about 24 million British thermal units per short ton (MMBtu/ston), central coal about 22 MMBtu/ston, and western coal about 18 MMBtu/ston. Lignite in the Dakotas and Texas has more moisture, with a still lower heat content. As explained in chapter 1, 1 Btu is 1,054.35 joules (J), or slightly more than 1 kilojoule (kj). This is equivalent to about one-fourth (0.252) of a kilocalorie (kcal). The ranges of energy content by general coal type are shown in table 3–4.

Table 3–4. Energy content by coal type

Coal Type	% Carbon	1,000 Btu/ston* Range	
Lignite	30	10,000	15,000
Subbituminous	40	16,000	20,000
Bituminous	50–70	22,000	30,000
Anthracite	90	> 28,000	

Source: Hinrichs (1996).
Notes: To convert to kJ/tonne, multiply the value in Btu/ston by 1.162. *British thermal units per short ton.

This same variation in energy content can be seen internationally in table 3–5, which includes coal production, consumption, reserves, and energy content of the world's most important coal producers and consumers. These values are for thermal, not metallurgical coal. Previously, we have seen the values for reserves in tonnes or short tons. In table 3–5, the coal has been adjusted for energy content and is given in million tonnes of oil equivalent (Mtoe).

Table 3–5. World coal production, consumption, and reserves, 2010

Region/ Country	Mtoe Prod. 2010	Mtoe Cons. 2010	Mtoe Net Exports 2010	Recoverable Reserves 10⁶ Tonnes Anthr. & Bitum.	10⁶ Tonnes Lignite Subbitum.	10⁶ Tonnes Total Coal	Heat (1,000 Btu/ tonne, 2006)
North America							
United States	552	525	28	108,501	128,794	237,295	22,392
Canada	35	23	11	3,474	3,108	6,582	23,099
Mexico	4	8	–4	860	351	1,211	19,315
Total	592	556	35	112,835	132,253	245,088	
South & Central America							
Colombia	48	4	45	6,366	380	6,746	27,087
Other	5	20	–15	524	5,238	5,762	
Total	54	24	30	6,890	5,618	12,508	

Region/ Country	Prod. 2010 Mtoe	Cons. 2010 Mtoe	Net Exports 2010 Mtoe	Anthr. & Bitum. 10^6 Tonnes	Lignite Subbitum. 10^6 Tonnes	Total Coal 10^6 Tonnes	Heat (1,000 Btu/ tonne, 2006)
Europe & Eurasia							
Germany	44	77	−33	99	40,600	40,699	10,905
Kazakhstan	56	36	20	21,500	12,100	33,600	18,297
Poland	55	54	2	4,338	1,371	5,709	17,083
Russia	149	94	55	49,088	107,922	157,010	20,980
Turkey	17	34	−17	529	1,814	2,343	9,381
Ukraine	38	36	2	15,351	18,522	33,873	21,778
United Kingdom	11	31	−20	228	—	228	24,709
Other	60	124	−64	1,857	29,285	31,142	
Total	431	487	−56	92,990	211,614	304,604	
Middle East							
Total	1	9	−8	1,203	—	1,203	
Africa							
South Africa	143	89	54	30,156	—	30,156	23,486
Other Africa	2	7	−5	1,362	174	1,536	
Total	145	95	50	31,518	174	31,692	
Asia Pacific							
Australia	235	43	192	37,100	39,300	76,400	22,526
China	1,800	1,904	−104	62,200	52,300	114,500	22,216
India	216	278	−61	56,100	4,500	60,600	18,130
Indonesia	188	39	149	1,520	4,009	5,529	25,628
Japan	1	124	−123	340	10	350	
South Korea	1	76	−75	—	126	126	
Taiwan	0	40	−40	0	0	0	
Thailand	5	15	−10	—	1,239	1,239	12,084
Vietnam	25	14	11	150	—	150	23,336
Total	2,509	2,385	125	159,326	106,517	265,843	
World Total	3,731	3,556	176	404,762	456,176	860,938	

Sources: US EIA (n.d.d); BP (2012).
Notes: tonne = metric tonne; Mtoe = million tonnes oil equivalent. Multiple by 1.5 to get a representative tonne of coal equivalent. Negative net exports are net imports.

From the table you can see the largest consumers (China, United States, India, Japan, and Russia), the largest producers (China, United States, India, Australia, and Russia), the largest exporters (Australia, Indonesia, Russia, South Africa, and Colombia), and the largest importers (Japan, China, South Korea, India, and Germany). Under net exports, positive numbers indicate net exporters and negative numbers indicate net importers.

Higher carbon coal has a higher energy or heat content and is more valuable. The heat contents are shown in the last column of table 3–5.

Electricity and heat generators worldwide buy about 65% of the coal used, and industry buys nearly another 20%. Table 3–6 shows the use of coal and its importance as a fuel in major sectors. About 40% of the world's heat and power is generated by coal, and about 20% of industrial energy use is coal. You can compute similar statistics for many countries for the latest year available from the IEA, as noted under table 3–6. (See chapter 5 for a discussion of electricity generation from coal and other sources.)

Table 3–6. Global coal use by major sector, 2011 (ktoe)

	ktoe	% of Coal Used in Each Sector	Coal's Share of Energy Use
Electricity & heat generation	2,365,638	64.8%	41.2%*
Other transformation & losses	362,722	9.9%	
Industry	728,932	20.0%	28.5%
Other uses	135,461	3.7%	2.4%
Non-energy use	39,223	1.1%	4.8%
Nonspecified	16,672	0.5%	13.4%
Total	3,648,648	100.0%	

Sources: IEA (n.d.f); IEA (n.d.d);
Note: ktoe = kilotonnes of coal equivalent

Energy Demand and Supply

I use our ideal market model for coal. First, divide the coal market into two groups. Buyers are represented by a demand equation and sellers by a supply equation.

Demand

The quantity purchased as a factor of production by one buyer will be influenced by the following:

- the price of coal (P_c)
- the price of substitutes to coal (P_{sb}), such as oil and natural gas
- the price of complements to coal (P_{cm}), such as coal boilers
- technology for coal use (T)
- the price of the output produced (P_{ot})
- energy policy (P_{ol})

Purchases by all buyers (Q_d) will also be influenced by the number of buyers $(\#buy)$.

Thus, we can write market demand as follows:

$$Q_d = f(P_c-, P_{sb}+, P_{cm}-, T+/-, P_{ot}+, P_{ol}+/-, \#buy+) \qquad (3-1)$$

The signs immediately to the right of the variables indicate the effect that the variable will have on coal purchases or the sign of the partial derivative of coal purchases with respect to that variable. For example, $\partial Q_d/\partial P_c$ tells us how much coal consumption will decrease or increase for an increase or decrease in the price of coal. This value may change depending on the price before the change. (For a review of derivatives and partial derivatives, see any standard calculus text. For a review of calculus applied to economic modeling, see Dowling [1992] or Chiang and Wainwright [2006]. You can test your calculus skills at http://dahl.mines.edu/courses/dahl/calc/.)

We expect that $\partial Q_d/\partial P_c$ would be negative. As the price of coal increases, electric generators and other users would try to economize on coal use and switch to other fuel sources, and this effect might be very small in the short run but much larger in the long run. $\partial Q_d/\partial P_{sb}$ should be positive. If the price of a substitute (such as natural gas) increases, we would expect a shift toward coal use. $\partial Q_d/\partial P_{cm}$ should be negative. If the price of coal boilers increases, coal becomes a less desirable fuel, and buyers may switch to competing fuels.

The sign of $\partial Q_d/\partial T$ is uncertain. It depends on the type of technological changes taking place. If the technology increases the productivity of coal, there should be a shift toward coal from other fuels. Alternatively, if the technology increases the productivity of other fuels, there may be a shift away from coal toward alternative fuels. The sign of $\partial Q_d/\partial P_{ot}$ should be positive. If the price of output goes up, we will want to increase production, and we will need more coal. Instead of the price of output, the quantity of output or some measure of economic activity is often used, which we will designate as Y. This is particularly appropriate if the buyer is a consumer rather than someone using the energy as a factor of production. Consumers will need to consider their income in making consumption decisions. The sign of $\partial Q_d/\partial P_{ol}$ depends on the policy.

For example, US energy law prohibited new under-the-boiler use of natural gas in 1978, favoring coal, as did German coal price subsidies. More recent environmental regulations are more likely to decrease coal use. The sign of $\partial Q_d/\partial \#buy$ should be positive, since more buyers will increase coal consumption or the quantity of coal demanded.

Supply

We represent suppliers by an equation in which we write quantity supplied (Q_s) as a function of the following:

- the price of coal (P_c)
- the price of factors of production for coal, such as labor and capital (P_f)

- the price of similar goods that coal miners could produce (P_{sim})
- the price of by-products or complements of coal production (P_b)
- coal production technology (T)
- government coal policies (P_{ol})
- the number of sellers $(\#sel)$

Thus, we can write the supply of coal as the following:

$$Q_s = f(P_c+, P_f-, P_{sim}-, T+, P_b+, P_{ol}+/-, \#sel+) \tag{3–2}$$

In equation 3–2, we expect $\partial Q_s/\partial P_c$ to be positive. As the price of coal increases, coal producers will want to produce and sell more coal. $\partial Q_s/\partial P_f$ should be negative. As the prices of factors of production increase, suppliers will produce less coal. $\partial Q_s/\partial P_{sim}$ should be negative. If the price of a similar good that coal producers could produce goes up, they might switch to the similar good. For example, they could acquire other property and switch to mining other minerals or producing gravel. The sign of $\partial Q_s/\partial T$ should be positive. Technical change should reduce costs and increase coal production.

The sign on $\partial Q_s/\partial P_{ol}$ depends on the policy. For example, laws passed to improve mine safety and reclamation have increased costs and decreased production, whereas policies encouraging research and development in coal mining have increased production. The sign of $\partial Q_s/\partial P_b$ should be positive. If the price of by-products that are complements to coal production increase, profits from coal mining increase, and coal production should increase. The sign of $\partial Q_s/\partial \#sel$ should be positive, since more sellers will increase coal production.

Equilibrium price and quantity

To show how markets work, let's develop an example. Let demand and supply be equal to the following linear functions:

$$Q_d = 100 - 2P_c + 3P_{sb} - 4P_{cm} + 0.10Y \tag{3–3}$$

$$Q_s = 6 + P_c - 1P_k - 0.2P_l - 0.8P_{nr} - 1.5P_{sm} \tag{3–4}$$

where

P_c is the price of coal,
P_{cm} is a complement to coal consumption, such as a boiler, set = 10,
P_k is the price of capital, set = 2,
P_l is the price of labor, set = 3,
P_{nr} is the price of other natural resources used in production of coal, set = 5,
P_{sb} is the price of a substitute to coal, such as natural gas, set = 6,

P_{sm} is the price of a similar product that a coal producer could produce, set = 4, and Y is a measure of economic activity, set = 954.

Standard procedure in economics requires that we hold all variables but price and quantity constant (called *ceteris paribus* or "all else equal") and develop demand and supply equations as follows. First, fix the ceteris paribus values in each equation, and you will get the following:

$$Q_d = 100 - 2P_c + 3 \times 6 - 4 \times 10 + 0.1 \times 954 = 173.4 - 2P_c \qquad \textbf{(3–5)}$$

$$Q_s = -6.6 + 1P_c \qquad \textbf{(3–6)}$$

(For a review of the algebra needed to solve linear equations, go to any standard algebra book or Speigel [1995]. You can test your algebra skills at http://dahl.mines.edu/courses/dahl/alg/.)

Next, invert the demand curves, or solve for price as a function of quantity, inverse demand, as follows:

$$Q_d = 173.4 - 2P_c$$

$$2P_c = 173.4 - Q_d$$

$$P_c = 173.4/2 - (1/2)Q_d = 86.7 - 0.5Q_d$$

Similarly solve for inverse supply:

$$P_c = 6.6 + Q_s$$

Note the above supply and demand curves are both linear functions. When $Q_d = 0$, $P_c = 86.7$, and when $P_c = 0$, $Q_d = 173.4$. We can graph these two points and connect them, as seen in figure 3–4.

Do the same for supply. When $Q_s = 0$, $P = 6.6$, and when $Q_s = 10$, $P = 16.6$. Again, graph these two points and connect them, as seen in figure 3–4.

To forecast equilibrium price and quantity in this model, we would find where quantity demanded equaled quantity supplied. The graph shows equilibrium to be at a price of $60, near the historical average, and at a quantity near 50. To be more precise, solve with our functions by setting quantity demanded equal to quantity supplied, as shown in the following:

$$Q_s = Q_d$$

$$Q_d = 173.4 - 2P_c = Q_s = -6.6 + 1P_c$$

$$173.4 + 6.6 = 1P_c + 2P_c$$

$$180 = 3P_c$$

$$P_c = 180/3 = 60$$

We can solve for equilibrium Q using either the demand or supply equation:

$$Q_d = 173.4 - 2(60) = 53.4 \text{ or } Q_s = -6.6 + 60 = 53.4$$

To check whether this is likely to be a stable equilibrium, suppose that the price is $70. Then quantity demanded would be as follows:

$$173.4 - 2 \times 70 = 33.4$$

Quantity supplied would be as follows:

$$-6.6 + 70 = 63.4$$

There is excess quantity supplied, and coal price is likely to fall. A similar argument can be used to show that equilibrium is stable from below as well.

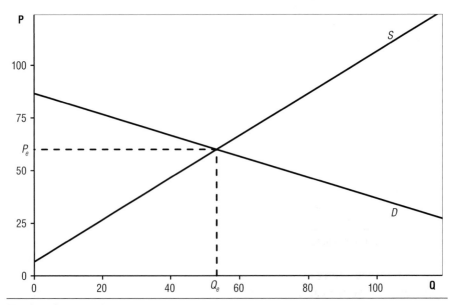

Fig. 3–4. Supply and demand

Shifts in Supply and Demand

Now suppose that one of our ceteris paribus variables (those variables held constant) changes. Let the price of natural gas go to 15. With gas more expensive, electricity generators might run their gas turbines less and their coal generators more. This would shift our demand curve to the right, as in figure 3–5, making the demand curve as follows:

$$Q_d = 200.4 - 2P_c$$

We call this a *change in demand*, since it shifts the whole demand curve. This shift would move us along the supply curve to a higher price and quantity. This movement along the supply curve to a different quantity in response to a higher price is called a *change in quantity supplied*.

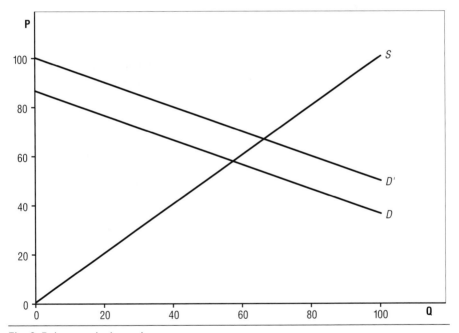

Fig. 3–5. Increase in demand

Sometimes we don't know the magnitude of variable changes or the specific functions, and so we can only do qualitative analysis. If large coal deposits for the export market are developed in Indonesia, supply shifts out, lowering price and raising quantity. If a financial crisis in the United States and Western Europe reduces demand, it lowers price and quantity. Some events might cause both curves to shift. Higher interest rates might shift demand away from coal toward other less capital-intensive fuels, which would lower price and quantity. Higher interest rates would also raise the costs to suppliers and reduce supply, thus raising the price and lowering quantity. Since both of these effects would lower quantity, we would expect quantity to fall. Because the reduction in demand lowers price, and the decrease in supply raises price, the change in price would depend on which effect is larger.

Often, more than one event at a time impinges on a market. For example, in 1973 and again in 1979, oil prices increased dramatically. We would expect this to increase the demand for coal, as in figure 3–5, and increase price and quantity of coal consumed. However, during this same period, increasing environmental and safety regulations in the United States, the largest coal producer at the time, caused productivity to decrease from 2.20 to 1.82 short tons per miner-hour (US EIA 2012d). More recently, China has been closing small, dangerous, and highly polluting mines. These types of events decrease the supply of coal to S' (fig. 3–6). At any given price, less coal would be supplied. Supply would move left, which would increase price and decrease quantity.

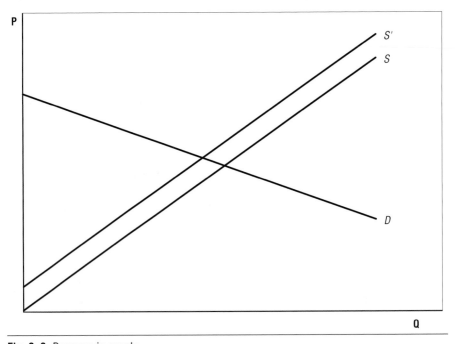

Fig. 3–6. Decrease in supply

What is the net effect of these two market changes? Since both increase coal price, we conclude that coal price will increase. The increase in demand increases quantity, while the decrease in supply decreases quantity. The net effect would depend upon which effect is larger, so the change in price is positive and the change in quantity is uncertain.

Subsequent to 1979, US coal mine productivity has shown dramatic improvements, with production per miner-hour increasing. Production increased from 1.9 tons per minor hour in 1980 to average more than 5.5 tons per miner-hour in the most recent decade (US NMA, n.d.). (For an Excel model that can be used to simulate supply and demand models similar to the one above, as well as other models in this chapter, go to http://dahl.mines.edu/ch03m.xlsx.)

Demand and Supply Elasticities

Often we want to measure how responsive quantities demanded and supplied are to prices or other variables, or both, to help design energy plans and policy. For example, if coal demand in Asia is very responsive to income growth and income rises or falls sharply, there will be a large effect on the coal market. If coal demand and supply are very responsive to price, only a small change in price will be needed to bring about equilibrium after demand or supply shocks. Economists use elasticities to provide such a measure of responsiveness. The price elasticity of demand is the percentage change in quantity divided by the percentage change in price:

$$\varepsilon_d = \frac{\text{\% change quantity}}{\text{\% change in price}} = \frac{\dfrac{\Delta Q_d}{Q_d}}{\dfrac{\Delta P_d}{P_d}} \tag{3–7}$$

If the price elasticity is –0.5, and price goes up by 100%, quantity demanded falls by 50%. If the price change is very large, we often get a different elasticity depending on whether the price and quantity in the respective denominators are those before (P_{d1}, Q_{d1}) or after (P_{d2}, Q_{d2}) the price and quantity changes. So arc elasticities are typically defined using the average of the respective denominators as follows:

$$\varepsilon_d = \frac{\text{\% change quantity}}{\text{\% change in price}} = \frac{\dfrac{\Delta Q_d}{(Q_{d1} + Q_{d2})/2}}{\dfrac{\Delta P_d}{(P_{d1} + P_{d2})/2_d}} = \frac{\dfrac{(Q_{d2} - Q_{d1})}{(Q_{d1} + Q_{d2})/2}}{\dfrac{(P_{d2} - P_{d1})}{(P_{d1} + P_{d2})/2_d}} \tag{3–8}$$

It is often convenient to rewrite equation (3–7) as follows:

$$\varepsilon_d = \frac{\Delta Q_d}{\Delta P_d} \frac{P_d}{Q_d}$$

If we take very small changes in price or take the limit as DP_d goes to zero, then we can rewrite the previous elasticity in terms of partial derivatives, as follows:

$$\varepsilon_d = \frac{\partial Q_d}{\partial P_d} \frac{P_d}{Q_d}$$

In this case, ∂Q_d represents a change in Q_d for a small change in $P_d (\partial P_d)$.

To compute demand elasticities, go back to the original demand equation (3–3): $Q_d = 100 - 2P_c + 3P_{sb} - 4P_{cm} + 0.1Y$. The price elasticity of demand is $(\partial Q_d / \partial P_d)(P_d / Q_d)$. In this example "price" is the price of coal represented by P_c, so the demand elasticity is $(\partial Q_d / \partial P_c)(P_c / Q_d) = -2(P_c / Q_d)$. This elasticity varies as P_c and Q_d vary. Thus, to evaluate this elasticity, we need to pick a price and values for our other right-side variables and compute a Q_d. Let $P_c = \$52.02$ and our other right-side variables be as above. Then at $P_c = \$52.02$, and from equation 3–3:

$$Q_d = 173.4 - 2P_c = 173.4 - 2 \times 52.02 = 69.36$$

and the elasticity is:

$$= -2 \times 52.02/69.36 = -1.5$$

This elasticity implies that if price declines 1%, quantity demanded goes up by 1.5%. If the demand elasticity is less than −1, quantity responds by a larger percent than the percent price change, and we call the demand *price elastic*. If the demand elasticity is between −1 and 0, the quantity responds by a percentage smaller than the percent price change, and we call the demand *price inelastic*.

If price changes, we can use the demand elasticity to estimate what would happen to sales. If a coal tax increased coal price 10%, and the demand elasticity was −0.2 in the short run, then the following expression applies:

$$\partial Q_d / \partial Q_d = \varepsilon_d \times \partial P_c = -0.2 \times 0.1 = -0.02$$

Coal consumption would fall by 2%. If coal consumption were 500 million tons before the price change, consumption after the price change would be the following:

$$(1 + \partial Q_d / Q_d) \times original\ demand = (1 - 0.02) \times 500 = 490$$

Price elasticities also reveal the relationship between price changes and total revenue. We know that total revenue equals price times quantity sold, and that the demand elasticity is as follows:

$$\frac{\dfrac{\Delta Q_d}{Q_d}}{\dfrac{\Delta P_d}{P_d}}$$

Suppose that this elasticity equals −2 for coal. If price decreases 10%, quantity demanded increases by −2(−0.10) = 0.2 or 20%. The price decrease causes revenue to decrease, but the quantity increase causes revenue to increase. Since the numerator or quantity effect is larger, total revenues increase.

We can see this same effect for different elasticities along our demand equation (3–5) in table 3–7. When price falls from 65 to 63, quantity rises from 43.4 to 47.4. Price elasticity computed from equation (3–8) is −2.819. Since the percentage quantity increase is larger in absolute value than the percentage price decrease, revenues increase. At the center of the demand curve when price falls from 44.35 to 42.35, the price elasticity is −1 or unitary, and the revenues do not change. The increase in revenue from the quantity increase is just offset by the revenue decreases from a lower price. Alternatively, if demand is inelastic, with elasticity at −0.503, the increase in quantity is more than offset by the decrease in price, and revenues fall. Notice how the price elasticity becomes less elastic as we move down the demand curve and is unitary at the center. Such an elasticity pattern holds for all linear demand curves.

Table 3–7. Revenues related to elasticities

Elasticity	P	Q	P*Q
−2.819	65	43.40	2,821.0
	63	47.40	2,986.2
−1.000	44.35	84.70	3,756.4
	42.35	88.70	3,756.4
−0.503	30	113.40	3,402.0
	28	117.40	3,287.2

Note: Elasticities are computed from equation (3–8). Q is computed from equation (3–5).

Income elasticity of demand (ε_y) tells us how sensitive sales are to income changes.

$$\varepsilon_y = \frac{\% \text{ change quantity}}{\% \text{ change in income}} = \frac{\dfrac{\Delta Q_d}{Q_d}}{\dfrac{\Delta Y}{Y}}$$

If $\varepsilon_y > 1$, demand is *income elastic*, and we have a luxury good. For a luxury good, sales increase at a faster percentage rate than income. If $0 < \varepsilon_y < 1$, demand is *income inelastic*, and sales increase at a slower percentage rate than income. If $\varepsilon_y > 0$, we have a *normal good*, but if $\varepsilon_y < 0$, we have an inferior good. For example, coal for household heating use has been an inferior good. As households got richer, they used less coal and substituted natural gas, fuel oil, and electricity for heating. Although coal is sometimes used for household heating in developing countries, very little coal is now used in this sector in the industrialized countries.

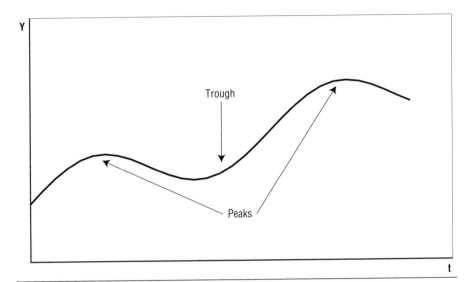

Fig. 3–7. Representative business cycle

Income elasticity also tells us how sensitive demand is to the business cycle. In industrial countries, income tends to change cyclically. Sometimes income is growing, but at other times, we have recessions, and income falls. Figure 3–7 shows a representative business cycle.

When income increases, we are said to be "going up the cycle," and when income falls, we are said to be "going down the cycle." A *peak* is the point where income growth changes from positive to negative. A *trough* is the point where income growth changes from negative to positive. A business cycle would be from one peak to the next peak, or one trough to the next trough. In the post–World War II era until the 1990s, US business cycles tended to average 8 to 10 years. We can discern such cycles for the United States and other countries in figure 3–8.

Activities that increase spending during peaks and decrease spending during troughs are said to be *procyclical.* That is, they tend to increase the cycle. Activities that decrease spending during peaks and increase spending during troughs are said to be *anticyclical.* That is, they tend to moderate or decrease the cycle by lowering the peak and raising the trough. If $\varepsilon_y > 1$, we have a luxury good, and sales are very sensitive to the business cycle, or sales are procyclical. If $0 < \varepsilon_y < 1$, we have a normal good with income inelastic demand. If $\varepsilon_y < 0$, when income increases, we buy less of the good. Such a good is called an inferior good, and sales of the good are anticyclical.

A cross-price elasticity (ε_{cross}) tells us how the quantity demanded of one good changes when the price of another good (P_o) changes:

$$\varepsilon_{cross} = \frac{\% \text{ change quantity}}{\% \text{ change in price of another good}} = \frac{\Delta Q_d / Q_d}{\Delta P_o / P_o}$$

For example, if the cross price elasticity of demand for coal with respect to natural gas is 0.5, then $\Delta Q_d / Q_d = 0.5(\Delta P_o / P_o)$. If the gas price goes up 10%, the percentage change in coal demand is $\Delta Q_d / Q_d = 0.5(\Delta P_o / P_o) = 0.5(0.10) = 0.05$ or 5%. Such a positive cross-price elasticity of demand indicates that the two goods are substitutes in demand. When the price of one good goes up, consumers switch to the other substitute good. If the cross-price elasticity of coal demand with respect to coal boilers is -1.2, and the price of coal boilers falls 20%, the percentage change in quantity of coal demanded is $\Delta Q_d / Q_d = -1.2(-0.2) = 0.24$ or 24%. Such a negative cross-price elasticity indicates the two goods are complements. If the price of the complement goes down, people consume more of the complement good and also more of the good itself.

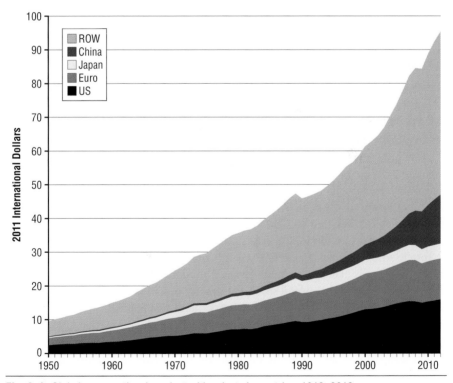

Fig. 3–8. Global gross national product with selected countries, 1913–2012

Sources: Computed from data in Maddison (2010) and World Bank (n.d.).

Notes: Euro-area data before 1980 was created by extrapolating with GDP for Western Europe. ROW = rest of the world.

Supply Elasticities

The responsiveness of quantity supplied to a variable is called the *elasticity of supply* with respect to that variable. It is the percentage change in quantity divided by the percentage change in the variable. We can write the elasticity of supply with respect to price (P) as:

$$\varepsilon_s = \frac{\% \text{ change } Q_s}{\% \text{ change } P} = \frac{\Delta Q_s / Q_s}{\Delta P / P}$$

where again Δ represents a discrete change in the variable.

If the supply price elasticity of coal is 0.89, then when the price of coal increases by 1%, the percentage change in quantity of coal supplied is $\Delta Q_s / Q_s = 0.89(0.01) = 0.0089$ or 0.89%.

As with demand, the cross-elasticity of supply indicates how quantity supplied is related to another price. For example, if the cross-price elasticity of gasoline supply with respect to the price of distillate is –0.2, and the price of distillate increases 1%, the quantity of gasoline produced decreases 0.2%. This negative cross-price elasticity indicates the two goods are similar in production. If the price of one goes up, the supplier shifts to producing this higher-priced product. If the cross-price elasticity of supply is positive, when the price of one goes up, we produce more of both, implying the two goods are complements in supply. For example, if the price of methane goes up, it might stimulate production of coal bed methane from certain deposits, as well as the coal itself.

The time period over which we measure supply and demand elasticities influences elasticity size. In the short run, if the price of coal goes up, coal mines may only be able to increase production a small amount. Since coal mining is very capital intensive, with specialized equipment, it takes time to buy new equipment, and it typically takes four to seven years to open a new mine. Thus, the short-run elasticity may be quite low. However, in the long run (the amount of time required to totally adjust to a price change), production may change much more. Elasticity in the long run is likely to be much larger. The more capital intensive the industry and the longer lived the capital stock, the greater the difference between long- and short-run elasticities. The same is true for demand elasticities.

Although the long run may vary from product to product, it is fairly well-defined. Simply put, it is the time required for total adjustment to take place. The short run is less well-defined and typically depends on the period of interest and is most often a year or less. If you are doing statistical analysis on real data, the short run is often the periodicity of your data. This could be daily or weekly energy prices, monthly data for natural gas in storage, quarterly data on economic indicators, or annual data for oil consumption. If your analysis focused on monthly data for coal supply, the short run would be a month. If your analysis concerned annual data, the short run would be a year, and you would expect the annual response to be larger than the monthly response. The long-run elasticities in both cases should be the same but would be more elastic than in the short-run. For example, Labys et al. (1979) found that the underground supply of US coal during the period 1955 to 1973 had a short-run annual elasticity of 0.07 and a long-run elasticity of 1.31. Thus, with time to adapt, coal producers were many times as price responsive as they were in a single year.

Using Elasticities to Forecast Supply

Supply elasticities are quite useful for policy and planning, as shown in the following examples. Suppose the world price of coal goes up. Government planners in Australia will want to know what will happen to domestic coal production. Because Australia is a large coal exporter, coal production will have implications for employment, GDP, tax revenues, balance of payments, and demand for mining equipment. Likewise, companies in South Africa and the United States will want such information about their own production, as well as that of their competitors. Let coal price go from $60 per tonne to $66 per tonne. Suppose the short-run (one-year) supply elasticity is 0.10, and the long-run supply elasticity is 1.1. Since

$$\varepsilon_s = \frac{\Delta Q_s / Q_s}{\Delta P / P},$$

in the short run:

$$\Delta Q_s / Q_s = \varepsilon_s (\Delta P / P) = 0.10 \times 6/60 = 0.01.$$

Thus, Australian production would go up 0.01 or 1%. Australian production was 415.5 million tonnes in 2011, so production would increase by $0.01 \times 415.5 = 4.155$ million tonnes. In the long run, the increase would be $\Delta Q_s / Q_s = e_s (\Delta P / P) = 1.1 \times 6/60 = 11\%$, for an increase of $0.11 \times 415.5 = 45.71$ million tonnes.

Price Changes from a Supply Disruption

Coal mining is often dirty and dangerous, and labor relations in coal mining have often been turbulent. If strikes lead to a production disruption, elasticities can tell us what happens to price. Demand elasticities imply the following for price changes:

$$\varepsilon_d (\Delta Q_d / Q_d) / (\Delta P_d / P_d) \rightarrow (\Delta P_d / P_d) = (\Delta Q_d / Q_d) / \varepsilon_d$$

Suppose a long strike reduces supply to world markets by 6%. What price change would we need to reduce demand by this same amount? If the short-run price elasticity of demand were –0.20, then world coal prices would increase by the following:

$$(\Delta P_d / P_d) = (-0.06)/(-0.20) = 0.3 \text{ pr } 30\%$$

With supply disruptions, a significant quantity is taken off the market in a very short time. Since elasticities tend to be very small over very short periods of time, the price spike may be rather large in the short run. However, in the long run, demand is more elastic. If the long-run demand elasticity were –0.50, the price would fall in the future until it was only $(-0.06)/(-0.50) = 0.12$, or 12% more than the current price.

Creating Demand and Supply from Elasticities

Elasticities tell us how responsive quantity demanded and quantity supplied are to various economic variables. They can also be used to create demand and supply equations for modeling and for forecasting. Because China is now the largest coal consumer, the second largest global economy, and has experienced relatively rapid growth, you may be interested in forecasting Chinese coal consumption. To do so, let's create a demand equation with the following information:

- Price elasticity $\varepsilon_p = -0.20$
- Income elasticity $\varepsilon_y = 0.90$
- Price per tonne in US dollars = $100
- Consumption in millions of metric tonnes a year = 4,000
- Income in trillions of US dollars = $10

To create a linear demand equation around the above values, use the following:

$$Q_d = a + bP + cY$$

We know the following:

$$-0.2 = \varepsilon_d = \partial Q/\partial P(P/Q) = b(P/Q) \rightarrow -0.2 = b(100/4,000) \rightarrow b = -8$$

Similarly for income:

$$0.90 = \varepsilon_y = c(Y/Q) \rightarrow 0.90 = c(10/4,000) \rightarrow c = 360$$

Next, b and c and current Q, P, and Y can be used to solve for a:

$$a = Q - bP - cY = 4,000 - (-8) \times 100 - 360 \times 10 \rightarrow a = 1,200$$

Finally, a is substituted into demand to get the following equation:

$$Q_d = 1,2008P_d + 360Y_d$$

If you want to forecast when only price changes, you would use the same procedure but only create a function: $Q = a - bP$. If you want to forecast for price, income, and cross-price changes, you would create a function with $Q = a - bP + cY + dP_{cross}$. You could add any number of variables, but you would need an elasticity and a value for each variable.

You could also measure price and income in Chinese yuan (¥), where $1 is typically between ¥6 and ¥7. Since the dollar and yuan are traded daily, the exchange rate changes over time. The current rate can be found at FXCM (2013) and other Internet sites. Although it does not matter what units you use to create your demand function, be sure to use those same units when you are forecasting.

Summary

Coal is our most abundant fossil fuel. It fueled the original Industrial Revolution in the industrial countries and is fueling the current industrial revolution in China and India. The competitive market model, with shifts in demand and supply, demonstrates how wood was surpassed by coal, and later coal was surpassed by oil. Although highly stylized, this powerful model offers clues as to how newer fuel sources will penetrate and supplant fossil fuels.

In a competitive market, we assume many buyers and sellers, a homogeneous product, and that both buyers and sellers have the necessary information relating to all market transactions and can freely enter into or exit out of the industry. To ensure that a competitive market allocates resources efficiently, we further idealize this model by assuming that property rights are well-defined, there are no externalities, and coal mining is not a decreasing cost industry. We will see why these assumptions are needed for efficiency and see the results of relaxing some of them in coming chapters.

In our competitive model, coal buyers are represented by a demand curve, and coal sellers by a supply curve. Movements along these curves as the result of a coal price change are called changes in quantity demanded or supplied. Shifts of the whole curve, from other variables changing, are called changes in demand and supply. By using basic demand and supply analysis, we can improve business decisions by analyzing the effects of many events on market price and quantity, including business cycles, changes in the prices of related goods, and government policy. We will consider two such policies—energy price controls and energy taxes/subsidies—in the next chapter.

Quantity of coal demanded is positively related to the number of buyers, the price of substitutes, such as oil or gas, and income or industrial activity. It is negatively related to the price of coal and complementary goods such as capital. The quantity of coal supplied is positively related to the number of sellers (producers), the price of coal, the price of by-products of coal production, and the price of output produced using coal. It is negatively related to the price of factors of production for coal, such as labor and capital, and the price of any similar product to which coal producers could switch. Technology and government policies also affect coal demand and supply, with the effect depending on the specific policy or technology.

Elasticities (the percentage change in quantity divided by the percentage change in another variable) measure how responsive quantity demanded and quantity supplied are to prices and other variables. Demand price elasticities indicate relations between price changes and revenues; income elasticities also tell us how sensitive the demand is to the business cycle, while cross-price elasticities indicate whether goods are substitutes or complements. Elasticities are important indicators of how markets behave and can also be used to create demand and supply curves for forecasting and policy analysis.

4

Energy Price Controls, Taxes, Subsidies, and Social Welfare

Taxes are what we pay for civilized society.

—Justice Oliver Wendell Holmes

Introduction

Governments sometimes prefer prices to be different from the market price. They might want to charge a higher price to consumers to discourage an activity. Thus, the government of a rich country might want high prices on cigarettes, alcohol, and activities that emit carbon dioxide. They might want to charge lower prices on activities that they want to encourage. In a developing country that is experiencing erosion and other problems from deforestation, the government might want to encourage the use of kerosene and liquefied petroleum gas (LPG) for cooking, heating, and lighting to reduce wood use. Alternately, governments might encourage or discourage a certain activity on the part of suppliers by raising or lowering the prices they receive. For example, European governments have wanted higher prices for electric power generated from renewable energy sources, with lower prices for electricity generated from fossil fuels.

In this chapter, we will consider two different mechanisms governments use for changing market prices—price controls and tax/subsidy policy. Before we discuss these mechanisms, let's develop the concept of social welfare to measure the effect such policies have on society.

Social Welfare

In any market, consumers benefit when they consume the good, and producers benefit when they sell the good. To develop a measure for these benefits, consider a competitive market with no externalities. Thus, neither the production nor the consumption of this energy product causes any pollution or other costs or any other benefits outside of this market. We will come back to the case with externalities later in the book. Suppose the inverse demand and supply curves in our coal market are the same as in chapter 3:

$$P_c = 86.7 - 0.5Q_d$$

$$P_c = 6.6 + Q_s$$

We can think of the demand curve as a marginal benefit curve. For example, if the quantity sold is 1, the market price $P_d = 86.7 - 0.5 \times 1 = \$86.20$. Then someone must value that first unit by at least $86.20. If two units are sold, the market price is $P_d = 86.7 - 0.5 \times 2 = \$85.70$. Someone must value that second unit by at least $85.70, and so on. At the market price of $85.70, the person who values that first unit at $86.20 is getting it at $85.70 and is getting a surplus of $0.50. We measure this consumer surplus for the market as the difference between a consumer's willingness to pay and what is actually paid. In the continuous case, we measure it as the area under the demand curve and above the price. This would be the triangle *abc* in figure 4–1, which is equal to $(86.7 - 60) \times 53.4 \times 0.5 = \$712.89$.

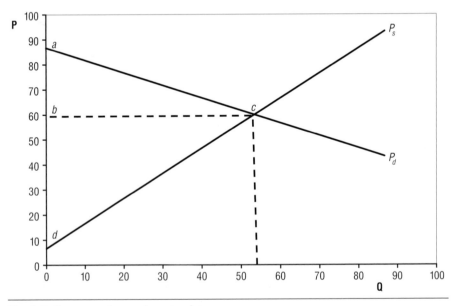

Fig. 4–1. Consumer plus producer surplus

On the supply side, suppose that we have a competitive market in which firms maximize profits. In such a market, each firm is small and takes the market price as given. Suppose their production costs are a function of their output, or $C = C(Q)$. Then profits (π) for the competitive supplier are revenues minus costs, as follows:

$$\pi = PQ - C(Q)$$

If the competitor chooses the Q that maximizes profits, first-order conditions are as follows:

$$\partial\pi/\partial Q = P - \partial C(Q)/\partial Q = 0$$

Since $\partial C(Q)/\partial Q$ is marginal cost, the first-order condition tells us that the supplier should produce up to the point where price equals marginal cost. Second-order conditions for a maximum are expressed as follows:

$$-\partial^2 C(Q)/\partial Q^2 < 0 \rightarrow \partial^2 C(Q)/\partial Q^2 > 0$$

Since $\partial^2 C(Q)/\partial Q^2$ is the slope of the marginal cost, second-order conditions are that marginal cost should slope up or increase, as in figure 4–2.

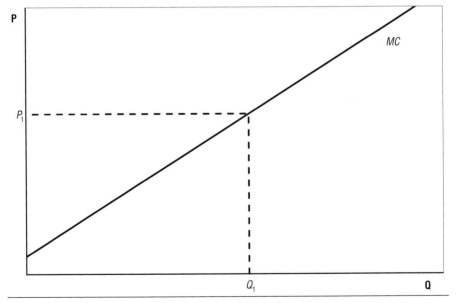

Fig. 4–2. Supply equals marginal cost in a competitive market

If price is P_1, the producer makes profits on all units up to Q_1. Beyond Q_1, the producer loses money on the last or marginal unit and so should not produce beyond Q_1. Thus, in a competitive firm maximizing profit, the supply curve is the marginal cost curve. For the total market in the short run, we get a supply curve at any point in time by adding together the production of all suppliers. In the long run, firms can enter and exit, giving us a somewhat different supply curve, but one that still relates to costs. So, the market supply curve is represented by the marginal cost curve for the whole industry.

To consider supplier welfare, take the inverse supply curve or the marginal cost curve to be $MC = P = 6.6 + Q_s$ as above. The marginal cost of the first unit $MC = 6.6 + 1 = \$7.60$, MC of the second unit $= 6.6 + 2 = \$8.60$, and so on. Recall at equilibrium, 53.4 units are sold at $60. However, the first unit only cost $7.60 to produce, leaving that supplier with a surplus of $60 − $7.60 = $53.40 for that unit. Similarly, for the second unit, which cost $8.60 to produce, the producer receives $60, yielding a producer surplus of $60 − $8.60 = $51.40. In the continuous case, we measure this producer surplus by the triangle bcd in figure 4–1 above. Thus, producer surplus at equilibrium in this market equals $(60 − 6.6) \times 53.4 \times 0.5 = \$1,425.78$. The sum of producer plus consumer surplus = 3, which is the area adc in figure 4–1, above. Economists refer to the total net benefits in the market, which are the consumer plus producer surpluses, as *social welfare*. In this case, social welfare at the competitive equilibrium is $712.89 + $1,425.78 = $2,138.67.

In the above case, we have assumed that there are no externalities so that the demand and supply curves represent all costs and benefits. If that is the case, note there are some interesting implications that can be seen from figure 4–1. First, the competitive equilibrium with price of $60 and quantity of 53.4 maximizes social welfare. At a higher quantity, marginal costs are greater than marginal benefits, reducing net benefits. At a lower quantity, marginal benefits are greater than marginal costs. Thus, we forgo the net benefits we could have had if we had consumed more. Second, the price is equal to the marginal benefit of the last consumer to enter the market. Third, the price is equal to the marginal cost of the last producer to enter the market. The implications are that the price is set by the marginal consumer and marginal producer, while maximizing social welfare. Economists call such a market *efficient* and idealize the competitive market and marginal cost pricing when there are no externalities with increasing unit costs in the market.

Government Price Controls

Governments often seek to influence markets. For example, after World War II, the European coal industry was contracting. To protect the industry and to protect jobs in the coal mines, the British, Belgium, Spanish, and West German governments all wanted higher coal prices. One possible policy would be to set a minimum price of P_{mn} above the market equilibrium, making it illegal to charge a lower price. If the coal industry obeyed the law, notice what would happen in figure 4–3. At P_{mn}, suppliers would want to bring more to market than consumers were willing to buy, and there would be surplus coal brought to market. If the minimum price were set at $70, as in the figure, the quantity demanded would be $Q_d = 173.4 - 2P_c = 173.4 - 2 \times 70 = 33.4$ and quantity supplied would be $Q_s = -6.6 + 1 \times 70 = 63.4$. Although the government might have been well intentioned, if the market were competitive, there would be downward pressure on price from the excess coal in the market. Such a policy would be hard to maintain without further interference from the government to buy up extra coal or provide subsidies to the coal industry, which we will consider in the next section.

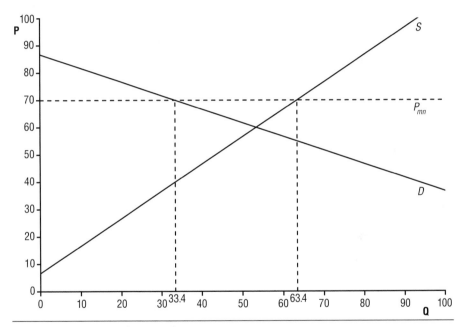

Fig. 4–3. Government price controls

More recently a number of governments have been trying to encourage renewable electricity by requiring electric utilities to buy electricity generated by renewables at set rates called *feed-in tariffs (FIT)*. Gipe (2012) and Ren21 (2013) contain examples for 66 countries as well as some states and provinces in Australia, Canada, India, and the United States. Programs include support for electricity generation from wind, small wind, offshore wind, geothermal, biomass, biogas, tidal waves, small-scale hydro, and solar photovoltaics. About 40% of the countries with FIT programs are in Eastern or Western Europe. Details of the programs vary from jurisdiction to jurisdiction, with sample programs and some of their features shown in table 4–1.

Table 4–1. Sample feed-in tariffs for electricity renewables, 2011

Jurisdiction	Contract Years	Renewable Source	FIT/ kWh Local Currency	P_El-i	Xrate LC/$	FIT US$/kWh	P_El-i US$/kWh
Japan < 20 kW	20	Small Wind	57.750	13.240	82.931	0.70	0.16
Switzerland < 10 kW	20	Small Wind	0.200	0.112	0.923	0.22	0.12
Germany	20	Offsh. Wind	0.130	0.080	0.748	0.17	0.11
Denmark (Maximum)	20	Offsh. Wind	0.620	0.567	5.571	0.11	0.10
China	NA	Onsh. Wind	0.610	0.505	6.732	0.09	0.08
Ecuador	15	Geothermal	0.132	NA	NA	0.13	NA
Italy < 1 MW	15	Geothermal	0.200	0.147	0.748	0.27	0.20
Canada, Ontario	20	Biomass	0.138	0.108	1.020	0.14	0.11
Czech Republic	15	Biomass	4.580	2.795	19.380	0.24	0.14
US, Vermont	20	Biogas	0.136	0.068	NA	0.136	0.068
Malaysia < 4 MW	16	Biogas	0.320	0.075	3.060	0.10	0.02
Ireland	15	Hydro	0.083	0.110	0.748	0.11	0.15
France	20	Hydro	0.055	0.074	0.748	0.07	0.10
Italy	15	Tidal Wave	0.340	0.147	0.748	0.45	0.20
Portugal 1st 5 MW	15	Tidal Wave	0.260	0.100	0.748	0.35	0.13
Israel	20	Solar PV	1.490	0.460	3.724	0.40	0.12
India < 1 MW	25	Solar PV	19.600	NA	49.124	0.40	NA

Sources: Gipe (2012) is the source for contract years and FIT in local currency. IEA (n.d.g) is the source for the price of electricity in industry, and US IRS (n.d.) is the source for exchange rates used to convert to US dollars. *Note:* FIT rate = the feed-in tariff rate in local currency; NA = not available; Offsh = offshore; Onsh. = onshore; P_El-i = the price of electricity in industry before tax in local currency; PV = photovoltaic; and Xrate LC/$ = the exchange rate measured as local currency per US dollar in 2011.

As seen in the table, these rates are usually considerably higher than the price of power to industry and, hence, cost more than power from nonrenewables. The reasons these programs do not cause excess quantity supplied, as in figure 4–3, is that the electricity market is typically not competitive. Utilities are able to pass along the costs of the feed-in tariffs to all of their customers, not just those that consume renewable power. However, these tariffs and programs are subject to continuing erosion as solar costs have been coming down rapidly. In some regions, cash-strapped consumers, who also vote, have become more resistant to higher electricity costs.

In other cases, governments may want to lower prices to consumers and set a maximum price, as in figure 4–4. For example, the United States had some sort of price controls on natural gas sold across state lines in the years 1954 to 1993. (See Natural Gas Supply Association [2011] for more details on these regulations, as well as chapter 8.) If the market is competitive, notice what will happen. At P_{mx}, consumers want to buy more (Q_d) than suppliers want to bring to market (Q_s), and there is a shortage. And that is exactly what happened in the 1970s. Nobel laureate Milton Friedman also pointed out that such price controls exaggerate a shortage and make it appear worse than it needs to be.

With only Q_s produced, there are social losses. For all output between Q_s and Q_e, the demand (marginal benefit curve) is higher than supply (marginal cost curve). For each of these units, marginal benefits are greater than marginal cost. By foregoing these units of output, society loses the area (*abc*). Thus, a price control in a competitive market may cause the market to be less efficient. In addition, since demanders are willing to pay more and suppliers are willing to produce more at a higher price, a black market might develop.

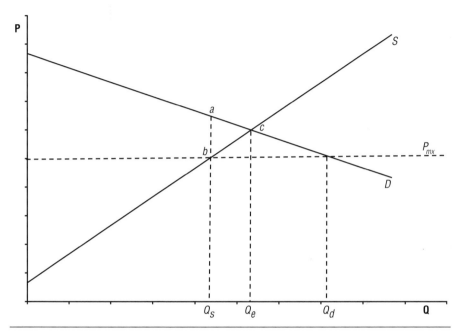

Fig. 4–4. A maximum price in a competitive market

What would happen if we switched the P_{mx} to be a P_{mn}? Now it is illegal to charge a lower price than P_{mn}. In this case, equilibrium is no longer an illegal price, and the price control would be nonbinding. Similarly, swapping the P_{mn} in figure 4–3 for a P_{mx} would make the price control nonbinding.

Whenever price controls in a competitive market are not set at equilibrium price and are binding and obeyed, they move us away from the competitive equilibrium. Thus, we can expect social losses and the market to be in disequilibrium, with economic forces encouraging surpluses and shortages. They may also encourage illegal trading tending toward the market clearing rate.

Governments also set quantity restrictions in markets. The US government set a quota on imported oil in effect from 1959 to 1973, restricting the amount that could be imported. Think about what such quantity restrictions (Q_r) would do if the market is competitive and the restriction is obeyed. Again, consider figure 4–4.

With Q_r below equilibrium quantity at Q_s, price will increase to above equilibrium price (a) to allocate the restricted product.

The European Union is requiring 20% of energy to be produced by renewables by 2020. If Q_r is above equilibrium quantity at Q_d in figure 4–4, the price will have to fall to below equilibrium price (P_{mx}) to allocate the extra product. However, at P_{mx}, producers are losing money, and the quantity restriction is not sustainable. Notice also that there are social losses from quantity restrictions.

Economists typically do not advocate price and quantity controls in competitive markets with no externalities and increasing costs. Such controls may cause losses in welfare, encourage law breaking, increase market instability, be unsustainable, and send the wrong price signals to producers and consumers.

Taxes and subsidies on energy products are other policies that might affect market prices and quantities. In the next sections, we consider such fiscal policies governments use to raise money and to influence markets.

Government Taxes and Subsidies

Governments have various reasons for taxing energy, and such taxes will influence energy markets. An energy-producing country or state may find energy taxation a politically expedient source of revenues, particularly if a large part of the output is exported. They may tax an energy source to provide a public good related to energy consumption, such as gasoline taxes that are collected to fund roadways. They may use taxes to discourage energy use, such as passing a carbon tax to discourage the burning of high-carbon fossil fuels that influence our climate. They may use differential taxes on energy products to influence industrial behavior, such as Europe's high import taxes on oil products and low import taxes on crude oil imports. Historically, this encouraged refineries to be built in Europe rather than in the oil-producing countries.

Alternately, governments may subsidize rather than tax an energy product to encourage use, such as California subsidies on solar energy panels, or to protect a domestic industry and employment, such as Germany's subsidies to protect its domestic coal industry. Governments may subsidize energy products as part of an income redistribution scheme, such as Indonesian subsidies on kerosene to help poor residents who use kerosene for cooking and lighting.

No matter what the reason for their adoption, energy taxes and subsidies are likely to distort prices and production in energy markets. Such distortions may decrease efficiency in efficient markets, or they may increase efficiency in inefficient markets with externalities present.

Types of Taxes

Governments around the world collect a variety of taxes, including personal income taxes, sales taxes, value added taxes, and corporate income taxes. Often, energy and natural resources are subject to taxes beyond corporate taxes. For example, Australia passed a mineral resources rent tax on coal, iron ore, and natural gas that went into effect July 1, 2012, with an effective rate of 22.5% (Allnutt and Lilly 2012). In 2013, 37 states in the United States had special taxes on fossil fuels and other minerals called *severance taxes*. These are taxes on producing the resource or "severing" it from the earth.

These taxes tend to more often be *ad valorem* taxes, which are taxed as a percent of price rather than a unit tax of a constant amount per unit. Thus, a natural gas tax of $0.10 per 1,000 cubic feet (Mcf) would be a unit tax, while a tax of 10% on natural gas price would be an ad valorem tax. The ad valorem tax would be $0.30/Mcf, if natural gas were $3/Mcf, but would be $0.60/Mcf if natural gas were $6/Mcf.

The top 13 fossil fuel–producing states or areas ranked in order by energy content of production in 2011 were Texas, Wyoming, the federal offshore area, West Virginia, Louisiana, Oklahoma, Pennsylvania, Colorado, Kentucky, New Mexico, Alaska, California, and North Dakota. All these states have oil and gas, while Wyoming, West Virginia, Kentucky, and Pennsylvania are the largest domestic coal producers. Texas had the largest oil production, followed by the federal offshore area, Alaska, and California. However, with recent production increases, by early 2014, North Dakota had risen from the number 6 to the number 3 spot. In 2011, Texas had the largest gas production, followed by Louisiana, Wyoming and Oklahoma (US EIA, n.d.f). However, with its new Marcellus Shale gas developments, Pennsylvania had risen from number 8 into the number 4 slot by 2012 (US EIA, n.d.f). All of these states but Pennsylvania have severance taxes and, Pennsylvania is also likely to pass such a tax. The main provisions of the tax laws for some representative US states are shown in table 4–2.

Total tax revenues from severance taxes were only 1.9% of all state tax revenues in 2013. The severance tax revenues vary considerably across states, and they contribute vastly different amounts to the coffers of the large fossil fuel–producing states. In 2013, Texas received the largest share of US state severances tax revenues (28%), followed by Alaska (24%) and North Dakota (15%). However, for some states they represent a significant portion of the state's total tax revenues. Severance taxes were the most important share of state revenues for Alaska (78%), North Dakota (46%), and Wyoming (40%), but contributed less than 10% in Texas (US Census Bureau 2013).

Table 4–2. Some representative severance tax rates for large fossil fuel–producing states, 2011

State	Oil (% or $/bbl)	Natural Gas (% or $/bbl)	Coal (% or $/bbl)	Uranium (%)	Percent of US Sev. Taxes	Percent of State Tax Rev.
Alaska					30%	74%
1st Five Years	12.25%	10%				
After Five Years	15%	10%				
Min. after Five Years	$0.80	No min.				
Mining License Tax			3%–7% varies by profits			
Texas					16%	4%
Severance Tax	4.60%	7.50%	None			
Min. Tax	$0.05	$0.08	None			
North Dakota					10%	43%
Gross Prod. Tax	5.00%	$0.04	$0.395			
Oil Extraction Tax	6.50%					
Louisiana					7%	9%
Severance Tax	12.50%	$0.16	$0.12			
Oklahoma					7%	11%
Gross Prod. Tax + Excise Tax		7.10%				
Oil Price ≥ $17/bbl	7%					
17 > Oil Price ≥ 14	4%					
Oil Price < 14	1%					
Wyoming					7%	34%
Severance Tax—Oil, Gas, & Uranium	6%	6%		4%		
Surface Coal			$0.57			
Underground Coal			$0.55			

New Mexico						
Severance Tax—Oil, Gas, & Uranium	7.09%	7.94%			6%	15%
Surface Coal			7%	3.50%		
Underground Coal			3.75%			
West Virginia						
Severance Tax—Oil & Gas	10.00%	10.00%			4%	9%
Severance Tax—Coal			5.00%			
37–45 in. Seam Thickness			2.00%			
Less than 37 in. Seam Thickness			1.00%			

Sources: Kent and Eastham (2011); Temte (2010); New Mexico Taxation and Revenue Department (2013); Alaska Department of Revenue Tax Division (2013).
Note: See the Alaskan government tax return forms for more information on how the mining license tax varies with mining net incomes. The % symbol refers to an ad valorem tax. See individual state government tax forms for the most complete and updated severance tax information.

Private owners of a resource receive a royalty if they lease their mineral rights to someone else to produce. The US situation, with private ownership of mineral rights allowed, is unique; in most of the rest of the world, the mineral rights are owned by the government. Where the US government owns the energy and mineral resources and leases them, as in the US offshore and national forests, it also receives royalty and other payments. Taxes in resource owning countries are highly variable and often are complicated to summarize. You can see details for many countries' oil and gas tax laws by consulting tax guides provided by consulting companies, such as Ernst and Young (2012).

Companies typically pay corporate income taxes plus a royalty or production tax. Figure 4–5 shows estimates of the percent of these taxes retained by the government relative to the company's share for a variety of countries in the years 1997 to 2007. There is wide variation across regimes, from a low of 30% government tax in Ireland to rates of more than 90%. Most countries received more than one-half of the per-barrel take. The majority of countries where the government take is higher than 85% are OPEC members. In some cases, the national oil company (NOC) is producing all the oil and paying into the government coffers. If private companies have concessions, projects in these countries must be quite profitable for companies to be willing to accept such a low share of the per barrel revenue. Countries with poorer quality, more expensive, less well-known reserves, or countries whose situation is more risky or less desirable in any way, will typically need to levy lower taxes in order to attract investment.

Severance taxes are applied at the production level, but energy taxes are also applied as tariffs, when energy is traded across borders, and as excise taxes, when energy is sold to final consumers, such as the gasoline tax you pay at the pump. The most heavily taxed energy sources are oil products. To illustrate the extent of this taxation, note the wide variation in price for oil products in table 4–3. Although some of the price differences reflect different transport distances, exchange rates, and distribution costs, much of the difference is from varying tax rates. The top rows of table 4–3 include wholesale prices of premium gasoline and light fuel oil (called *distillate* in the United States and gas oil in other places) at three world oil market centers—New York Harbor, Amsterdam-Rotterdam-Antwerp (ARA), and Singapore. Diesel fuel has tighter specification and is a bit more expensive but is otherwise similar to light fuel oil. Prices for light fuel oil (called distillate in the United States and gas oil in other places) is not available, so the spot wholesale diesel price is used.

Heavy fuel oil, also called residual fuel oil or resid in the United States, varies in sulfur content. The wholesale prices given here are for high sulfur fuel oil (3.5% sulfur). Much of this fuel is used in international shipping, as many countries restrict sulfur emissions from domestic fuel burning, which requires burning low sulfur fuel oil or capturing the emissions from burning high sulfur fuel oil. However, those restrictions are being tightened for shipping zones around countries as well. As high sulfur fuel oil is still the most frequently reported price in the data available, it is used here, but where a high sulfur fuel oil price is not reported, a low sulfur (<1% sulfur) fuel oil price is given if available.

Notice that wholesale prices for gasoline and diesel are all between $0.75 and $0.85 per liter with heavy fuel oil on a per liter basis varying between $0.60 and $0.75. However, retail prices vary considerably more. For example, they are considerably below the wholesale prices for Saudi Arabia and Iran, where these fuels were subsidized in 2012. In the United States, the excise tax on gasoline and diesel is around $0.10 per liter, and the retail prices are around 20% higher than wholesale prices. In Russia, the retail prices for these two fuels are roughly similar to those in the United States. In heavily taxed Europe, the gasoline and diesel prices are typically between two and three times these wholesale prices. Prices for fuel oils are typically not as heavily taxed, and their prices vary less across countries, particularly for heavy fuel oil. All these excise taxes distort markets; in the next section we will see how.

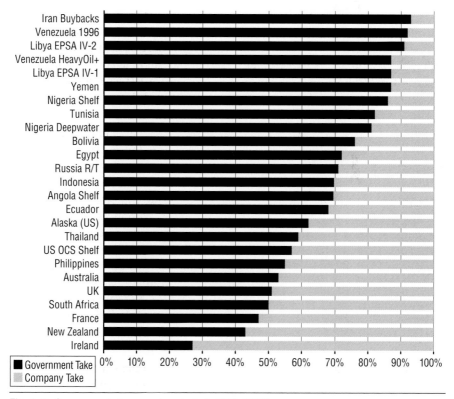

Fig. 4–5. Government share per barrel of oil, 1998–2007
Source: Johnston (2008).

Table 4-3. World survey of selected petroleum product prices, 2012

Region/Country Spot Wholesale Prices	Gasoline $ per liter	Diesel Fuel $ per liter	Light Fuel Oil $/1000 liter	Light Fuel Oil $/1000 liter	Heavy* Fuel Oil $/tonne
New York Harbor	0.83	0.79	788.84 !	788.84 !	665.45
ARA	0.77	0.81	805.24 !	805.24 !	704.33
Singapore	0.78	0.81	805.51 !	788.84 !	773.41

Retail Prices including Taxes	Automotive Fuels		Residential	Industrial	
	Premium Gasoline $ per liter	Diesel Fuel $ per liter	Light IEA $/1000 liter	Light IEA $/1000 liter	Heavy IEA $/tonne
North America					
Canada	1.32	1.23	1154.89	929.61	779.76
Mexico	0.86	0.85	NA	662.24	600.85
United States	0.97	1.05	1040.73	800.78	727.21
Central & South America					
Chile	1.56	1.24	1253.85	NA	NA
Brazil	1.39	1.02	NA	NA	NA
Colombia	1.28	1.18	NA	NA	NA
Ecuador	0.58	0.29	901.89~	901.89~	599.65
Venezuela	0.02	0.01	685.53~	685.53~	14.74
Western Europe					
France	1.91	1.78	1245.57	1007.46	772.97
Italy	2.28	2.18	1872.84	1547.81	825.21
Netherlands	2.33	1.95	NA	NA	718.25
Norway	2.53	2.35	1739.11	1391.29	NA
Spain	1.75	1.75	1220.63	1026.24	772.82
Sweden	2.10	2.16	2032.35	1040.63	1434.90
Turkey	2.54	2.33	1817.53	NA	1224.28
United Kingdom	2.17	2.27	1121.24	1040.56	NA

Eastern Europe & Former Soviet Union

Czech Republic	1.93	1245.27	949.94	541.97
Poland	1.74	1273.10	999.93	706.97
Russia	0.99	NA	NA	NA
Middle East				
Iran	0.33	800.00~	800.00~	1704.48~
Saudi Arabia	0.16	115.72~	115.72~	NA~
United Arab Emirates	0.47	NA~	NA~	NA~
Africa				
Nigeria	0.62	310.06~	310.06~	NA~
South Africa	1.38	NA	NA	NA
Far East & Oceania				
Australia	1.57	NA	NA	NA
China	1.37	NA	NA	NA
India	1.25	NA	NA	NA
Japan	2.00	1148.99	1003.46	1021.42 #
Korea, South	1.80	1238.00	NA	1001.09 #

Sources: Spot Wholesale Prices: IEA n.g.d. Retail gasoline and diesel prices: GIZ (2014). All other OECD prices: IEA n.g.d. All other OPEC prices: OPEC (2013).
Note: NA indicates price not available. ! spot diesel price used as proxy for price of light fuel oil. * Industrial heavy fuel oil represents high sulfur fuel oil unless otherwise specified; #Indicates low sulfur. ~For OPEC countries, kerosene price is used for light fuel oil price for both households and industry and fuel oil price is used for heavy fuel oil price. For OPEC countries prices have been converted from $/bbl using 159 liters per barrel and 6.7 barrels per metric tonne of fuel oil. To convert heavy fuel oil price to $ per 1,000 liters, divide the price in tonnes by 1.065. 1 liter = 0.0264 US gallon.

Modeling Taxes in a Competitive Market

Let us go back to our competitive model. Let demand and supply and their inverse functions be as follows:

$$Q_d = 18 - 2P_d \rightarrow P_d = 9 - 0.5Q_d$$

$$Q_s = -3 + 3P_s \rightarrow P_s = 1 + (1/3)Q_s$$

We can see equilibrium in this market in figure 4–6. Solving for the price that makes supply equal to demand, $18 - 2P = -3 + 3P \rightarrow$ equilibrium P = 4.2 and Q = 9.6.

Now suppose that we put a unit tax of three on this product. The demand price will have to include the payment to the supplier plus the tax. The easiest way to see the intuition of the tax is to add three to the supply price, as in figure 4–7. Where this new function crosses demand is where demand price (P_d') equals supply price plus tax $(P_s' + t)$. Both suppliers and demanders are happy with the quantity (Q_t') at their price.

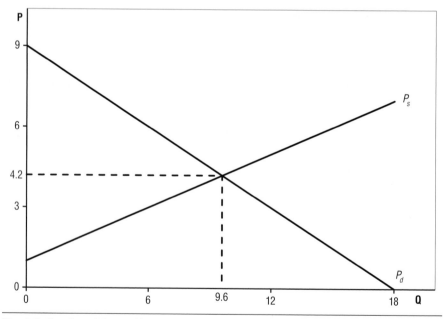

Fig. 4–6. Supply and demand in an energy market

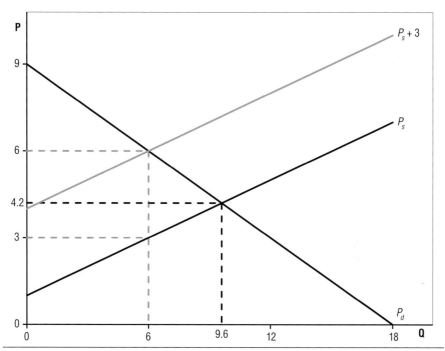

Fig. 4–7. Supply and demand with a unit tax

From the diagram, we can see that if we added the tax at the original quantity of 9.6, demanders would not want to buy 9.6 at the supply price plus the tax. At the new higher demand price, there would be excess quantity supplied, which would push supply price down. Supply price would keep falling until the demand price equaled the supply price plus tax, and demanders and suppliers wanted to exchange the same quantity. This would be at a quantity of 6, a demand price of $6, and a supply price of $3. To see how to determine those prices and quantities, go back to the model. If we invert the demand and supply curves and set $P_d = P_s + t$, we get the following:

$$P_d = P_s + t$$

$$P_d = 9 - 0.5Q_d = P_s + 3 = 1 + (1/3)Q_s + 3$$

At equilibrium $Q_d = Q_s = Q_t'$, as the following:

$$9 - 0.5\,Q_t' = 4 + (1/3)Q_t'$$

Solving for equilibrium Q_t', we determine that $Q_t' = 6$.

To get the demand price, substitute Q_t' back into the inverse demand to get the following:

$$P_d' = 9 - 0.5 \times 6 = 6$$

To get supply price, substitute Q_t' back into the inverse supply, as follows:

$$P_s' = 1 + (1/3)6 = 3$$

The above solution matches the economic "intuition" of how a tax would work in a competitive market. An alternative solution technique would be to substitute the relationship $P_d = P_s + t$ into the direct demand curve and solve as follows:

$$Q_d = 18 - 2(P_s + t) = -3 + 3P_s$$

Substituting in the tax and solving will yield the same solution as above.

Who pays this tax? As demand price goes up from \$4.20 to \$6.00, consumers of this energy product pay \$1.80 of the tax. Supply price falls from \$4.20 to \$3.00, so producers pay \$1.20 of the tax. Government revenues from the tax equal $Q_t' \times t = 6 \times 3 = 18$.

Suppose that we collect the entire tax from the consumer. The consumer pays the tax, and what the consumer is willing to pay to the supplier is $P_d - t$. We represent $P_d - t$ in figure 4–8. It is easy to see that the effect would be the same. Thus, it does not matter whether the unit tax in a competitive market is collected from the consumer or producer, provided the costs of collecting the tax are similar.

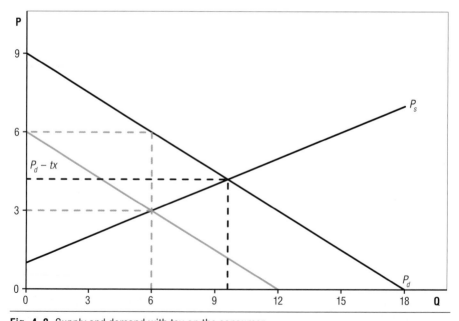

Fig. 4–8. Supply and demand with tax on the consumer

Incidence of Tax Depends on Demand and Supply Elasticities

Who pays the tax, which is called the *incidence* of the tax, depends on the shape of demand and supply. For example, on the left side of figure 4–9, there is a perfectly elastic demand curve. Notice that the demander is unwilling to pay a higher price and will cut consumption to zero if a higher price is imposed. In this instance, the supplier pays the whole tax at the new equilibrium quantity of Q_e'. On the right side of figure 4–9, the consumer has a perfectly inelastic demand. Since the consumer will not take a lower quantity but will pay any price for the given quantity, the consumer now pays the whole tax at the same equilibrium quantity.

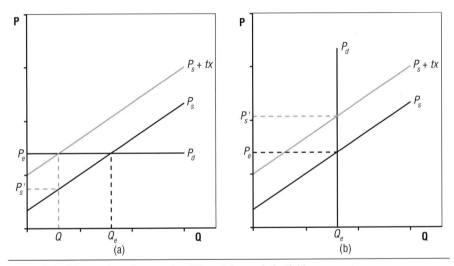

Fig. 4–9. Incidence of a unit tax under different demand elasticities

Figure 4–9 illustrates that the more responsive the demanders, the more they can pass the tax on to suppliers. Similarly, you will find that the more responsive the suppliers, the more they can pass on the tax to demanders. Indeed, it can be shown that the ratio of the changes in the demand and supply price is inversely related to the ratio of the supply and demand price elasticities, as follows:

$$\frac{\Delta P_d}{\Delta P_s} = \frac{\varepsilon_s}{\varepsilon_d}$$

(See http://dahl.mines.edu/st04.pdf, question 14 and answers for a proof.) You will notice that the tax typically moves the equilibrium price and quantity away from the competitive price and quantity and thus causes social losses. These losses are typically called *deadweight losses*, and we will consider them in the next section.

Consumer and Producer Surplus Show Deadweight Loss from a Tax

The above analysis tells us that in most cases, energy taxes decrease quantity consumed, increase demand price, and decrease supply price, causing a market distortion. To measure this distortion, we consider our two measures of social welfare: consumer and producer surplus.

In figure 4–10, the loss in consumer surplus in the market from a tax is *hbcg*.

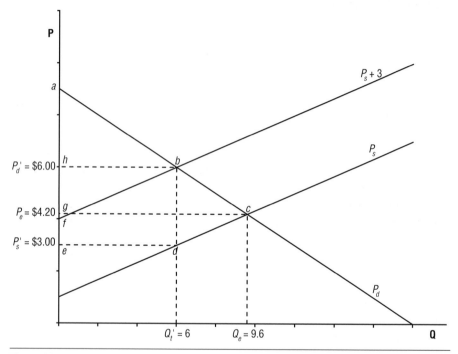

Fig. 4–10. Deadweight loss from an energy tax

The loss in producer surplus is *gcde*. Part of the consumer and producer surplus goes to the government as revenues *hbde*. We consider this a transfer. It is not lost; whoever benefits when the government spends the tax revenues gets this amount. What no one gets is the area *bdc*. This area we call the deadweight loss or the welfare loss from the tax, which is equal to $(9.6 - 6)(5.40 - 5)0.5 = \$5.40$, in this case. It is because of this distortion that unit and ad valorem taxes are thought to be less efficient than income taxes. Taxes with smaller deadweight losses are considered more efficient.

Energy Subsidies

The opposite of an energy tax would be an energy subsidy. The IEA (2011b) estimates global energy subsidies for fossil fuels to be $400 billion. The analysis for a subsidy would be similar, but instead of adding the tax to supply, you would subtract it. Thus, consumers would pay a lower price than suppliers receive. Alternatively, instead of subtracting a tax from supply, you could add the subsidy to demand. In addition, the government would pay out the subsidies instead of receiving tax revenues. Remember to always read the demand and supply price after the subsidy from the original demand and supply curves.

Summary

Governments often want to control the price of a good or the quantity purchased. One way to do so is to set maximum or minimum prices or quantities. If the market is competitive, binding constraints will put the market in disequilibrium, giving economic incentives to break the law. In addition, if there are no externalities, such constraints will cause social losses. We can measure such losses as consumer surplus (area under demand and above price) plus producer surplus (area above supply and below price).

Taxation of energy products is another way that governments influence prices and quantities in markets. Governments have various reasons for taxing products, with the most predominant being to collect revenues. Energy and natural resources are often subject to special taxes beyond corporate taxes. Severance taxes are taxes on the production of energy, while royalties are payments to the owner of the resource for exploiting it. In the United States, this owner may be a private citizen or the government. In most other countries, the government owns the mineral rights, and the share the government takes is highly variable across countries.

Energy taxes are also applied as tariffs when energy is traded across borders, and as excise taxes when energy is sold to final consumers. Most governments tax energy products, with petroleum products, particularly gasoline and diesel fuel, typically being the most heavily taxed, followed by diesel fuel.

Excise taxes and subsidies have distribution effects, which include who pays the tax (also called the incidence of the tax) and how much revenue the government collects. The elasticities of demand and supply determine who pays the tax, with the more price-elastic side of the market paying less of a tax or receiving less of a subsidy. Excise taxes in general are inefficient because they distort economic decisions, creating losses in welfare referred to as deadweight losses. The deadweight losses may be losses in consumer or producer surplus, or both. However, later we will see that taxes and subsidies may correct distortions caused by externalities.

5

Natural Monopoly and Electricity Markets

Communism is socialism plus electricity.

—V. I. Lenin

Introduction

Ever since the first public electric supply was provided in London and New York in 1881 and 1882, electricity use has grown to make our lives more comfortable and interesting. We can see this growth in the United States since 1940, and for the whole world since 1960, in figure 5–1.

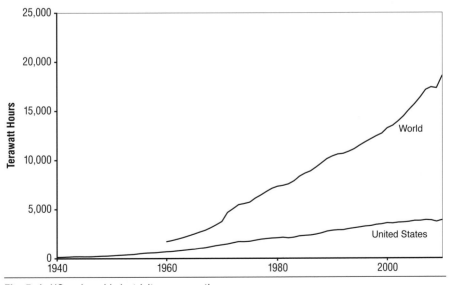

Fig. 5–1. US and world electricity consumption
Sources: United States, 1960–1948: US DOC (1975); 1949–2012: US EIA (2012d); World Bank (n.d.).

The underlying unit of measurement in electricity generation is the watt (W), and fortunately, SI units are used worldwide for electricity. A watt equals 1 joule (J) per second. Since this is a small unit, often watts are converted into larger units. For example, some of the most common standard conversions and abbreviations are as follows:

1 kilowatt (kW) = 1,000 watts
1 megawatt (MW) = 1,000,000 watts
1 gigawatt (GW) = 1,000,000,000 watts
1 terawatt (TW) = 1,000,000,000,000 watts

In standard SI unit abbreviations, the W is capitalized in honor of James Watt, as well as the J in honor of James Joule. If a kilowatt operates for an hour, it is called a kilowatt hour (kWh), which is usually the basic unit of measurement for household electric bills. One kilowatt hour equals 3,412 Btus, 3,600 kJ, or 860 kilocalories (kcal). One kilocalorie is the same as the Calorie that we use to measure energy in food, which is designated with a capital C. Thus, a kilowatt hour, which typically costs less than $0.20 in most countries, has the amount of energy in three to four chocolate bars. Costs per kilowatt hour in the United States are sometimes measured in mills, which are tenths of a cent.

Some care needs to be taken when interpreting electricity quantity data. We need to know where we are measuring the electricity along the supply chain. I will use the terminology used by US Energy Information Administration (2012d) in their electricity flow diagrams and note some differences in terminology and conversions for other major databases. If we begin with the total electricity produced, it is typically measured in kilowatt hours or some larger denomination in watts. US EIA denotes this total as *gross electricity generation*. The International Energy Agency (n.d.d) calls this *total production* in their electricity/heat statistics. The amount of power used by the power station is called *plant use* by US EIA and *energy own use* by IEA. If we subtract plant use, including that used for pumped storage, from the gross, we have *net electricity generation*. (Plant use is between 5% to 6% of gross electricity production in the United States.) If we are interested in power use by domestic consumers, we subtract electricity exports and add in imports. The IEA calls this value *domestic supply*; Eurostat and the United Nations call this value *gross inland consumption*. US EIA does not provide a separate category at this point. (See IEA [n.d.e] for imports and exports of power by country for OECD countries.)

Power is moved to domestic customers through high-voltage transmission lines and then stepped down to local distribution networks for delivery to local customers. Take away these transmission and distribution losses (IEA simply calls them losses) and we have *end use*. (IEA calls this value *final consumption*). These losses are about 7% of gross generation for the United States. Alternatively, if we want to measure energy flow along the whole supply chain (see fig. 5–2 for a US example), we start with the energy in the coal and other power sources going into the power plant. This flow might be measured in British thermal units, kilojoules, metric tonnes of oil equivalent (toe), or other standardized energy units. US EIA calls this *energy consumption to generate electricity*. For thermal or nuclear power, when the power is converted to electricity, a significant amount of heat is generated. Unless some of this heat is reused, such as with a combined cycle power plant, it all becomes a conversion loss. For the United States, the average conversion loss is about 63%.

The US EIA indicates such conversion losses through a concept called a heat rate, or the number of British thermal units needed to produce 1 kWh. Thus, if the heat rate of a coal power plant is 9,000 Btus, then 9,000 Btus of coal produce 1 kWh or 3,412 Btus of electricity. The efficiency of this plant is energy out divided by energy in. In this case, efficiency is 3,412/9,000 = 0.379, or 37.9%. Average heat rates for US coal plants in 2013 are 10,444, with similar rates for oil and nuclear, while natural gas plants have a more efficient average heat rate of 8,152 (US EIA 2013d).

US EIA (2013c) gives the following estimated heat rates for a number of new electricity plants: advanced pulverized coal (8,800), advanced combined cycle natural gas (6,430), conventional gas turbine (10,850), combined cycle biomass (12,350), and municipal solid waste (18,000).

When computing gross energy used to produce electricity for international comparison, US EIA (n.d.d) also attributes a heat rate, which they call *heat content* in their international database, to hydro and other renewables, even if no heat is lost in the conversion process. For example, their heat rate for all renewables for power generation in 2011 was 9,756. This differs from the IEA, which only converts the kilowatt hours into a larger heat content if heat is lost in the process (IEA 2005). As a result, the IEA counts hydro, wind, and solar by the energy in the electricity output and thus computes smaller total energy consumption for countries with these sources than does the US EIA.

Gross Energy (PJ) coal, oil, gas, hydro, nuclear, geothermal, wind, biomass	−	Conversion Loss	=	Gross Electricity Generated	−	Power Plant Use	=
41.3	−	25.9	=	15.4	−	0.8	=
Net Generation	−	Transmission and Distribution Loss	+	Net Imports = Imports − Exports	=	Net Consumption	
14.6	−	1.0	+	0.2	=	13.8	

Fig. 5–2. Electricity energy balance in the United States, 2013
Source: US EIA (2013d).
Note: PJ = petajoules. Divide by 1.055 to convert to quads.

Electricity Market Evolution

From 1920 to 1970, electricity consumption growth in the United States averaged around 7%. At this annual rate, electricity consumption roughly doubled every decade ($1.07^{10} = 1.97$). However, in the 1970s and early 1980s, the US economy experienced higher fuel prices and maturing electricity markets. In addition, the service sector became an increasing share of the economy. As a result, electricity growth rates slowed to an average of 4.1% per year in the 1970s, 3.0% in the 1980s, 2.4% in the 1990s, and less than 1% thereafter. A similar pattern has occurred in the European Union since 1960, with growth rates falling in each successive decade.

Electricity was not only important to the development of the United States and other developed capitalist countries but was also a cornerstone of the development strategy in the former Soviet Union, as demonstrated by the quote at the beginning of this chapter. It is equally important in the current plans for the developing countries. China, while still building up its electricity infrastructure, had growth rates averaging 7%–8% in each decade from 1970 to 2000. China's growth rate accelerated to an average of more than 11% from 2000 to 2009 (World Bank, n.d.). The status of current electricity consumption by major region of the world, along with population, is given in figure 5–3.

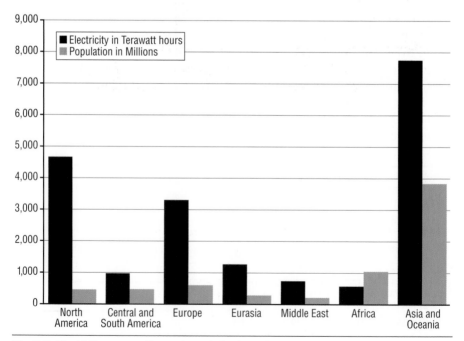

Fig. 5–3. Electricity consumption and population by major world regions, 2011
Source: US EIA (n.d.d).

The oil-rich Middle East has the highest rates of growth, followed by Asia and Oceania. However, if we include only the developing countries for Asia and Oceania, rates of growth have averaged 7% or more in the decades since 1980, matching earlier growth rates in the United States. African per capita consumption is currently near that of the United States in 1925. Central and South American per capita consumption is near that of the United States in 1950.

Chemical reactions from burning coal, oil, and natural gas to produce thermal energy account for more than two-thirds of the world's electricity generation, with coal being the dominant fuel. Around another quarter is generated from nuclear power or from gravitation in the form of hydropower. Less than 5% comes from other renewables, including biofuels, waste, geothermal, solar, wind, and tidal waves. Although this latter share is now small, it will no doubt grow with the implementations of policies to reduce global carbon emissions. The shares from each of these sources for the world are shown in figure 5–4, panel (a). For the United States, shown in figure 5–4(b), the patterns are somewhat similar. Coal dominates; fossil fuels provide around two-thirds of US power; hydro and nuclear provide slightly more than one-fourth; and other renewables around 5%.

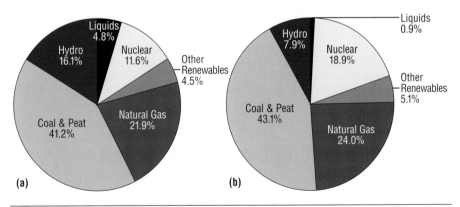

Fig. 5–4. World (a) and US (b) electricity production by fuel type, 2011
Source: IEA (n.d.d).

Table 5–1 compares regional electricity generation patterns for the world from the IEA. Thermal electricity, which uses the heat generated from the burning of fossil fuels, dominates generation in all major regions except South and Central America, where hydropower accounts for about two-thirds of generation. In non-OECD Europe and Eurasia, which is dominated by the Russian Federation, natural gas is the dominant fuel. The Middle East, with its huge reserves of oil and gas, is the most heavily dependent on oil, with most of the remainder from natural gas. The European Union has the highest share of nuclear and other renewables, while China, which recently passed the United States to become the largest power producer, is very heavily dependent on coal. There is also considerable variation within each of the regions across countries. The IEA database offers information about electricity generation by fuel for more than 100 countries (IEA, n.d.d).

Table 5–1. Share of electricity and heat generation by fuel and total generation, 2011

Region	Coal & Peat %	Oil %	Ngas %	Nuc %	Hydro %	Oth. Renew. %	Total tWh	Heat
OECD Americas	37	2	25	17	14	5	5,348	2.7
United States	42	1	25	18	8	5	4,350	3.1
European Union -27*	27	3	25	23	8	14	3,279	17.3
OECD Asia & Oceania	37	10	30	13	7	4	1,931	3.2
Non-OECD Europe & Eurasia	23	4	54	8	7	4	1,729	56.5
Brazil	2	3	5	3	80	7	532	0.2
Middle East	0	37	60	0	2	0	845	0.0
Asia exclude China	49	6	25	4	12	4	2,202	0.5
China	81	1	2	2	12	2	4,716	15.8
Africa	38	10	33	2	16	1	695	0.0
India	68	1	10	3	12	5	1,052	0.0
Russia	18	4	61	6	6	5	1,055	62.7
World	41	5	26	10	14	5	22,201	15.3

Sources: IEA (n.d.d.)
Note: Ngas = natural gas, Nuc. = nuclear, Oth.Renew = other renewables than hydro power. Asia exclude China = developing Asian countries excluding China. *European Union -27 includes all members prior to Croatia joining in 2013: Austria, Belgium, Bulgaria, Cyprus, Czech Republic, Denmark, Estonia, Finland, France, Germany, Greece, Hungary, Ireland, Ireland, Italy, Latvia, Lithuania, Luxembourg, Malta, Netherlands, Poland, Portugal, Slovakia, Slovenia, Spain, Sweden, and United Kingdom.

Modeling Electricity Markets

Costs

To model electricity markets, a good starting point is production costs. Total cost is composed of *fixed costs*, which are not related to electricity production and must be paid whether electricity is produced or not, and *variable costs*, which are related to quantity of production. Fixed costs can include leases on office space, insurance premiums, and equipment that has already been paid for, such as an electric generator or a stack gas scrubber to remove sulfur dioxide. If fixed costs cannot be recovered, they are referred to as *sunk costs*, which tend to be large in capital-intensive industries such as energy. Some fixed costs may not be sunk. Although you may want to keep an insurance contract even if you do not produce, if you go out of business and can cancel your insurance contract, it is not sunk. Any amount returned is the nonsunk portion. Resale or salvage value can also reduce the sunk cost of capital. For example, in July 2012, ConocoPhillips received $180 million from the state of Pennsylvania and Delta Air Lines for its refinery in

Trainer, Pennsylvania (Mouawad 2012). The cost of the refinery is a fixed cost. If ConocoPhillips produced anything and kept the refinery, they would not receive any sale value. However, the $180 million minus transaction costs would not be a sunk cost. Any capital costs above this amount that are not recouped while in business are unrecoverable and would be sunk costs.

Power plants are typically large and require infrastructure to acquire fuel and to deliver power to end-use customers. These capital costs are typically fixed in the short run, since such equipment has a long life. Plant life for an electric power plant is considered to be 40 years. However, in the long run, we can mothball or sell any piece of equipment, or even the whole plant, and we do not have to replace them. Thus, in the long run, all costs are variable.

Suppose the following equation represents costs for electricity generation:

$$TC = FC + VC(Q)$$

where

TC is total cost,
FC is fixed cost, and
$VC(Q)$ is variable cost as a function of quantity of production (Q).

These costs should include both out-of-pocket costs and opportunity costs. For example, if a generator invests its own funds in building a new unit, the cost of these funds is the value of the funds in their next best alternative. For a hydro unit, the out-of-pocket cost of the water might be zero, but the true opportunity cost of the water would be the foregone value of the water in its next best alternative. If the next best use is for irrigation, then the revenue you could have earned from selling the water for irrigation would be the opportunity cost for the water.

The above is our total cost outlay. We may also want to know some unit production costs. The two unit production costs of interest to us are average cost and marginal cost.

Consider average cost first. If we want an overall view of our cost per unit, we use average cost, which we get by dividing total cost by output, as follows:

$$TC/Q = FC/Q + VC(Q)/Q$$

Average total cost (TC/Q) equals average fixed cost (FC/Q) plus average variable cost [$VC(Q)/Q$]. Average fixed cost (which is just a constant divided by Q) has a consistent structure. It falls as Q increases. Average variable costs, however, may take on a variety of structures depending on the production process. Knowing your variable cost structure is critical for making sound business decisions. In what follows, we will consider some of the possibilities.

Suppose a power plant owns a coal deposit of 800 tons that has a fixed cost of $1,000 and costs $10 per ton to produce. To mine the deposit, total and average costs are as follows:

$$TC = 1,000 + 10Q \quad \text{and} \quad TC\!\big/\!Q = 1,000\big/Q + 10 \tag{5-1}$$

In this case, average variable cost (10) is constant, but average fixed cost $(1,000/Q)$ and average total cost decrease as production increases until Q is 800. However, at 800, variable cost jumps to infinity; since we can produce no more coal at any cost, our average total cost curve becomes vertical. What happens to average cost in equation (5-1) as Q increases? We know that FC/Q decreases over time. Since average variable costs are constant, average total cost falls as Q increases until the mine runs out of reserves. For an example, see figure 5-5, panel (a).

Costs structures can vary considerably across industries and technologies. The costs in figure 5-5, panel (a) demonstrate economies of scale, with average costs decreasing as the operation gets larger. Firms with such a cost structure are called *decreasing cost* firms.

Let's consider some other cases before discussing the electricity market. Take the following general case, which allows a wide variation in cost structures depending on our choice of the parameters for a, b, c, d, and e:

$$TC = a + bQ^c + d \times \exp(eQ) \tag{5-2}$$

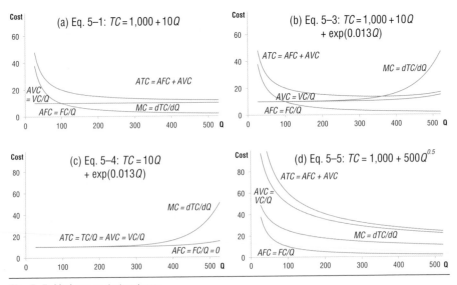

Fig. 5-5. Various cost structures

Case (a) is equation (5–2), with $a = 1,000$, $b = 10$, $c = 1$, and $d = 0$, which removes the last term from the equation.

In case (b), assume that costs increase faster than in case (a). Thus, the first tons in our mine are cheaper to produce than later tons, which are buried deeper in thinner seams. Represent this case with the following function, which is equation (5–2) with $a = 1,000$, $b = 10$, $c = 1$, $d = 1$, and $e = 0.013$. Now total and average costs are as follows:

$$TC = 1,000 + 10Q + \exp(0.013Q) \quad \text{and} \quad \frac{TC}{Q} = \frac{1,000}{Q} + 10 + \frac{\exp(0.013Q)}{Q} \quad \textbf{(5–3)}$$

This average total cost and its components are shown in figure 5–5, panel (b). In this case, the falling fixed costs dominate and pull average total costs down until about 417, and then the increasing variable costs dominate, pulling up costs.

Consider the next panel in figure 5–5, case (c), which is the case in panel (b) but with no fixed costs, or $a = 0$. Now total and average total costs are as follows:

$$TC = 10Q + \exp(0.013Q) \quad \text{and} \quad \frac{TC}{Q} = 10 + \frac{\exp(0.013Q)}{Q} \quad \textbf{(5–4)}$$

In this case, total costs equal variable costs. Average costs increase as the operation gets larger over the whole range of potential production. A firm that displays such diseconomies of scale is called an *increasing cost* firm.

Panel (d) is the last case we will consider, and the one of most interest in this chapter. Suppose the firm in question is capital intensive, with a large proportion of fixed costs. Variable costs may display economies of scale, as larger production may allow quantity discounts. For example, the bulk of coal used by utilities is transported by rail. If railroads give lower rates per ton of coal moved to a utility for large shipments than for small shipments, marginal costs may fall when the utility buys more coal and produces more electricity. Variable costs may also display economies of scale as larger production may allow specialization and learning by doing. This case is represented in panel (d) with $a = 1,000$, $b = 500$, $c = 0.5$, $d = 0$, and $e = 0$. Total and average total costs are shown in equation (5–5):

$$(d)\, TC = 1,000 + 500Q^{0.5} \rightarrow \frac{TC}{Q} = \frac{1,000}{Q} + \frac{500Q^{0.5}}{Q} \quad \textbf{(5–5)}$$

Average costs are a useful measure of the average resources needed to produce products. We can easily determine our total costs, if we know our average costs by multiplying average total costs by output, as follows:

$$TC = ATC \times Q = \frac{TC}{Q} \times Q$$

Thus, if average cost is $15, and we produce 20 units, our total cost is $300.

The other important unit cost to know for making good economic decisions is the cost of the last unit, which economists call marginal cost. This, along with *marginal revenue*, will help us determine whether a particular unit should be produced or not.

To demonstrate marginal cost, take the cost curve in equation (5–3) and measure the marginal cost of the 401st unit. This can also be seen by measuring marginal cost at two output levels rounded to the nearest three decimal points. If output is 400, our total cost is as follows:

$$TC(400) = 1,000 + 10Q + \exp(0.013Q) = 1,000 + 10 \times 400 + \exp(0.013 \times 400)$$

$$= 5,181.272$$

If production is 401, our total cost is the following:

$$TC(410) = 1,000 + 10 \times 401 + exp(0.013 \times 401) = 5,193.944$$

Our marginal cost is the change (Δ) in total cost for that last unit:

$$\Delta TC = TC(401) - TC(400) = 5,193.944 - 5,187.272 = \$12.672$$

Notice that since fixed costs are included at both 400 and 401 units, they cancel out when we take the difference. Thus, marginal costs only include variable costs (VC), not fixed costs.

If production occurs in 5-unit batches, the marginal cost of the 401st to 405th units is computed as follows:

$$MC = \Delta VC/\Delta Q = (VC(405) - VC(400))/(Q_2 - Q_1)$$

$$= (5,243.446 - 5,181.272)/(405 - 400) = \$12.435$$

For small changes in Q, or if we take the limit as ΔQ goes to zero, marginal cost can be approximated as the derivative of variable cost with respect to Q:

$$\lim_{\Delta Q \to 0} = \Delta VC/\Delta Q = dVC/dQ$$

Marginal cost for small changes in the general case is as follows:

$$dTC/dQ = dFC/dQ + dVC(Q)/dQ$$

where dTC/dQ is the derivative of total cost with respect to Q, or it is the change in total cost divided by the change in total quantity. Also, dFC/dQ is the derivative of fixed cost with respect to output, and $dVC(Q)/dQ$ is the derivative of total cost with respect to output. (Review derivatives for various mathematical functions at http://dahl.mines.edu/courses/dahl/calc/.)

If all fixed costs are sunk, then FC is constant, and $dFC/dQ = 0$. If some fixed costs are not sunk, they are incurred once we start producing and should be added to marginal costs for the first unit of production. Thus, marginal cost comes from changes in variable cost as we change production, and its sign depends on whether we have decreasing, constant, or increasing costs for our short-run variable costs.

Marginal, not average, costs are important for determining whether we should produce a given unit or not. The graphical examples in figure 5–5 give us some intuition on the importance of considering the cost of an additional unit and show how marginal cost is related to average cost. The convenience of derivatives allows computation of marginal cost functions in panels (a), (b), (c), and (d), corresponding to the four cost functions, which are as follows:

$$(a)\ TC = 1{,}000 + 10Q \rightarrow MC = \frac{dTC}{dQ} = 10$$

$$(b)\ TC = 1{,}000 + 10Q + \exp(0.013Q) \rightarrow MC = \frac{dTC}{dQ} = 10 + 0.013\exp(0.013Q)$$

$$(c)\ TC = 10Q + \exp(0.013Q) \rightarrow MC = \frac{dTC}{dQ} = 10 + 0.013\exp(0.013Q)$$

$$(d)\ TC = 1{,}000 + 500Q^{0.5} \rightarrow MC = \frac{dTC}{dQ} = 0.5 \times 500Q^{-0.5}$$

For panel (a), marginal cost is constant, average variable cost is constant, and they are equal. Thus, adding an additional unit at the same cost does not change the average variable cost. In panel (b), notice that marginal cost is everywhere above average variable cost, and average variable cost is rising. Thus, as we produce more units, the additional units cost more and drag average variable cost up. Another interesting feature in panel (b) is the relationship between marginal and average total cost. For the first units, marginal cost is below average total cost, pulling down the average. After marginal cost crosses average total costs and continues on up, it drags average variable costs up. As a result, marginal cost goes through the minimum of the average variable cost curve as well as the average total cost curve.

In panel (c), there are no fixed costs, and average variable cost equals average total cost. Marginal cost is everywhere above average total cost, pulling them both up as production increases. In panel (d), marginal cost is everywhere below average variable cost and average total cost, pulling them both down as production increases. The above examples show how costs may vary with output. They may also vary by time of day and year, as explored in the next section.

Load Cycle

The pattern of electricity consumption across time is called the *load cycle*. Daily on-peak consumption tends to occur during the day, shoulder consumption occurs early in the morning and later in the evening, and off-peak consumption takes place during the night. Figure 5–6 shows typical daily load curves for Israel, Jordan, and Egypt. Israel, the most developed country of the three, has its peak during the day, as is common in industrial countries. Jordan and Egypt, which are less industrialized, show peak loads during the evening hours.

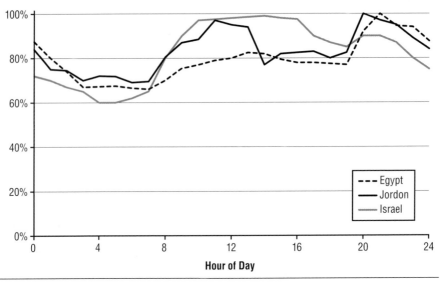

Fig. 5–6. Typical daily electric load curves for Israel, Jordan, and Egypt
Source: Used with the permission of Professor Amnon Einav, Tel Aviv University.

Electricity consumption also varies seasonally. In much of the United States, the summer air conditioning season is the peak of this consumption, as seen in figure 5–7. In Canada (also in the US Pacific Northwest), where summers are cooler, and a higher percentage of homes are heated with electricity, the winter heating season is the peak, also shown in figure 5–7. In other parts of the world, the load curve depends on the climate, how much electricity is used for heating and cooling, and whether commercial and industrial activity follows a seasonal pattern or not.

To satisfy electricity demand, utilities have a merit order for dispatching their units. Baseload plants, such as coal, nuclear, and hydro typically have a higher proportion of fixed or sunk costs relative to operating costs, and their operating costs tend to be lower. They also may have a longer lead time to ramp power production up or down. Baseload plants tend to run at all times, except when offline for maintenance and repair. Peaking units are flexible and may have lower capital costs but higher fuel costs. Historically gas turbines were often a preferred choice. Hydro is also flexible. Although it has high capital costs and no fuel costs, lending itself to baseload production, it may make sense to hold water back if peak price is high enough to more than recoup any losses from reduced power sales off peak. A small amount of pumped storage (water pumped up into a reservoir during off-peak periods to be released to produce energy when needed during peak demand) provides some peak power worldwide. Older, dirtier plants with higher operating costs are also held in reserve. Thus, during on-peak periods, the variable cost may increase as more electricity is produced.

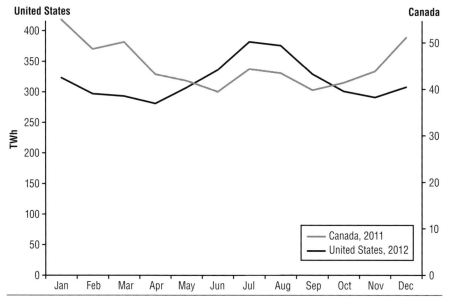

Fig. 5–7. United States' and Canadian electricity end use by month
Sources: US EIA (n.d.b); Statistics Canada (2012).

With newer technology, combined cycle gas plants (with efficiencies of up to 60%) have become more flexible, and in many instances, may even be the cheapest choice for both on-peak and off-peak power.

In the next sections, we develop some models for understanding this decreasing cost industry with seasonal and daily load patterns.

Monopoly in a Decreasing Cost Industry

If average costs decrease over a wide range of values, as in figure 5–5, panel (d), we call such industries with economies of scale *decreasing cost industries.* Historically, this has often been the case for electricity generation, which is very capital intensive. Let's see what happens in such an industry using a simple example.

Suppose the annual inverse demand and cost curves for electricity in a village are as follows:

$$P = 75 - 4Q$$

$$TC = 19Q - 0.25Q^2 \rightarrow ATC = 19 - 0.25Q \text{ and } MC = 19 - 0.50Q$$

Since there are no fixed costs, this is a long-run model. Price (P) is measured in cents (¢) per kilowatt hours and quantity (Q) is measured in kilowatt hours per year. For example, total cost for producing 20 units is the following:

$$TC = 19 \times 20 - 0.25(20)^2 = 280¢ \text{ or } \$2.80$$

Average total cost for producing 20 units is given as follows:

$$ATC = 280/20 = 14¢ \text{ or } \$0.14$$

The marginal cost of the 20th unit is the following:

$$MC = 19 - 0.50Q = 19 - 0.5 \times 20 = 9¢ \text{ or } \$0.09$$

You can visualize this model in figure 5–8. P is the inverse demand curve, which slopes down. Note that there are economies of scale. Thus, as we produce more, average unit costs fall. Since unit costs are falling, marginal cost (or the cost of the last unit) must be below the average, pulling the average down. Such cost curves also imply that the largest producer of electricity will have the cheapest unit cost and will be able to undercut producers with smaller generating units. In such an instance, a monopoly is likely to evolve. Now consider how a profit-maximizing monopolist could maximize profits, which requires we develop the marginal revenue (MR) curve (the dotted line) in figure 5–8.

A producer that is a profit-maximizing monopolist enjoys the whole market demand. Instead of competing, the producer picks the point on the demand curve that maximizes profits. We can easily work out what that quantity would be. Let the inverse demand curve be $P = P(Q)$, which slopes downward ($P' < 0$). In other words, if she reduces quantity, she can sell at a higher price. Let the monopolist's output be Q and the monopolist's cost be $TC(Q)$. We assume that TC slopes upward ($TC' > 0$), which means that marginal costs are positive, but $TC'' < 0$, which means that marginal costs are decreasing. The monopolist's profits (π) are total revenues minus total costs, as follows:

$$\pi = P(Q)Q - TC(Q) \tag{5-6}$$

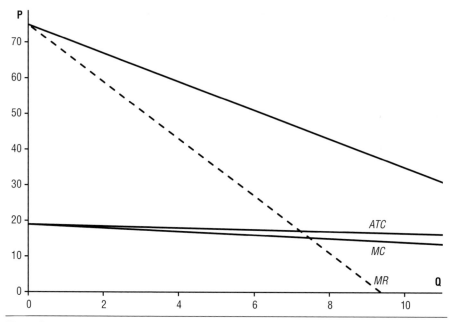

Fig. 5–8. Inverse demand and cost curves in a decreasing cost industry

To maximize profits with respect to output, take the first derivative of the function with respect to output, and set it equal to zero. Using the product rule for derivatives, the first-order condition for profit maximization is the following:

$$\frac{d\pi}{dQ} = P + \frac{dP}{dQ} \times Q - \frac{dTC}{dQ} = 0$$

The first two terms on the right-hand side of the equation are $P + (dP/dQ) \times Q$ (which is also written as $(P + P'Q)$, and they constitute marginal revenue (MR). Let's see its composition. P is the demand curve, or the average price at each quantity. The second expression is dP/dQ (the slope of the demand curve) times output. dP/dQ is the reduction in price required to sell an extra unit of output. Since this reduction is for all units, we multiply the reduction times the number of units sold, Q. Thus, what we get for an additional unit is the price minus the revenue reduction on all previous units. Then marginal revenue, which is the dotted line in figure 5–8, is less than or below the demand curve. The third term, dTC/dQ, which is also written as TC', is the marginal cost curve (MC). The first-order condition tells us a profit maximizing monopolist is to produce at the point at which the following is true:

$$MR - MC = 0 \rightarrow MR = MC \qquad (5\text{--}7)$$

Find the quantity where equation (5–7) holds in figure 5–8. Then consider second-order conditions to confirm whether we have a maximum or not. Taking the derivative of the first equation in (5–7) with respect to Q results in the following:

$$\frac{dMR}{dQ} - \frac{dMC}{dQ} < 0$$

Since the first expression above is the slope of marginal revenue, and the second is the slope of marginal cost, the second-order condition requires the following:

$$\text{slope } MR < \text{slope of } MC$$

Since the slope of MR and MC are both negative in a decreasing cost industry, this result means that MR must be more negative or steep than the MC. Note that MR is more steep than MC in figure 5–8.

Before we apply this result to the above problem, let's develop one additional general result that will be useful. Suppose we face an inverse linear demand curve $P = a - bQ$. Total revenue for this demand curve is as follows:

$$TR = PQ = (a - bQ)Q = aQ - bQ^2$$

Marginal revenue equals the following:

$$\frac{dTR}{dQ} = a - 2bQ$$

Thus, marginal revenue is downward sloping and is twice as steep as the demand curve for linear demand.

We are now ready to work out how much the monopolist should produce in the above example to maximize profits. We represent the monopoly market in figure 5–9, including marginal revenue, which in this case is linear and twice as steep as the demand curve. Thus, it bisects the Q axis halfway between the zero and where the demand crosses the Q axis.

The profit maximizing monopolist would produce Q_m, which is where marginal revenue equals marginal cost, and sell at a price of P_m. The monopolist's profits (π_m) could be computed in a variety of ways. Substitute all the functions and the optimal quantity (Q_m) into equation (5–6). Alternatively, note that average profit per unit is ($P_m - AC_m$). This average profit is earned on Q_m units, yielding the area of the grey rectangle equal to ($P_m - AC_m)Q_m$. Last, since marginal cost is the cost of the final unit, if we add marginal cost of each unit to the marginal costs of all the preceding units, we get total variable cost. With no fixed costs, this sum also represents total costs. In the continuous case, we can represent this total cost as an integral and write profits yet a third way, as follows:

$$\pi_m = P(Q_m)Q_m - \int_0^{Q_m} MC(Q_m)dQ_m$$

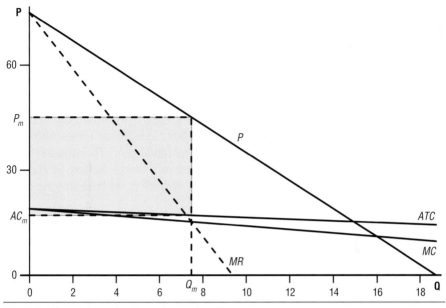

Fig. 5–9. Monopoly production, price, and profit

For the above example, the numerical solution for P_m and Q_m can be computed as follows (note answers are rounded to three decimal places):

$$TR = (75 - 4Q)Q = 75Q - 4Q^2$$

$$MR = \frac{dTR}{dQ} = 75 - 8Q$$

Setting marginal revenue equal to marginal cost and solving for Q results in the following:

$$MR = 75 - 8Q = MC = 19 - 0.5Q \rightarrow Q = 7.467$$

Second-order conditions are as follows:

$$\frac{dMR}{dQ} - \frac{dMR}{dQ} = -8 - (-0.5) < 0$$

So, we have a maximum. Monopoly price is as follows:

$$P_m = 75 - 4(7.467) = 45.132$$

Monopoly profits computed the three different ways are the following:

$$\pi_m = PQ - TC = 7.467(45.132) - 19(7.467) + 0.25(7.467)^2 = 209.067$$

$$\pi_m = (P_m - AC_m)Q_m = (45.132 - (19 - 0.25 \times 7.467))7.467 = 209.067$$

$$\pi_m = P(Q_m)Q_m - \int_0^{Q_m} MC(Q)dQ = 45.132 \times 7.467 - \int_0^{7.467} 19 - 0.5QdQ$$
$$= 337.001 - 19 \times 7.467 + 0.5 \times 7.467^2/2 = 209.067$$

Now let's consider the effect on society of the monopoly result. In figure 5–9, notice that at Q_m, consumers are willing to pay a higher price than the marginal cost of that unit, implying social losses. Furthermore, these losses accrue to each unit of output up until demand crosses marginal cost. To formalize these losses, remember from chapter 4 that we represent society's welfare by the sum of producer plus consumer surplus. Using this definition, we measure the social optimum by again turning to calculus.

The sum of consumer surplus plus producer surplus can be represented by the area below demand and above price plus the area above marginal cost and below price. We can represent these areas up to output Q_o and price P in integral notation as follows:

$$W = \int_0^{Q_o} P(Q)dQ - PQ_o + PQ_o - \int_0^{Q_o} MC(Q)dQ = \int_0^{Q_o} P(Q)dQ - \int_0^{Q_o} MC(Q)dQ$$

We are in effect maximizing the area between the inverse demand P and MC in figure 5–9. If we integrate or add up marginal costs, we get total variable costs, so consumer surplus can also be written as follows:

$$W = \int_0^{Q_o} P(Q)dQ - TVC$$

Optimizing this function with respect to output Q_o gives us the following:

$$\frac{dW}{dQ_o} = \frac{d\int_0^{Q_o} P(Q)dQ}{dQ_o} - \frac{dTVC}{dQ_o} = P(Q_o) - MC(Q_o) = 0$$

Remember in this case with no fixed costs that our total variable costs are equal to our total costs. The calculus tells us we should operate where the price equals marginal cost or where the demand curve crosses the marginal cost curve. We can see this same result in figure 5–9. At outputs before marginal cost crosses demand, price is above MC, so we can increase social welfare by increasing output. However, when producing outputs beyond this point, the price is below the marginal cost. Consumers value the extra output by less than the cost of the extra output, and welfare is diminished.

Using the above example, we can solve for the social optimal P_o and Q_o as follows:

$$P_o = 75 - 4Q_o = MC = 19 - 0.50Q_o \rightarrow Q_o = 16$$
$$P_o = 75 - 4(16) = 11$$

Notice that at Q_o, average total costs are above price, and the utility would lose money. The amount lost would be determined as follows:

$$\pi = PQ_o - TC(Q_o) = 11 \times 16 - 19(16) + 0.25(16)^2 = -\$64$$

Calculus does not tell us how to cover these losses, but only how to price each unit of the product and the total amount that should be produced. In the next section, let's consider policy in this decreasing cost case.

Government Policy for a Natural Monopoly

Since the social optimum in a decreasing-cost industry is at a loss-making quantity, it would not be sustainable in a competitive market. Also, without intervention in such a market, we would expect a monopoly to evolve. In such a case, economists think the market will fail and there is room for government intervention. From the time it was generally agreed that electric utilities were natural monopolies, governments have historically responded with government ownership or regulation. In the United States and Japan, more often, public utilities have been privately owned (referred to as investor-owned utilities or IOUs) and regulated. Elsewhere, public utilities have more often been owned by the government (publicly owned).

In the United States, these utilities have been subject to price regulation by regulatory commissions since 1907. The early agencies had a vague mandate to "maintain just and reasonable prices" and to limit price discrimination that was unrelated to variation in cost. When a utility wanted a rate increase, it would present a case to the commission in the state in which it operated. Prices would be set in these rate cases and would generally be fixed until the next case. The benefits of such regulation are the avoided social losses associated with monopoly, while the costs are those associated with running the regulatory agencies and any other unintended side effects of the regulation.

A 1944 US legal decision established that regulated prices should not only cover costs but include a return high enough to attract capital and compensate investors for risks. Next, let's consider three possible regulatory price models: rate of return, fully distributed cost, and peak-load pricing.

Rate of Return Regulation

In rate of return regulation, the utility is allowed a rate of return based on its capital stock or rate base (RB). With economies of scope, the typical utility produces n products ($q_1, q_2, q_3, \ldots, q_n$) and charges a price ($p_i$) for the ith product. (Economies of scope arise when unit costs fall when more than one type of product is produced.) Total revenue is $\sum_{i=1}^{n} p_i q_i$. The firm is allowed to charge pi such that it covers its noncapital expenses and earns a normal rate of return (s) on its rate base. The basic accounting equation for rate of return regulation is given in equation (5–8):

$$\sum_{i=1}^{n} p_i q_i = \text{expenses} + s(RB) \qquad \text{(5–8)}$$

where

p_i is the price of the ith service class,
q_i is the quantity of the ith service class,
n is the number of service classes,
s is the allowable or "fair" rate of return, and
RB is the rate base of the regulated firm's investment.

Equation (5–8) requires that the company's revenues equal its costs, including a normal rate of return. Economic profits are zero, but economically efficient prices are not required.

Let's apply equation (5–8) to a simple rate case. Suppose that a utility has a rate base of $2,000. It expects to sell 4,000 kWh to industrial customers and 2,500 kWh to residential customers. Operating costs are $200. The regulator believes that 10% is a normal rate of return for the utility. The utility has asked the public utility commission for a rate increase to $0.05/kWh for industrial users and to $0.10/kWh for residential users. Should the commission approve the rate increase?

The utility's revenues would be 0.05 × 4,000 + 0.10 × 2,500 = $450. Its expenses plus return on capital would be 200 + 0.10 × 2,000 = $400. Since revenues are larger than expenses, the commission would not approve the rate increase.

Problems with Rate of Return Regulation

There are a number of difficulties in implementing rate of return regulation. For the most part, it is straightforward to compute expenses. There may be some difficulties with transfer prices, which are prices at which one part of a company sells products to another. For example, if a utility owned a coal mine and sold itself coal, the book or accounting value recorded may be distorted for tax or rate of return regulation purposes. Thus, the value of the coal may not represent an arm's-length transaction from one independent firm to another. Such cost distortions may arise from tax laws as well as from cost pass-through. Also, since costs can be passed on to the ratepayers, there may be less incentive to hold the line on costs than in a competitive industry.

There is also ambiguity over a normal rate of return and the rate base. A normal rate is required to attract financial capital to the market. Rates on capital depend on how the investment is financed. Suppose the utility finances capital purchases using the three conventional sources, bonds, preferred stocks, and common stocks, in the proportions shown in table 5–2. Any required rate of return could be computed as the weighted average of the three financing sources.

Table 5–2. Financing for a representative utility

	% of Capitalization	Required Return
Bonds	48	8.22
Pref. Stocks	14	9.34
Common Stocks	38	12.5
Weighted Average	$0.48 \times 8.22 + 0.14 \times 9.34 + 0.38 \times 12.5 = 10.0$	

Annual rates of return for a bond are relatively easy to compute. First, take the easiest case of a bond that lasts forever, called a *perpetuity*. A utility bond usually pays coupons quarterly or as stipulated on the bond. The quarterly rate of return for such a long-term bond is the quarterly coupon divided by the bond price. For example, suppose the utility sells you a bond for $1,000 and pays four quarterly coupons of $25 each. Then the quarterly rate of return on the bond is $(25)/1,000 = 0.025$ or 2.5%.

The rate of return will be influenced by other rates of return in the market. For example, suppose that the Fed tightened monetary policy to reduce inflation, and the market interest rate went up to 5% on items of risk similar to the utility bond. Now if the utility bond only paid 2.5%, it would not look very attractive. If you wanted to sell the bond, you would have to lower the price so that it paid the same rate as other similar assets. Thus, for a perpetuity, the equation $25/P = 0.05$ must hold. This equation implies that the price of the bond would have to fall to $P = 25/0.05 = 500$ for the bond to be competitive. This relationship, where bond prices fall when interest rates on similar assets increase, also holds in the case of an asset that has a finite life. However, the numerical value of a finite life asset is more complicated and requires us to value money across a finite time period.

Valuing Money across Time

Suppose we know the future flow of dividends and the stock price at issue. To see how to develop a precise measure of rate of return, let us first review how to value money across time.

Suppose you have $1 today. The interest rate at the bank is 5%, and the bank compounds interest annually. After one year you will start to receive interest on the interest accrued. If you put the money in the bank you will have $(1 + 0.05) \times \$1 = \$1.05$ at the end of the year. If you leave this $1.05 in the bank for another year, you now get interest on the principle and the interest, so at the end of two years you will have $(1.05)(1.05) \times \$1 = 1.1025$. If you hold it for t years, you will have $(1.05)^t \times \$1$. If you have A dollars and the interest rate is r, at the end of t years, you will have its compounded value: $(1 + r)^t \times \$A$.

Similarly, we can see what money in the future is worth today. If I offer you a dollar in a year, what is it worth to you today? In other words, how much money (B) would you have to put in the bank today to have a dollar at the end of the year? The answer depends on the interest rate and compounding. If the interest rate is 5%, then we know that $B(1.05) = 1$, the dollar I am offering you in a year. Solving for B, we get $B = 1/1.05 = \$0.9524$. We call this process *discounting future income*, and the value today is called the *present value*.

What is the value now (B = present value) of $1 received in two years, if the interest rate is 5% and interest in compounded annually? $(1.05)^2 B = 1$, so $B = 1/(1.05)2$. The present value of $1 received in t years $= 1/(1.05)t$. The present value of D_t dollars received in t years at an interest rate of r is $D_t/(1 + r)^t$.

We can now value a stream of income quite easily. If I offer you D_1 dollars at the end of one year and D_2 at the end of two years, the present value of this stream of income with interest rate r is the sum of the present values (PV) of each amount as follows:

$$PV = \frac{D_1}{(1 + r)} + \frac{D_2}{(1 + r)^2}$$

If I offer you a stream of income D_i from time period i = 1...k, its present value is as follows:

$$PV = DCF = \sum_{i=1}^{k} \frac{D_i}{(1 + r)^i} \qquad \text{(5–9)}$$

If you have a flow of income (D_i) and a flow of costs (C_i) for k years starting at the end of the year, the net present value (NPV) of your income steam (also called your discounted cash flow or DCF) is the present value of your income minus the present value of your cost, as follows:

$$NPV = DCF = \sum_{i=1}^{k} \frac{D_i}{(1 + r)^i} - \sum_{i=1}^{k} \frac{C_i}{(1 + r)^i} = \sum_{i=1}^{k} \frac{D_i - C_i}{(1 + r)^i}$$

If the cash flow starts immediately, start i in the summations at zero instead of one.

Utility Rate of Return on a Bond or Stock

Discounting enables us to compute a utility's rate of return (also called an *internal rate of return* or *irr*) on an asset. To do so, note that the stock's price should be equal to the discounted cash flow of its future dividends and any resale value or:

$$P = \sum_{i=1}^{k} \left[\frac{D_i}{(1 + irr)^i} \right] + \frac{P_k}{(1 + irr)^k} \qquad \textbf{(5–10)}$$

where

P is the current cost of the bond/stock,

D_i is the coupon/expected dividend in year i,

k is the number of years coupons or dividends are received,

irr is the cost of capital for assets with similar risk characteristics, and

P_k is the bond redemption price or any resale value of a stock when it is sold.

The discounted cash flow of the future income is what the future flow of income is worth today, if it pays you the required return or discount rate. This return is your opportunity cost plus any adjustment for risk. If the project is more risky, you will require a higher rate of return; if it is less risky, you will require a lower rate of return.

To illustrate that an asset price should be equal to its discounted cash flow, take a simple example of a project that lasts a year. Suppose the project pays $100 in a year. Other projects that are equally risky pay 10%, therefore $DCF = 100/1.10 = \$90.91$. Thus, if you have $90.91 now and invest it for a year, you would have $100. If you invested your $90.91 in another equally risky asset with the same cash flow, you would also have $100. Now suppose you only have to pay $85 for this asset. This would be a good buy, since it would pay $100 at the end of a year, whereas other projects that yielded $100 would cost $90.91. Because this is a desirable project, everyone would want to buy it, instead of the other more expensive assets. As investors switched out of more expensive assets into the cheaper one, the price of the cheaper asset would be bid up, and the price of the more expensive ones would fall until they were equally desirable or their prices equaled their discounted cash flow.

Assuming efficient markets, notice that *irr* in equation (5–10) equals the rate of return required by investors for this type of asset. This required return is also called the *opportunity cost of capital*. Its value depends on how risky the asset is. Here risk is taken to mean variability in profits from the capital or investment. The causes of the risk include business, financial, and regulatory risk, among others. The more risky the firm, the higher the required return investors will demand to invest in the firm.

Now let's compute the internal rate of return from price and coupon information. You buy a bond for $100 today, the bond pays $10 in dividends at the end of the year for two years, and the bond redeems for its face value of $100 after two years. The formula to solve for *irr* would be as follows:

$$100 = \frac{10}{(1 + irr)} + \frac{10}{(1 + irr)^2} + \frac{100}{(1 + irr)^2}$$

Solving:

$$(1 + irr)^2 100 = \frac{(1 + irr)^2 10}{(1 + irr)} + \frac{(1 + irr)^2 10}{(1 + irr)^2} + \frac{(1 + irr)^2 100}{(1 + irr)^2}$$

$$(1 + 2irr + irr^2)100 = 10(1 + irr) + 10 + 100$$

Simplifying:

$$100 irr^2 + 190 irr - 20 = 0$$

The solution for *irr* in the above equation can be found using the quadratic formula, as shown here:

$$irr = \frac{-190 \pm (190^2 - 4 \times 100(-20))^{0.5}}{2 \times 100}$$

Solutions yield the following positive and negative roots:

$$irr+ = \frac{-190 + (44,000)^{0.5}}{2 \times 100} = 0.10 \text{ and } irr- = \frac{-190 - (44,000)^{0.5}}{2 \times 100} = -2.00$$

Although the second negative root makes mathematical sense, it does not make economic sense, and we ignore it. Notice in the above case, the redemption price is the same as the price we paid for the bond, and the return is the same as the D_i/P. This will always hold for a bond that is bought at face value, pays out a constant stream of income, and is held to maturity and then redeemed at face value, no matter how far out the maturity. Thus, computing an internal rate of return for a bond is relatively straightforward.

For preferred stocks, it is a little more complicated to compute a rate of return. Although the terms of various preferred stocks vary, a preferred stock typically pays a maximum return or maximum dividend. Companies do not have to pay dividends on preferred stock, but if they pay any dividends on common stock, they must pay the maximum dividend on preferred stock. Take the case where the dividend rate is a constant D_i. Then we know D_i in equation (5–10). However, we do not know *k* since the stock does not have a fixed redemption time. If we picked a time (*k*) we were going to sell, we would still not know the price at *k*. However, if we intended to hold the stock in perpetuity, and it continued to pay the same dividends year upon year, the return would be D_i/P.

Both common and preferred stocks represent equity or ownership, but computing the rate of return on common stocks poses even more of a challenge.

As with preferred stocks, there is no specific maturity (k) for the stock. Although companies go in and out of business every day, in reality, some company stocks may last for a very long time. For example, Royal Dutch Shell (now called Shell Oil) has been in business since 1907. If you bought Shell stock for P_1 and held it for k years, received dividends of D_i at the end of each year i, with $i = k$, and sold the stock for P_k in year k, your internal rate of return would be found by solving the following equation:

$$P_1 = \sum_{i=1}^{k}\left(\frac{D_i}{(1 + irr)^k}\right) + \frac{P_k}{(1 + irr)^k}$$

Notice that if k is very large, such a formula is cumbersome to solve. However, you can go to Microsoft Excel and use the IRR formula. Put the values for $(-P_1, D_1, D_2, \ldots, D_k + P_{kj})$ in cells A1 to Ak + 1. Values are negative for cash outlays and positive for cash receipts. Years with no dividends would have values of 0. For $k = 20$, type in cell B1 = IRR(A1:A21,α). The expression just before the comma in the internal rate of return formula is the address for the cash flows, and the expression just after the comma is a starting value for Excel to use to solve this nonlinear equation. Thus, if k were 200 and α were 10%s, the internal rate of return formula would be IRR(A1:A201,0.10). You can leave a blank after the comma, but a well-chosen starting value may speed up the computation.

Another difficult problem with computing the internal rate of return for common stock is that values for D_i in the future are hard to measure. Our choices are to turn to an electricity market analyst to estimate D_i or use some historical value, such as the last value in the market or some average over time. If we can come up with good estimates for the D_i values, and we have computed the rates of return on stocks, bonds, and preferred stocks, and their shares of financing, then we can compute s, the required rate of return, to be able to raise capital.

A more pragmatic approach would be to raise s if investments were deemed insufficient and decrease s if there was overinvestment in electricity.

Utility Rate Base

The last thing we need for rate of return regulation is the rate base (RB). It is usually original cost minus depreciation, but this value understates actual investment if inflation has occurred. To see why, consider how to convert values measured in one year's dollars to those measured in another. Suppose you purchased your plant in 1980 for 500 million 1980 dollars. You depreciate the plant over 40 years, so that annual depreciation is 500/40 = 12.5. Let's compute the rate base in 2000: RB (2000) = 500 − 20 × 12.5 = $250 million. However, if prices have doubled from 1980 to 2000, the value of the dollar has fallen in half. The rate base related to 2000 dollars would be 2 × 250 = $500 million. Original cost understates the rate base.

We can easily adjust the rate base if we have the appropriate index. For example, suppose we have capital stock (K) measured in 1985 dollars, and we want to know the value in 2000 dollars. We need an index that captures the value of a dollar in 2000 divided by the value of a dollar in year 1985. Call this index 2000$/1985$. The denominator is called the base year. It is computed by dividing the value of a basket of goods in year 2000 by the value of the same basket of goods in 1985. For example, if the basket of goods cost $200 in 1985 and cost $350 in 2000, the I2000$/1985$ = (350/200) = 1.75. This index tells us that prices for the goods in the basket in 2000 are 0.75 times higher than their prices in 1985. To use our index to inflate 1985 dollars up to 2000 dollars, multiply by our 1985 capital stock, as follows:

$$\frac{K(\text{in } 1985\$)(I2000\$)}{1985\$} = K(\text{in } 2000\$)$$

For example, if we paid $200 for a piece of capital in 1985, and the wholesale price index for year 2000 with base year, 1985 = (I2000$/1985$) = 1.75, the price of capital in 2000$ is as follows: 200(1.75) = 350.

Alternatively, if we bought capital in 2000 and wanted to deflate these 2000 dollars back to 1985 dollars, we would divide by the index in decimal form (remember, to divide by a fraction, you invert the fraction and multiply) as follows:

$$\frac{K(\text{in } 2000\$)}{(I2000\$/1985\$)} = K(\text{in } 1985\$)$$

A piece of capital purchased for $350 in 2000 would be worth 350/(175/100) = $200 (in 1985$).

What if we need to know the index I2000$/1985$ with base year 1985 but only have indices with base year 1990? Use the index I1985$/1990$ and the index I2000$/1990$ to construct the appropriate index as follows:

$$\frac{(I2000\$/1990\$)}{(I1985\$/1990\$)} = \frac{I2000\$}{1985\$}$$

Thus, if the wholesale price index for year 2000 with base year 1990 = (I2000$/1990$) = 1.61, and the wholesale price index for the year 1985 with base year 1990 = (I1985$/1990$) = 0.92, then the wholesale price index for year 2000 in 1985 dollars is (1.61/0.92) = 1.75. When indices are reported in the literature, they are usually reported in percent, which means the above numbers would have to be multiplied by 100. Thus, (I1985$/1990$) = 0.92 would be reported as 92. However, to use the indices to deflate or inflate, you would need to divide them by 100.

The financial approach to valuing the utility's capital would be to ascertain the value of the company's stocks and bonds. However, this is not an independent value. It depends on s, the allowed rate of return. To see why, take a simple example. Suppose that our allowed rate of return is 10%, and we have stocks equal to $100 and no bonds. Thus, our dividends are $10. If equally risky alternatives in the market paid 9%, then our stock would look like a good investment. People would keep buying our stock, bidding up the price until it was paying only 9% as well. Assuming a perpetuity, the new price could be determined as follows:

$$\text{Dividend}\Big/\text{Price} = 0.09 = {}^{10}\!\big/_{P}$$

Solving, we get the following:

$$P = {}^{10}\!\big/_{0.09} = \$111.11$$

However, if the rate of return (s) changed, dividends would change, and the price would change.

Rate of return regulation usually has a regulatory lag. Once a utility decides that its prices are too low, it must present a case and get the rate increased by the public utility commission. These new prices stay fixed until the next approved rate case. The utility then has an incentive to reduce costs through technical improvement, since it gets to keep the extra savings until the next rate case.

Rate of return regulation does not dictate a rate structure but approves a suggested rate structure. Most utilities price discriminate and charge different prices to different customer classes. Some sample US average prices and consumption per customer class are shown in table 5–3.

Table 5–3. US average electricity prices and consumption by customer class, 2012

Sectors	Electricity Prices (¢/kWh)	Electricity Consumption (terawatt hours)
Residential	11.9	1,374.6
Commercial	10.12	1,323.8
Industrial	6.7	980.8

Source: US EIA (2013b).

Since utilities are only allowed a nominal rate of return on their capital, they may have a tendency to overinvest in capital relative to the least-cost input mix of production. This tendency has come to be known as the *Aversch-Johnson* (AJ) effect. (Crew and Kleindorfer [1979] mathematically analyze how this might distort the choice of inputs. For a summary of their analysis see http://dahl.mines.edu/b0501.pdf.)

Although economic theory leads us to suspect an AJ effect, empirical studies investigating the issue have found mixed results. However, if the AJ effect exists, it is raising the costs of generating electricity.

Utility Cost Allocation

Utilities typically are multiproduct firms supplying different customer classes. For example, they supply high- and low-voltage customers as well as on-peak and off-peak services. A utility's fixed cost may contribute to services for more than one product or customer class. Fully distributed cost (FDC) deals with the issue of how to distribute these fixed costs over customer or product classes.

One way to allocate fixed costs is to distribute them in the price across all consumers. This, however, distorts consumption decisions and causes losses in social welfare, since price will no longer be equal to marginal cost. Since markets with less elastic demand will cut consumption the least, there is greater economic efficiency in allocating more costs to consumers with less elastic demand. For more information on allocating cost in this way, known as *Ramsey pricing*, see Viscusi et al. (1995).

A more efficient way to allocate fixed costs is in the form of fixed charges. As long as you do not drive consumers out of the market by charging them more than their consumer surplus, you can allocate fixed costs over consumer groups in any way you like, and it will be economically efficient. For more information on distributing costs as fixed charges and issues of fairness, again, see Viscusi et al. (1995).

Peak-Load Pricing

Peak-load pricing means charging different prices for electricity depending on the load factor. Since it is expensive to store electricity, capacity is usually made large enough to satisfy the on-peak demand. This means, however, that during much of the time some capital is sitting idle.

In the example of Duke Power Company, given in Viscusi et al. (1995), Duke produces 55 billion kWh per year. Its average single-customer demand is 6,300 megawatts (MW), peak demand is 11,145 (MW), and installed capacity is 13,234 (MW), with the extra required as a mandated reliability margin. If a utility can move some of the on-peak demand to off-peak, it can decrease the amount of total capital needed and its cost by using existing capital more intensely.

To develop the efficiency criteria for peak-load pricing, we use a simple model (Dreze 1964). Suppose daily demand on peak is D_{pk} and off-peak demand is D_{ofpk}. Demands are independent, so the price in the on-peak period does not affect the quantity demanded in the off-peak period, and vice versa (fig. 5–10).

Operating costs are C_{oppk} for on peak and C_{opof} for off peak. Capital costs per unit are constant at C_k, and we assume capital can be added in small increments. We want to pick prices in the two markets (on peak and off peak) to maximize social welfare.

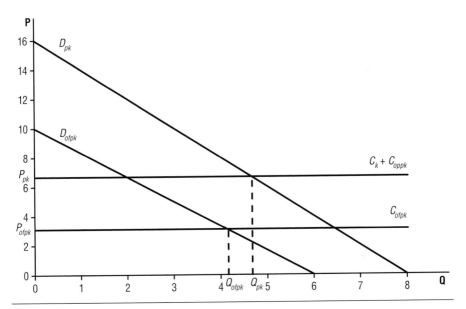

Fig. 5–10. Peak load model

We know that the social welfare is the area under the demand curve minus cost. We can use either prices or quantities as our choice variable, since choosing one determines the other. It is easiest to optimize in terms of quantity. Capacity will be assumed to be at the peak with no margin for safety. Total consumer benefits (area under the demand curve) minus total cost or social welfare for this case is expressed as follows:

$$SW = \int_{0}^{Q_{pk}} P_{pk}(Q_{pk})dQ_{pk} + \int_{0}^{Q_{ofpk}} P_{ofpk}(Q_{ofpk})dQ_{ofpk} - C_{oppk}Q_{pk} - C_{opof}Q_{ofpk} - C_{k}Q_{pk}$$

First-order conditions are the following:

$$\frac{\partial SW}{\partial Q_{pk}} = P_{pk} - C_{oppk} - C_{k} = 0 \longrightarrow P_{pk} = C_{oppk} + C_{k}$$

$$\frac{\partial SW_{pk}}{\partial Q_{opk}} = P_{ofpk} - C_{opof} = 0 \longrightarrow P_{ofpk} = C_{opof}$$

The above first-order conditions require that on-peak load pays its operating plus capital costs, but off-peak load only pays its operating costs. Since off-peak has idle capacity, we increase welfare by increasing consumption off-peak as long as we are covering our costs. However, before such a scheme is implemented, analysis is necessary to ensure costs of metering do not negate the gains from peak-load pricing.

The above result holds when there is no peak switching. Thus, the on-peak period has higher consumption after implementing the pricing scheme. (For the peak-switching case, see question 35 at http://dahl.mines.edu/st05/st05.pdf.)

Summary

Electricity is produced from a variety of sources, including fossil fuels, nuclear, and renewable sources. Fossil fuels tend to dominate power production worldwide, but fuel sources vary considerably across countries and regions. Coal dominates in the United States, China, and a number of other countries. Hydropower dominates in Latin America, while Europe has the highest proportion of nuclear power and other renewable energy sources. The Middle East has the largest share of oil generation. North America remains the most electricity-intensive region of the world, followed by Europe, the former Soviet Union, and the industrial countries of Asia and Oceania. The developing countries lag behind, especially Africa. Such countries, however, are eager to catch up and provide very promising markets for new generation capacity.

Costs are an important consideration in determining which fuels to use to generate electricity and how much electricity to generate. Total costs include fixed costs, which are unrelated to production, and variable costs, which are related to how much is produced. In the long run, all costs are variable. However, in the short run, fixed costs include existing plant and equipment costs, leases, and insurance. If they cannot be recouped, they are also sunk. Average unit cost tells us what it costs us to produce a unit on average and is useful in helping us compute total costs, total profits, and overall performance. Marginal cost tells us what the last unit costs to produce and is an important input in deciding how much to produce.

We also see that through the day and the year, the load cycle for electricity varies. We have daily on-peak and off-peak demand. On-peak demand tends to occur during the day and early evening; shoulder demand occurs early in the morning and later in the evening. Off-peak demand occurs during the night. There are also seasonal on- and off-peaks, depending on the climate and daylight hours. To satisfy electricity demand, utilities continually use their most efficient, least-cost, or baseload plants; these are usually coal, nuclear, and hydro. During on-peak periods, less efficient reserve capacity is used, such as turbines with lower capital costs but higher fuel cost, or pumped storage and older plants. Thus, during on-peak periods, the variable cost is usually higher.

Revenues are also an important factor in deciding how much electricity to produce. Price is the average revenue for each unit and is important in determining profits. Marginal revenue is the revenue from the last unit and is important in helping determine how much to produce. If a plant has market or pricing power, its marginal revenue will be below its price.

Historically, increasing economies of scale gave the largest plant the lowest marginal cost. This largest, most efficient plant is referred to as a natural monopoly. It would be able to drive other plants out of business. If it became sole producer, it could maximize profits where marginal revenue equaled marginal costs and make monopoly profit. However, the social optimum would be at the larger output, where price equals marginal cost. The resulting social losses from a monopoly have historically caused governments to intervene in this market to either regulate or have the government own the electricity generation.

We considered three different types of regulatory approaches to regulating prices: rate of return, fully distributed cost, and peak load. Under rate of return regulation, revenues must cover variable costs plus a fixed rate of return on capital necessary to attract the appropriate amount of capital to the industry. This required rate depends on the rates of return on bonds, common stock, and preferred stock used to finance capital and may be difficult to compute. Valuing the rate base is also fraught with difficulties. Inflation means that using original cost may understate the rate base, whereas using the current cost is likely to overstate the rate base. Rate of return regulation may cause utilities to invest in too much capital and may not be efficient.

Economic efficiency requires that price be set equal to marginal cost. This leaves the problem of allocating the fixed costs across consumer classes. If we are constrained to allocating the costs across each unit of production, we will not have the economically efficient outcome. However, economic theory suggests social losses are smaller if we allocate more costs to groups with less elastic demand.

If we maximize social welfare, we should set price equal to marginal cost in each customer class and allocate the fixed costs as fixed charges rather than dividing them up into price. Efficiency criteria do not tell us how to allocate the fixed charges, except that the charges should not distort consumer decisions on the margin. Thus, no fixed charge should be greater than the consumer surplus in that market.

If the consumer classes are on peak and off peak, and there is no peak shifting, then theory indicates that the most efficient way to price electricity is to charge all capital costs to the on-peak users and only marginal costs to the off-peak users.

6

Restructuring in the Electricity Sector

As has long been noted, the key resource of governments is the power to coerce.
Regulation is the use of this power for the purpose of restricting the decisions
of economic agents.

—Viscusi et al. (1995)

Introduction

Natural monopoly arguments traditionally led most governments to either regulate the electricity industry or produce electricity themselves in vertically integrated monopolies. Thus, large publicly or privately owned companies owned much of the generation, the high-voltage transmission lines, and the local distribution companies. They were responsible for all aspects of the power market. They built and maintained generating facilities and assured the quality of the electricity supply. They built and maintained the transportation network, making sure that power was dispatched and transported when and where it was needed. They built and maintained local distribution companies, distributed power, and billed customers.

Problems with Regulated and Government-Owned Utilities

The regulatory theory applied in the last chapter was that regulation was implemented for the public good. However, markets are not the only things that can fail; government regulators or government owners may fail also. Proper regulation of electricity requires expertise in a variety of areas, including engineering knowledge of the electricity industry to get the power reliably and safely generated, transported, and distributed. Accounting, finance, and economic expertise are also needed to ensure that the power system is economically viable and efficient. Since regulators are often political appointments, they may be swayed by powerful interest groups, such as producers, consumers, environmental groups, property owners near the power plant, and even possibly the executive and legislative branches of government. These groups will likely have a variety of competing interests with varying lobbying abilities (Becker 1983). Even nonpolitical bureaucrats may be subject to the same political pressure as their political counterparts (Winston 1993).

Regulation can invite the dangers of regulatory capture, where the regulator tends to take the point of view of the regulated industry. Often, regulators retire from government and move into the regulated industry in a process referred to as a *revolving door*. Regulations may even be invited by industry and serve to protect them rather than the consumer (Stigler 1971). The Aversch-Johnson (AJ) effect mentioned in the last chapter may raise costs by encouraging too much investment in real capital. Furthermore, guaranteed rates of return may not be conducive to cost minimization, and regulators may err when setting the rates.

In the case of government ownership, you may get what Leibenstein calls *X-inefficiency*, or higher costs than would prevail in a competitive market with cost minimization. Governments may also have other goals than cost minimization. For example, in New Zealand before privatization, the government-owned electricity monopoly employed far more people, and at higher costs than privately owned utilities elsewhere. Presumably, this was a government employment policy. In developing countries, electricity prices are often held low to subsidize development, while costs are often high, resulting from poor management, electricity theft, and corruption. This leads to capital shortages for developing new generation capacity in many poor countries, and as a result, electricity brownouts and blackouts are routine occurrences. For example, a Gallup poll of workers in electrified workplaces in 17 sub-Saharan African countries reported an average of more than 2.5 days of power outage in the previous week (Tortora and Rheault 2012).

Kumbhakar and Hjalmarsson (1998) found municipal electricity distributors to be significantly less economically efficient than privately owned distributors and cited the following reasons:

- The government may guarantee municipal debt, shielding municipal distributors from bad decisions and bankruptcy.
- There may be a conflict of interest when governments own distribution while also regulating private distribution.
- A public rather than private rate of return may distort municipal investment and production decisions.
- Municipal procurement and hiring rules may be more stringent than for the private sector.
- Municipals may have added political objectives not faced by private sector plants.
- Municipals may have fewer incentives for profit maximization and cost minimization than in the private sector because of the lack of discipline from shareholder pressure and the stock and debt markets.

The electricity market is comprised of a number of segments, including generation, long-distance transmission, local distribution, and marketing. These segments have different cost curves, with different evolutions across time. Transmission and distribution enjoy economies of scale over a wide range of capacities. Coordination and management of the grid are likely natural monopolies. This is less likely the case for retail marketing if retailers have open access to the grid.

Large economies of scale in generation were once the rule, leading governments to treat them as natural monopolies. As markets have gotten larger and technical changes have lowered costs for smaller generators, economists have also questioned whether the electricity market is really a natural monopoly, particularly at the generation level. Christensen and Greene (1976), in their classic cost study for US generators, found that in 1963 most electricity firms in the United States could exploit significant economies of scale, while by 1970, a large share of firms were in a relatively flat portion of the long-run average total cost curve. Hunt and Shuttleworth (1996) noted that the optimal electricity generation plant size had fallen from 1,000 MW to approximately 100 MW in the previous 30 years. Nor did Feibelman and Britt (2012) find more recent statistical evidence of economies of scale in generation for dozens of US utilities. All these changes led to new work that considers how to transform the electricity supply sector into a more competitive industry with minimal regulation.

Models for the Electricity Sector

We will consider four electricity models, combining the model types from Hunt and Shuttleworth (1996) and Tenenbaum, Lock, and Barker (1992). The four models are distinguished by the type of competition at each step in the supply chain rather than by ownership:

- *Model one.* No competition at any stage, or monopoly utilities. Often, these companies are vertically integrated, and they may be publicly or privately owned.
- *Model two.* Model one, but with competition in generation. A single buyer such as a distribution company may buy from a number of different producers to encourage competition in generation. The United States started moving to this model with the Public Utilities Regulatory Policy Act (PURPA) of 1978 that required US utilities to purchase output from independent power producers (IPPs) at avoided costs. *Avoided costs* are the costs a utility would have to incur to generate an equal amount of power. This model is sometimes called the *single-buyer model.*
- *Model three.* Model two, but with common or contract carriage of high-voltage transmission lines offered to all wholesale sellers and buyers. Often distribution companies (DISCOs) own the distribution

wires and can choose their suppliers, with competition in generation and in the wholesale market.

- *Model four.* Model three, but retail customers also choose their suppliers in *full retail competition.* There is open access in both transmission and distribution, which are still considered natural monopolies and are government owned or regulated. In the British model, there is also complete separation of generation, transmission, and distribution, with an independent company to own the high-voltage transmission and perform the dispatch function.

The important differences in these models are whether there is competition among generators, whether retailers or distribution companies can choose the generator to buy from, and whether the final consumer can choose who to buy their power from. Hunt and Shuttleworth (1996) argue that model four is the most economically efficient, given the following conditions:

- a well-established electricity retailing system
- mature market institutions
- constant vigilance against market power
- appropriate methods of dispatch

Models two and three seem to be the most popular for countries starting the restructuring process. The United Kingdom, New Zealand, Russia, and others have had model four as their ultimate restructuring goal. European Union Directive 96/92/EU in 1996 required full retail competition by 2007. However, Blumsack and Perekhodtsev (2009) note that in practice for model four, usually only a small percent of retail customers switch to new suppliers from the incumbent utility when offered the chance. They site inertia and high transaction costs relative to financial saving among the likely reasons.

With privatization and restructuring, the need for dispatch and coordination becomes crucial, particularly where formerly vertically integrated companies have been broken up. Often, an independent system operator (ISO) coordinates the whole physical system based on a wholesale market for electricity or a power pool that links together wholesale buyers and sellers on the supply chain. The ISO may manage the financial aspects of the wholesale market, such as matching bids and offers and dispatching, or a separate company may do it.

Examples of Electricity Restructuring

One of the early moves toward restructuring electricity markets in the United States, the Public Utility Regulatory Policies Act of 1978, allowed qualifying small producers using renewables and combined heat and power facilities to access the grid. More extensive moves toward restructuring began a bit later in Latin America. The electricity industry there was more often developed under the initiative of the private sector. However, as happened elsewhere, concerns about natural monopoly led to governments regulating, building power plants, and eventually nationalizing existing private sector facilities. Most notably, this was the situation in Chile, Argentina, Brazil, and Peru. In the alternative Latin American model, the power sector was largely developed by government initiative (e.g., Bolivia, Colombia, and Venezuela). Thus, by the 1980s, most of the electricity in Latin America was generated and delivered by governments. However, the usual problems with government ownership and the debt crisis led many Latin American countries to move toward increasing efficiency beginning in the 1980s (Henisz and Zelner c. 2011; Economic Consulting Associates 2010).

Chile was the leader of the electricity restructuring pack, both in Latin America and for the rest of the world. The restructuring was a continuation of economic reforms in 1975, moving toward a freer market. Beginning in 1978, the two large government-owned utilities were broken up and subsequently privatized by 1988. Chile unbundled generation, transmission, and local distribution. New generating companies were allowed to enter and sell to large customers. Regulated prices were maintained for small consumers that bought through local distribution companies (Sol 2002). Transmission remained regulated, and a club of generating firms operating the system dispatched based on marginal cost (Rudnick and Zolezzi 2001).

This early attempt has been judged as one of the most successful reforms in the developing world, and it has served as a model for numerous other countries in Latin America (Pollitt 2008). As with most reforms, it continues to evolve in response to encountered problems, such as drought or the disruption of gas supply from Argentina. The most significant problem in Latin American restructuring has resulted from the high percent of hydropower in the system (66% overall, but even higher for some countries). Insufficient quantities of water have been stored during wet periods, leading to power crisis during periods of drought (e.g., Brazil, 2001; Chile, 1998–1999, 2011; and Venezuela, 2009–2010).

Freed (1997) has summarized efforts in the United Kingdom, New Zealand, Norway, and Sweden, other early leaders in electricity restructuring. Their changes have been significant over time, and all programs have provided lessons for later restructuring efforts. The latter three are unusual for industrial countries because their largest power source is hydro, not fossil fuels. Because inexpensive fuel prices result from hydropower, these countries have reasonably cheap electricity prices, though they have relatively high electricity taxes on household use. (See table 6–1 for prices in these early movers, along with the United States.)

Each of these four countries began its deregulation under a different set of circumstances, which we will consider in turn.

Table 6–1. Electricity prices and taxes, $/kWh, 2013

	Chile		United Kingdom		New Zealand		Norway		Sweden		United States	
	Price	Tax	Price	Tax	Price	Tax	Price	Tax	Price	Tax	Price	Tax
Industrial	0.11	0.00	0.13*	0.00	0.09*	NA	0.07	0.01	0.09	0.00	0.07	NA
Households	0.17	0.03	0.23	0.01	0.22*	0.03*	0.15	0.05	0.23	0.09	0.12	NA

Source: IEA (n.d.g).
Note: NA indicates number not reported. * indicates the data is for 2012.

United Kingdom

UK electricity restructuring began with the Electricity Act of 1983, aimed at the three vertically integrated national companies in England and Wales, Northern Ireland, and Scotland. The act was part of the Thatcher revolution to remove government controls and ownership and to move toward a more competitive environment. Independent power companies (IPCs) were allowed open access to the national grid, with their power purchased by the Central Electric Generating Board (CEGB) at avoided costs. However, CEGB's low interest rate cost advantage (5% real) kept independent power producers (IPPs) from entering.

From 1957 until restructuring, the English and Welsh electricity supply industry (ESI) had a publicly owned CEGB, with a vertically integrated monopoly over generation and transmission. CEGB supplied 12 area boards that held local monopolies over distribution. Both CEGB and the area boards were allowed to pass on costs to captive consumers. ESI's mandate was to operate for the public good with autonomy over the electricity industry. However, CEGB was often called upon to alter plans in the interest of wider economic policy, including price decreases to lower inflation, ordering plants ahead of time to stimulate employment, and limiting gas use (because it was considered a premium fuel). CEGB also bought British nuclear generators to support the local nuclear industry and supported the local coal industry. Since it was unable to keep a lid on investment costs, CEGB power stations took longer to build, with costs up to 100% higher than similar privately owned systems.

In 1988, the government began massive restructuring, with a two-year goal to set up a structure for privatization, along with accompanying regulatory and licensing schemes. It sought to protect its nuclear industry and have a successful public share offering, with expectations of rising electricity supply industry share prices. It proposed a horizontal and vertical deintegration of the industry. The area boards, called regional electricity companies (RECs), would each be sold off intact. The National Grid Company (NGC), owned by the RECs, would operate the high-voltage transmission grid and a new power pool. On Vesting Day in 1990, generation went to two new companies, National Power and PowerGen, with pumped storage going to NGC's First Hydro subsidiary.

Nuclear power was not initially privatized because of weak economics from higher private sector discount rates and uncertainty about possible liabilities and decommissioning costs. Nuclear power received government support through requirements that a certain percent of electricity must be generated from sources other than fossil fuels, called a *non-fossil fuel obligation* (NFFO), with support from a 10% fossil fuel levy (FFL) on power sales. Nuclear power remained under state ownership, with newer plants being privatized into British Energy in 1996. The old Magnox plants remained in government ownership. The government has been decommissioning these old plants, and 9 out of 11 of them had been decommissioned by 2012 (World Nuclear Association 2012).

The power pool was to have been based on bidding from both sides of the market, as is the case with most wholesale markets. However, because of software constraints prior to Vesting Day, CEGB's one-sided dispatching algorithm was used. The first bidders in the pool were Power Gen, National Power, NGC, Eléctricité de France (EdF), and the Scottish generator. In this process, demand by half-hour periods was forecast from models, while suppliers were to bid their marginal costs. From the cost bids, a system marginal cost was constructed. Dispatch was based on the marginal costs and transmission constraints.

For example, suppose that forecast load is 100 kW for the next hour. There are bids from five generators as follows:

- National Power bids $0.05/kWh for 75 kWh.
- Power Gen bids $0.06/kWh for 25 kWh.
- Scottish utility bids $0.07/kWh for 50 kWh.
- EdF bids $0.075/kWh for 10 kWh.
- NGC bids $0.08/kWh for 50 kWh.

Assume there is a current transmission capacity constraint of 65 kW from National Power to market but no other constraints. In this case, you would dispatch 65 kW from National Power, 25 kW from Power Gen, and 10 kW from the Scottish utility. The system marginal price for this case would then be $0.07. All generators who bid the system marginal price or lower would be paid the pool purchase price, calculated as $PPP = SMP + CC$, where SMP is the system marginal price and CC is the capacity charge. Capacity charge signals how much need there is for new generation capacity. It is computed as $CC = (LOLP \times VOLL)/L$, where $LOLP$ is the loss of load probability, which is the likelihood of a load interruption because of capacity constraints, and $VOLL$ is the value of the lost load. This expected value of lost load ($LOLP \times VOLL$) is distributed over load (L).

For example, suppose there is a 5% probability of a 10 kWh shortfall. The loss of output from a 10 kWh shortfall is estimated at $15. Multiplying the loss of load probability by the value of the lost load results in the following:

$$LOLP \times VOLL = 0.05 \times 15 = \$0.75$$

Dividing this over all kilowatts consumed results in the capacity charge, as follows:

$$CC = \$0.75/100 = \$0.0075/kWh$$

From this we can determine the value for the pool purchase price as follows:

$$PPP = SMP + CC = 0.07 + 0.0075 = \$0.0775$$

An alternative to a capacity charge is to require a certain amount of excess capacity. Demanders pay the pool selling price (*PSP* in the equation below) equal to the pool purchase price plus an uplift charge (*U* in the equation below). The uplift charge is used to recover costs from extra electricity or spinning reserve dispatched to cover transmission constraints and demand forecast errors. For example, if a plant is dispatched to provide 10 kWh of spinning reserve at \$0.06/kWh, but there is no market for the power, the producer must still be paid for being ready. If the payment is divided over all power users, then $U = 0.06 \times 10/100 = \$0.006$. The pool selling price, or the total charge to cover energy, capacity, and uncertainty, would then be determined as follows:

$$PSP = SMP + CC + U = \$0.07 + \$0.0075 + \$0.006 = \$0.0835$$

The basics of the initial UK market offer one example of how the wholesale market can be organized. The numerous electricity wholesale markets that have developed over the last two decades vary somewhat from market to market and are often much more complicated than the basic features discussed above. Markets are more likely to allow two-sided bidding, with demands proving bids as well. The price received may be the price each generator bids (price as bid) rather than a uniform price from the marginal bid (last price bid). Markets may be for an hour ahead, a day ahead, or even longer. There may also be markets for ancillary services that could include voltage regulation, spinning reserve, balancing services, and load shedding. Pricing may be uniform across the market (postal pricing) or vary by zone (zonal pricing) or node (nodal pricing). With zonal pricing, everyone in a geographic zone pays the same price; with nodal pricing, prices vary by node within geographic zones. There can also be markets for new capacity. For more discussion of power exchanges with some international examples, see Bichpuriya and Soman (2010) and Su (2007).

In the United Kingdom, the director general (DG) heads up the Office of Electricity Regulation (Offer), established in 1990. The DG's task is to make sure demand is satisfied, encourage competition, protect customer interests, issue licenses to generators and RECs, and regulate transmission and distribution using the price-cap methodology. The price cap $(RPI - X)$ is the price-cap rate of inflation (RPI) minus the target productivity factor (X). X is reset every four to five years. This lag is to encourage utilities to reduce costs, since they are allowed to keep any savings greater than X. Utilities are free to choose how to reduce costs. Disagreements between the regulator and regulated companies are referred to the Monopolies and Mergers Commission.

At Vesting Day, large customers could choose their supplier, and by 1999, all customers could choose their supplier. From the beginning of the restructuring, the wholesale or pool market was plagued by generators trying to game the system, especially during peak periods. Regulatory responses followed. To reduce market power, the DG ordered National Power and PowerGen to dispose of about 15% of their capacity, and the DG discouraged attempts of generators to buy RECs. First Hydro was sold to a new generator. In 1995, the National Grid was separated from the RECs. From 1995, there was a flurry of takeovers and mergers of the RECs, largely by British and US interests. During 2001, electricity distribution and generation were broken up into separate companies as required by the Utility Act of 2001. Electricity price was totally deregulated in April 2001. In July 2002, German-based E.on took over PowerGen to become the largest energy service provider in the world.

From Vesting Day to 2002, dozens of new generation units came on the market. Much of the new capacity was provided with combined cycle gas turbines (CCGT). This flurry of construction was dubbed the *dash for gas*. In 1990, less than 5% of UK power was generated from gas. By 2002, it had risen to 30%, and by 2011, it had risen to about 40% (IEA, n.d.d).

In 2001, the electricity and natural gas regulators were combined into the Office of Gas and Electricity Markets (OFGEM), and the power pool was abandoned for the New Electricity Trading Arrangements (NETA). Now instead of being required to bid into the power pool with central dispatch, participants can sign bilateral contracts, and power is self-dispatched. There is a separate balancing market for 24 and 4 hours ahead, with different balancing price depending on whether participants are long (need to sell power) or short (need to buy power) (Thomas 2001). Although NETA essentially remains in effect, Newbery (2005) has suggested that the pool problems were being ironed out and criticized NETA on a number of grounds, including a lack of price transparency, increasing barriers to entry, increased price risk in the balancing market, and implementation expense.

There is a push toward renewables and a lower carbon footprint in many places, and the United Kingdom is no exception. However, higher costs for renewables and their intermittency, such as wind and solar, will likely prevent them from entering the market soon without government intervention. In the United Kingdom, the NFFO was replaced by a renewable obligation after 2000, and the FFL was largely replaced by the climate change levy (CCL) in 2001. The UK Energy Acts of 2008 and 2010 include provisions for feed-in tariffs for renewable electricity generation, regulation allowing carbon capture and sequestration, and requirements for smart metering. The UK Climate Act of 2008 targets an 80% reduction in greenhouse gas emissions from 1990 by 2050.

New Zealand

Another early electricity industry restructuring took place in New Zealand. It too arose out of a desire to reduce government involvement in the economy. In the 1980s, New Zealand went from one of the most heavily regulated industrial economies in the world to one of the least. The impetus for this change was a weak economy, unemployment, inflation, and balance of payments problems. The major reforms began with the 1986 Commerce Act, aimed at the removal of price controls. It prohibited activities that restricted competition and strove to make state-owned entities (SOEs), such as electricity, as efficient as they could be in the private sector.

Prior to 1987, the central government owned most of the generation and transmission system, with local governmental Electricity Supply Authorities (ESAs) owning local distribution and retailing. In 1987, New Zealand's government chose not to privatize but to corporatize government generation and transmission into the Electricity Corporation of New Zealand (ECNZ). The government owned the corporate shares. ECNZ had to pay taxes and was to be governed by commercial rather than political considerations. Although others were allowed to enter generation, excess capacity prevented new generation from coming online for many years. Transmission was still considered a natural monopoly, and Transpower was set up in 1988 as an ECNZ subsidiary to run the transmission grid, becoming a separate crown (government) corporation in 1994.

The Electricity Act of 1992 required information disclosure and removed the monopoly franchises, first for small and then for large customers of the 61 ESAs. The ESAs were allowed to compete with each other. Obligation to supply was to be phased out, and electricity distribution was to be ring fenced from other activities. (*Ring fencing* is to have separate financial accounts for regulated distribution from other nonregulated activities. It prevents a regulated sector from subsidizing a nonregulated sector of the business.) The ESAs were corporatized as local government trusts called Municipal Electricity Departments (MEDs).

In 1993, the Electricity Marketing Company (referred to as *M-co* or *EMCO*), was set up by the electricity industry to develop a market for wholesale trading, with small customers able to choose suppliers for full retail competition. In 1996, a competitive wholesale market commenced. EMCO, as market manager, set prices to clear the market, and Transpower dispatched the power. The market in this case had two-sided or double-sided bidding. In such a market, the system operator lines up the supply bids from lowest to highest, and the demand bids from highest to lowest, as in figure 6–1. The system marginal price is determined where the supply and demand bids intersect, and there are no price caps.

Nodal pricing was instituted, in which half-hour prices are made at 244 grid connection points or nodes. These prices, which reflect available electricity, transmission losses, and grid constraints, provide signals to potential investors. The government was no longer responsible for setting bulk tariffs. Contact Energy was broken off from ECNZ to provide more competition in generation, and restrictions were placed on ECNZ until its market share fell below 45%. (For more on the ongoing New Zealand wholesale or spot electricity market, see Electricity Authority [2013].)

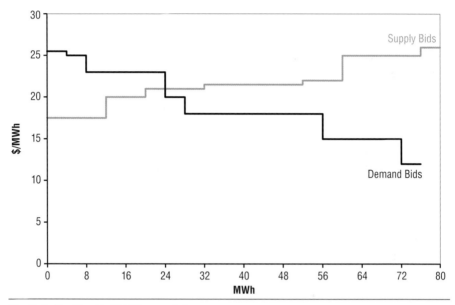

Fig. 6–1. Double-sided bidding market

The Electricity Industry Reform Act of 1998 mandated the separation of distribution from retailing and generation. (This mandate was relaxed somewhat in later acts.) Thus, generators, which can own their own retail arms and were called *gentailers*, could not own transmission or distribution lines. Nor could distributors own generation or retailing. This led to a lot of reorganization in the subsequent years. Some of the distribution companies remained as local government trusts, while others were privatized. Their retailing arms had to be spun off, and some were acquired by the generating companies (*gencos*) (New Zealand Ministry of Economic Development 2001). (See Switchme [n.d.] for recent information on power retailers in New Zealand.)

In 1999, the government privatized Contact Energy, ECNZ was broken into three state-owned corporations, EMCO was sold to the Rand Merchant Bank, and retail customer supplier switching was made easier.

Winter 2001 (June, July, and August) was cold and dry, which led to a reduction of hydropower production. At the same time, demand was high, which brought price increases. Winter 2003 was also dry, leading to more shortages. The Electricity Commission was set up to regulate the electricity market, taking over from EMCO. It would contract with generators to provide reserve power in dry years, and it set a maximum fixed charge for domestic customers. As in the past, much of the commission's ensuing work related to ensuring adequate capacity and a competitive spot market. After another dry winter in 2008 and more study, generators were required to set up a hedge contract market, allowing fixed prices as of 2010. EMCO, with its market clearing and settling role, was taken over by the New Zealand Exchange (NZX) in 2009. The commission was replaced by the Electricity Authority, with electricity reform in 2010, which seems to have continued the same themes as earlier regulatory efforts.

Regulation in New Zealand for the most part has not taken the form of price caps, as in England and Wales, but has been lighthanded. Regulators seeking to restrict anticompetitive behavior are allowed to award damages or implement temporary price controls. In addition, information disclosure rules require that information such as prices, energy and line charges, and condition of supply be made available to customers and investors. More information must be divulged by the natural monopoly sectors of transmission and distribution than by the more competitive generation and retailing sectors.

As of 2011, about 75% of New Zealand's electricity was generated by renewable power (IEA, n.d.d), with a government target of 90% by 2025. Nevertheless, with the high percent of hydro power, in 2010 the Electricity Authority recommended that smart metering not be regulated or required as current technology, as it did not create sufficient benefit to warrant the cost. (For more on the chronology of the New Zealand electricity market restructuring, see Energy Markets Group [2012].)

A climate change act required purchase of a permit to emit carbon dioxide or equivalent gases (CO_2e) starting in 2010, with an effective price of NZ$12.50 per tonne through 2012. Covered sectors are being phased in through 2015, but there is no permit trading yet.

Norway

Norway's electricity reforms began in 1991, without the same degree of economic reform as in England or New Zealand. Norway is probably only second to the United Kingdom in Europe in its efforts to promote more competition, although the Norwegian reforms have occurred in a very different setting. More than 90% of Norway's electricity was hydropower, with more than 90 producers and 200 distributors/suppliers that retailed electricity. Ownership was mixed, allowing *yardstick* comparisons between public and private companies. Thus, the private firms (which had to compete) or other best practice firms were the yardstick. Public performance was compared to these benchmark firms. If the public sector did not do as well as the benchmark firms, they were pressured to do better.

The largest generator, the Norwegian State Power Board (Statkraft), has continued to produce roughly 30% of the total energy supply. Another 30% is produced by county and municipal governments. The second largest generator,

Norsk Hydro, is about 40% owned by the government. Between 80% and 90% of distribution is by publicly owned companies, some of which are vertically integrated utilities.

Prior to 1991, Statkraft was part of a government ministry. It was the price leader, owning 80% of the transmission network and maintaining an import-export monopoly. Price regulation was light, with prices approved by the parliament under a principle of marginal cost pricing, with a rate of return constraint. There was a 20-year-old market operated by the Norwegian Power Pool (NPP). The market was voluntary, with buy and sell bids used to set the price on a weekly basis. Statkraft voluntarily acted as a swing producer to balance supply and demand when the market did not clear.

During the reorganization, Statkraft was incorporated as a state-owned generating company, and the transmission network was transferred to a new state-owned enterprise, Statnett. Statnett ran a spot market (intraday and day-ahead) and a six-month futures market, which was later extended to three years.

Bids and offers are aggregated, and prices in the spot market are made for the following day at hourly intervals, determining a system price. If there are transport capacity constraints, separate prices are computed within respective areas based on the local bids and offers. This will lower price in the surplus areas and raise the price in a deficit area. The Norwegian Water Resources and Energy Directorate (NVE) became and remains the electricity regulator.

Since there is third-party access (TPA) for all networks, anyone is allowed to buy in the spot market, even households. Generation, still largely government owned (90%) (NVE 2012), must be ring fenced from distribution for any vertically integrated utilities. Brokers, traders, and domestic importers and exporters can bid into the pool. The regulators grant licenses for operation, provide regulatory oversight of the transmission and distribution networks, set principles for transmission and distribution access charges, investigate monopoly power, and mediate network price disputes. Network services or distribution companies (159 as of 2011) are considered natural monopolies, are dominated by local municipalities, and are under income frame regulation, which is now a combination of rate of return, price cap, and yardstick regulation (NVE 2012).

Sweden

With the energy intensity of Swedish industry and high per capita electricity consumption, the Swedish government felt that electricity restructuring would stimulate a weak economy, and adopted a competitive electricity market in 1992. However, Sweden had a much more concentrated industry than Norway. Prior to restructuring, the Swedish State Power Board, Vattenfall, a state-owned limited liability company, produced one-half of Swedish power and owned and operated the high-voltage transmission grid. The next nine largest companies produced another 40%. The two largest, Vattenfall and Sydkraft, dominated the import-export trade. Ownership of distribution was mixed, with 60% municipally owned, 22% privately owned, 14% owned by Vattenfall, and 4% owned by cooperatives. Sweden's power pool was less open than Norway's and was more like an exchange, with bilateral contracts. Prices in the pool were set halfway between the marginal costs of the buying and the selling firms. Weak price regulation existed. Vattenfall was required to break even using the government bond rate.

In generation, Vattenfall was the price leader and yardstick. Competition pressured private firms to keep prices in line. Yardstick competition was also used to keep distributor prices in line between areas and municipalities, which were not permitted to earn a profit on electricity distribution. With reform, generation was separated from the transmission and the international interconnection network. These network activities were transferred from Vattenfall to a new SOE, Svenska Kraftnät, with more transparent network prices. Wholesale and retail power wheeling were allowed. In 1994, the government slowed liberalization over concerns that it would discourage a planned nuclear plant phase-out (passed in 1980) and increase rural prices.

Liberalization was resumed in 1996 with third-party access to the network. Local distribution companies (LDCs) have an obligation to supply existing customers, but customers can switch suppliers for full retail competition. Distribution is ring fenced from transmission and generation, with no formal price controls. Since reform, there has been consolidation among distributors, with large local and foreign power producers buying municipal distributors, and municipals combining into intermunicipals.

In 1996, Sweden joined Norway's power pool, Nord Pool, making it the first multinational power pool. Nord Pool included Finland in 1998, part of Denmark in 1999, and the rest of Denmark in 2000. These four countries became an open market, and Nord Pool came to be jointly owned by the grid companies in all four countries: Energinet, Fingrid, Statnett, and Svenska Kraftnät. These four countries have different fuel mixes for power generation, with the share of the top fuels by country in 2011 (IEA, n.d.d) as follows:

- Norway—hydro 95.3%, natural gas 3.4%
- Sweden—hydro 44.3%, nuclear 40.2%
- Finland—nuclear 31.6%, hydro 16.9%
- Denmark—coal 39.7%, natural gas 16.5%, and wind 27.8%

The large hydropower capacity in this market provides flexibility except in dry years, when prices are likely to spike. By 2010, Nord Pool was also trading spot power in Germany, Estonia, and Lithuania (Nord Pool Spot 2011).

In the futures and options markets, Nord Pool Financial launched power contracts in Germany and the Netherlands, along with some environmental products, including green certificates, and European Carbon Allowances. By 2010, Nord Pool Spot and Nord Pool Financial Markets had separated, with Nord Pool

Financial taken over and becoming NASDAQ OMX. For more information on the history of Nord Pool and NASDAQ OMX, see NASDAQ OMX (2014). Nord Pool is widely believed to be one of the more successful wholesale markets and has served as a role model for other new power pools around the world (Economic Commission for Africa 2009). For more information on power pool features, see Ku (1997).

Evaluation of Early Reforms

Freed (1997) considered the early performance of the reforms in New Zealand, Norway, Sweden, and the United Kingdom. Cost overruns in investment were prevalent in government-owned utilities in New Zealand and the United Kingdom prior to reform, but were relatively rare in Sweden and Norway due to their mix of public and private ownership and yardstick comparisons across utilities. Overruns in the former countries occurred because of government-designated contractors and design. For example, when Sweden allowed nuclear generators to choose their own technology, they chose boiling water and pressurized water reactors and developed low-cost nuclear power relative to the United Kingdom with its Magnox plants. Other cost overruns included the use of public rather than private sector discount rates, and cost pass-through that left all the risk with captive customers. With privatization, these reasons have been eliminated, as demonstrated by the prevalence of new UK generators choosing combined cycle gas turbines, which can be built in less than two years, over coal or nuclear.

Efficient electricity pricing requires that prices reflect costs by time and location to send appropriate signals to producers and consumers. Generally, costs are higher for sparsely populated areas than for those that are more densely populated. Costs are also generally higher for smaller users than for larger consumers and for on-peak rather than off-peak consumption. However, prices have not often reflected such costs. Prior to deregulation, cross-subsidization was prevalent in three of the four countries. New Zealand subsidized household and rural consumers from business and urban consumers. Norway subsidized electricity-intensive heavy industry such as aluminum production. The United Kingdom subsidized the largest consumers. With competition, third-party access, and ring fencing, it is hard to maintain such cross-subsidization.

Efficiency in the electric power market after the reforms is contingent upon creating competition in those sectors that are not to be subject to regulation. Free entry or at least contestability is needed for a market to be competitive. Barriers to entry and market power remained a concern in all four of these early reforming countries. Excess capacity and the dominance of hydropower generation in New Zealand by ECNZ and Contact led to barriers to entry in that market. National Power and PowerGen in England and Wales reputedly manipulated SMP. Their low-cost plants were bid in at low, even loss-making, levels to ensure they were dispatched. Their high-cost plants were bid in above marginal costs to ensure high SMPs to earn rents on all their production. Governments responded in both cases by further breaking up companies to improve competition.

One simple measure of concentration in an industry with n firms is the Herfindahl–Hirschman Index of concentration, or HHI in the following:

$$HHI = \sum_{i=1}^{n} \alpha_i^2$$

where
α_i is the ith firm's percent of the market.

If there is one participant or a monopoly in the industry, its market share is 100%. Its $\alpha = 100$, and $HHI = 100^2 = 10,000$. If there are two equal-sized firms in the industry, then market share is one-half or 50% each, and $HHI = 50^2 + 502 = 5,000$.

Perfect competition is represented by a large number of firms with small shares, as follows:

$$\lim_{\alpha_i \to 0, n \to \infty} HHI = \sum_{i=1}^{n} \alpha_i = 0$$

Thus, the closer to zero, the more expected competition, and the closer to 10,000, the more expected market power. Sometimes α is used to measure share, and then the measure falls between 0 and 1.

With divestitures in the late 1990s and early 2000s in England and Wales, HHI fell from 3,000 to 1,600, which would be roughly equivalent to going from three to six equal-sized firms. Vattenfall had a large share of the Swedish electricity market (HHI equaled roughly 3,333), with a virtual monopoly on gas imports. However, competition across Nord Pool pushed HHI below 2,600 for the pool, with connections and competition likely increasing further since then. Even so, occasionally regulated distribution companies in vertically integrated utilities have found ways to subsidize production, and market participants have found ways to collude and manipulate supply in the power pool.

Norway would seem to have the structure most conducive to competition, but even Norway has environmental barriers to entry for new hydropower plants and production externalities for new entrants on water courses with existing plants. These externalities exist because one plant's use of water output will affect that of another plant downstream.

Deregulation requires that regulation be revamped for the natural monopoly elements of the system (transmission and distribution) and that regulators keep a watchful eye on generation and retailing where competition is thought to be the most economically efficient form of organization. This requires either ring fencing or unbundling of competitive operations from the natural monopoly elements and an independent regulator with access to current and accurate information to make sure that no one is cheating. Regulatory approaches vary, and it remains to be seen whether the United Kingdom's price cap, Norway and Sweden's yardstick-performance-based regulation, or New Zealand's light-handed approach will prove the most effective.

Ownership is another issue to be considered when deregulating the electricity supply industry. Three types of ownership have been prominent around the world, as follows:

1. *Direct government ownership.* Ownership is usually run by a ministry, often with a mix of political and economic goals.
2. *A government-owned corporation.* The company has separate accounts, but all shares are owned by the government.
3. *Private ownership.*

Each of the four countries discussed started from a different point, and each has followed a different pattern of change. All of these countries now allow private ownership. The United Kingdom and New Zealand started with direct government ownership. The United Kingdom attempted to jump to private ownership as quickly as possible. New Zealand gradually moved to a government-owned corporation and then toward private ownership in generation. Norway was a mix of the first and third choices, and Sweden a mix of all three. Norway corporatized its state-owned generation and transmission, but there has not been a strong push to privatize at any level. Sweden has seen a bit of privatization with large power companies, both domestic and foreign, purchasing municipal and pension-owned distribution systems.

United States and California

Although the countries profiled above show early representative examples of restructuring, they are not unique. Electricity restructuring is ongoing in many areas, including the United States. In 1996, the US Federal Energy Regulatory Commission (FERC) issued Orders 888 and 889 to encourage competition in the interstate wholesale power market by requiring nondiscriminatory open access in interstate transmission, independent system operators (ISO), and a real-time information network. In 1999, FERC Order 2000 amended the orders to have these wholesalers form and belong to regional transmission organizations (RTOs), similar to ISOs but with more specifically designated properties. FERC has regulatory jurisdiction over power sales in interstate markets, whereas states typically regulate generation and local distribution. Currently there are seven RTO/ISOs in the United States, as follows:

- California Independent System Operator (CAISO)
- Electric Reliability Council of Texas (ERCOT) (within Texas; not regulated by FERC)
- Midwest Independent Transmission System Operator (MISO) (RTO in 2001)
- ISO New England (ISO-NE) (RTO in 2005)

- New York Independent System Operator (NYISO) (ISO)
- Pennsylvania, New Jersey, Maryland Interconnection (PJM) (RTO in 2001)
- Southwest Power Pool (SPP) (RTO in 2004)

More information on RTOs and their performance can be found at US FERC (2012). With the FERC orders, a number of states began electricity restructuring. California and Pennsylvania were the pioneers beginning in 1996, followed by others. Although all had hiccups as the restructuring progressed, for the most part, their efforts were reasonably successful. US FERC (2007) summarized many of these changes, and Willrich (2009) describes the hodgepodge features of the US electricity industry. Although the details of the whole system and the restructuring successes and failures are too numerous to consider, the most exceptional case is worthy of more careful scrutiny.

The most notorious US restructuring has been in California, the 10th largest economy in the world in 2010 (US BEA, n.d.; IMF, n.d.). Throughout the 1990s, its economy grew quite rapidly with the boom in information technologies. In 1996, at the beginning of the restructuring, its electricity prices were one-third above the US average and among the highest in the country.

To help correct high rates, which were at least partially blamed on rate-of-return regulation, California passed a law effective in March 1998. Its goal was model four, discussed earlier in this chapter: to begin with competition in the wholesale market, followed by full retail competition after investor-owned utilities had divested their generating capacity.

With restructuring and opening up of generation to new producers, while allowing consumer choice, existing higher cost plants may be driven out of business. Since generators put in these higher cost assets (called *stranded costs*) in good faith with a guaranteed rate of return, they asked to be compensated for these now-noncompetitive assets. Each state that restructures must decide who pays the costs for these stranded assets, whether generators, rate payers, or tax payers.

In California, stranded costs were handled through a competitive transition charge (CTC) assessed on all retail service, accompanied by a rate freeze for larger consumers and a 10% rate reduction on small consumers until the stranded costs were recovered. With the rate reduction, few small consumers switched suppliers. Utilities were allowed to fund the freeze and rate reduction by tax-free bonds. (With tax-free bonds, since the interest on the bonds is not taxable, it is subsidized by the government.)

The California law required mandatory nondiscriminatory open access to transmission and distribution, with the existing utilities owning the grid. Two new entities were formed. The California Power Exchange (CalPX), an independent nonprofit organization, was created to perform power trading in the wholesale power market. It had auctions for electricity in hourly blocks on the same day and in the day-ahead market. The California Independent System Operator (CAISO), a nonprofit independent system operator, was set up to operate the transmission system and manage congestion. CAISO, in turn, was governed by representatives of stakeholders, including customers, government, independent power producers (IPPs), and environmentalists. Since stakeholders managed, rather than owned, the asset, they battled over rent distribution instead of wealth generation. The utilities originally had to buy from the exchange on the day-ahead spot market and could not enter into long-term contracts. For the IPPs, exchange use was voluntary.

All worked well until June 2000, when shortages occurred. Between 1990 and 1999, electricity consumption had increased 11.3%, while aggregate generating capacity had fallen by 1.7%. Environmental agitation had caused early decommissioning of two nuclear generators, and no new power plant applications were filed from 1994 to 1998. Power production increased in California by using existing capacity more intensively, with the shortfall of power, as in the past, made up by imports.

Natural gas provided more than one-third of California's power. This was an economical arrangement from 1985 to 1999, since US gas prices fell in real terms. Gas consumption remained relatively flat from 1995 to 1999, and gas stocks were relatively low entering the cold winter of 1999 to 2000. An explosion on the El Paso natural gas line from Texas to California further exacerbated the gas shortage. Spot gas prices, which had been around $2.25/Mcf (multiply by 35.3 to get the price per thousand cubic meters), shot up to $10/Mcf by late 2000. (A $1/Mcf increase in natural gas price translates into roughly a $10/MWh increase in electricity prices.) *Wash trading*, in which a company buys and sells power simultaneously to the same client, is suspected to have increased natural gas prices as well. Another source of cost increase was the cost of NO_x emission credits in the Los Angeles Basin. As fossil-fuel generation increased, the permits increased from $6 to $45 per pound, adding almost another $40 to the cost per megawatt hour.

Drier weather and salmon management restrictions reduced hydropower in California and the Pacific Northwest. The hot summer of 2000 increased power needs. At the same time, there were transmission constraints from neighboring regions, as well as constraints between Northern and Southern California. Forest fires reduced transmission capacity in the western states, and scheduled outages in British Columbia reduced power exports to California. Such constraints take time to ease, since it takes six years to install new transmission lines in California, with three years to plan and site, and three years to build. Also, more than one-half of California's plants were more than 20 years old at the time. During 2000, when these older plants were being run hard, sometimes up to 10 GW of power was out during high demand periods.

With high demand and restricted supply, wholesale prices, normally set by the cost of the marginal producer in a competitive market, shot up. Capacity constraints gave more market power to the generators, allowing them to set prices even higher than the cost of the marginal producer. Meanwhile, retail prices were capped. In the ISO real-time market, the governing board of CAISO set price caps that were reduced over the summer from $750 to $500 to $250. The CalPX day-ahead market and other states had no price caps. With the CAISO cap, buyers "underscheduled" in the CalPX market, causing a flood of excess demand in the CAISO real-time market. This disequilibrium destabilized operations and forced the CAISO to make out-of-market (OOM) buys at any price. This, in turn, exaggerated gaming and market manipulation. Unpaid generators stopped deliveries, leaving a mismatch between the retail and wholesale market. Brownouts and blackouts were the result.

Things calmed down following the 2000 summer period. However, by December a cold snap and growing fears of drought caused hydropower producers to panic, and wholesale prices once again soared. That month, the US Department of Energy ordered some generators and marketers to supply power to the California market if there was a danger of an outage. In addition, FERC eliminated the requirement that the three large utilities buy their power on the spot market through the CalPX, which subsequently ceased operation in January 2001. By January 2001, wholesale power prices in California remained the highest in the country. They were $313/MWh in California compared to $74/MWh in the New England Power Pool (NEPOOL), $53/MWh in the New York Power Pool (NYPP), and $39/MWh in the PJM Interconnection.

When the lights went out, blame started to fly. California's Governor Davis blamed deregulation and greedy power generators and traders. The conservatives blamed environmentalists and antigrowth groups, the prohibition of long-term contracts, and a centralized spot market that discouraged investment. The power generators blamed the California Public Utilities Commission (CPUC), residential customers blamed the government, and the CPUC blamed FERC, which did not allow utilities to obtain capacity at avoided cost.

So why were prices so much higher in California than elsewhere? One important factor was difference in fuel costs. In California, gas-fired plants were approaching one-half of the generation capacity, whereas in NEPOOL (now ISO-NE), NYPP (now NYISO), and PJM, the share was only 19%, 18%, and 4%, respectively. The much higher dependence on nuclear and coal for East Coast generators helped shield them from the gas price run-ups. Further, in the East, fuel cost pass-through provisions were allowed, so prices went up and the lights did not go out.

The three eastern power pools had installed capacity requirements. For example, NYPP had a penalty of three times the cost of peaking capacity if peak power needs were not met. Thus, utilities in these power pools had strong incentives to make sure there was enough peaking capacity. Such regulation is likely to make average prices higher and peak prices lower. However, it may be inefficient, since peak supply is increased rather than reducing peak demand.

California and Ontario both allowed peak wholesale price changes to ration demand with no capacity requirements. However, California subsequently put on price caps in the wholesale market. Further, retail prices were also capped, and there was no peak load pricing. There was overreliance on the spot day-ahead market, and there were no capacity payments built into the design. Thus, except briefly for SDG&E's customers, consumers did not see any power price increases during the periods of shortages. Other states also have had shortages, but of a lesser magnitude, and they have allowed rates to increase. To illustrate the two different cases, one where price is allowed to increase, and the other where it is constrained and shortages develop, see figure 6–2.

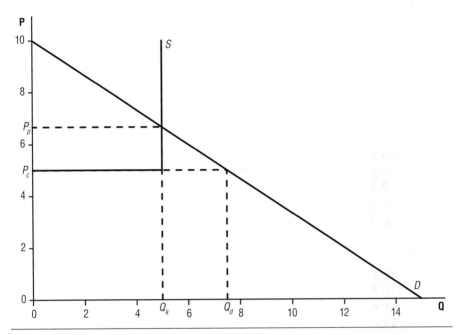

Fig. 6–2. Peak load demand and supply

Suppose the diagram represents the summer peak load market in California. Assume supply is perfectly elastic until capacity is reached at Q_k. If price is allowed to allocate this capacity, the price will jump up to P_p. There will be no shortages, but existing generators will make rents of $P_p - P_c$. Thus, there will be a transfer of wealth from consumers to producers. However, if the price is constrained at P_c, there will be excess quantity demanded of $Q_d - Q_k$. Forced blackouts will occur, and there will be political fights over the outputs.

Although there will be no wealth transfer, price sends the wrong signals to both consumers, who will try to over consume, and to producers, who will not see the advantage of putting in expensive capacity that will only operate during peak periods. An alternative is to put a price control on existing capacity but not on new capacity. This gives more incentive to invest but signals that producers will not receive rents during shortages as in a nonregulated market.

If prices are allowed to run up and a region is interconnected to other regions, power will flow into the shortage region. The US lower-48 states is divided into 10 North American Electric Reliability Corporation (NERC) regions. For more information and a map of the NERC regions, see NERC (2013). They are nonprofit organizations responsible for overall security and planning for the grid within their regions. They are connected to adjacent regions for power trading, and each of these regions is also included in three larger regions with more limited connections between them: the Western Interconnect, the Texas Interconnect, and the Eastern Interconnect. These interconnects also contain parts of Canada, Alaska, and Mexico. The Texas Interconnect, also known as the Electricity Reliability Council of Texas (ERCOT), is only connected to the other two interconnects by high-voltage direct current. Since it is wholly within one state, it is free from federal regulation within its borders.

With connections between the reliability regions, prices should be the same after adjustments for transportation costs unless there are transportation constraints. For example, suppose that wholesale electric power prices are P_c = $150/MWh in California, but only P_e = $75/MWh in ERCOT, with transport costs and losses from ERCOT to California of $10/MWh. If you are buying in the wholesale market in California, your cost of power per megawatt hour is $150, but if you buy from ERCOT, it will cost you $75 + $10. Since it is cheaper to buy from ERCOT and transport it, a buyer will want to buy from ERCOT instead of California. If the market is open and there are no transport constraints, consumers will buy from ERCOT. As more power is bought from ERCOT and less from California producers, the price will increase in ERCOT and decrease in California. The power purchases will continue to shift until the cost of purchasing from California or from ERCOT becomes the same on the margin or when P_e + $10 = P_c.

However, with transport constraints into California, power prices could diverge from surrounding areas. Throughout 2001, both the CPUC and FERC moved to ease the shortage. In the first part of the year, CPUC allowed retail rate increases for Pacific Gas and Electric (PG&E) and Southern California Edison (SCE). The State of California's Department of Water was authorized to sign long-term power contracts with generators for resale to cash-strapped PG&E and SCE. These long-term contracts allocated the risk from consumers to generators. Bids for nonpeak 3- to 10-year contracts were expected to be $55/MWh but were almost $70/MWh. Peak price bids were around $250/MWh, with exact contracted prices not known. These prices may have been higher for the state because of the risk that the state might renege on the contract if electricity prices subsequently fell.

In addition, laws were passed to shorten permitting times for new power plants in California and to ease environmental requirements. Another move that put the state squarely in the power business was the purchase of SCE's transmission grid to ease SCE's financial difficulties. This move, along with subsidies for energy conservation, the authority to build and operate electricity facilities, and the suspension of retail choice, signaled that California's power reform was far from turning the market over to the private sector.

Meanwhile, in the first half of 2001, FERC improved market signals in the wholesale power market by allowing more fuel cost pass-through, better data reporting, simplified regulation for the wholesale power market, and easier permitting for new natural gas transport projects. In addition, the Western Area Power Administration, managed by the US Department of Energy (DOE), increased electricity transportation capacity on Path 15 from Southern to Northern California, which was completed in 2004.

Although summer 2001 was as hot as the summer of 2000, peak demands were approximately 10% less than for the previous summer from higher prices, media attention, and a weaker economy. By October 2001, 2,236 MW of new generation capacity had become available, and the Bonneville Power Administration had increased hydropower output. By July 2003, 4,470 MW of new generation capacity became available. With the passing of time, the three major utilities became somewhat more integrated as they put in new capacity, and bilateral trading became the norm. The transmission grid has remained in the ownership of local utilities (some public, some private), with CAISO operating most of the transmission in California and making sure the market is balanced.

From 2002 to 2009, CAISO moved toward more reliance on the market to dispatch and balance the market rather than regulation and also moved from zonal to nodal pricing. With zonal pricing, the market price is determined by the clearing price of the last megawatt generated to meet load at any point in the geographic zone. When localized congestion occurs within a zone, it increases prices for all parties within the zone, even if they are not located near the congestion and even if they are not contributing to the congestion in any way. This results in overall higher system costs because everyone in the zone is paying the higher congestion costs, whereas with nodal pricing, only the parties at the congested location would be paying the increased cost. Nodal pricing is a method in which market prices are calculated for a number of locations on the transmission network (nodes) that represent physical locations on the system. These locations can include both generators and loads. The price at each node represents the incremental cost of serving one additional megawatt of load at that location subject to system constraints. Using this method, it is easier to determine which parties are responsible for transmission congestion and charge them accordingly.

In April 2009, the fully restructured day-ahead and real-time market had nodal marginal pricing. The day-ahead integrated forward market, with simultaneous operation of energy and ancillary markets with real-time dispatch at five-minute intervals, commenced. It is similar to the markets in other RTO/ISOs and seems to be working well. (For the history of the reform and more details on how the markets actually are administered, see the Department of Market Monitoring, Annual Reports on Market Issues and Performance at CAISO [2013].) California has always tended to take the lead on environmental policy and energy conservation, and state legislation will continue to provide challenges to its electrical system. Examples include its Global Warming Solutions Act of 2006, requiring a reduction in CO_2 emissions to 1990 levels by 2020, and its Renewable Portfolio Standard, most recently expanded in 2011, requiring 33% of electricity from renewables by 2020 (California Public Utilities Commission 2013). In 2009, California passed a law requiring planning for smart grid deployment to deal with the legislated onslaught of renewables. A smart grid uses digital and other communication technology to more efficiently coordinate generators, transmission, distribution, and end use.

Although many states have moved forward with restructuring, the negative California experience caused a number of states to suspend restructuring activities as shown in figure 6–3.

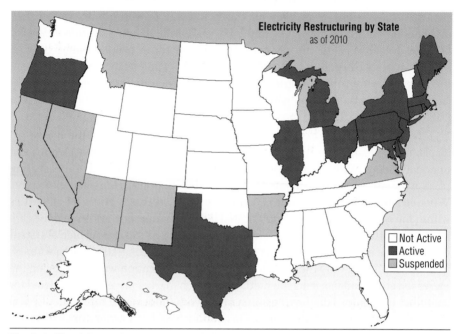

Fig. 6–3. Electricity restructuring in the US electricity sector, 2010
Source: US EIA (2010a).

In both the eastern and western regions of continental Europe, as in the United Kingdom, electricity generation was nationalized in many countries after World War II. Germany and Spain were exceptions. Electricity restructuring on the continent, for the most part, was encouraged by three successive EU directives relating to electricity restructuring, each replacing the previous: 96/92/EC in 1996, 2003/54/EC in 2003, and 2009/72/EC in 2009. All were in line with EU goals of freedom of movement of goods, services, persons, and capital across national borders. The first called for common rules for generation, transmission, and distribution. Further, it required full competition in generation (with some exceptions), consumer choice of supplier, unbundling, and breaking up of the existing monopolies.

Given the slow progress of market opening in some countries, the second directive focused on speeding up the process and increasing cross-border competition. National regulatory authorities were to oversee tariff rates that should reflect cost, ensure nondiscriminatory open access to transmission and distribution operated by ISOs, and oversee some sort of balancing mechanism between buyers and sellers. Full retail access was required by mid-2007, and a European Regulatory Agency for Electricity and Gas was set up—Council of European Energy Regulators (CEER).

Still, progress toward competition was slow and uneven. The third directive is aimed more specifically at unbundling and enhancing regulatory supervision in member states. Although the directive's preferred model is ownership unbundling between transmission, distribution, and generation, other models include ring fencing of these regulated activities from generation. Transmission operators are to be independent, with some independent supervisory body to monitor operator behavior. It also calls for modernizing the distribution networks along the lines of smart grids to encourage decentralized generation and energy efficiency. This provision aligns with the European 20-20-20 targets of 20% reduction in CO_2, a 20% share of low-carbon energy, and a 20% improvement in energy efficiency by 2020.

Europe differs from the United States as it has focused on retail choice, and there are no formal EU requirements for real electricity trade (power pools or exchanges) or financial markets (futures and options). Instead, the directive calls for market-based mechanisms to allocate power. In continental Europe, the day-ahead price is the main reference price for trading and for trading financial products. Wholesale trading is done in exchanges and over the counter (OTC) in bilateral trades. Pricing is zonal rather than nodal pricing as adopted in the United States (WEC 2010).

Although many of the European markets have not restructured as much as the RTO/ISOs in the United States, or as much as the regulators desire, they are gradually getting more open. A number of companies have expanded to become Pan European, including E.on, Vattenfalls, and Électricité de France (EDF). Nord Pool has been successfully trading for more than a decade. Other regional and national markets exist as well. For example, the European Power Exchange (EPEX Spot) is the exchange for trading contracts in intraday and day-ahead power in France, Germany, Austria, and Switzerland. Elexon provides balancing services in the United Kingdom. The Italian Power Exchange (IPEX) is the electricity futures market in Italy.

Africa is struggling to provide electricity for its inhabitants. More than one-half the population still lacks access to electricity (IEA 2011b). Three-fourths of the electricity generated in Africa is in Egypt, South Africa, Libya, Morocco, and Algeria. Common tendencies in the last decade have been to privatize power companies, attract private capital, and develop regional power connections to allow bilateral trading between countries. Build-own-operate-transfer (BOOT) and build-operate-transfer (BOT) projects have been built by Électricité de France, AES (United States), and Siemens (Germany), among others (MBendi 2013).

India's power generation has been owned and operated by electricity boards in each state using model 1. They had tended to suffer large operating losses, both physical and financial, with frequent power outages. Government attempts to draw in private capital in the 1990s were not particularly successful. Electricity Acts in 1998 and 2003 created central and state regulatory commissions and required unbundling of generation, transmission, and distribution. The distribution companies were to be privatized and a wholesale market developed. Most states have unbundled, along with corporatizing or privatizing distribution companies (Cropper et al. 2012). There is a wholesale market that has been operating reasonably well since 2008 (IEX, n.d.). Cropper et al. (2012) found some improvement in operating rates with the unbundling, with highest improvements in states that unbundled sooner. However, the authors suggest that tariff reform has lagged behind other reform and would likely improve operation more. (See Central Electricity Regulatory Commission of India [n.d.] for more details on the Indian power reform.) China and Russia have implemented some reforms, with China moving toward model 2 and Russia targeting model 4. More details for China and Russia can be found in Xu and Chen (2006) and Belyaev (2011).

Summary

Natural monopoly considerations have led most governments to either regulate or produce electricity themselves in vertically integrated monopolies. However, regulatory inefficiencies, government inefficiency, changes in economies of scale, and changes in market size have led many governments to consider restructuring their markets. Industrial countries are often doing so in the interests of reducing electricity prices and promoting economic efficiency. In contrast, developing countries are often doing so to attract much-needed capital to provide adequate electricity for their population.

A number of issues must be considered when restructuring: where we allow competition, where we require regulation, what market structure will be permitted, and who owns what. We can have competition at three levels:

1. Allowing new generators to enter the market and sell to the existing grid.

2. Allowing generators and distributors to choose trading partners from across the entire grid, which requires access to the transmission system, an efficient way of pricing and dispatching electricity across the transmission grid, such as a power pool, and an independent system operator. It often allows large customers to bypass the distribution companies if they wish.

3. The most extensive form of competition is full retail competition, in which large and small consumers can choose suppliers. In such an environment, there needs to be open access to the distribution and transmission network. Thus, a distributor can sell power that includes

distribution costs, or consumers can buy power from someone else and only pay distribution costs to the local distributor.

A number of examples, including Chile, the United Kingdom, New Zealand, Norway, Sweden, California, and India, show how restructuring evolves. These examples also reveal how restructuring depends on the political and economic climate within the country and existing structure of the electricity supply industry. Some common threads and lessons come from these and other examples:

- Transmission and distribution are often still considered natural monopolies, with some sort of regulatory restraint or oversight.
- Price regulation may take varying forms, such as rate of return, price cap, light-handed, or yardstick, with some combination of these being the most prominent.
- Generators, marketers, and retailers are typically considered potentially competitive, and these are the areas gradually being opened up.
- Usually generation is opened up first.
- When opening up the electricity market, existing generators, which had been monopolies, often have market power. This may result in government intervention to promote more competition.
- Open access to transmission allows generators, wholesalers, and retailers to compete.
- System operators need to be independent of the stakeholders in the system, with constant surveillance of the wholesale market.
- Nodal rather than zonal prices are thought to better signal capacity constraints.
- Real-time balancing works best if energy and ancillary services are dispatched simultaneously but requires the technology to support it.
- Well-informed independent regulators should be in place, with a well-designed transition plan, before entering into the restructuring.

The extent of vertical integration is another important issue. With some parts of the supply chain competitive, and other parts natural monopolies, it is clear that the competitive activities should be ring fenced or otherwise separated from the monopoly portions.

Ownership structure is a final important issue to consider in restructuring. The three most prominent types of ownership are government ownership, a government-owned corporation, and private ownership. Some believe that a competitive model does not depend on ownership but rather on the degree of competition at each stage of the supply chain. However, with a drive for economic efficiency, government ownership appears to be losing favor to government-owned corporations or privatization. Government-owned corporations are likely to be less economically efficient for several reasons:

- Government debt guarantees provide barriers to exiting the industry for failing companies.
- Public, rather than private, interest rates and required rates of return distort decisions.
- Political objectives may conflict with economic objectives.
- A lack of market discipline reduces incentive to maximize profits and minimize costs.

Although the majority of economists would likely argue that private corporations are more economically efficient than government corporations, many countries have opted to not wholly privatize but to keep parts of their system as public corporations. Whether these will outperform the many utilities worldwide that are privatizing remains to be seen.

An important lesson from California is that provision must be made to ensure adequate capacity. If market prices are to allocate electricity and signal the need for more capacity, they must be allowed to do so. Prices should be varied by time of day to reflect the real level of scarcity. If prices are not to allocate electricity, then some other means must be developed to meet capacity requirements. California also shows that poorly designed restructuring is often worse than no restructuring at all.

Economists are inclined to dislike government ownership and recommend letting the market operate except where market failures are persistent. However, markets will not work if regulatory restraints and uncertainty prevent entry and exit from the industry and market, and proper price signals are not being sent to producers and consumers. Although both transmission and distribution are thought to be natural monopolies, the operation of the grid should be returned to the utilities and governed by economic principals rather than political wrangling. Here, open access and some sort of rate control for grid usage is likely to be warranted.

Electricity market deregulation and restructuring have proven far more difficult than for natural gas, coal, or oil, largely because electricity cannot be cheaply stored and must be balanced in an interconnected grid. An active system operator must dispatch generators to meet actual load, while financial transfers must balance as well. Managing dispatch through market mechanisms of bids and offers becomes complex in the presence of transmission congestion, which creates pockets of market power. Managing increasing requirements for intermittent renewables further increases complexity. Distributed energy with electricity produced and consumed at the source (e.g., solar homes or companies generating their own power) can avoid the grid management problems but will lose any advantages of scale and scope offered by a centralized grid.

7

Monopoly, Dominant Firm, and OPEC

Go directly to jail. Do not pass go. Do not collect $200.

—From the game *Monopoly*, created by Charles Brace Darrow
in 1931, Colombia World of Quotations

Introduction

John D. Rockefeller developed the first oil monopoly out of the oil boom pandemonium that began with the first US oil strike in Titusville, Pennsylvania in 1859. He established the Standard Oil Company in Cleveland, Ohio in 1870. It focused on refining and transportation and proceeded to swallow up competitors or drive them out of business. Standard's tactics included negotiating freight discounts from railroads, undercutting prices of competitors, and controlling pipelines. By 1880, it controlled about 90% of the US oil product market through the refining sector, although it was not integrated backwards to oil production.

This market contrasts sharply with the many buyer and seller competitive models in chapter 3, in which each takes the market price as given. The marginal cost curve is then the supply curve. In the absence of externalities, economists believe that such a market is efficient because it maximizes social welfare as measured by consumer plus producer surplus, shown in figure 7–1.

Such a situation may approximately prevail in the coal market or in the market for windmills, but not in all energy and energy equipment markets. In chapter 5, we considered how electricity markets for many years were considered natural monopolies and were typically regulated or government owned. Now generation is considered amenable to competition, and many countries are moving toward competition in this sector.

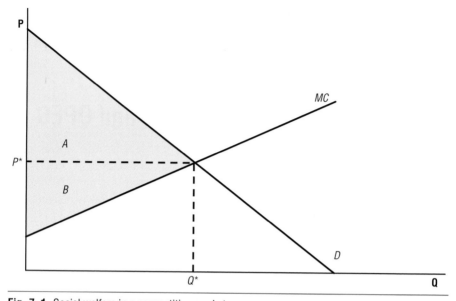

Fig. 7–1. Social welfare in a competitive market

Note: A is consumer surplus in a competitive market, B is producer surplus in a competitive market, D is the demand curve, MC is the marginal cost curve, P is price, Q is quantity, and S is the market supply curve.

Oil is another market in which monopolies or some other dominant groups have often formed, starting with the Standard Oil Trust. It was broken up in 1911 by a US Supreme Court ruling. Since the companies could no longer merge horizontally, many merged vertically. Three of the old Trust companies, joined by four others, subsequently dominated the world oil market. These large Anglo-American multinational oil companies—Esso, Mobil, Socal, Gulf, Texaco, British Petroleum (BP), and RD Shell—came to be pejoratively called the Seven Sisters.

In the United States, oil and gas reserves could be privately owned, and initially the law of capture generally prevailed. If the oil came out of your well, it was yours. Big oil strikes often resulted in overdrilling, overproduction, and waste. This led states to step in over the years with conservation laws and regulations to curtail overproduction and shore up prices (Lovejoy and Homan 1967).

A voluntary organization of nine states, called the Interstate Oil and Gas Compact Commission, formed in 1935 to assist member states to develop sound regulatory conservation practices. It now has more than two dozen member states and a number of foreign affiliates (IOGCC, n.d.). Some regulatory considerations could include minimum distance well spacing, requirements to unitize fields, with each producer receiving a designated share no matter from whose well the oil came, or even having the state prorate or designate how much each well could produce.

The most famous, long-lasting, and successful of the prorationing schemes was by the Texas Railroad Commission. After some initial tries, legislative reversals, and court challenges, this government regulatory agency prorated Texas oil from 1934 to 1972. The federal Connally Hot Oil Act (1935) supported prorationing by making it illegal to move nonprorated oil across state borders. Since Texas produced between 35% and 45% of US oil during this period, it served as a dominant player in the market. Later, the US government passed a mandatory oil import quota restricting foreign oil. All these government activities had the effect of supporting domestic prices (Kalt 1981).

Global oil markets in more recent years have been influenced by OPEC, and particularly by its core producers, such as Saudi Arabia. OPEC has functioned with a role similar to that of the Texas Railroad Commission in managing the oil markets. In these cases, market power is displayed, whether as a result of actions by prominent companies or due to government regulations. Unlike the case for natural monopolies, these monopolies seem to wax and wane.

Monopoly Model

Let's review how a firm can maximize profits when it has market power. Recall from chapter 5 beginning with equation (5–6) that the profit function and first-order profit maximizing condition for the monopolist are as follows:

$$\pi = P(Q)Q - TC(Q)$$

$$d\pi\big/dQ = P + dP\big/dQ \times Q - dTC\big/dQ = 0$$

Recall also that this second equation can be interpreted as marginal revenue (MR) equals marginal cost (MC), giving the second-order condition as follows:

$$dMR\big/dQ - dMC\big/dQ < 0 \rightarrow \text{slope } MR < \text{slope } MC$$

A useful result for monopoly, which relates price to the demand price elasticity, is easily developed. Start with the first-order condition and rearrange as follows:

$$P + \frac{dP}{dQ}Q - MC = 0 \rightarrow P\left[1 + \frac{dP}{dQ}\frac{Q}{P}\right] = MC$$

Note that the second expression inside the parentheses can be rewritten as follows:

$$\frac{dP}{dQ}\frac{Q}{P} = \frac{1}{\varepsilon_p} \text{ which can be rearranged to } P(1 + \frac{1}{\varepsilon_p}) = MC$$

Since it is easier to relate to positive elasticities than negative elasticities, take the absolute value of ε_p and change the + sign in the parenthesis to a minus sign, rearranging to get equation (7–1) as follows:

$$P = \frac{MC}{1 - \dfrac{1}{|\varepsilon_p|}}$$

(7–1)

Equation (7–1) suggests that higher costs raise optimal price, and more elastic demand lowers optimal price. In a competitive market, $|\varepsilon_p| = \infty$, $1/|\varepsilon_p| = 0$, and we have the competitive profit maximization rule, $P = MC$ or $P/MC = 1$. Remember that MC is not the accounting measure of cost. Rather, it is the economic measure including a normal rate of return, which is the required cost of doing business. Officials looking for evidence of monopoly power may note whether P/MC, which can be referred to as a *markup*, is substantially greater than 1.

Now suppose you are Rockefeller, and you have managed to monopolize the petroleum product market. You face the linear demand in figure 7–2. Marginal cost slopes up as assumed in the above example; marginal revenue is linear and twice as steep as the linear demand curve. Thus, it bisects the Q axis halfway between the intercept and where the demand crosses the Q axis.

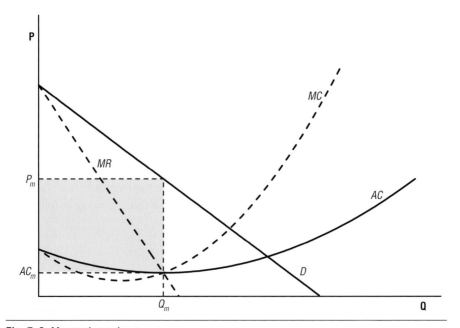

Fig. 7–2. Monopoly producer
Note: AC = average cost; *D* = demand; *MC* = marginal cost; and *MR* = marginal revenue

In figure 7–2, we can see that MR crosses MC at Q_m. It is easy to see that this is a maximum. If we produce less than Q_m, the marginal revenue is greater than marginal cost. Thus, it is profitable to produce extra units until we get to Q_m. After that point, the marginal cost, or the cost of an extra unit, is greater than the marginal revenue, and increasing production decreases profits. Also, since the slope of the marginal revenue is negative and the slope of the marginal cost is positive, the second-order conditions are satisfied, confirming that we have a maximum. The price that relates to Q_m is P_m, which can be read from the demand curve. Any excess of profits over economic costs is called producer surplus but is often referred to by economists as profits. It is the area $(P_m - AC_m)Q_m$. Such surplus arising in a competitive market is also called Ricardian rent. It arises when the marginal unit produced costs more than the preceding units. The excess surplus garnished by the monopolist above Ricardian rents is often called *monopoly profit*.

Now let's review a numerical example. Let the demand curve be as follows:

$$Q = 250 - 2.5P$$

Q is measured in barrels, and price is measured in dollars per barrel. The inverse demand curve is the following:

$$P = 100 - 0.4Q$$

Let the total cost curve be as follows:

$$TC = 0.15Q^2 + 75$$

Then profits are the following:

$$\pi = PQ - TC = (100 - 0.4Q)Q - (0.15Q^2 + 75)$$
$$= 100Q - 0.4Q^2 - 0.15Q^2 - 75$$

First-order conditions are expressed as the following:

$$\frac{d\pi}{dQ} = -1.100Q + 100 = 0$$

Solving, we find that $Q = 90.909$ barrels. Checking second-order conditions, we find the following is true:

$$\frac{d^2\pi}{dQ^2} = -1.100 < 0$$

So, we have a maximum. Substituting production back into the inverse demand curve gives us the monopolist's price as follows:

$$P = 100 - 0.4 \times 90.909 = \$63.636$$

Total cost, average cost, and profit for the monopolist at the optimal output are the following:

$$TC = 0.15Q^2 + 75 = 0.15(90.909)^2 + 75 = \$1,314.667$$

$$AC = \frac{TC}{Q} = 0.15Q + \frac{75}{Q} = 0.15(90.909) + \frac{75}{90.909} = \$14.461$$

$$(P - AC)Q = (63.636 - 14.461)90.909 = \$4,470.450$$

This is the gray area in figure 7–3.

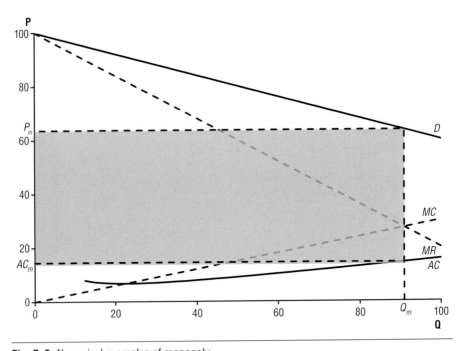

Fig. 7–3. Numerical examples of monopoly

We compare this area to what might be expected in a competitive market in the next section.

Monopoly Compared to Competition

How does the previous result compare to the results in a competitive example? It depends on the cost curves for all the additional firms that enter the market. Suppose the cost curves are the same for each additional firm and there are no barriers to entry. If a monopoly existed in this market and the monopolist was maximizing profits, he would be making excess profits, as shown above. This would cause other firms to enter into the industry. See AC_2 in figure 7–4.

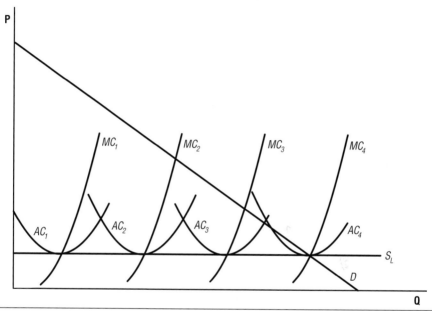

Fig. 7–4. Competitive supply in a constant cost industry

If another firm entered and the two firms did not collude, the short-run supply curve would be the horizontal sum of the two marginal cost curves, as in MC_2. Notice price would still be above costs. As there would still be profits in the industry, more firms would enter. For this constant-cost industry, notice that with the entry and exit of identical firms, a horizontal long-run supply curve is traced out, S_L, as seen in figure 7–4. There would be four firms in the industry in long-run equilibrium. The social losses in this market from monopoly power would then be the area under the demand curve and above the supply, between the monopoly output and the competitive output, as shown by the shaded area in figure 7–5.

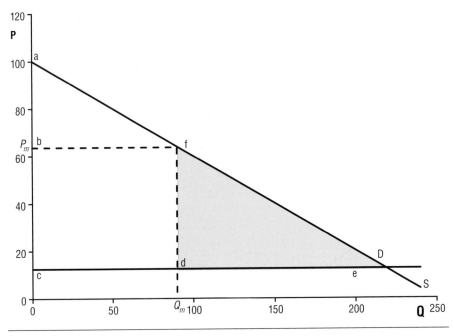

Fig. 7–5. Social losses from monopoly

If you have market power, you can optimize profits if you exploit it. Thus, the monopoly makes more money, because it can pick the price or the quantity and does not have to accept the market price, as in the competitive case. However, note that the monopoly can pick price or quantity, but not both. If it picks the price, the market dictates the quantity. If it picks the quantity, the market dictates the price.

Price Controls in a Monopoly Market

A policy that governments might consider in the monopoly case is maximum price control. It is easiest to see the effect of price controls by considering the four panels in figure 7–6. A price control makes operating on the demand equation above the price control illegal. Thus, the operational demand curve for a monopoly facing a maximum price control of P_{mx} is P_{mx} for all Q up to the demand curve and is the demand curve below P_{mx} as shown in figure 7–6(a). The monopoly should still maximize at $MR = MC$, but since part of the demand curve has changed, part of the marginal revenue curve has also changed. Now MR (the heavy dotted line in the figure) is flat along the new flat P_{mx} portion of the demand curve, but then becomes discontinuous to follow the marginal revenue for the market demand curve. With the new MR curve, it becomes easy to see what the monopoly should do to maximize profits when we know its marginal cost curve.

If we set the maximum price at P_{mx} above the monopoly price, P_m, as shown in figure 7–6(b), the control is nonbinding, and the monopoly will still produce at the monopoly price and quantity (P_m,Q_m). Next, pick a maximum price between P_m and where MC crosses the demand. In a competitive market, we expect price to equal marginal cost, so I label this price and corresponding quantity P_c and Q_c. The monopoly will produce at a higher level than Q_m at the controlled price, and interestingly, Q increases as the price control is lowered. If we can pick the price control just right, we can get the monopoly to produce Q_c, which mathematical optimization suggests will maximize social welfare under ideal conditions.

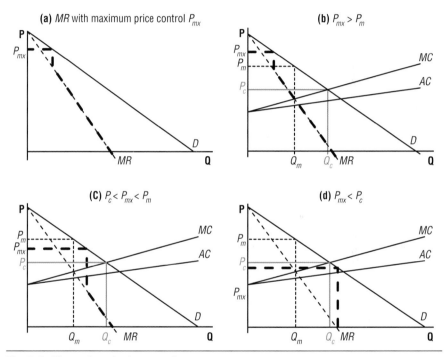

Fig. 7–6. Monopoly and price controls

Now take a price P_{mx}, which is below P_c as in figure 7–6(d). Now the marginal revenue curve crosses marginal cost to the left of the demand curve. Although the monopoly produces more than at its desired price of P_m and charges a lower price, there is excess quantity demanded in the market or a shortage. If the price control falls below where marginal revenue from the original demand curve crosses marginal cost, the monopoly produces less than Q_m. If the price control falls totally below the average cost curve, the monopoly would not cover costs and would go out of business in the long run. Thus, a price control could improve the monopoly allocation by lowering price and raising output, but in some cases, it could cause a shortage in the market. For this reason, price-controlled public utilities with monopoly franchises, discussed in chapter 5, were required to satisfy market demand at the controlled price.

We could also tax the monopolist. Just add a unit tax (t) to the marginal cost and solve for Q as follows:

$$MC(Q) + t = MR(Q)$$

However, note that such a tax would actually raise price above and lower quantity below the monopoly solution and move us even further from a social optimum. (For a numerical example of taxing a monopolist, see question 10 in http://dahl.mines.edu/st07/st07.pdf.)

Antitrust Laws

Although price controls would lower the price, they would not give us the socially optimal amount of output in the case of monopoly, if the cost structure resembled that in figure 7–4. As a result, governments often have relied on antitrust laws. In the United States, the Sherman Antitrust Act of 1890 made monopoly and restraint of trade illegal. It was under this act that the most famous energy antitrust case in the United States was conducted against the Standard Oil Trust. This case led to the breakup of Standard Oil in 1911 as shown in table 7–1. In 1914, the Clayton Act supplemented the Sherman Act and made mergers to restrain competition also illegal. The US Federal Trade Commission (FTC) Act prohibited unfair competition and set up the FTC to supervise trade and enforce the antitrust laws.

A more recent case of the dissolution of a monopoly occurred due to the breakup of the Soviet Union (USSR) in 1991. The oil industry in the USSR, then the Russian Empire, started in Azerbaijan. The first well drilled was in 1846. The Swedish Nobel family and the French Rothchild family invested in this very oil-rich spot in the world, making the Russian Empire the largest oil producer in 1900. By the time the Bolsheviks came to power in 1917, Standard and Royal Dutch Shell had also entered the fray. All lost their properties to nationalization after the Russian Revolution (Tolf 1976; Hassmann 1953). Soon thereafter the oil industry in the USSR came to be owned and centrally managed by the state under various ministries (Ministry of Geology, Ministry of Oil, Ministry of Gas, Ministry of Oil Refining and Petrochemical Industry, and Ministry of Energy and Electrification). (For a wealth of information on the oil industry in the USSR, see Considine and Kerr [2002].)

Table 7–1. Sample of oil company mergers, acquisitions, and restructuring, 1910–2012

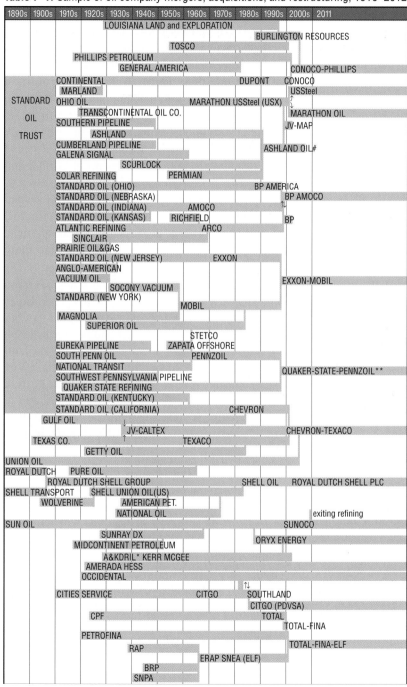

Source: Berger and Anderson (1992) updated from company home pages and other Internet sources.
Note: *A&KDRIL = Anderson and Kerr Drilling; **Taken over by Shell US in 2002; JV = joint venture.
See Thompson (n.d.) for histories of numerous oil companies.

From 1987 to 1990, various reform restructuring measures (*perestroika*) gave more decision-making authority to managers. The hope was that the reforms would rejuvenate a moribund Soviet industry, especially oil, a major foreign currency earner. Between 1989 and 1991, ministries were replaced by *konserny* (resembling holding companies or government corporations) to be managed for commercial purposes. The Ministry of Natural Gas became Gazprom in 1989. The Ministry of Oil became the association Rosneftgas, with less power than a konserny, in 1991.

With the breakup of the USSR in 1991, Russia continued to restructure its oil industry. In 1992, there were some 300 state-owned oil-related enterprises that were centrally controlled. A decade later, there were six large privately owned oil companies, Lukoil, Yukos, Surgutneftegaz, Sibneft, Tyumen Oil (TNK), and Slavnest, producing 75% of Russia's oil. Three state-owned companies, Rosneft, Tataneft, and Bashneft, were producing another 8%. Transneft, a state-owned company, still controls most of the huge Russian oil pipeline network. In this case, privatization and restructuring probably had less to do with increasing social welfare and more to do with lining the pockets of oligarchs (Sim 2008). For more on this enigmatic transition, see Grace (2005) and Gustafson (2012).

The restructuring continues under Putin's leadership, starting when he served as Russia's prime minister from 1999 to 2000 and then 2008 to 2012, and also as Russia's president from 2000 to 2008 and beginning again in 2012. In the ensuing struggle for power, the state regained some of the assets that had been privatized. Gazprom, with the state owning controlling shares, acquired Sibneft in 2005 and one-half of the stock of Slavneft. Rosneft, 75% owned by the state, continually acquired assets. It took over Yukos, bankrupted after a tax dispute in 2006. Rosneft is now the largest Russian oil company, with more than 20% of Russian production (Rosneft 2013). TNK and BP formed a partnership in 2003, and they own the other one-half of the shares of Slavneft. BP sold its shares of TNK-BP to Rosneft in 2013 (Gosden 2013).

Another possible policy to counteract privately owned monopoly is nationalizing the industry. In this case, the company would probably remain big, since many small companies would be less manageable for the government. However, there is no economic rationale for having one big company, as in the case of a natural monopoly. With no competition and no discipline from the market to control cost, we might expect X-inefficiency and costs to go up. The current solution in many constituencies is to promote competition with antitrust (or threats of antitrust) actions, and to regulate where competition is not feasible.

Some governments have national oil companies, which are some of the largest oil companies in the world. However, as with electricity, some national oil companies have been privatized. Earlier, Canada and the United Kingdom privatized their national oil companies, and Norway has partially privatized. Argentina privatized in 1993 but renationalized in 2012. The majority of the Russian oil industry was privatized in the decade after the breakup of the Soviet Union, but more recently there has been some reconsolidation and return of assets to state control.

Brief History of Oil Markets

Let's briefly consider the history of the oil market and then go on to monopoly models for our analysis. For more detailed information, see Yergin (1991), Yergin (2011), Adelman (1972), and Sampson (1975); for a chronology of oil market events, see Ratnikas (c. 2013).

From its inception, we can see the boom and bust of the oil market. US prices fell from more than $200 per barrel in 1860 (measured in 2013 dollars) to less than $15 (about $0.50 in money of the day) in 1861 as drillers produced as fast as possible (see fig. 7–7.) Such rapid exploitation largely resulted from the law of capture, which is relatively unique to the United States. Whoever drilled down and pumped out the oil got to keep it, even if the pool of oil was physically located under more than one owner's well. Unless producers were colluding, they had incentive to capture as much as possible from their own and neighboring plots. In this competitive foray, once drillers put in their well, it was a sunk cost. Then drillers had an incentive to produce where the price equaled their marginal cost, which was relatively low in this capital-intensive industry.

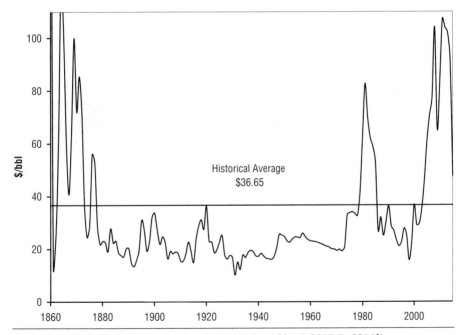

Fig. 7–7. Real US oil prices to refineries, 1861–2014 and March 2015 (in 2014$)
Sources: API (1971); US EIA (n.d.b); US DOC (1975); and US BLS (n.d.).

Rockefeller consolidated control over the US product market to stabilize it, and he did so by means fair and foul. Meanwhile, half a world away, Royal Dutch made oil finds in Indonesia beginning in the 1890s, while Shell Transport and Trading commissioned its first tanker to transport kerosene from Russia to the Far East in 1892. These two companies came together in 1907 to become Royal Dutch Shell.

Oil production in the United States was concentrated in Appalachia until 1901, when the huge Spindletop well in East Texas came in, and Texaco and Gulf Oil were born. Prior to the turn of the century, Standard Oil and Shell were the major players on the world oil scene. As noted above, the Nobels and the Rothchilds soon joined them, producing oil in Baku. Standard Oil was broken up under antitrust action in 1911. During World War I, a need for oil to fuel the British fleet induced that country to buy a controlling share in Anglo-Persian to later become British Petroleum.

By 1920, postwar oil prices had slumped. Attempts by Standard Oil of New Jersey (later Exxon), Royal Dutch Shell (RD Shell), and Anglo-Persian to shore up prices were thwarted by newcomers Gulf and Texaco and two other oil companies. These companies were out of the old Standard Trust (Standard of California [Socal] and Standard of New York [Socony]), which evolved into Chevron and Mobil. These seven companies later managed to stabilize world oil prices for many years.

Executives of three of the so-called Sisters (Anglo-Persian, RD Shell, and Jersey Standard) met at Achnacarry, Scotland in 1928 and agreed to share world markets. In the same year, RD Shell, Anglo-Persian, Companie Française Petrole (CFP), Jersey Standard, Mobil, Standard of Ohio (Amoco), Gulbenkian, and others agreed to cooperate through the Turkish Petroleum Company in much of the old Ottoman Empire. This cartel-like agreement, later known as the Red Line Agreement, took its name from the red line marking the agreement area on a map.

By the end of World War II, the Seven Sisters controlled world crude oil trade. Markets in the United States were more competitive, but states stepped in as noted above.

During the 1950s, new companies such as Getty and Occidental produced oil in North Africa, again putting downward pressure on world oil prices. Taxes levied on the companies were set at 50%, a rate initially established in Venezuela, which spread to all the major producing countries. Oil companies paid these taxes to the countries based on posted prices but did not immediately reduce posted prices when spot prices fell. However, falling demand from a European recession and rising world supply finally caused the major multinationals to cut market and posted prices in the late 1950s. This reduced taxes paid to producing countries and prompted Venezuela, Iran, Iraq, Kuwait, and Saudi Arabia to form OPEC in September 1960. Qatar joined in 1961, Libya and Indonesia in 1962, the United Arab Emirates in 1967, Algeria in 1969, and Nigeria in 1971. Two other countries were members for a time: Ecuador from 1973 to 1992, and Gabon from 1975 to 1994. In 2007, Angola joined and Ecuador rejoined. Indonesia dropped out in 2009 because it had become a net importer of oil, and since then, OPEC membership has remained at 12.

OPEC did not manage to increase prices in the 1960s, but it did manage to hold the line on taxes. However, when oil production in the United States peaked in 1970 and markets tightened, prices started to rise. With the Arab Oil Embargo and production cuts associated with US and Netherlands support of Israel during the Yom Kippur War of 1973, market prices rose, and OPEC prices followed. Prices rose again in 1979 during the Iranian revolution, when Iranian production was cut dramatically. These rapid price increases from supply shocks can be seen in figure 7–7. High prices caused a reduction in oil consumption and an increase in production outside of OPEC. OPEC production fell by about one-third from 1973 to 1985, while its share of the world oil market fell from more than 50% to less than 30% in 1984 (OPEC 2013). (See fig. 7–8.)

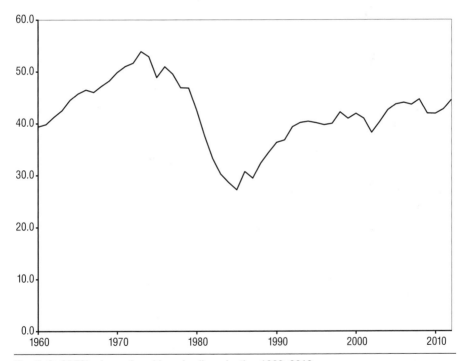

Fig. 7–8. OPEC's share of world crude oil production 1960–2012
Source: OPEC (2013).

In 1982, OPEC moved to a quota system (called a *production allocation* by OPEC), which has been in place since then. (See OPEC [2013] for quotas by country.) Iraq, although a member of OPEC, has often been outside of the quota system. Figure 7–9 compares OPEC quotas and OPEC production.

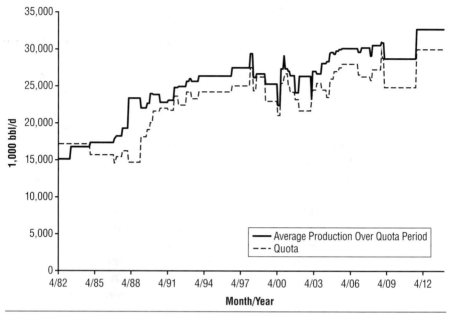

Fig. 7–9. OPEC monthly production and quotas, 1982–2012
Source: OPEC (2013).

In 1983, OPEC reduced posted prices from $79.54 to $67.84 in 2013 dollars per barrel ($34 to $29/bbl in money of the day). Since late 2007 with relatively high prices, OPEC as a group has had an overall quota or production allocation, but individual countries have seldom been given a separate allocation.

Saudi Arabia, in particular, took a large hit before quotas were implemented, with its production falling from a peak of more than 9 million barrels a day in the late 1970s and early 1980s to less than 3.5 million barrels a day in 1985. As a result of this role as a swing producer, in late 1985, the Saudis began using netback pricing, and in 1986, they raised production to more than 4.5 million barrels a day. (*Netback pricing* means that the Saudis charged companies an oil price equal to the value of product from a barrel of oil minus transport and refinery cost.) As a result, world prices plummeted, with prices to US refineries falling from $52.16 to $26.57 per barrel through 1986 (US EIA, n.d.b; US EIA, n.d.d). (All prices hereafter in this chapter refer to US refinery prices as graphed in figure 7–10 in 2013 US dollars unless otherwise indicated.)

Following the 1986 market collapse, crude oil prices fluctuated with an overall modest decline, except for the brief spike during the Gulf War against Saddam Hussein, bottoming out in 1998 during the Asian financial crisis. At lower prices, consumption increased and non-OPEC production decreased through the early 1990s. OPEC sales climbed, as did its share of world oil production. Oil demand growth remained strong, and discipline within OPEC remained reasonably good until after 1996. However, production continued to grow in 1997 and early 1998, largely the result of quota violations, especially from Venezuela. Prices again fell averaging below $20 per barrel for 1998, leaving large budget deficits and a need for funds.

In early 1999, the tide started to turn. Under the leadership of a new president in Venezuela, OPEC pulled together to reduce production. Gradually, the Asian economy picked up. Low oil prices had taken their toll on higher-cost producers, knocking out some of the competition. (Note the company consolidations from 1999 to 2001 in table 7–1.) The market tightened, pushing real oil prices during 2000 to about $36. They fell after the US World Trade Centers in New York were attacked on September 11 and averaged less than $28 for the year. Third-quarter US GDP fell, and US GDP income growth rate for the year fell from 4.1% in 2000 to 1.1% for 2001 but rebounded somewhat to 2.2% in 2002. World income growth rates also fell, but not as much as in the United States. Real oil prices remained below $30/bbl. OPEC's share fell during the weaker economy.

From 2004 to 2007, world income growth exceeded 4% every year, fueled by high growth rates in India, China, and other emerging markets. With this demand shock and little excess capacity, real oil prices rose sharply from 2003 through 2008, roughly tripling. Cost of oil production, oil taxes, wages for oil industry workers, metals, and other raw material prices gyrated upwards as well. OPEC's share increased until the world financial crisis, beginning in the US housing sector in the second half of 2008, put a damper on an overheated world economy. Growth in world GDP in constant international dollars fell by about half in 2008 (5.1% to 2.5%) and became negative in 2009 (–0.9%) (World Bank, n.d.). Real oil prices fell to an average of about $64/bbl in 2009. World GDP rebounded in 2010 to an almost pre-crisis growth rate of 5%, and real oil price went up to almost $82. Although world GDP growth slowed in 2011, it remained positive. A revolution also removed some Libyan production from the market for a while. Libyan production dropped from 1.65 million barrels a day in January 2011 to 0 in August 2011, before beginning to rebound. Increasing consumption and decreasing production had the expected effect, and oil averaged more than $100 per barrel for the year (US EIA, n.d.d).

OPEC regained its pre-Libyan production levels by November 2011. Libya was producing around 82% of its pre-revolution production by early 2013. However, production again fell later in the year and remained less than one million barrels a day through early 2014 (US EIA, n.d.).

Although global GDP growth was lower than expected in 2012, and the oil markets seemed well supplied, prices remained relatively high. Confrontation over Iran's nuclear program and other political unrest may have contributed to these prices and we can see the resulting nominal prices for three market crudes by month in figure 7–10. WTI nominal price averaged close to $95 per barrel for the year, while nominal Brent averaged more than $110 per barrel. This uncharacteristically wide spread between these two light crudes persisted over much of 2011 and 2012, averaging more than $15 a barrel. When Libyan light crude was knocked out of the market, it put pressure upwards on light crude prices. However, increasing US production of light tight oil put downward pressure on light oil in the United States. With no capacity or authorization to export light oil from the United States, the WTI/Brent price differential persisted into 2013. However, with market changes and increasing US product exports, the price differential generally trended down, especially with the dramatic plunge in prices in the last half of 2014, averaging around $11 a barrel in 2013 and $6 in 2014.

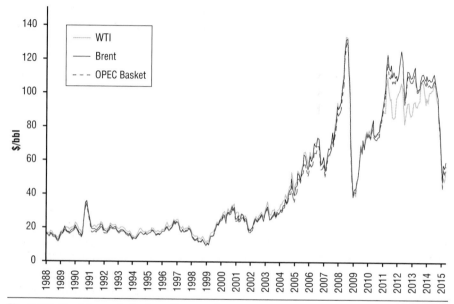

Fig. 7–10. Monthly nominal prices, three marker crudes, January 1988–April 2015
Sources: US EIA (n.d.b); OPEC (n.d.).

What happens to price and supplies in the coming years, of course, depends on a variety of factors. IMF (2015) projects world annual growth in GDP from 2016 through 2020 to average 3.9%, 0.3% faster than it was from 2011–2015. Advanced economies are projected to grow at an average of 2.1%, slightly higher than the previous decade and a half, while emerging markets and developing countries are projected to grow at an average of 5.1%, not matching the steamy average of 6.2% of 2000–2010. How US tight light oil production and other fringe production responds to the lower prices in 2015 as well as OPEC's response will also influence if and when oil prices recover. Political and ethnic strife in oil producing countries such as Iraq and Iran or strife that may spill over from nearby countries would put some pressure up, while settling problems that increased production would put some pressure down.

Some argue that the United States should become self-sufficient in oil and energy to isolate themselves from all these price shocks. The problem with the self-sufficiency argument is that oil is fungible and can move around. With arbitrage, high prices in one area of the world will attract crude supplies from other areas. Pockets of high prices will be temporary as crude oil moves to the highest bidder. We can clearly see this phenomenon in figure 7–10, which shows monthly nominal spot prices for three marker crudes: WTI, UK Brent Blend (Brent), and the OPEC basket, a weighted average of representative crudes for each OPEC member.

Despite the fact that these three crude streams are produced in different parts of the world and have slightly different characteristics, their prices have tracked each other quite well, decade after decade. The only exception is the recent break starting with the Libyan crisis in February 2011. Normally such a spread would cause less crude to flow to the US Gulf and more to flow to areas with higher prices, thus lowering Brent and OPEC crude prices and raising the price of WTI. However, increasing tight light oil from shale in the United States has flowed from North Dakota into Texas. With no infrastructure to enable export yet, prices have diverged more than has been the case historically.

Multiplant Monopoly Model

OPEC decisions are made using a combination of political bargaining and individual country decisions, with perhaps a variety of economic and political goals and subgoals. In this section, we consider how OPEC should behave if it wanted to focus on the economic goal of maximizing its total profits. First, OPEC is not a single country but consists of 12 countries, so we use a multiplant monopoly model. If OPEC wants to maximize profits as a group, it must find its combined marginal cost and demand function. The monopolist's costs depend on costs in each of the countries. To see how, consider marginal costs in the following diagrams.

To simplify the exposition, we assume that OPEC consists of two countries, a lower-cost and a higher-cost country, whose costs are represented by the two MC curves on the left in figure 7–11.

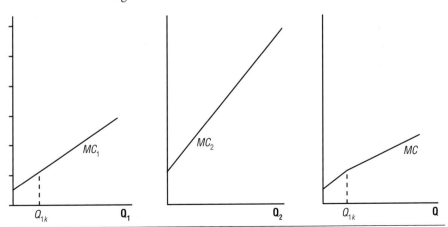

Fig. 7–11. Marginal cost for a two-country OPEC

Their total marginal cost curve is the horizontal sum of the separate cost curves. We can think of this summation as adding up the Qs for a given marginal cost. For example, suppose the two marginal cost functions are as follows:

$$MC_1 = 5 + 0.4Q_1$$

$$MC_2 = 11 + 0.8Q_2$$

For market prices below $11, only country 1 can afford to produce, so the group's MC curve would coincide with country 1's curve. Q_1 could produce 15 at an MC of $11 ($11 = 5 + 0.4Q_1 \rightarrow Q_1 = 15$), after which country two would start producing, and OPEC's marginal cost curve would be the horizontal sum of the two curves. Since we have to sum quantities for a given marginal cost, we need to invert these two functions, yielding the following:

$$Q_1 = -12.5 + 2.5MC_1$$

$$Q_2 = -13.75 + 1.25MC_2$$

Summing the right side, we see market quantity (Q) as follows:

$$Q = -12.5 + 2.5MC_1 + (-13.75 + 1.25MC_2)$$

OPEC should allocate production so that the following is true:

$$MC_1 = MC_2 = MC$$

Otherwise, it should reallocate production to the country where it is cheaper to produce. With MC the same across all countries, we can substitute in MC for MC_1 and MC_2, which yields the following:

$$Q = -26.25 + 3.75MC$$

Inverting gives us marginal cost after the kink as follows:

$$MC = 7 + 0.267Q$$

Next we need the demand for OPEC oil. Since OPEC has a large share of oil reserves, it has market power, and its demand curve is downward sloping. We know that the monopolist should produce where marginal revenue equals marginal cost. Let the demand for OPEC oil be as follows:

$$Q = 70 - 0.45P \rightarrow P = 155.556 - 2.222Q$$

Then total revenues are the following:

$$TR = PQ = (155.556 - 2.222Q)Q$$

And marginal revenues are as follows:

$$MR = \frac{\partial TR}{\partial Q} = 155.556 - 4.444Q$$

We set marginal revenue (MR) equal to marginal cost (MC) as follows:

$$155.556 - 4.444Q = 7 + 0.267Q$$

Solving for Q yields the following:

$$Q = 31.534$$

We need to check that we are assessing the portion where both countries are producing. Since Q is greater than 15, both countries are producing. If Q were less than 15, only the low-cost country would be producing, and we would have to use the marginal cost curve to the right of the kink to solve for Q.

We solve for market price by substituting Q into the demand equation as follows:

$$P = 155.556 - 2.222Q = 155.556 - 2.222 \times 31.534 = 85.487$$

The solution is illustrated in figure 7–12.

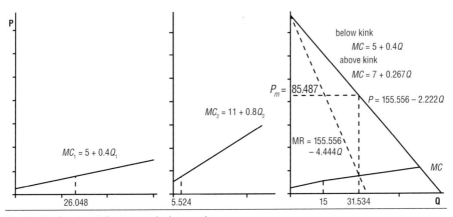

Fig. 7–12. Dominant firm numerical example

Now let's see how much each country produces. Marginal revenue at profit maximizing output gives us marginal cost as follows:

$$MR = 155.556 - 4.444Q = 155.556 - 4.444 \times 31.534 = 15.419 = MC$$

Marginal cost lets us solve for each member's production allocation as follows:

$$Q_1 = -12.5 + 2.5MC_1 = -12.5 + 2.5 \times 15.419 = 26.048$$

$$Q_2 = -13.75 + 1.250MC_2 = -13.75 + 1.250 \times 15.419 = 5.524$$

Thus, the high-cost country is producing much less than the low-cost country. Only if costs are very similar should each country get a similar production quota. Right away, we can see political problems could evolve trying to allocate production across producers.

OPEC's Demand Curve and Marginal Revenue Curve

In the above example, we assumed a demand curve for OPEC. To derive a demand curve for OPEC, we divide the world into two groups: OPEC (the dominant firm) and the rest of the world, which we call the competitive fringe. Since the fringe is competitive, its supply curve is equal to its marginal cost curve. Now see world demand and the fringe supply in figure 7–13.

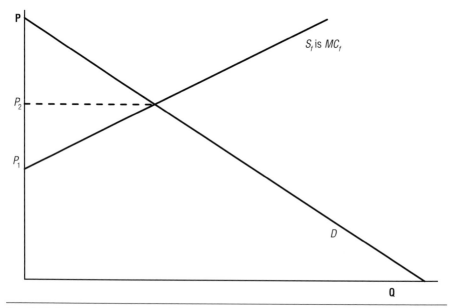

Fig. 7–13. Developing demand for OPEC's oil

In this model, if oil prices are below P_1, the fringe does not produce, and OPEC faces the whole world demand curve. If price is P_2 or above, the fringe satisfies the whole market, and OPEC does not produce. At any price between P_1 and P_2, OPEC demand is world demand minus the production of the fringe. Then OPEC demand is the darker kinked line in figure 7–14.

To develop the marginal revenue curve for OPEC's oil, we need the marginal revenue for the two parts of the demand curve. For the flatter portion of the demand curve to the left of the broken line, we need to take the marginal revenue from the flatter portion, which is world demand minus the fringe supply. For the steeper portion of OPEC demand to the right of the broken line, we need to take the marginal revenue curve from the steeper portion, which is the total world demand. This yields the discontinuous marginal revenue function as pictured in figure 7–14. Once we have MC and MR for OPEC, the optimum is the quantity where marginal cost equals marginal revenue. The price is then the price on the demand for OPEC's oil, the darker line.

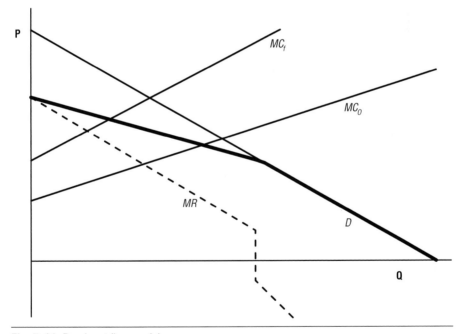

Fig. 7–14. Dominant firm model

To better see how this works, let's work through another numerical example. Let world demand (Q_w), marginal cost of the fringe (MC_f), and marginal cost of OPEC (MC_o) be as follows:

$$Q_w = 230 - 2P \rightarrow P = 115 - 0.5Q_w$$
$$MC_f = 60 + 0.7Q_f$$
$$MC_o = 20 + 0.5Q_o$$

The first step is to find the demand for OPEC's oil, which is world demand minus supply of the fringe. To get Q_f of the fringe, we know that fringe supply is equal to its marginal cost curve, as follows:

$$Q_f = -85.714 + 1.429MC = -85.714 + 1.429P$$

The kink in demand is where the fringe does not produce or where the following applies:

$$Q_f = 0 = -85.714 + 1.429P \rightarrow P = 60$$

So the fringe produces nothing when the price is 60 or lower. At this price world demand is expressed as follows:

$$Q_w = 230 - 2(60) = 100$$

For any given price above the kink, the demand for OPEC's oil is world demand minus the fringe as follows:

$$Q_o = Q_w - Q_f$$

We have world demand from above. OPEC demand to the left of the kink is expressed as follows:

$$Q_o = Q_w - Q_f = 230 - 2P - (-85.714 + 1.429P)$$
$$Q_o = 315.714 - 3.429P$$

Inverting this demand yields the following:

$$P = 92.072 - 0.292Q_o$$

Marginal revenue for this demand curve can be derived as follows:

$$TR = PQ = (92.072 - 0.292Q_o)Q_o$$
$$MR = \frac{\partial TR}{\partial Q} = 92.072 - 0.584Q_o$$

Remember, it has the same intercept but is twice as steep as the inverse demand curve.

As shown in chapter 5, this is always the case for a linear demand. For any price below the kink, we are to the right of the kink and on the world demand and world marginal revenue curves, which are the following:

$$P = 115 - 0.5Q$$

$$MR = 115 - Q$$

We know that MC_o could cross the MR to right or left of the kink. First, we will try to the left. Setting marginal revenue equal to marginal cost results in the following:

$$92.072 - 0.584Q = 20 + 0.5Q$$

$$72.072 = 1.084Q$$

$$Q_o = 66.487$$

The solution is to the left of 110, so MC crosses MR before the kink. (If Q had been greater than 110, we would have been on the wrong MR curve, and we would need to go back to MR for world demand [$MR = 115 - Q$], set it equal to OPEC's marginal cost [$20-0.5\ Q_o$], and resolve. See http://dahl.mines.edu/st07/st07.pdf, questions 23, 40, and 41 for worked-out examples of the various cases.) Now that we know OPEC's optimal quantity, we can get price from the demand for OPEC's oil to the left of the kink, which is as follows:

$$P = 92.072 - 0.292Q_o = 92.072 - 0.292(66.487) = \$72.658$$

The quantity supplied by the fringe easily follows from its supply curve:

$$Q_f = -85.714 + 1.429P = -85.714 + 1.429 \times 72.658 = 18.114$$

OPEC's producer surplus or profit is as follows:

$$\pi_o = PQ_o - \int_0^{Q_o} MC_f dQ = 72.658 \times 66.487 - 20 \times 66.487 - 66.487^2 \times 0.25 = \$2395.942$$

Fringe producer surplus is the following:

$$\pi_f = PQ_f - \int_0^{Q_f} MC_f dQ = 72.658 \times 18.114 - 60 \times 18.114 - 18.114^2 \times 0.35 = \$114.446$$

This example is summarized in figure 7–15.

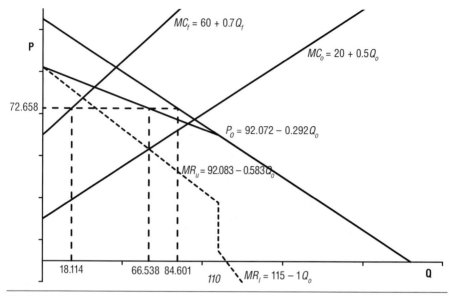

Fig. 7–15. Dominant firm numerical example

Price Elasticity of Demand for OPEC's Oil

Recall from equation (7–1) that optimal price and price elasticity are related. So, what is the demand price elasticity (ε_o) for the dominant firm? OPEC's demand is world demand minus supply of the fringe:

$$Q_o = Q_w - Q_f$$

Take the derivative with respect to P to get the following:

$$\frac{\partial Q_o}{\partial P} = \frac{\partial Q_w}{\partial P} - \frac{\partial Q_f}{\partial P}$$

Multiply by P/Q_o to get OPEC's elasticity, ε_o:

$$\varepsilon_o = \frac{\partial Q_o P}{\partial P Q_o} = \frac{\partial Q_w P}{\partial P Q_o} - \frac{\partial Q_f P}{\partial P Q_o}$$

We can rewrite this as follows:

$$\varepsilon_o = \frac{\partial Q_o P}{\partial P Q_o} = \frac{\partial Q_w P Q_w}{\partial P Q_o Q_w} - \frac{\partial Q_f P Q_f}{\partial P Q_o Q_f}$$

Remember that world price elasticity of demand, ε_w, and fringe price elasticity of supply, ε_f, are defined as follows:

$$\varepsilon_w = \frac{\partial Q_w}{\partial P}\frac{P}{Q_w}$$

$$\varepsilon_f = \frac{\partial Q_f}{\partial P}\frac{P}{Q_f}$$

Using these two definitions, OPEC's elasticity can be rewritten as follows:

$$\varepsilon_o = \varepsilon_w \frac{Q_w}{Q_o} - \varepsilon_f \frac{Q_f}{Q_o}$$

This equation tells us that OPEC's price elasticity of demand is the world price elasticity of demand weighted by world production, divided by OPEC production, minus the fringe price elasticity of supply, weighted by fringe production, and divided by OPEC production. Thus, the greater the elasticity in world demand and world supply, the greater the elasticity of OPEC demand.

Now let's investigate the elasticity for one country within OPEC, if it is the only country to change price. Let's say Venezuela changes its price and the rest do not. Suppose that Venezuela has α percent of the market $(0 < \alpha < 1)$ or Venezuelan production is $Q_v = \alpha Q_o$. Then Venezuela's price elasticity is the following:

$$\varepsilon_v = \frac{\partial Q_v}{\partial P_v}\frac{P_v}{Q_v} = \frac{\partial \alpha Q_o}{\partial P_v}\frac{P_v}{\alpha Q_o}$$

But if only Venezuela lowers price, Venezuela gets the entire increase in OPEC demand. Writing the total change in output in partial differential notation, we get $\partial Q_v = \partial Q_o$. Substituting this result into Venezuela's elasticity, we find that Venezuela's elasticity can be rewritten as follows:

$$\varepsilon_v = \frac{\partial Q_o}{\partial P}\frac{P}{\alpha Q_o} = \frac{\varepsilon_o}{\alpha}$$

The above shows that the smaller the share, the more elastic (more negative) the demand. We know that the greater the elasticity, the lower the price. Thus, the smaller the country's market share, the more tempting it is to cheat by producing over quota. Also, the smaller the country's share, the less likely the country will be caught if it cheats.

Alternatively, all countries could follow Venezuela's lead and lower their prices. In that case, it is likely that Venezuela would get its normal share of the increase, or $\partial Q_v = \alpha \partial Q_o$. Substituting into Venezuela's elasticity results in the following:

$$\varepsilon_v = \frac{\partial Q_v}{\partial P_v}\frac{P_v}{Q_v} = \frac{\partial \alpha Q_o}{\partial P_o}\frac{P_o}{\partial Q_o} = \frac{\partial Q_o}{\partial P_o}\frac{P_o}{Q_o}$$

Thus, Venezuela's elasticity is the same as OPEC's elasticity.

From the equation $P = MC \big/ \left(1 - \dfrac{1}{|\varepsilon_p|}\right)$, you can conclude that higher cost producers want higher prices by restricting production, and smaller countries with more elastic demand will put downward pressure on prices. Thus, some of the tension within OPEC during weak markets may come from different oil reserves amounts and their cost of production.

Non–Profit Maximization Goals for OPEC

Profit maximization is not the only goal for OPEC. With development in mind, Cremer and Isfahani (1980) suggest that some countries may have a target revenue that they want to use to invest in domestic industry. To see how countries might behave under this goal, take a simple example. Assume that all oil revenues are used for investment purposes. Suppose that we have two types of OPEC countries depending on the quantity of investment funds they are able to absorb and invest. The first is comprised of high absorbers with low income and large populations, such as Nigeria. The second is comprised of low absorbers with high income and low populations, such as Kuwait and the United Arab Emirates (UAE). See table 7–2 for OPEC country's population and GDP per capita. The high absorbers may have a larger target revenue for investment than the low absorbers.

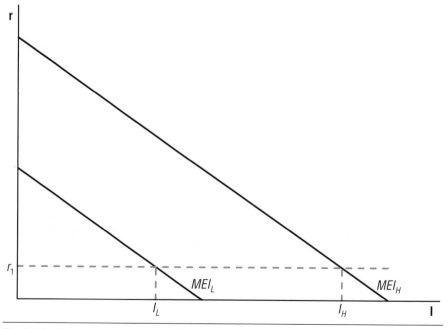

Fig. 7–16. Marginal social efficiency of investment

To see how an OPEC country might choose such an investment target revenue, suppose that curve MEI_L in figure 7–16 represents the marginal efficiency of investment curve for a low absorber like the UAE, and MEI_H represents the marginal efficiency of investment for a high-absorber country like Nigeria. Note the marginal efficiency of investment is the rate of return on the last unit of investment.

Assuming that the social discount or interest rate in both countries is r_1, the UAE would want to invest I_L, since investment up to that point brings in a higher social return than it costs. The high-absorber country with a large population, Nigeria, has a higher marginal efficiency of investment and, consequently, a higher target revenue for investment, I_H. If each country wants to raise target investment revenues from its oil industry, then investment expenditure must equal oil revenues or $I_L = PQ_L$ and $I_H = PQ_H$. These two functions along with oil demand are shown in figure 7–17.

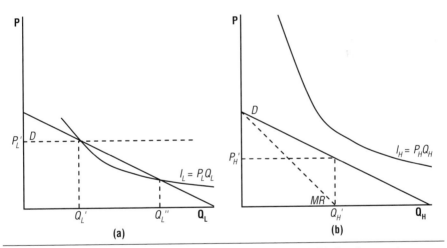

Fig. 7–17. Target revenues and price increases for low (a) and high (b) absorber country

Let D in this same graph be the demand for oil for each group, which is assumed to be identical for easier comparison. Also assume zero costs to keep the diagrams less cluttered. We can see that for the low-absorber countries, a variety of points on the demand curve would satisfy or more than satisfy their investment demand, but no price would put the higher absorber on its target investment. The low absorbers would prefer to get their revenue with the least sales and so would want to produce Q_L' and charge price P_L. The high absorber would want to get as close as possible to the target revenue. In this simple no-cost case, that would be where $MR = 0$ at (P_H', Q_H'). Thus, we can see the tension between the two groups. Now suppose that demand shifts out to D' for both countries, as shown in figure 7–18.

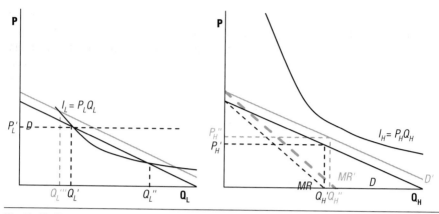

Fig. 7–18. Target revenue and price increase for high-absorber countries

The low absorber will want to remain on I_L and would want to decrease output to Q_L'' at a higher price. Since it produces less at a higher price, it has a downward sloping supply curve. The high absorber would want to move closer to I_H and would shift consumption out to Q_H'. Since the high absorber increases output as price increases, it has an upward sloping supply curve. It would like to increase output even more, but the market cannot absorb that output at the given price. Again, there is tension between the targets of the different groups.

Table 7–2. OPEC petroleum, income, and population statistics for 2012

Country	Joined OPEC	Prod. 1,000 bbl/d 2012	R 10⁹ 2012	R/P 2012	New Well Depth Feet 2009	Wells (W) 2012	Prod. Share 2012	Pop. in 10⁶ 2012	P/W/d 2012 bbl/d	Per Capita GDP 2012
Algeria	1969	1,200	12.2	27.8	9,547	2,061	0.04	37.8	582	5,204
Angola	2007	1,704	9.06	14.5	9,060	1,554	0.05	18.58	1,097	6,391
Ecuador	2007	504	8.24	44.7	9,730	3,177	0.02	15.5	159	4,621
Iran	1960	3,740	157.3	114.9	7,224	2,199	0.12	76.52	1,701	7,173
Iraq	1960	2,942	140.3	130.3	5,435	1,700	0.09	34.21	1,731	6,419
Kuwait	1960	2,978	101.5	93.1	8,181	1,831	0.10	3.82	1,626	45,351
Libya	1962	1,450	48.47	91.3	6,583	1,910	0.05	6.41	759	12,777
Nigeria	1971	1,954	37.14	51.9	9,095	2,168	0.06	167.7	901	1,535
Qatar	1961	734	25.24	94.0	10,039	517	0.02	1.77	1,420	108,458
Saudi Arabia	1960	9,763	265.85	74.4	8,061	3,407	0.31	29.2	2,866	24,911
UAE	1967	2,652	97.8	100.8	8,679	1,640	0.08	8.39	1,617	45,726
Venezuela	1960	2,804	297.74	290.1	5,757	14,959	0.09	29.52	187	12,956

Source: OPEC (2009), OPEC (2013).
Note: bbl/d = barrels per day, Pop. = population, Prod = production, R/P = Reserves over production, P/W/d = barrels of production per well per day, Wells = producing wells.

The quantity of an OPEC member's reserves might also influence its long-term views on prices. High-reserves Gulf countries have a longer run interest in reserves and are more worried that oil substitutes will be found. This, and the fact that high-reserves countries often have lower costs, may lead them to prefer lower prices. Shallower well depth and greater production per well result in lower production costs. Thus, population, revenue needs, production costs, and reserves all may influence a member country's preference for oil price. Table 7–2 shows such variables for OPEC countries.

Summary

Market power has been a dominant feature of the global oil market since its inception. High capital costs and low operating costs have led to price instability as the market lurched from boom to bust. Economies of scale were prevalent and led to large firms with an incentive to collude. Governments highly dependent on oil revenues have also sought to stabilize and support higher prices.

With market power, firms should operate where marginal revenue equals marginal costs. They will produce less than is socially optimal and charge a higher price. A binding maximum price control should reduce social losses if set between the monopoly and competitive price. If set below the competitive price it could cause shortages or even reduce welfare if set low enough.

A monopolist will make more profit than in the competitive case with the total loss in consumer-plus-producer surplus being greater than the monopolist's gain in profits. This net loss is called a deadweight loss and is why many governments try to prevent monopoly. In a corrupt government, a rent-seeking corrupt official is likely to seek a share in the monopoly profit or rent.

With a normal monopoly, the arguments for government ownership and regulation are weaker. Economists are more likely to recommend increasing or maintaining competition through breaking up monopolies, fining or punishing monopoly behavior, or denying mergers and takeovers when undue market power would ensue. As in electricity, there have also been some moves to privatize national oil companies.

If OPEC's 12 members want to maximize total revenue, they should behave as a dominant firm, multiplant monopolist. Their demand curve is world demand minus the supply of the competitive fringe, and their marginal cost curve is the horizontal summation of all 12 members' marginal costs. As with any producer with market power, they should operate where marginal revenue equals marginal cost. Each member country would produce where its marginal cost equaled OPEC's marginal revenue, so high-cost countries would produce less than low-cost countries. The pricing rule is $P = MC/(1 - 1/|\varepsilon_p|)$. Thus, the higher the marginal cost or lower the price elasticity, the higher the price. OPEC's overall price elasticity is as follows:

$$\varepsilon_o = \varepsilon_w \frac{Q_w}{Q_o} - \varepsilon_f \frac{Q_f}{Q_o}$$

Thus, the more elastic world demand is, or the more elastic the fringe supply is, the more elastic the demand for OPEC's oil is, and the lower the desired price. If one OPEC country with market share α changes its price and no other country changes its price, then that country's elasticity is ε_o/α. Smaller countries, therefore, have higher price elasticities and a bigger incentive to lower prices. However, if all countries change their prices, then each is likely to have OPEC's elasticity.

Revenue needs, costs, and reserves vary from country to country and may cause different desired price patterns. Yet, despite the economic and cultural differences among member nations, OPEC has managed to stay together, often influencing prices in its own interests.

We live in exciting times pricewise. But such prices are not unique. Initially, turbulent prices stabilized when Rockefeller gained monopoly power, and they remained more stable when the Seven Sisters and Texas Railroad Commission reigned, from 1930 to 1970. There was turbulence as OPEC took over until 1986, when market pressures moved OPEC to a quota system, and prices reverted to historic levels. Recent demand shocks and $100/bbl oil have again focused our attention on the price of oil, as has the price plunge of later 2014 and early 2015. Whether supply or demand shocks have been the cause, spiking prices lasting decades or more have been followed by periods of lower prices and relative calm.

Whether prices return to more historical levels in the coming decade, of course, depends on many market variables. One great benefit is that emerging markets are dragging billions of people out of dire poverty and increasing oil demand. However, failure to increase supply as quickly as this demand has pushed up prices. Thus, price in the coming decades will depend on world economic growth and growth in oil production. Oil production will depend on economic and political factors. Hydrocarbon reserve scarcity is not yet a problem; rather, there is the need to provide infrastructure to find, process, and deliver the hydrocarbons. High prices may encourage consumers to find alternative sources of energy and could facilitate emerging conservation technologies. At $100/bbl oil, gas to liquids, coal to liquids, heavy oils, tar sands, tight light oil, and shale gas all look attractive and have contributed to downward pressure on price. Global warming concerns and more regulations or taxes may be just around the corner, putting even more downward pressure on price and providing further challenges to producers only so recently awash in cash.

8

Market Structure, Transaction Cost Economics, and US Natural Gas Markets

The choice between the firm and market organization is neither given nor largely determined by technology but mainly reflects efforts to economize on transaction costs; the study of transaction costs is preeminently a comparative institutional undertaking; and this very same comparative contractual approach applies to the study of economic organization quite generally—including hybrid forms of economic organization, externalities, and regulation.

—Oliver Williamson, Edgar F. Kaiser Professor of Business Administration, Professor of Economics, Professor of Law, University of California Berkeley

Introduction

Market structure is an important determinant of how firms behave. The market structures of various energy industries have evolved in different ways across time and countries. The coal industry in the United States and in world markets has remained fairly competitive. The majority of coal is sold under long-term contracts, which is the predominant form of market governance. Electricity has evolved in a far different way. Once private or government-owned, vertically integrated monopolistic firms were the predominant form of market organization, but they are now privatizing and deregulating, particularly at the generation level.

Historically, dominant entities have arisen in oil until new suppliers have weakened their grip. The Texas Railroad Commission, the Seven Sisters, and OPEC have all had formal or informal agreements to restrict output to maintain price. Natural gas has evolved in different ways in different markets. In this chapter, we will broadly survey world natural gas markets, including consumption, production, reserves, and natural gas production technology. We will then use transaction cost economics to look more specifically at the evolution of US natural gas markets. In chapters 9 and 10, game theory techniques will be applied to the Asia-Pacific LNG market and European natural gas market.

Natural Gas Consumption and Production Worldwide

From 1980 to 2012, natural gas experienced the fastest consumption growth globally of all fossil fuels, averaging 2.6% a year, compared to 2.2% per year for coal and less than 1.3% a year for oil (BP 2013). Of the major energy sources, only primary electricity production in the form of hydroelectric, nuclear, geothermal, and other renewable sources has seen faster annual growth, averaging 3.3% from 1980 to 2012 (US EIA n.d.d.). The net result is that by 2011 natural gas constituted about 21% of global commercial energy supply. Coal was a bit higher at 29%, followed by oil at 31.5%. Primary electricity accounted for much of the rest (IEA 2013a).

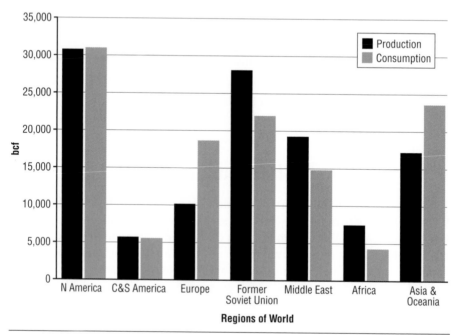

Fig. 8–1. Natural gas world consumption and production by major region, 2012
Source: US EIA (n.d.d).
Note: bcf = billions of cubic feet. One cubic meter equals 35.314 cubic feet. N America = North America, C&S America = Central and South America including the Caribbean, Europe = Europe outside the former Soviet Union.

Figure 8–1 is a regional summary of natural gas production and consumption. For example, North America is the largest producer and consumer of natural gas, followed by the former Soviet Union. The Western Hemisphere (North, Central, and South America) is almost self-sufficient in natural gas except for small and declining amounts of LNG imports to the United States dominated by those from Trinidad and Tobago. That may change in the coming decade as shale gas continues to develop in the United States and Canada, particularly if export licenses are issued. The former Soviet Union and Africa are net exporters to Europe, while Asia supplements internal consumption with relatively small shipments of LNG from Alaska and the Middle East.

For more detail on how this gas consumption is distributed across countries, see table 8–1, which contains a summary of natural gas production, consumption, and reserves by major regions and for selected countries.

Table 8–1. World dry natural gas consumption, production, and heat content

Area	Production (bcf)				Consumption (bcf)				Res. tcf	Heat Cont.
	1980	1990	2000	2012	1980	1990	2000	2012	2012	2012
N. America	23,061	22,562	26,966	30,812	22,559	22,470	27,723	31,015	412	NA
Canada	2,759	3,849	6,470	5,070	1,883	2,378	2,991	3,057	61	1,034
Mexico	900	903	1,314	1,684	799	918	1,398	2,424	17	1,074
USA	19,403	17,810	19,182	24,058	19,877	19,174	23,333	25,533	334	1,022
C&S America	1,227	2,014	3,430	5,738	1,241	2,024	3,304	5,577	270	NA
Argentina	280	630	1,321	1,329	359	717	1,173	1,641	13	1,045
Bolivia	79	107	117	644	14	30	44	131	10	1,045
Brazil	42	97	257	598	42	97	333	1,071	15	1,058
Trin&Tob	81	177	493	1,428	81	177	354	787	13	1,045
Venezuela	517	761	961	803	517	761	961	869	195	1,191
Other	228	242	281	936	228	242	438	1,078	23	NA
Europe	9,144	8,596	10,982	10,183	11,193	13,360	17,394	18,684	147	NA
France	280	101	66	19	981	997	1,403	1,559	0	1,122
Germany	NA	NA	779	434	NA	NA	3,098	3,080	6	944
Italy	443	611	587	304	972	1,674	2,498	2,646	2	1,023
Netherlands	3,398	2,693	2,559	2,840	1,493	1,535	1,725	1,624	46	895
Norway	917	976	1,867	4,155	35	80	140	150	71	1,063
Spain	0	49	6	2	56	192	588	1,148	0	1,092
Turkey	0	7	23	22	0	122	524	1,598	0	1,026
UK	1,323	1,754	3,826	1,323	1,702	2,059	3,373	2,641	9	1,068
Eurasia	15,370	28,782	25,426	28,125	13,328	24,961	20,532	22,014	2,165	NA
Russia	NA	NA	20,631	21,685	NA	NA	14,130	15,437	1,680	1,026
Turkmenistan	NA	NA	1,642	2,492	NA	NA	261	868	265	1,012
Ukraine	NA	NA	636	694	NA	NA	2,779	1,856	39	993
Uzbekistan	NA	NA	1,992	2,222	NA	NA	1,511	1,861	65	1,017
Other	NA	NA	525	1,031	NA	NA	1,850	1,992	116	NA

Area	Production (bcf)				Consumption (bcf)				Res.tcf	Heat Cont.
	1980	1990	2000	2012	1980	1990	2000	2012	2012	2012
Middle East	1,424	3,716	7,570	19,292	1,311	3,599	6,822	14,826	2,800	NA
Iran	250	837	2,127	5,649	232	837	2,221	5,511	1,168	1,056
Oman	28	99	322	1,035	28	99	221	715	30	1,020
Qatar	184	276	1,028	5,523	184	276	532	1,257	890	1,111
S. Arabia	334	1,077	1,759	3,585	334	1,077	1,759	3,585	284	1,020
UAE	200	780	1,355	1,854	105	663	1,110	2,235	215	1,047
Other	428	647	979	1,646	428	647	979	1,523	213	NA
Africa	686	2,459	4,440	7,489	735	1,351	2,038	4,275	510	NA
Algeria	411	1,787	2,940	3,053	460	681	726	1,323	159	1,127
Egypt	30	286	646	2,141	30	286	646	1,882	77	1,020
Nigeria	38	131	440	1,190	38	131	238	244	180	1,020
Other	207	255	414	1,105	207	253	428	826	94	NA
Asia&Oceania	2,462	5,660	9,582	17,227	2,577	5,865	10,517	23,625	505	NA
Australia	313	723	1,159	1,902	322	625	797	1,718	28	1,120
China	505	508	962	3,811	505	494	902	5,181	107	1,045
India	51	399	795	1,460	51	399	795	2,056	41	1,034
Indonesia	654	1,602	2,237	2,559	248	630	958	1,329	141	1,090
Japan	78	211	205	169	903	2,028	2,914	4,617	1	1,187
Korea, S.	0	0	0	37	0	107	669	1,764	0	1,119
Malaysia	56	654	1,600	2,176	56	315	820	1,104	83	1,053
Pakistan	286	482	856	1,462	286	482	856	1,462	27	867
Thailand	0	208	658	1,458	0	208	705	1,796	11	977
Other	520	872	1,111	2,193	205	577	1,103	2,597	67	NA
World	53,375	73,788	88,396	118,866	52,943	73,629	88,330	120,017	6,809	NA

Source: US EIA (n.d.d).
Note: bcf = billion cubic feet; tcf = trillion cubic feet; Heat C = heat content measured in British thermal units per cubic foot (Btu/cf); NA = not available; C&S = Central and South. There are about 35.3 cubic feet per cubic meter, and 1 Btu = 1.055 kilojoules.

Let's first consider from the table the largest players in the global gas market. In order of magnitude, Russia, Iran, Qatar, the United States, and Saudi Arabia have the largest gas reserves. The United States, Russia, Iran, China, and Japan are the largest consumers; the United States, Russia, Iran, Qatar, and Canada are the largest producers. Russia, Qatar, Norway, Canada, and Algeria are the largest exporters; Japan, Germany, Italy, South Korea, and Turkey are the largest importers.

With higher transportation costs, less natural gas is traded across countries, and natural gas trade patterns tend to be more regionally oriented than for oil. There are nice statistical summaries of gas trade flows in British Petroleum (BP 2014), as well as earlier editions. About 22% of natural gas was traded across national borders by pipeline in 2011. Some of the prominent flows are as follows: Canada is a net exporter to the United States; the United States is a net exporter to Mexico; Bolivia exports to Brazil and Argentina; and Columbia exports to Venezuela. Major suppliers to Western Europe are the Netherlands, Norway, Russia, and Algeria, with smaller amounts of LNG from Africa and the Middle East. (See chapter 10 for more discussion of the European natural gas market.) Russia supplies Eastern and Western Europe, as well as some FSU countries and Turkey; Kazakhstan, Turkmenistan, and Uzbekistan export gas to Russia; Iran exports to Turkey; Qatar exports to the United Arab Emirates and Oman; and Mozambique exports to South Africa. More of the Asia-Pacific gas trade is LNG, but about 20% is by pipeline as follows: Timor-Leste exports to Australia; Indonesia exports to Malaysia and Singapore; Malaysia exports to Singapore; and Myanmar exports to Thailand.

About 10% of global natural gas consumption is traded as LNG. About 63% of these LNG exports in 2011 went to the Asia-Pacific market, and another 27% went to Europe. Japan is the largest importer, with around 30% of the global LNG import market. The largest exporter is Qatar, with around 30% of the market. The bulk of Qatari exports are divided between Europe and the Asia-Pacific market. The next three exporters, Malaysia, Indonesia, and Australia, supply a combined 27% of the market, mostly to Asia-Pacific countries. The majority of African exports go to Europe. New Russian LNG exports all go into the Asia-Pacific market; the majority of Trinidad and Tobago's exports go to the Americas. Expectations of huge US imports have evaporated with new shale gas production. The LNG market will be considered further in chapter 9. Gas transportation will be discussed further in chapter 17.

Natural gas is a fairly clean fuel. When burned, it produces less carbon dioxide and nitrogen oxides (NO_x) than the other fossil fuels, little sulfur dioxide (SO_2), and no particulate matter. Its current known reserves, shown in table 8–1 by region and for selected countries, suggest that it is more plentiful than oil on the basis of energy content. (Roughly 6,000 cubic feet of gas equals a barrel of oil; roughly 1,300 cubic meters is a metric tonne of oil.) Oil and gas are both found at shallow depths. However, the greater heat and pressure at lower depths break down the hydrocarbons into smaller molecules like methane. This suggests that natural gas is likely to become relatively more plentiful as exploration takes the search deeper within the earth.

Natural Gas Conversions

As with oil, natural gas is measured in various ways. In the United States, the volumetric measure is given in cubic feet (cf or ft^3), 1,000 cubic feet (Mcf), million cubic feet (MMcf), billion cubic feet (bcf) and trillion cubic feet (tcf), usually measured at a pressure of 1 atmosphere, or 14.73 pounds per square inch. (One inch = 2.54 centimeters, and 1 pound = 0.45 kilograms.) The energy unit measure is a British thermal unit. One British thermal unit is roughly the amount of energy emitted by burning one kitchen match. Sometimes British thermal units are aggregated in units of 100,000 called *therms*; 10 therms = a *dekatherm*, and 10^{15}

Btus = a *quad* (for quadrillion Btus). In metric units, the volume measure is cubic meters (m^3), and the energy unit tends to be kilocalories (kcal = 3.97 Btu) or the SI unit of energy kilojoules (kJ = 0.95 Btu).

Natural gas is composed primarily of methane (CH_4), which contains 1,012 Btu/cf. Natural gas in the reservoir, called *wet gas*, typically contains small amounts of heavier hydrocarbon gases such as ethane (C_2H_6), propane (C_3H_8), and butane (C_4H_{10}), and even heavier liquid molecules, which raise the energy content. These heavier hydrocarbons are typically more valuable and so are separated out and sold separately. Wet gas can also contain other gases, such as carbon dioxide (CO_2), nitrogen, helium, and sulfur dioxide (SO_2), which lower the energy content.

When gas is produced, heavier molecules that are liquid at normal atmospheric pressure and temperatures condense at the wellhead and are separated out. This liquid, called *lease condensate*, consists of pentane (C_5H_{12}) and heavier molecules. Liquefied petroleum gas (LPG), which consists of propane and butane, is stripped out at gas processing plants. These two gases liquefy easily under pressure, facilitating transport. Ethane is also taken out at gas plants. This product remains a gas and is largely used in the petrochemical industry as a feedstock to produce ethylene. All these heavier products lumped together are referred to as *natural gas liquids* (NGLs). The marketed natural gas is called *dry* or *marketed natural gas*.

Table 8–1 includes representative energy contents for natural gas from various countries. In the table, energy content for gas varies from a low of 867 Btu/cf in Pakistan to 1,191 Btu/cf in Venezuela. The Netherland's huge Groningen field also produces fairly low calorific gas, giving it an average heat content of 895 Btu/cf. Russia, which is the second largest producer and largest exporter of natural gas, produces gas averaging 1,026 Btu/cf. The United States, which is currently the largest producer, produces gas with a similar energy content, averaging 1,024 Btu/cf. Where precise values are not needed, the energy content is often rounded to a more convenient 1,000 Btu/cf or 1,000,000 Btu/Mcf. Other major exporters, except for the Netherlands, tend to produce gas of a bit higher energy content than the United States or Russia.

In 1971, about one-third of global energy use of natural gas was in the industrial sector, and another one-third was used in the electricity and other transformation sector. The residential sector accounted for around 15%, and the commercial and public services another 7%. Figure 8–2 traces the evolution of energy natural gas use from 1971 to 2011. The average annual growth over this period was about 2.8%. The electricity and other energy transformation sector has seen the most stunning growth, and it now accounts for roughly one-half of natural gas consumption worldwide. In 2011, about 80% of this category was for electricity and heat. The industrial sector's share fell by about one-half, and residential, commercial, and public services roughly maintained their share during this period. From the graph, you can also see the sensitivity of natural gas to the economy, with dips or stagnant growth coinciding with weakness in the world economy in 1975, the early 1980s, the early 1990s, the end of the 1990s, and in 2009.

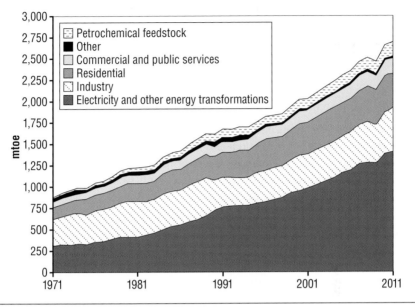

Fig. 8–2. Global natural gas use by sector, 1971–2011
Source: IEA (n.d.f).

Gas production and use has changed over time and varies across regions. As the market has changed and evolved, the structure of the industry has changed as well. In the coming sections, we will see how transaction cost economics helps us to better understand these changes.

Transaction Cost Economics

Neoclassical economics views the corporation as a production function, with output as a function of the various inputs, which include capital, labor, and materials. It has paid little attention to such philosophical questions as, "Why do corporations exist?" or "What determines their function and their relationship to each other?"

Transaction cost economics focuses on these more fundamental issues. It suggests that corporations are the integrating force for the increasing complexity and division of labor, and the market structure evolves that minimizes transaction costs between entities. After a brief discussion of transaction cost economics, we will consider how the US natural gas market has evolved historically in response to changes in transaction costs.

Cost is an extremely important parameter in optimal economic decision making. The costs we have considered so far in this book are production costs, or what we pay for all the resources used in transforming inputs into outputs.

They should include both direct and opportunity costs. Total costs (TC) can be written as

$$TC = \sum_{i=1}^{n} p_i q_i$$

where

p_i and q_i are the price and purchase of the ith factor.

Price can be either the market price of purchased resources or the opportunity cost of owned factors of production, and n is the number of factors used. Looking at costs in this way helps us choose the appropriate factor mix. Another way to consider costs is to break total costs into fixed and variable costs to help make short- and long-run economic decisions.

In macroeconomics, costs are viewed in a somewhat different way. When goods are produced, their costs are payments to factors, but they also generate income for these factors. If we tally the costs of all end-use products, we get a measure of total income. Aggregate income (Y) represents both the supply of goods available and the income generated. Thus, sufficient income is generated to buy back all goods produced.

In a simple Keynes model, the economy is divided into consumers, business, government, and the foreign sector. Aggregate purchases of goods from each of these sectors, respectively, is represented by C = consumption, I = investment, G = government, and $Ex - Im$ = net exports, or exports minus imports. Equilibrium in this model comes when income equals purchases:

$$Y = C + I + G + (Ex - Im)$$

Looking at costs in this framework will help in forecasting macro aggregates, an important input into business decision making.

Once goods are produced, most are exchanged in a modern capitalist economy, which leads to *transaction costs*, or the costs of conducting such exchanges. Coase (1937) divides them into the following:

- search costs
- information costs
- bargaining costs
- decision costs
- policing costs
- enforcement costs

We call the institutional arrangements that govern such transactions the *governance structure*. Transaction cost economics considers four models of transaction governance. From the least to the most formal, they are the following:

- *Market governance.* Trading in a spot market.
- *Bilateral governance.* Trading using long-term contracts.

- *Trilateral governance.* Trading with long-term contracts, with a third party facilitating the transaction.
- *Unified governance.* Transactions are internal to the company.

Survival of the economically fittest suggests that the chosen mode of governance should minimize transaction costs.

Underlying these costs are some basic economic realities. Economic agents have bounded rationality, and they may be opportunistic. Bounded rationality assumes that agents cannot acquire, assimilate, and use all information, nor can they anticipate all future contingencies in a contractual relationship. Opportunism assumes that some agents may take advantage of information asymmetries and make promises they do not intend to keep, or that they may be intentionally misleading in their economic dealings for personal advantage.

Given these two facets of economic behavior, termed *bounded rationality* and *opportunism*, transaction cost economics focuses on three institutional factors: uncertainty, asset specificity, and frequency of transactions. It considers how they influence transaction costs, and the choice among the four alternative modes of governance. Uncertainty comes from not being able to predict volatility in future states of the world and from complexity, which bounded rationality may prevent us from understanding. If we cannot anticipate the future and mitigate opportunism, we need an adaptive, sequential decision-making process to handle new situations as they occur. Complexity requires the flexibility to respond efficiently to possibly conflicting stimuli. An increase in uncertainty will increase bargaining or noncompliance costs as opportunistic parties employ strategic behavior to their advantage.

Asset specificity refers to investment in relationship-specific assets, which may be physical, locational, dedicated, or human. A *nonspecific* asset is a piece of standard equipment that can be used in many applications and places by many different operators. A general utility van is a nonspecific asset. A *mixed asset* is more specific and could be a piece of equipment that is not totally specific but is customized in some way. A utility van that has been modified to carry radioactive waste would be a mixed asset. An *idiosyncratic investment* is very specific to a particular transaction. A conveyor belt from a mine to a power plant in an isolated region would be an idiosyncratic investment. For an idiosyncratic investment, the value or opportunity cost of the asset in its next-best use is often quite low and may even be zero or negative if there are disposal costs. For example, it might be prohibitively expensive to tear up the conveyor belt and move it to another mine.

Now, if you owned the conveyor belt with large sunk costs and were selling to an unscrupulous power plant owner, he might be able to push the price for your services below your total costs. For example, if your total cost for conveying coal is \$0.50 per ton, your variable costs are \$0.05 per ton, and the conveyor has no alternate value, price (*P*) for conveying coal would have to be greater than \$0.05 for you to keep producing in the short run. As long as you are covering your variable costs, you are better off producing. Any amount above what the producer requires to be willing to produce is called a *rent* (R), which would be $R = P - \$0.05$. If price were \$0.25, your rents would be $R = \$0.25 - \$0.05 = \$0.20$.

In the long run, price would have to be greater than \$0.50 per ton for you to keep producing. (Your rents would be $P - \$0.50$ for $P > \$0.50$.) At prices less than \$0.50, you would not be willing to produce, as you would be making avoidable losses, and you would not buy new capital to produce.

Thus, a return to cover your fixed cost that is not required in the short run is considered rent in the short run; if it is required in the long run, it would not be considered a rent in the long run. Such a return that is required in the long-run but not the short run is called *quasi-rent*. If price were \$0.75 in the short run, then total rents would be $\$0.75 - \$0.05$, but $\$0.50 - \$0.05 = \$0.45$ (the amount of fixed costs) would be quasi-rents. If the price is \$0.45 in the short run, you would make quasi-rents of $\$0.45 - \$0.05$, but no long-run rents, only losses. The difference between rents and quasi-rents is shown in table 8–2.

Table 8–2. Rents and quasi-rents

Short Run					
$0 <$	P	$< AVC$	shut down		
$AVC <$	P	$< ATC$	quasi-rent	$=$	$P - AVC$
	$ATC <$	P	total rent	$=$	$P - AVC$
Long Run					
$0 <$	P	$< ATC$	shut down		
	$ATC <$	P	total rent	$=$	$P - ATC$
Example					
Short Run					
$AVC = 0.05$			ATC	$=$	0.50
$0 <$	P	< 0.05	shut down		
$0.05 <$	P	< 0.50	quasi-rent	$=$	$P - 0.05$
	$0.50 <$	P	total rent	$=$	$P - 0.05$
Long Run					
$0 <$	P	< 0.50	shut down		
	$0.50 <$	P	total rent	$=$	$P - 0.50$

Now complicate the problem a bit: Suppose you can sell the conveyor at any time for an equivalent return of \$0.10 per ton of coal. Now your variable costs are your variable production costs of \$0.05 per ton plus your opportunity cost of \$0.10 per ton. In the short run, your rents would now be $R = P - 0.05 - 0.10$. Quasi-rents would be those rents that would disappear in the long-run. They would be \$0.50 – \$0.15, if $P > \$0.50$, but $P - \$0.15$ for $0.15 < P < \$0.50$.

Although quasi-rents are not rents in the long run, in the short run they may be reduced or eliminated through strategic bargaining, if assets are very specific. Such a reduction is called a *holdup*. For example, suppose that you own a pipeline and are selling gas to a single power plant. Your fixed costs are $0.50/Mcf, your variable costs are $0.75/Mcf, and your total costs are $1.25/Mcf. Your pipeline is far from any other buyer. Suppose the power plant refuses to pay you more than $0.80/Mcf. You can refuse to sell, but with no recourse, you are better off getting $0.80/Mcf than nothing. With $0.80/Mcf, you are paying variable costs and have $0.05/Mcf toward your fixed costs, so you are losing $0.45/Mcf. If you stop selling, you have nothing toward your fixed costs and are losing $0.50/Mcf. In the long run, however, you would not replace the pipeline unless you were reimbursed for your total costs. With such idiosyncratic investment and the threat of hold up, there is a need to establish formal guidelines for a trading relationship.

Others variables may also influence how you conduct business. The frequency of transactions varies, and purchases may be recurrent or occasional. For example, food for the cafeteria may be purchased weekly or even more often, but paper and supplies may be purchased less often, perhaps monthly or quarterly. Large capital equipment may be purchased every decade or even less often. Frequent transactions of nonspecific assets in a competitive market are likely to be governed by the market. Someone who acts opportunistically in one period may be hurt in the next transaction, which should discourage opportunism. With frequent transactions, informational asymmetries are less likely. If uncertainty is low in such a case, there is little incentive to lock in transaction arrangements by more formal forms of governance. If uncertainty is very high, there is the fear of locking in transaction provisions that may be disadvantageous as market conditions change. However, infrequent transactions are more likely to require more formal governance, especially when the assets are very specialized.

Market governance, represented by a spot market, is adequate for transactions involving generalized or nonspecific assets, regardless of transaction frequency, when there are alternative trading partners available to protect against opportunism by either party.

With bilateral governance, two parties contract with each other but remain separate entities, as is the case with long-term natural gas contracts. This is appropriate when recurring transactions involve medium to high degrees of asset specificity, information asymmetries, and there are few buyers and sellers in the market. Products are more likely to be customized with a medium level of uncertainty in the market. Transactions are now frequent enough to recover the contract setup cost. The degree of specificity increases the length and comprehensiveness of contracts, while uncertainty may decrease the length of the contract or cause the inclusion of clauses that allow for specified contract changes as market conditions change. Thus, many natural gas contracts have price escalation clauses, where price is tied to another energy product, such as fuel oil or the spot or futures price at a gas market hub. Bargaining typically sets prices under bilateral governance.

Trilateral governance also involves two firms that trade with each other, but a third firm may be involved in the transaction to make sure one firm does not take unfair advantage of the other. When firms provide for mandatory arbitration by an impartial third party in cases of dispute, there is trilateral governance. Although contracts hold parties to particular activities, it may be expensive to enforce contracts through the courts. When assets are mixed, and transactions are only occasional, trilateral governance may be desirable to reduce the cost of enforcing contracts. With frequent transactions and mixed assets, there is less need for a third party in the transactions, and bilateral governance may suffice. Neither side is as likely to try to take advantage of the other, since they know that such behavior may jeopardize future transactions. An example of trilateral governance was the US Export-Import Bank's threat to call in loan guarantees and pressure arbitration between Enron and India's government over the unfinished Dabhol power plant.

Organized exchanges that trade physical natural gas (e.g., NGX) and financial natural gas contracts (e.g., ICE and CME Group, see chapters 18 and 19) also provide a form of trilateral governance by guaranteeing both sides of the trade. Margin accounts are required to protect against default. However, typically the exchange pools all the trades, so traders do not know the counter party to a trade unless delivery is taken.

Unified governance, also called *vertical integration*, may be appropriate with highly specific assets or idiosyncratic purchases, large information asymmetries, and complex customized products. For example, labor that becomes very specific to the firm is likely to be hired rather than outsourced through a long-term contract for labor services. For instance, you may outsource for building custodial services but have in-house employees to operate your firm-specific software.

The choice between unified and other forms of governance for purchasing inputs is termed the *make-or-buy decision*. Purchasing a service that was formerly performed in-house is referred to as *outsourcing*. With unified governance, the need for adaptive, sequential decisions resulting from uncertainty is satisfied without the cost of involving another party by removing the transaction from the market. However, internal transaction or control costs also exist and influence the choice of governance. With infrequent idiosyncratic transactions, there are less likely to be scale economies to unified governance, and firms that go to the market use trilateral governance to insure against holdup.

Williamson (1993) summarizes the likely types of governance depending on specialization of the assets and frequency of transactions, as shown in table 8–3.

The predominant form of governance in the US natural gas industry has evolved in response to the market and government interventions, as considered in the next section.

Table 8–3. Likely governance structure matrix

		Type of Asset		
		Nonspecific	Mixed	Idiosyncratic
Frequency of Transaction	Occasional	Market	Trilateral	Trilateral
	Frequent	Market	Bilateral	Unified

Source: Williamson (1993).

Evolution of the US Natural Gas Industry

Dahl and Matson (1998) trace the history of the US natural gas industry and show how it has adapted as regulatory frameworks have changed. In its infancy, the natural gas industry inherited the regulatory structure of the existing town gas (also called *coal gas*) systems. In the 1800s, distributors of synthetic gas, produced from coal, actively sought municipal and state assistance, as well as protection of their investment in coal gasifiers and distribution lines in the form of eminent domain and franchises. Lacking incorporation laws, corporate status was granted on a case-by-case basis in the form of charters. Local rate regulation was practiced until out-of-town natural gas supplies became available, which resulted in the formation of state public utility commissions (PUCs).

Short-term contracting was common in the first quarter of the 20th century because the natural gas fields were relatively close to markets. Investment in the specialized assets needed to transport gas to market was fairly low, and the existence of multiple pipelines allowed competition. Uncertainty came from the capability of individual wells to sustain production, encouraging the use of a spot market where buyers could switch suppliers as necessary.

The development of welded steel pipelines allowed for the long-distance transmission of natural gas. These pipelines introduced the industry into interstate commerce and required a large investment in specialized assets. State regulators were not allowed to control matters beyond their borders, but no federal regulatory authority was able to grant monopoly status to protect investments from competition. Therefore, vertical integration was adopted as the common governance structure. Groups of producers and distributors combined into holding companies to build pipelines that brought the gas to market, thereby internalizing the transactions. Fear that such holding companies in the electricity industry would abuse their market power (as they eventually did) led to enactment of the Public Utilities Holding Company Act (PUHCA) in 1935 (Thakar 2008). This act, which also covered natural gas utilities, allowed only two layers of integration, largely splitting off distribution from production and transportation. Such a breakup was less common in the intrastate market.

The Natural Gas Act (NGA) of 1938 required the Federal Power Commission (FPC) to control interstate transport and sale-for-resale of natural gas. Interstate natural gas pipeline companies were then, and still are, subject to federal oversight of their rates. Although interstate oil pipeline companies were common carriers controlled by the Interstate Commerce Commission (ICC) to prevent monopolies, gas pipeline companies were initially allowed to remain private carriers. US common carriers are required by law to provide transportation on a nondiscriminatory basis, whereas a private carrier can restrict carriage to its own products. Common carriage is often called *third-party access*.

Gas purchased from others was recognized at cost or the price paid. However, affiliated producing companies were regulated on the same basis as the pipeline company using the original cost of capital and a regulated rate of return. Therefore, vertical integration was discouraged by regulation. There was little competition in the early history of the interstate transportation of natural gas, and regulation rather than vertical integration may have curbed opportunistic behavior. Private regulated monopoly pipeline companies purchased gas from producers and sold gas to distributors under long-term contracts. These contracts often were for 20 years and longer rather than the more typical five-year contracts in nonregulated industries of similar capital intensities. Since the right to abandon sales was typically not granted by regulators, sellers had no reason to request shorter contracts. Because prices were typically the highest allowable by law, neither producers nor pipelines saw any advantages to shorter term contracts. Regulatory stability also reduced uncertainty and lengthened contracts.

Long-term contracts contained provisions relating to payment, pricing changes, and dedication of reserves, such as the following:

- *Take-or-pay clauses.* Such clauses guaranteed payment by pipeline companies to producers of a contracted payment whether product was taken or not.
- *Minimum-billing clauses.* Local distribution companies (LDCs) were obligated to take or pay for contracted volumes made available to them.
- *Most-favored-nation clauses.* MFN clauses gave the producer the right to the highest price paid by the purchaser within a specified geographic area.
- *Price renegotiation.* Price renegotiation was allowed on specified dates or under certain conditions.
- *Automatic price adjustments.* Price adjustments occurred either at specified intervals or were triggered by specific events.

- *Exclusivity or sole-source clauses.* Such a clause could indicate dedication of reserves to a specific party.

Embedded in their regulated rates were all the products and services associated with the merchant function, including the commodity, transportation, storage, and all complementary goods. Pipelines bundled these services and were required by certificate and contract obligations to maintain sales capacity and provide it on demand without notice.

Regulation from wellhead to burnertip

It is debatable whether US gas producers had monopoly power that would have required price regulation except in isolated fields with few producers. Nevertheless, beginning in 1954, all producer sales-for-resale transactions involving the interstate market were regulated. This regulation disrupted economic allocation by distorting the signaling function of market-determined prices. The result was supply shortages when controlled natural gas prices were depressed below market clearing levels. Supply shortages in the interstate market in the early 1970s prompted the FPC to raise price ceilings for producers. Continued shortages encouraged industrial users and utilities with interruptible contracts to develop dual-fuel-burning capability, allowing them to switch between gas and other sources, such as fuel oil.

In 1977, the US Federal Energy Regulatory Commission (FERC) replaced the FPC in regulating interstate gas pipelines. To increase gas going into the interstate market, the Natural Gas Policy Act (NGPA) of 1978 instituted a phased decontrol of prices for new gas and extended price controls to gas sold in the intrastate market. With uncertain supplies and expected high demand growth, interstate pipelines negotiated high-priced contracts with producers. The contracts contained more stringent than usual take-or-pay provisions.

The worldwide recession of 1981 and 1982, energy conservation, and fuel switching spurred by high energy prices reduced the consumption of oil and gas, causing oil product prices to fall sharply. However, inflexible pricing provisions in long-term contracts, price regulations, and large take-or-pay obligations on higher-priced gas prevented gas prices from falling to compete with oil products. Pipelines took expensive take-or-pay gas at the expense of lower cost gas, with the anomalous result that prices rose in a surplus market. Producers shut in gas supplies, and pressure for political reform mounted.

In 1983, FERC authorized special marketing programs (SMPs) to allow producers to deal directly with end users, who could switch to competing fuels. This less expensive gas prevented the further loss of market share, increased the throughput on pipelines, and opened the door for contract carriage and the development of a spot market in natural gas. Some pipeline companies began reneging on their take-or-pay obligations, and producers filed suit against them. However, producers faced the prospect of long court battles with no revenue in the interim, and so regulation by court order was found not to be an effective means of enforcement. In 1984, FERC responded with Order 380, offering regulatory relief for pipeline customers (local distribution companies or LDCs) by eliminating minimum billing or their take-or-pay obligations to the pipeline companies. This change shifted the cost burden from their customers to the pipeline companies, particularly those with take-or-pay obligations with producers. Minimum billing had allowed pipeline companies to transfer risk associated with their investment in physical assets to their customers. Price risk was also passed on to residential and commercial users by purchased gas adjustment (PGA) rules at the state level, which allowed the cost of gas to an LDC to be passed on to its customers.

The new economic burdens on pipeline companies led them to attempt renegotiation of contract terms with producers. However, uncertainty of supply and demand, limited court precedents, and potential regulatory relief made this a difficult process in the existing institutional environment. After this period of uncertain supply, demand, and regulatory interventions, contracts in the mid-1980s generally extended no longer than three years.

Open access: development of a spot market

The courts found SMPs discriminated against end users who were not able to switch fuels. In 1985, FERC Order 436 allowed interstate pipelines to voluntarily offer nondiscriminatory open access transportation capacity. Pipeline customers were allowed to convert their bundled, firm-contract sales volumes to firm (i.e., noninterruptible) transportation. Contract carriage reduced asset specificity and the need for long-term contracts, and the spot market rose from 4% of consumption in 1983 to more than 70% of the gas market in 1987 and 1988. Pipelines challenged Order 436 because it did not reduce their take-or-pay liabilities. These challenges resulted in Order 500 in 1989, which allowed pipeline companies to offset take-or-pay liabilities with the transportation of gas. Sales and transportation were becoming separate economic activities.

FERC Order 636, issued in 1992, carried on the deregulation. The order essentially restructured the pipeline industry by mandating contract carriage, including storage in the definition of transportation, unbundling transportation from sales, and generally encouraging a more competitive atmosphere based on private contracts. Competition was to play a major role in the conduct of business, with regulation used only where competition was not present (i.e., transmission and distribution).

The growth of intermediaries

As marketers gained experience and network effects increased, assets became more specific. A number of marketers have moved toward more vertical and horizontal integration within the industry. There were only about 50 marketers in

1986 that purchased gas for resale. That number increased rapidly to more than 350 companies in 1991 before falling back to about 260 in 1995, which is more or less the same number of marketers as in 2000. In 1993, the top 20 marketers moved 46.2 bcf/d. In 1996, they moved 84.8 bcf/d, but by 2000, they moved 149.7 bcf/d. In 2011, they moved 124.8 bcf/d (*NGI* 2013). Since gas often changes ownership more than once, the marketed values are larger than the total consumption for North America (83.6 bcf/d) (BP 2012).

The *NGI* list of top marketers in table 8–4 is certainly not exhaustive. The exact number of marketers is constantly in flux as companies enter and exit the industry. Fielden (2012) suggests that the *NGI* list represents about 70% of natural gas transactions.

Table 8–4. Top North American natural gas marketers, 2011

Rank	Company	2011 (bcf/d)
1	BP	23.00
2	ConocoPhillips	15.45
3	Shell Energy NA	13.20
4	Macquarie Energy	10.23
5	EDF Trading NA	7.22
6	J.P. Morgan	6.68
7	Chevron	6.35
8	Tenaska	6.25
9	Louis Dreyfus	5.58
10	Sequent	5.21
11	ExxonMobil*	4.33
12	Citigroup	3.57
13	J. Aron & Co.	3.42
14	Encana*	3.33
15	Chesapeake*	2.75
16	Devon*	2.61
17	Anadarko	2.33
18	ONEOK	2.32
19	Hess	2.17
20	Southwestern Energy Co.	1.68
	Total	**124.78**

Source: NGI (2013).
Note: *Indicates NGI took the value from North American gas production and does not include any third-party trading. Divide by 35.3 to convert to billion meters per day.

Eleven of the marketing affiliates above (BP, ConocoPhillips, Shell Energy, Chevron, ExxonMobil, Encana, Chesapeake, Devon, Anardarko, Hess, and Southwestern Energy) are associated with 11 of the top 20 natural gas producers shown in table 8–5. Three are banks (Macquarie, J.P. Morgan, and Citigroup);

four are energy trading companies (EDF Trading, Tenaska, Louis Dreyfus, and J. Aron & Co.) and the other two (Sequent and ONEOK) are affiliated with local distribution companies.

There are more than 6,000 natural gas producers in the United States. They range in size from the multinationals and independents to a single low production well operated by a single person with multiple owners. Sheer numbers suggest a competitive market, with the top 20 producers in table 8–5 accounting for less than one-half of US natural gas production in 2011.

Table 8–5. Top United States natural gas producers, 2012

Rank	Company	bcf
1	Exxon Mobil	1,518.0
2	Chesapeake Energy Corp.	1,129.0
3	Anadarko Petroleum Corp.	916.0
4	Devon Energy Corp.	752.0
5	ConocoPhillips	685.0
6	BP*	602.6
7	Encana*	592.0
8	Southwestern Energy Co.	564.5
9	Chevron Corp.	440.0
10	WPX Energy Inc.	407.0
11	Royal Dutch Shell*	397.0
12	EOG Resources, Inc.	380.2
13	Fidelity Exploration and Production Co.	379.8
14	Apache Corp.	312.6
15	Occidental Petroleum Corp.	300.0
16	EQT Production	259.0
17	Cabot Oil and Gas Corp.	253.0
18	Ultra Petroleum	249.3
19	QEP Resources	249.3
20	Exco Resources Inc.	156.3
	Total US	**24,000.0**

Source: *OGJ* (2013b).
Note: *Non-US-based companies with production numbers from their respective 2012 annual reports.

Gas Consumers

The gas produced is sold to four main types of customers: residential, commercial, industrial, and electricity generation. Figure 8–3 shows how consumption by these US sectors has changed over time. Through 1970, natural gas consumption grew rapidly in all sectors, with flat, regulated prices and growing income. With price run-ups in the 1970s and early 1980s, consumption in the industrial and electricity generation sectors fell, and consumption stagnated in the residential and commercial sectors. Lower prices over the 1990s caused industrial consumption to rebound. Electricity rebounded also, but with a longer delay. Spiking prices beginning around 2003 again caused industrial consumption to drop precipitously. However, residential and commercial sector consumption remained fairly flat, while electricity consumption continued to grow dramatically. A small amount of natural gas is also used for transportation. If natural gas prices remain low and oil prices remain high, the transport sector is likely to see significant growth and a rising share. A number of countries have quite successfully promoted natural gas for transport, with more than 15% of road energy transport supplied by natural gas in 2011 in a number of countries, including Argentina (15.8%), Bangladesh (30.3%), Bolivia (21.0%), and Pakistan (79.0%) (IEA, n.d.h).

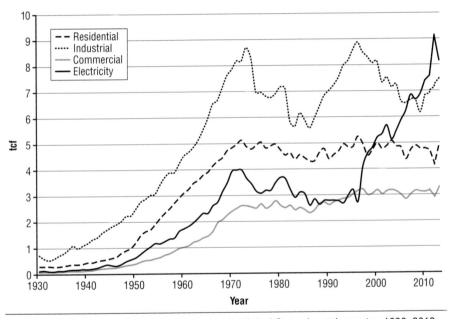

Fig. 8–3. Historical natural gas consumption in the United States by major sector, 1930–2013
Source: Compiled from information in US EIA (n.d.b).

Gas Transmission

As with electricity, restructuring is leaving transmission and distribution regulated with the wholesale marketers providing large consumers a choice of sticking with the local distribution companies or buying directly from producers. As of 2009, 21 states and Washington, DC, allowed full retail access (US EIA 2010b).

Through the process, there has been a strong consolidation among interstate pipelines. From 1997 to 2001, some of the largest companies acquired systems that spanned much of the United States. The top 10 companies owning interstate pipelines then owned almost 90% of the market, and the majority had marketing affiliates as well (table 8–6). Since 2001, the restructuring has continued. Only 4 of the largest pipeline companies are still recognizable in the top 10 (Kinder Morgan, Williams, NiSource, and Dominion). The others are no longer the same companies. Many of the assets of Enron and El Paso are now controlled by Kinder Morgan. Duke's pipeline assets have been spun off into Spectra. CMS's assets are now controlled by Energy Transfer; Reliant's assets are with CenterPoint; and Koch's assets are owned by Boardwalk.

Table 8–6. Top ten interstate pipeline companies by mileage, 2001 and 2011

Company	Pipeline Mileage 2001	Company	Pipeline Mileage 2011
El Paso	45,805	*	
Williams	27,002	Williams, LP	15,000
Enron	23,372	*	
Kinder Morgan	18,694	Kinder Morgan	63,000
NiSource	15,620	NiSource	19,000
Duke	14,596	Spectra	14,300
CMS	11,071	Energy Transfer	15,700
Dominion	9,950	Dominion	7,800
Reliant	8,204	Centerpoint	8,200
Koch	7,278	Boardwalk	14,300
		TransCanada Affiliates	16,561
		MidAmerica	16,580
Other	24,408	Other	28,965
Total	**206,000**		**219,406**

Sources: Tobin (2001, 20); OGJ (2001); US FERC (2013); supplemented with company home pages.
Note: *Much of these companies' assets are now controlled by Kinder Morgan.

Regulated pipelines have largely been broken off into separate entities, with some form of limited partnerships the most popular organizational form. In 2012, FERC indicated there were 162 regulated natural gas–related entities, with 116 owning interstate pipelines. The remainder included natural gas storage companies, liquid natural gas (LNG) companies, and natural gas hubs. More than 70% of these interstate pipelines were in some form of limited partnership. As regulated entities, interstate pipelines have lower commercial risks, and many owners apparently do not feel the reduced risk from lower corporate liability worth the extra taxes a corporation has to pay. (A corporation pays taxes on its profits, and shareholders in turn have to pay taxes on their income, whereas a partnership itself does not pay taxes, only the partners themselves pay taxes on their income from the partnership.)

Volatility in the Natural Gas Market

Natural gas consumption is highly seasonal, with weather being a strong driver, as can be seen in the monthly consumption data in figure 8–4. The most prominent feature in figure 8–4 is the strong winter peak for heating, with a much smaller summer peak for natural gas to supply electricity generation for the cooling season. Prices also show a jagged tendency, with peaks often during the winter and troughs during the off season. However, there is also a strong trend from the economy, with low prices during the weak economies in 2002 and 2009. This suggests that price volatility often comes from the demand side of the market, with two important determinants of natural gas being the state of the macro economy and the season. Further price reductions occurred through 2012 as the shale gas revolution spurred gas production. However, prices have trended up through early 2014 as producers reduced drilling and production. High storage costs, with large distances between storage and market in some cases, have also contributed to volatility and shortened contract lengths.

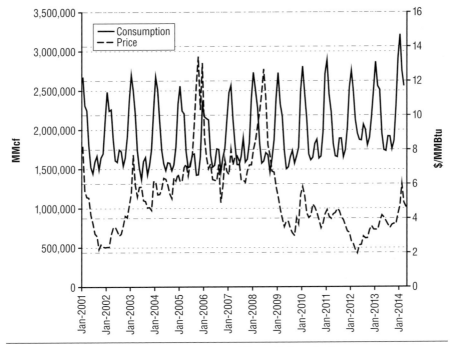

Fig. 8–4. Monthly US natural gas consumption and Henry Hub spot prices
Source: US EIA (n.d.b).

To see how seasonal demand shifts relate to prices under different contract regimes, see figure 8–5. In panel (a), assume that price is fixed (P_c), and in panel (b), assume that quantity is fixed (Q_c). Notice that seasonal demand shifts cause a variation in production under fixed prices but a variation in prices under a fixed quantity.

Seasonal effects are not the only source of volatility. Consider figure 8–6, which shows annual historical prices. Price controls kept prices fairly flat and stable from 1939 to the early 1970s. With shortages in the 1970s and the beginning of wellhead price decontrol in 1978, which was completed in 1990, prices have become much more volatile. This volatility prompted the creation of an organized futures market for natural gas on the New York Mercantile Exchange (NYMEX) in April 1990 for delivery at Henry Hub, Louisiana. By buying or selling a futures contract, you can essentially lock in a price for your physical product and avoid price risk. (Futures and other financial derivatives will be considered further in chapters 18 and 19.)

In 1993, NYMEX increased access to futures markets by instituting after-hours electronic trading with NYMEX Access. In 2008, NYMEX was acquired by CME Group (formerly the Chicago Mercantile Exchange). Natural gas futures contracts on the exchange still trade in New York in open outcry auction (9:00–14:30) and trade almost 24 hours a day electronically on CME Globex. Over-the-counter contracts, which are bilateral and not guaranteed through the exchange, are traded electronically almost 24 hours a day on CME Clearport (CME Group 2012b). The Intercontinental Exchange (ICE), founded in 2000 as an over-the-counter market and headquartered in Atlanta, Georgia, is also a key player in gas trading. With offices around the world, it now offers exchange and over-the-counter financial gas trading, as well as physical trading on its electronic platform WebICE (ICE 2013b). A daily spot market exists, with business transacted over the phone or electronically on ICE, electronic bulletin boards, and e-commerce sites. This market can be tapped to create market balance. ICE provides the trading platform for the Calgary-based Natural Gas Exchange (NGX), the largest North American physical exchange for natural gas and electricity. Spot natural gas was estimated to be about 16% of gas consumption in 2009 (US FERC 2010).

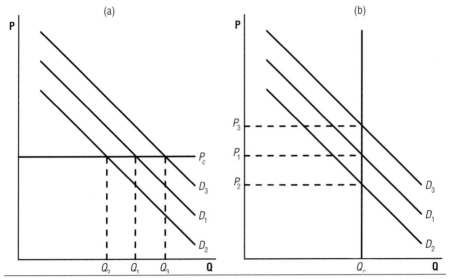

Fig. 8–5. Price and quantity changes under fixed price and fixed quantity regimes

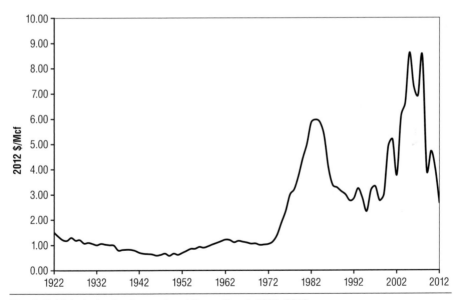

Fig. 8–6. Historical natural gas price at the wellhead, 1922–2012
Source: US EIA (n.d.b).

Contracts

Contracted gas typically takes three forms: swing contracts, base contracts, and firm contracts. Swing contracts are also called *interruptible contracts*. With *base contracts*, buyers and sellers make a best effort to satisfy contract amounts, but neither side is legally obligated to do so. With *firm contracts*, both sides are legally obligated to satisfy contracted volumes. These contracts can be long-term (greater than a year) or short-term (less than a year). The bulk of natural gas production is contracted during *bid week*, which is the last five business days of the month. Buyers typically keep a portfolio of contracts expiring at different times. For example, Portland General Electric had future gas purchase obligations of $136 million dollars on December 31, 2011. Slightly more than one-third was for 2012; about 15% was contracted for each year from 2013 to 2015; and about 9% was contracted for 2016. The remaining 7% was contracted even further out (Portland General Electric 2013).

Initially, there was no standard long-term contract for gas marketing, and contractual relationships were highly personalized. However, with huge increases in volume, long-term gas contracts have become more standardized to include quantity, contract term, price, delivery point, payment schedule, and performance obligations. This has lowered transaction costs by saving time on confirmation of verbal agreements and lessening the potential for misunderstanding. In response to the problem of buyers defaulting if they can get a better price, marketers began to include language about nonperformance. For example, if prices fell, and the buyer turned back the gas, then the buyer would owe the difference in price to the original seller. A similar penalty would be invoked if prices rose and the seller would not deliver.

Since the market demands flexibility in pricing, few contracts contain fixed prices. By 1995, 90% of marketers' purchase contracts were indexed; slightly more than 80% of their sales contracts were indexed. To maintain allocative efficiency, these prices are usually indexed to factors that reflect current market conditions, such as Henry Hub near-term futures or published spot price indexes. In 2009, more than 90% of the contracted amounts still had prices pegged to a price index.

Prior to restructuring, pipelines took possession of gas and then resold it. Interstate pipelines have continued to phase out their gas sales in favor of transportation sales as required, and contracting for pipelines has changed as well. All have bulletin boards, or open access same-time information systems (OASIS), as required by FERC, to post capacity information. FERC subsequently required that pipelines use Internet-compatible messaging and business formats, which have been established by the North American Energy Standards Board (NAESB 2010). While the many market changes outlined above had reintroduced some degree of asset specificity, these information standards have been reducing it.

Some of the capacity devoted to pipeline-owned gas became *no-notice transportation*, which means shippers can have the capacity on demand. This category is similar to the service provided to noninterruptible customers when pipelines were merchants and transporters. *Firm transportation* means that the buyer pays for the capacity.

Under FERC Order 637 in 2000, released pipeline capacity could be traded in a nonregulated market. *Released capacity* is transportation capacity that was under long-term contracts that have expired, or capacity wanting release from long-term contracting. The released capacity can be traded in the spot market. Interruptible transport service is available as well. The ability to shift fuels and take advantage of interruptible and released capacity can be quite advantageous. The largest decreases in transportation costs have been for the electricity generation sector, followed by the industrial sector, both of which have traditionally used the bulk of interruptible service.

On-system consumers are those who still buy in the traditional way from the local distribution companies, whereas *off-system consumers* are those who have left the old system. Off-system customers do not buy from LDCs but may buy from marketers or even producers. By 2010, most deliveries to electric power companies were unbundled, with pipelines delivering gas directly to power plants off-system and not through local distribution companies. The EIA refers to gas delivered on behalf of a third party as gas delivered for the account of others. This unbundled gas was about 82% of industrial sales in 2010, 42% of commercial sales, and 11.5% of residential sales (US EIA, n.d.b).

On-system commercial and residential customers did not fare as well price-wise for transportation services because they do not use much interruptible service. Off-system commercial, industrial, and electricity generators saw larger declines. Part of the off-system price declines also resulted from the change from modified fixed variable transportation rates, in which all customers share in paying the fixed costs, to straight fixed variable rates, in which peak customers pay all of the fixed costs, as discussed in chapter 5.

Residential, small commercial, and small industrial customers have stayed on-system. These customers are often using gas for heating and require reliable peaking service. No-notice delivery under Order 636 allows these users the kind of pipeline service previously provided to firm sales customers. Capacity release of firm transportation also encouraged this movement, as it has made firm transportation cheaper and further spurred a gray market, in which reliability is being offered by the bundling of sales, transportation, and storage. This ability to bundle recreates some asset specificity not realized in the separate markets. This change in transport cost structure and increased competition is reflected in the narrowing of the price spread between the industrial and electricity sectors. It is also reflected in a divergence of the price spread between those two sectors, which largely moved off-system, and the commercial and the residential sectors (especially the latter), with more remaining on-system (see fig. 8–7). The residential sector was especially hesitant to move off-grid.

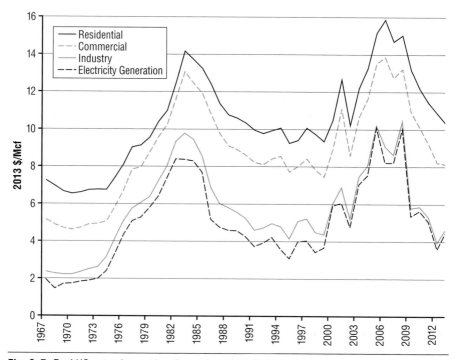

Fig. 8–7. Real US natural gas prices by sector, 1967–2013
Sources: US EIA (2012d), for prices from 1967 to 2002; US EIA (n.d.b) for 2003 to 2013. Deflated by consumer price index from US BLS (n.d.).

Natural gas markets are not as complicated as electricity markets because gas can be stored and everything does not have to happen in real time. Figure 8–8 shows seasonal storage changes and their relationship to price. Storage is a complementary good to interruptible transportation and sales. If you have gas stored, an interruption in gas or transportation can be filled from storage. Storage is therefore a substitute for firm transportation and can be purchased under varying contract arrangements. If your gas deliveries are secure, you do not need storage. Earlier marketers did not have access to storage or firm transportation, hence their reliance on interruptible services.

With Order 636, storage also received open access status in 1993. Open access, along with an 11.6% increase in storage capacity between 1992 and 2011, allowed marketers to package gas in ways similar to traditional pipelines (US EIA, n.d.b). Players have tended to hold portfolios of gas and transportation with contracts of varying lengths, which can be complemented with storage.

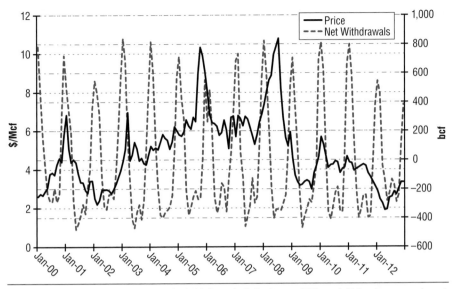

Fig. 8–8. Natural gas net withdrawals (withdrawals [+] minus additions [−]) to storage and spot price

Source: US EIA (n.d.b).

With restructuring, market hubs developed quite rapidly, usually around areas where a number of gas pipelines came together or where storage was available. Prior to 1994, 12 gas market hubs existed in the United States and Canada. In 1996, 39 market hubs were operational, and by 2012, ICE was quoting prices at more than 100 market points. About 20 of the major hubs are called *market centers.* Gas loans, along with balancing and parking services, are now offered at many market hubs. These loans and services help avoid imbalance penalties and improve reliability. Prices of these and other services, such as title transfer and real-time tracking, are market based. However, transportation and storage services are open access at controlled prices unless given special dispensation to operate in an environment that FERC considers competitive. Typically, quantities are contracted with the delivery point specified and the price tied to a price indexed at a specified market hub (US EIA 2009a). See figure 8–9 for a map of major North American gas hubs and gas flows.

Open access, the increasing interconnectedness of the natural gas grid, and increased storage suggest reduced asset specificity for existing pipelines. The reduction in specificity and the regulated rates provide some degree of protection from opportunism. Along with increased uncertainty in the natural gas industry and shorter contracts in the marketing of gas, these factors have also led to a reduction in length of transportation contracts. For a pipeline relatively isolated from the rest of the network, implying more asset specificity, or where markets are more stable, we can expect contracts to be longer. Also, where new gas assets are being built, we expect contracts to be longer. FERC requires proof of a 10-year commitment for new projects. Traditional funding of large capital projects is highly leveraged, which also requires that contracts be very long. For example, cogeneration facilities often require a firm fuel source, and long-term contracts average 15 years.

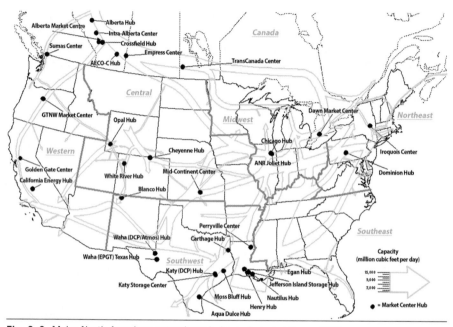

Fig. 8–9. Major North American natural gas hubs and market flows, 2009
Source: US EIA (2009b).

Shorter term contracts make gas prices more volatile, but in a world where prices are indexed, contract length is irrelevant to volatility. In a competitive world, even fixed-price contracts do not ensure stable prices, as shown by the take-or-pay debacles of the 1980s. If we move from governance by regulation to governance by market, we can expect to see price volatility continue. Ultimately, increased price volatility is the cost of the improved efficiency flowing from increased competition.

North and South of US Borders

Regulatory structure for natural gas has followed a similar path in Canada, with the National Energy Board (NEB) being the counterpart to FERC. TransCanada dominates an extensive interprovincial pipeline network, which begin operating in 1957 and extends into the United States. The price of natural gas, which had been tied to crude oil price, was deregulated in 1985 in an agreement between the

Canadian government and the three major producing provinces: Alberta, British Columbia, and Saskatchewan. Sales and transport services were subsequently unbundled. However, there is no released capacity market in Canada. Released capacity remains under the control of the pipeline and is not traded competitively (Makholm 2012).

Gas market restructuring in Mexico has been more restrained than in its more dynamic neighbors to the north. Government-owned Pemex and certain subsidiaries have the right to explore for, produce, and first market all the natural gas in Mexico. But there have been changes in the midstream and downstream markets. In 1995, Mexican law was changed to allow private investment in gas pipelines (previously under the sole purview of Pemex). Private investment in storage and local distribution also became possible. A regulatory commission, Comissión Reguladora de Energía (CRE), was created to oversee privatization of Pemex assets in these midstream and downstream segments of the market. CRE also regulates pipeline tariffs under RPI-X, with a five-year review horizon. CRE required open access of Pemex's pipelines, which resulted in some private purchase of local natural gas LDCs, private investment to build new LDCs, and private investment in pipelines (*OGJ* Editors 1999; Center for Energy Economics c. 2004).

For example, Sempra Pipelines and Storage has a natural gas LDC and a pipeline system in Mexico. TransCanada owns two natural gas pipelines in Mexico: the 193-mile Guadalajara, completed in 2011, and the 81-mile Tamazunchale, completed in 2006, with a contract to build, own, and operate a 146-mile extension granted in 2012. The 25-year transportation service contracts for TransCanada are reminiscent of the long contracts in the United States before the market was restructured (TransCanada 2007, 2011, 2012).

Summary

Since 1980, natural gas has exhibited the fastest consumption growth of all fossil fuels, and currently constitutes about 22% of global energy consumption. Natural gas is a fairly clean fuel, and it is more plentiful than oil on the basis of energy content.

North America is the largest producer and consumer of natural gas, followed by the former Soviet Union. Russia contains one-third of the world's proven reserves of natural gas; the Middle East holds another one-third. With higher transportation costs than oil, natural gas trade patterns has tended to be more regionally oriented, with regional markets in North America, South America, and Western Europe. Natural gas traded in the Asia-Pacific region mostly takes the form of LNG imports into China, India, Japan, South Korea, and Taiwan. As the LNG markets have been extended, the markets have become more interconnected.

Transaction and transportation costs have influenced the evolution of the natural gas industry. Transaction cost economics assumes bounded rationality and opportunism. Given these assumptions, transaction cost economics focuses on three institutional factors that influence costs: uncertainty, asset specificity, and frequency of transactions. Transaction cost economics also involves four types of governance: spot market, bilateral, trilateral, and unified governance. With nonspecific assets, spot markets are likely to dominate. More idiosyncratic assets, coupled with more uncertainty and less-frequent transactions, make it more likely that unified governance will occur. Less idiosyncratic assets, coupled with more uncertainty and more frequent transactions, mean it is more likely that spot market governance will occur.

Although the frequency of transactions has not changed much over the history of the US natural gas market, uncertainty and asset specificity *have* changed as markets, technologies, and regulations have evolved.

Early production of natural gas tended to occur close to end-use markets, with relatively inexpensive transportation assets. Thus, governance was a competitive spot market at the turn of the 20th century. With the development of welded pipe and distribution systems, assets became more specific, and companies moved into the unregulated interstate market. Producers adopted vertical integration to protect against holdup in supply. However, with public utility regulation in the 1930s, companies separated production from transportation and dismantled vertically integrated systems. Contracts with very long terms at regulated prices came to dominate in a highly regulated environment. Interstate pipeline companies became merchants, buying gas from producers, reselling the gas to local distribution companies, and providing all services required in between.

Natural gas shortages in the 1970s led to a messy price deregulation beginning in 1978 and largely completed by the early 1990s. Pipeline companies were converted from merchants to open access regulated transport providers, and numerous marketers entered the industry. The spot market became larger during periods of high uncertainty but now tends to vary between 10% and 20% of gas sales.

Contracts have been streamlined and standardized to some extent, although contract lengths vary considerably. Decreased asset specificity and increased uncertainty have led to much shorter contracts than the standard 20-year contracts preceding deregulation. Current contracts of a year or more are considered long-term, and contracts with a term from 30 days to a year are considered short-term.

High price volatility led to the development of a gas futures market to moderate price uncertainty, and marketers handled much of their risk management this way. Price uncertainty is also being handled by indexing price in some way, usually to some market basket. This basket could include the Henry Hub near-term future price, gas prices at another hub, averages of two or three hubs, fuel oil prices, or some combination of these factors. Organized exchanges and bulletin boards allow trading of spot natural gas, pipeline capacity release, and natural gas futures contracts.

Residential, small commercial, and small industrial customers have stayed on the system and have not seen the degree of price reduction in transportation that larger users have experienced. Larger users have been able to take advantage of interruptible service.

Although the shale gas revolution will require new infrastructure, the existing governance appears to be working quite well. I do not foresee any major changes to asset specificity or uncertainty that will change the domestic governance structure. However, the development of an export market will be highly contingent upon the regulatory response, and increased natural gas in the transportation market will also require a governance structure to be developed.

Similar natural gas regulatory changes have occurred in Canada, except Canada has no spot market for capacity release. Mexico has allowed private capital to enter the midstream and downstream sectors, but competitive markets are not in place for either natural gas or pipeline service sales. It will be interesting to see whether Mexico follows its northern neighbors down the restructuring path.

9

Monopsony: Japan and the Asia-Pacific LNG Market

In an idealized competitive market, the market sets the price. Market power is the ability of smart players to influence market prices. Single sellers, monopolists, will want to raise prices. Single buyers, monopsonists, will want to lower prices.

Introduction

While almost two-thirds of the world's oil is exported, this is true for only one-third of the world's natural gas. For these gas exports, transport by pipeline is typically cheaper unless gas is far from its market or is separated from its market by an ocean. (Chapter 17 will have more on energy transport costs by various modes.) Thus, pipelines trumped newer LNG exports by more than 2 to 1 in 2012 (BP 2013). The first commercial shipments of LNG traveled from Algeria to the United Kingdom in 1964. In contrast, long-distance natural gas pipelines were first commissioned in the USSR in 1946 and even earlier in the United States (Stern 2005; Makholm 2012).

Although LNG has a relatively small share of the total natural gas market (around 10%), LNG growth rates have been higher than for piped gas for the last 25 years (BP 2013). Lowered gas exploration and production costs, along with new storage tank design, larger compressors, larger tankers, and larger LNG trains, had decreased costs considerably by the early 2000s, making this fuel more competitive (Fletcher 2002). Although costs vary considerably from project to project, the general consensus seems to be that costs generally fell through about 2004, but then rose steeply along with the worldwide economic boom. Wood Mackenzie estimates the capital costs of 36 liquefaction projects between 1965 and the present. They found that real capital costs for liquefaction per tonne fell by around 57% per tonne between 1980 and 2000, but rose by 300% from 2000 to 2013 (Songhurst 2014). IGU (2014) estimates regasification costs of new facilities rose by 230% from the average of 2000–2006 to 2007–2013.

For tanker purchases in AVRHTS (2012), after adjustment by the US producer price index, the real average price of new LNG tankers per unit of capacity fell by more than 50% from the mid-1990s to 2008. However, with expectations of large imports into the United States, new tanker orders increased. With increasing purchases and higher capital costs, new LNG average tanker prices rebounded by 19% from 2008 to 2010. However, with overbuild of tankers, these real prices have trended down thereafter for tankers delivered and on order through 2016.

Spot charter day rates fell much more precipitously (Zhou and Holloway 2011). After spiking to more than $120,000 per day in 2006, they bottomed out to less than $50,000 in mid-2010, when nearly one-third of the LNG tankers were out of service. Several factors contributed to a turnaround, including a strong Asian economy, closures of nuclear power plants in Japan after the Fukushima nuclear power plant accident in 2011, and some German nuclear power plant closures related to their nuclear phaseout. As a result, spot day rates soared to more than $140,000 per day by early 2012, and most LNG tankers were back in service (Royal Institution of Naval Architects 2013). However, new tanker deliveries and slower-than-expected completions of new gasifaction terminals had again reduced day rates to $70,000 by early 2014 (Ship and Bunker 2014).

Whereas rising steel prices and tight markets for engineers and engineering firms have raised capital costs for capital-intensive industries such as LNG, technical progress continually helps to reduce costs. For example, it is technically feasible to produce, store, and regasify LNG from a ship. Excelerate Energy, Golar LNG, and Höegh Energy already build or convert ships to transport, store, and regasify LNG, called *floating storage and regasification units* or *FSRUs*. Excelerate had nine such vessels in 2012, Golar had five, and Höegh had two. In 2005, the first FSRU ship in the United States began offloading offshore of Louisiana. Argentina was able to build a floating import terminal in less than a year to receive such ships, which is considerably shorter than for a land-based terminal. These units are increasing rapidly with a 34% increase in 2013 (IGU 2014).

Currently, Exmar is taking the technology a step further and investing in the first floating liquefaction, regasification, and storage unit (FLRSU) for stranded gas offshore of Colombia in 2014 (Singh 2012). Shell has also made a final investment decision (FID) to build the largest floating liquefied natural gas (FLNG) system. Depending on how costs evolve, such offshore production of LNG could further enhance this market by making smaller and remote gas fields economically viable.

Typically, reserves must be large enough to produce for at least 20 years to justify the high upfront investment. As LNG facilities are idiosyncratic, market governance has typically been long-term contracts. Further, investors financing LNG contracts often require long-term contracts to avoid any possibilities of holdup. A checklist of other items bankers and financiers may consider before granting a loan for an LNG project, or other large-scale energy project, is contained in table 9–1.

Table 9–1. What your banker wants to know

I. Loan Request
    A. Amount and type
    B. Term: long run/short run
    C. Purpose
        1. Working capital, inventory, equipment, real estate
        2. Merger/acquisition, refinancing
    D. Project plan and what it will accomplish
    E. Collateral
        1. Accounts receivable, inventory
        2. Fixed assets, real estate
        3. Marketable securities
        4. Guarantor
        5. Letters of credit
    F. Demonstrate credit worthiness and when and how the money will be repaid
II. Business Information
    A. Purpose of business and when established
    B. Number of employees
    C. Accountant/insurance agent/attorney
    D. Years at present location (own/lease)
    E. Ownership and other commitment of funds
III. Management
    A. Names/credentials/responsibilities
    B. Compensation
    C. Hierarchy
    D. Reporting policies
    E. Role of board of directors
    F. Bank access to board and management
IV. Environmental
    A. Philosophy
    B. Real or potential liabilities
    C. Accident response preparedness
    D. Precautions when purchasing new properties
    E. Environmental reports
V. For Oil and Gas Reserves
    A. How do you add the following?
        1. Exploration, development
        2. Reservoir management
        3. Acquisitions (properties or companies)
    B. Your inventory of reserves and replacement ratios
    C. Performance statistics
        1. Finding costs
        2. Replacement ratios (find vs. acquire vs. extensions)
        3. Level of effort (LOE)
    D. Three years of engineering reports
    E. Reserve breakdown by
        1. Proved developed producing reserves (PDP)
        2. Proved developed nonproducing reserves (PDNP)
        3. Proved undeveloped reserves (PUD)
    F. Major fields

VI. Financial Information and Controls
    A. Your bank and credit relations
    B. Fiscal statements for the last three years
        1. Balance sheet/profit and loss/cash flow
        2. External audit report
        3. Federal tax returns
        4. State and local tax returns
    C. Your dividend policy
    D. Contingent liabilities
        1. Lawsuits
        2. Tax audits
        3. Pension fund commitments
        4. Environmental liabilities
    E. Any history of bankruptcy
    F. Implementation and control of hedging strategies
    G. Organizational agreement
        1. Articles of incorporation, partnership and trust agreements, etc.
    H. Information on affiliates and off-balance-sheet activities
VII. Performance Statistics and Plans
    A. Production costs
    B. General and administration costs
    C. History of stock trading prices
    D. Forecast of three-year cash flow with all planned investments
    E. Brief history of your company
    F. Analysis of the market and business outlook
    G. Your competition
    H. Future drivers in your industry

Sources: US SBA (2013); Lee (2013); Kramer and Fusaro (2010).

LNG Production and Trade

IGU (2014) and BP (2014) indicate that 17 countries exported from 92 LNG trains by the end of 2013 with total exports at 325.3 billion cubic meters, equivalent to about 243 million tonnes per annum (MTPA) of LNG. Total operating nominal liquefaction capacity was about 291 MTPA. Capacity in Alaska was temporarily out of service for much of 2013. Its capacity would put the number of exporters at 18 and capacity a little higher. Twenty-nine countries imported LNG through 94 regasification terminals with regasification capacity having reached 688 MTPA by the end of 2013 (IGU 2014). Floating capacity of 44.3 MTPA is included in this total. The Netherlands, Norway, Sweden, and Thailand added their first ever regasification capacity during 2011, Indonesia during 2012, while Israel, Malaysia, and Singapore joined their ranks in 2013 (IGU 2013, 2014). (See table 9–2 for some of the trade flows; other minor importers and exporters are included in the regional import and export totals) (IGU 2014).

As the number of importers and exporters has increased, the spot market has also grown. In 2000, it was less than 5% of world LNG trade. By 2012, it had risen to more than 30%. As contracts have been getting more flexible, the number of countries that re-export had been increasing, rising to eight by 2013—Belgium, Brazil, France, Netherlands, Portugal, South Korea, Spain, and the United States (IGU 2014).

Construction and interest in LNG remains high. More than a dozen new tankers were added in 2013, increasing the fleet to 357 LNG vessels holing more than 18,000 cubic meters of LNG each, with another 108 on order, 31 of which were scheduled for delivery in 2014. IGU (2014) lists 29 liquefaction projects with a total capacity of more than 100 MTPA under construction that should come on stream between 2014 and 2019. The larger share of this will be floating, bringing the floating share up to almost a quarter by 2018. They also list 26 gasification projects with a total capacity of more than 70 MTPA that should come on line over this same period.

The size of trains has steadily increased over time. New Qatari megatrains have a nameplate capacity of 7.8 MTPA. These huge trains are more than 20 times as large as the initial Algerian trains in 1964, the oldest of which has been decommissioned (IGU 2012). The number of completed greenfield projects that have commenced deliveries is also increasing. As opposed to brownfield projects, these totally new projects are not additions to existing projects.

Algeria, the first to export commercial quantities of LNG, is still among the countries with the largest number of liquefaction trains, but it no longer has the largest capacity, having dropped to fifth place. Qatar, which only began shipping in 1997, now has about one-fourth of world capacity, followed by Indonesia and Australia (see table 9–3). One-half of the 17 countries have begun exporting LNG since 1999. Atlantic LNG in Trinidad and Tobago and Nigeria's Bonny Island began shipments in 1999; Qalhat in Oman began shipments in 2001. SEGAS and Egyptian LNG began shipments from two Egyptian Mediterranean ports in 2005. Norway's Statoil began shipments from Snøhvit, and Equatorial Guinea LNG began shipments in 2007. Sakhalin II, controlled by Russia's mighty Gazprom, began shipping in 2009, and Peru LNG and Yemen LNG began shipping in 2010. Libyan liquefaction capacity was damaged in the civil war and decommissioned in 2012, but was offset by Australian Pluto coming onstream. Additions in 2013 included Angola's delayed T1 project using previous flared gas, adding capacity of 5.2 MTPA with first shipments in June of 2013 to Brazil (*OGJ* Editors 2013a), and a rebuild in Skikda, Algeria, adding another 4.5 MTPA. Papua New Guinea (PNG) LNG began shipping in 2014 (IGU 2014).

Table 9–2. Natural gas: trade movements by LNG 2013 (billion cubic meters)

To \ From	T&T	Peru	Nor	Rus	Oman	Qatar	UAE	Yemen	Algeria	Egypt	EqGuin	Nig	Aus	Brun	Indo	Mal	Other*	Total Imports
Mexico	0.4	2.5	0.4	–	–	1.6	–	0.5	–	–	–	1.6	–	–	0.4	–	0.5	7.8
Argentina	3.6	–	0.1	–	–	0.9	–	–	–	0.2	–	0.5	–	–	–	–	1.6	6.9
Brazil	2.5	–	0.3	–	–	0.3	–	–	0.1	0.1	–	0.9	–	–	–	–	1.1	5.1
Chile	3.5	–	–	–	–	0.2	–	0.4	–	–	–	–	–	–	–	–	–	4.1
Ot Americas	5.6	–	0.2	–	–	1.0	–	0.3	–	–	–	0.2	–	–	–	–	–	7.2
Americas	15.6	2.5	0.9	–	–	3.9	–	1.3	0.1	0.2	–	3.1	–	–	0.4	–	3.2	31.2
France	–	–	0.3	–	–	1.8	–	0.1	5.3	–	–	1.2	–	–	–	–	0.1	8.7
Italy	–	–	–	–	–	5.2	–	–	0.0	–	–	–	–	–	–	–	0.3	5.5
Spain	2.0	1.5	1.1	–	0.2	3.5	–	–	3.2	0.0	–	3.1	–	–	–	–	0.3	14.9
Turkey	–	–	0.2	–	–	0.4	–	0.1	3.8	0.2	–	1.3	–	–	–	–	0.1	6.1
UK	0.1	–	0.1	–	–	8.6	–	–	0.4	0.1	–	1.3	–	–	–	–	–	9.3
Ot Eur/Eurasia	0.1	–	0.6	–	–	4.0	–	–	0.7	0.1	–	1.3	–	–	–	–	0.2	6.9
Eur/Eurasia	2.2	1.5	2.3	–	0.2	23.4	–	0.2	13.5	0.4	–	6.9	–	–	–	–	1.0	51.5
China	0.1	–	–	–	–	9.2	–	1.5	0.1	0.6	0.5	0.5	4.8	–	3.3	3.6	0.2	24.5
India	–	–	0.1	–	–	15.3	–	0.7	0.1	0.4	–	0.9	–	0.1	–	–	0.1	17.8
Japan	0.4	1.0	0.4	11.6	5.5	21.8	7.4	0.7	0.6	0.8	3.0	5.2	24.4	6.9	8.5	20.3	0.4	119.0
South Korea	0.7	0.7	0.1	2.5	5.9	18.3	–	4.9	0.2	0.8	0.2	3.8	0.8	1.6	7.7	5.9	0.3	54.2
Ot Asia/Pacif	0.7	–	0.1	0.1	–	13.6	–	0.4	0.4	0.5	1.3	2.0	0.2	0.9	2.6	4.0	0.4	27.2
Asia-Pacific**	1.6	1.7	0.7	14.2	11.4	75.0	7.4	8.2	1.3	3.0	5.1	12.1	30.1	9.5	22.1	33.8	1.1	238.1
Total Exports	**19.8**	**5.6**	**3.8**	**14.2**	**11.5**	**105.6**	**7.4**	**9.6**	**14.9**	**3.7**	**5.1**	**22.4**	**30.2**	**9.5**	**22.4**	**33.8**	**5.6**	**325.3**

Source: BP (2014).

Notes: T&T = Trinidad and Tobago, Nor = Norway, Rus = Russian Federation, UAE = United Arab Emirates, EqGuin = Equitorial Guinea, Nig = Nigeria, Aus = Australia, Brun = Brunei, Indo = Indonesia, Mal = Malaysia, UK = United Kingdom, Eur/Eurasia = Europe and Eruasia. Imp = Imports, Exp = Exports, Ot = Other. *includes reexports. **Asia-Pacific and Ot Asia/Pacific (Asia/Pacif) includes the Middle East. – indicates no trade flow. BP used conversion 1 billion cubic meters (bcm) natural gas = 0.74 million tonnes of LNG.

Table 9–3. Data on global liquefaction capacity in 2013

Rank	Country	First Shipments	Capacity (MTPA)	# Trains	Average Train Size (MTPA)
1	Qatar	1997	77.0	13	5.9
2	Indonesia	1978	34.1	17	2.0
3	Australia	1989	24.2	6	4.0
4	Malaysia	1984	23.9	8	3.0
5	Algeria	1964	23.8	16	1.5
6	Nigeria	1999	21.9	6	3.7
7	Trinidad	1999	15.5	4	3.9
8	Egypt	2005	12.2	3	4.1
9	Oman	2001	10.8	3	3.6
10	Russia	2009	9.6	2	4.8
11	Brunei	1972	7.2	5	1.4
12	Yemen	2010	7.2	2	3.6
13	UAE	1977	5.8	3	1.9
14	Angola	2013	5.2	1	5.2
15	Peru	2010	4.5	1	4.5
16	Norway	2007	4.2	1	4.2
17	Equatorial Guinea	2007	3.7	1	3.7
18	United States	1969	1.3	2	0.7
	Total		292.1	94	3.1

Sources: Olaya Morales (2006), GUI (2013), supplemented with company home pages.
Note: MTPA = million tonnes per annum. US capacity was offline for most of 2013 for upgrading.

One-half of the eighteen countries have begun exporting LNG since 1999. Atlantic LNG in Trinidad and Tobago and Nigeria's Bonny Island began shipments in 1999; Qalhat in Oman began shipments in 2001; Segas and Egyptian LNG began shipments from two Egyptian Mediterranean ports in 2005; Norway's Statoil began shipments from Snovhit, and Equatorial Guinea LNG began shipments in 2007; Sakhalin II controlled by Russia's mighty Gazprom began shipping in 2009; Peru LNG and Yemen LNG began shipping in 2010. New expected greenfield projects include Angola LNG, which used formerly flared gas as feedstock and began small shipments in 2013; Papua New Guinea (PNG) LNG, which was expected to begin shipping in 2014; and a number of Australian projects, which are expected online in 2014–2015.

Given the huge capital costs for LNG projects, they are often undertaken by a consortium of companies, and contracts tend to be long-term, on the order of 20 to 25 years. Take-or-pay clauses are often included, meaning that the gas must be paid for whether taken or not. Although projects are not typically vertically integrated from gas production through to final consumers, buyers often take a stake in the liquefaction facilities.

Contract prices are usually indexed to some benchmark crude oil or crude oil product prices, often with minimum price provisions. For example, Japan's prices are indexed to a basket of crudes called Japan's Crude Cocktail (JCC), and continental Europe more often indexes to fuel oils. However, there has been some continental movement toward gas hub pricing, as is the case with prices linked to Henry Hub in the United States and the National Balance Point (NBP) in the United Kingdom. With considerable new capacity coming onstream and global imbalances, the spot market has also become more prominent, accounting for about one-fourth of sales by 2011 (IGU 2011; Jensen 2011).

Table 9–2 shows most of the LNG importers as of 2013. Japan has a long history of imports and is the largest importer, with around a third of the market. It began importing with small shipments from Alaska in 1969 and has continually diversified its supply portfolio as new suppliers have come on stream. It receives at least some shipments from each of the 17 suppliers, with Australia being its largest suppliers followed by Qatar and Malaysia. South Korea is the second largest importer at about half the volume of Japan. Together these two countries account for more than half of the LNG market. Korea's two largest suppliers are Qatar and Indonesia. China and India are the two largest recent importers. Their imports account for another 13% of global LNG importers. Long-time importers, Spain and the UK, followed by France are the largest European importers. Their suppliers are coming mostly from Qatar and Africa.

Historically, Japan has been even more prominent in the LNG market. It typically constituted more than one-half of the global LNG market, and from 1969 to 1986, it was the only LNG buyer in the Asia-Pacific market and the only buyer of LNG from the United States, Brunei, United Arab Emirates, Indonesia, and Malaysia for a number of years as they came onstream.

Japan has had three types of energy policies at various times since World War II. Before the oil crisis in 1973, Japan's energy policy focused on acquiring energy supplies to fuel economic growth. The second period lasted from just after the energy crisis to about 1985, and policy focused on energy security and diversifying out of oil. In the third period, after the Chernobyl nuclear plant disaster in 1986, policy focused on environmental safety. These last two periods have seen the buildup of the LNG trade and are currently being reinforced after the nuclear tragedy at Fukushima. All of these policies have increasingly favored LNG (Nei 2000).

Although a number of Japanese electricity and natural gas companies buy the LNG, the contracts are negotiated through the Ministry of Economy, Trade, and Industry (METI), formerly known as MITI—the Japanese Ministry of International Trade and Industry. This suggests that Japan might have some market or monopsony power as a buyer. A strict *monopsonist*, which is one buyer in a market, is the counterpart of a *monopolist*, which is one seller in a market. Japan was a monopsonist in the Asia-Pacific market prior to 1986. It was also quite dominant thereafter, still accounting for more than three-fourths of the Asia-Pacific market in 2000. In the next section, we will consider how Japan could optimize when it was the only buyer.

LNG Monopsony on Input Market, Competitor on Output Market

To analyze how Japan should behave if it were the only buyer of LNG in the Asia-Pacific market, consider the following simple monopsony model. In all examples, we will abstract from any contracting issues and assume price and quantities are flexible. First, assume that suppliers of LNG to Japan are competitive and that Japanese companies are competitors on their output market, but they band together through METI to become a monopsonist buyer of LNG. The majority of Japan's LNG purchases go to produce electric power, and the second largest are LDCs that sell to the residential sector. Both gas and electric utilities were regulated prior to 1986. Assume that the regulations simulate what would have happened in a competitive output market. To make the exposition easier, assume that all of Japan's LNG goes to produce electricity and that LNG (L) is the only input into the production of electricity for these plants. If Japan maximized profits, its objective function would have been:

$$\pi = P_e E(L) - P_L(L)L$$

where

P_e = the price of their output or the price of electricity,
$E\,(L)$ = the production function for electricity as a function of LNG,
$P_L\,(L)$ = the supply of LNG, which should be the combined marginal cost curve for the competitive producers of LNG, and
L = the quantity of LNG purchased to produce electricity.

The first-order condition for Japan's optimization problem is given in equation (9–1):

$$\frac{\partial \pi}{\partial L} = P_e \frac{\partial E}{\partial L} - P_L - \frac{\partial P_L}{\partial L} L = 0 \longrightarrow P_e \frac{\partial E}{\partial L} = P_L + \frac{\partial P_L}{\partial L} L \tag{9-1}$$

The expression on the left after the arrow in equation (9–1) has the price of output or electricity times the extra electricity produced by an extra unit of LNG (the marginal product of LNG). The whole expression represents the extra revenue gained from buying an additional ton of LNG and is referred to as the *marginal revenue product* of LNG (MRP_L). To understand this expression, take a simple example. Suppose regulators set the price of electricity at P_e = \$0.11, and utilities can sell all they want at that price. Assume that the thermal efficiency for the power plants that burn LNG is about 0.35 (for every 100 Btus of energy input, there is an output of 35 Btus of electricity). Assume constant marginal product up to generation capacity of 150,000 kWh. Suppose 1 Mcf of gas has 1,012,000 Btus and 1 kWh of electricity has 3,412 Btus.

At this thermal efficiency, the marginal product of 1 Mcf of gas = 1,012,000 × (0.35/3,412), or about 103.8 kWh/Mcf. The revenue per thousand cubic feet is then

0.11 × 103.8 = \$11.418. Gas use at capacity in million cubic feet = 150,000/103.8, or about 1,445 Mcf. Once you reach capacity, the *MRP* falls to zero, since you can produce no more electricity.

In equation (9–1), the right-hand side expression after the arrow represents the extra cost of buying an additional unit of LNG, referred to as the *marginal factor cost* of LNG (MFC_L). It is composed of two components. The first expression is the price of LNG, which is a function of LNG. This function is the competitive supply curve for LNG and also the average cost of LNG for the buyer. The second expression is the slope of the supply curve for LNG (or the change in price for a given change in quantity) times the quantity of LNG purchased. It represents the increase in cost for all previous units consumed. Thus, as we move up the supply curve, we pay a higher average price, not just for the last unit, but for all units.

It is easy to see the relationship of marginal factor cost to price by taking a simple supply function. Suppose that supply of LNG is $L = -20 + 100P_L$. When the price of LNG (L) is 5, then 480 units of LNG are supplied. When the price of LNG is 5.25, then 505 units of LNG are supplied and so on, as shown in table 9–4.

Table 9–4. Developing marginal factor cost for LNG supply function

P_L	$L = -20 + 100P_L$	$TFC = P_L \times L$	$MFC = \Delta(TFC)/\Delta L$
5.00	480	2,400.00	
5.25	505	2,651.25	(2,651.25 – 2,400)/(505 – 480) = \$10.05
5.50	530	2,915.00	(2,915 – 2,651.25)/(530 – 505) = \$10.55
5.75	555	3,191.25	(3,191.25 – 2,915)/(555 – 530) = \$11.05
6.00	580	3,480.00	(3,480 – 3,191.25)/(580 – 555) = \$11.55

The total LNG bill, or total factor cost (*TFC*) at any point on the supply curve, is the price of LNG times the amount purchased ($P_L L$) as seen in the third column in table 9–4. Now, as we move up the supply curve, it is easy to see $MFC = \partial(TFC)/\partial L$. In a discrete problem, we can approximate $\partial(TFC)/\partial L$ by $\Delta(TFC)/\Delta L$, as seen in table 9–4, column 4. For example, the change in total factor cost when going from price \$5 to \$5.25 is (2,651.25 – 2,400) and the change in L when going from price \$5 to \$5.25 is (505 – 480). Then

$$\Delta(TFC)/\Delta L = (2,651.25 - 2,400)/(505 - 480) = \$10.05.$$

So at a price of \$5.25, the average factor cost is \$5.25, but the marginal factor cost is \$10.05. Though we pay \$5.25 for each of the 505 units, on the margin we are paying more, since we pay \$5.25 not only for the additional 25 units, but also for each of the first 480 units. Recall that we only paid \$5 each when we bought only 480 units.

The first-order conditions tell us that Japan should buy LNG up to the point where the marginal revenue product equals marginal factor cost. Since marginal revenue product is \$11.42, we can see from table 9–4 that it crosses *MFC* somewhere between a price of $5.75 and $6.00. We can solve for the exact amount using the continuous functions $MFC = \partial(TCF)/\partial L$.

First, we need total factor cost, $TCF = P_L L$. To get P_L, invert the above supply curve as follows:

$$L = -20 + 100P_L \rightarrow P_L = 0.2 + 0.01L$$

Then $P_L L$ can be determined as follows:

$$P_L L = (0.2 + 0.01L)L = 0.2L + 0.01L^2$$

To get MFC for a change in LNG, take the derivative as follows:

$$MCF = \partial(P_L L)\big/\partial L = \partial(0.2L + 0.01L^2)/\partial L = 0.2 + 2 \times 0.01L$$

Note that when the supply curve is linear, the MFC has the same intercept but is twice as steep as the inverse supply curve. The units of measurement for MFC are dollars per thousand cubic feet. This must be set equal to the marginal revenue product of LNG, which is equal to the price of electricity times the marginal product of LNG, as follows:

$$\$11.418 = 0.2 + 0.02L$$

Solving for L yields the following:

$$L = 561 \text{ Mcf}$$

To buy 561 Mcf, the generator will need to pay a price of $P_L = 0.2 + 0.01 \times 561 = \$5.81$. This solution can be seen in figure 9–1.

Now let's sum up what we have learned about pricing power on the demand side of the market. As with all economic decisions, we need to look at the marginal benefits and costs of undertaking an activity. In this market, the benefits are the extra revenues brought in by another unit of the factor (MRP), and the marginal costs are the extra payments we need to hire an additional unit of the factor (MFC). With pricing power in demand, the costs of an extra unit are higher than the average cost, which is measured on the supply curve. To optimize, we buy the factor up to the point where the MRP is equal to the MFC.

In a more realistic example, a utility may be large enough to face a downward sloping demand for electricity (E) of $P_e(E)$ instead of a fixed price. In addition, if the utility has more than one plant, with some plants more efficient than others, the marginal product of LNG may fall with increased electricity production and LNG use ($E_L < 0$). Thus, the utility puts plants in merit order and uses the most efficient plants first. Then the equation $MRP_L = P_e E_L$ slopes down as in figure 9–2. Again find the purchases (L^*) that equate MRP_L and MFC_L. For this amount of purchase, the monopsonist has to pay the amount on the supply curve, P_L^*.

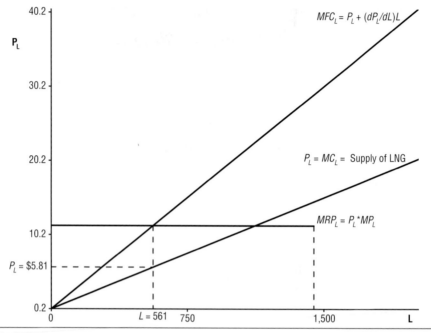

Fig. 9–1. Monopsony purchases of LNG for constant marginal product up to generating capacity

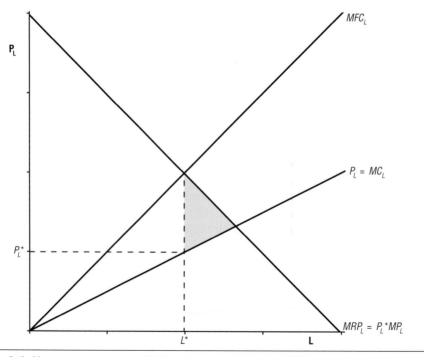

Fig. 9–2. Monopsony purchases of LNG with downward sloping MRP_L

Monopsony Model Compared to Competitive Model

Suppose that METI no longer bargains for the utilities, so utilities lose their monopsony power and must compete for natural gas. To see the value of monopsony power in this simple example, compare the above result to a competitive case. In the monopsony case, the optimization conditions are as follows

$$\partial \pi \big/ \partial L = P_e \left(\partial E \big/ \partial L \right) - P_L - \left(\partial P_L \big/ \partial L \right) \times L = 0$$

In the competitive case, the utilities do not have any pricing power, so $(\partial P_L / \partial L) = 0$ and the competitive solution is the following:

$$\partial \pi \big/ \partial L = P_e \left(\partial E \big/ \partial L \right) - P_L = 0$$

Note this is where marginal revenue product equals price. Thus, the demand for the factor is the marginal revenue product curve, and the new equilibrium is where the marginal revenue product curve crosses the supply curve.

Monopsony Model with Price Discrimination

In the above examples, we have assumed that each supplier receives the same price. However, contract prices are not usually public information. If each supplier does not know what other suppliers are receiving, it might be possible to pay each supplier a different price. In such a case, with monopsony power, Japan should pay each supplier its reservation price, which is the lowest price the supplier would be willing to accept, assumed here to be the supplier's marginal cost. At a price lower than the reservation price, suppliers should drop out of the market. We represent this situation in figure 9–3.

Japan pays the first supplier P_1 for L_1 units of output. It pays P_2 for $L_2 - L_1$ units, and P_3 for $L_3 - L_2$ units. If Japan can perfectly price discriminate and pay the second supplier a higher price than the first, and the third a higher price than the first two, the marginal factor cost is not above the supply curve. The marginal cost of a factor is just its price. In this case, Japan should buy L_3 units and pay each supplier its reservation price. The marginal price, or the highest price paid, would be the same as in a competitive market, but Japan would be able to garner the entire producer surplus.

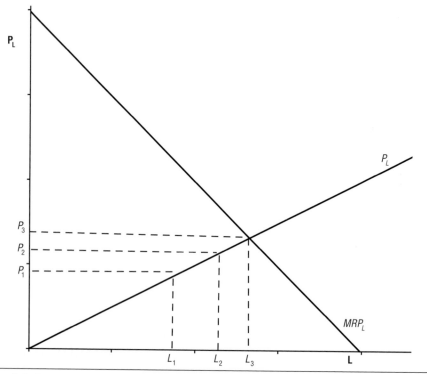

Fig. 9–3. Perfectly price-discriminating monopsonist

Monopoly and Bilateral Monopoly

If Japan has more competitors, it will not have as much market power, and the LNG suppliers should do better with higher prices and larger quantities. Another way for LNG suppliers to counteract monopsony power would be for them to band together and behave as a monopolist. Let's see what would happen in such a case.

Assume that Qatar, the largest seller of LNG, organizes a cartel called the Organization of LNG Exporting Countries (OLEC). Again, to keep the model simple, assume that they are selling to a competitive market of buyers. In a competitive buyer market with flexible price and quantity, the monopoly exporter or seller should sell up to the point where marginal revenue equals marginal cost. The demand for LNG in Japan is the marginal revenue product of LNG, while the marginal revenue for LNG exporters is below this demand if exporters have market power. The solution is L^* in figure 9–4.

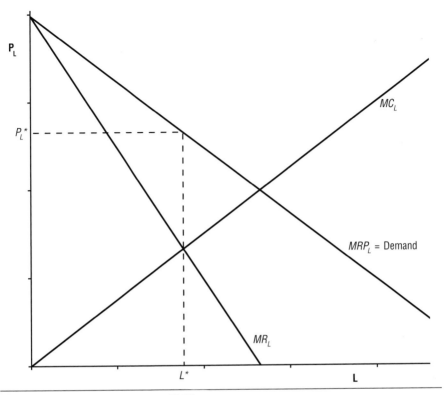

Fig. 9–4. OLEC as monopoly seller of LNG

Compare this to figure 9–2. The price for LNG is considerably higher when the supplier has the market power than when the buyer has it. Whether quantity is larger or smaller is uncertain and depends on the slope of the *MRP* and the *MC* curves.

Now put the two markets together into a bilateral monopoly, as in figure 9–5. To simplify the explanation and exposition, assume that both players want the same quantity and contract it to be L_c. Both players would desire the same quantity if the slope of the marginal revenue product were equal in absolute value to the slope of the marginal cost curve.

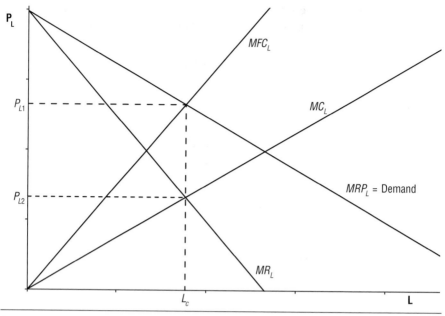

Fig. 9–5. Bilateral monopoly in the Asia-Pacific LNG market

The seller wants to receive P_{L1} and the buyer wants to pay P_{L2}. Where the price settles depends on the bargaining power of the two parties. We can, however, determine the range in which we might expect the price to fall. First, the absolute minimum price a monopoly seller would be willing to accept would be the point at which they received no producer surplus or are just covering all their costs. (Remember that an economist includes in *cost* a normal rate of return.) Thus, the absolute minimum price if volume is fixed at L_c would be somewhat below P_{L2} in figure 9–5. Mathematically, it would be as follows:

$$\text{Producer Surplus} = P_{min}L_c - \int_0^{L_c} MC(L)dL = 0$$

Solving yields the following:

$$P_{min} = \frac{\int_0^{L_c} MC(L)dL}{L_c}$$

At this point, losses on the last units just cancel the economic profits on the first units sold.

The absolute maximum price that the consumer would be willing to pay, if volume is fixed at L_c, is where consumer surplus is zero. This would be a price somewhat above P_{L1}. Mathematically, consumer surplus is the area under the demand curve minus the total amount paid, which is price times quantity. The maximum price they would pay is where this value is zero, as follows:

$$\text{Consumer Surplus} = \int_{0}^{L_c} MRP(L)dL - P_{max}L_c = 0$$

Solving yields the following:

$$P_{max} = \frac{\displaystyle\int_{0}^{L_c} MRP(L)dL}{L_c}$$

At this price, the consumer surplus gains on the first units just cancel out the consumer surplus losses on the last units bought. Price must fall below P_{max} and above P_{min}, but its exact location is determined by the party with the stronger bargaining power. These maximum and minimum prices are shown in figure 9–6.

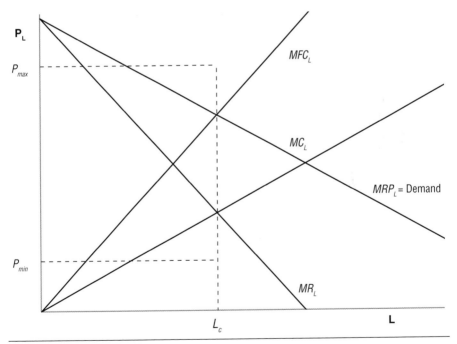

Fig. 9–6. Reservation prices in a bilateral monopoly

In reality, the situation is much more complicated because producers and consumers are making huge upfront capital investments facing an uncertain future. There is some but not complete market power on either side, and negotiation is on long-term contracts.

Bargaining and Negotiation

Bargaining and negotiation skills are important for success in your personal and professional life, since there is no shortage of disputes. However, it is necessary to have some mechanism for solving disputes. Methods to settle disputes include traditions, regulations, and courts. More decentralized methods of settling disputes include markets and negotiations. Honing your negotiating and bargaining skills improves the chances that disputes will be settled in your favor.

In *distributive games*, the negotiation is over how payments are distributed between the players. Negotiating over the purchase of an item is the classic distributive game. Your opponent's reservation price should be her opportunity cost. If you do not meet or exceed her reservation price, she will walk away. However, sometimes parties walk away strategically hoping you will make a better offer. In other cases, when you have not met their reservation price, they may walk away for good. Thus, if the players are too greedy, a bargain may not be struck.

In *distributive negotiations*, an unreasonable offer with concessions tends to work better than a reasonable offer with no concessions. It is also better to get your opponent to make the first offer. This tends to be good advice in most types of negotiation. Caution is advised in choosing your opening offer. If you are too conservative, you may give away too much. If you are too extreme, you may insult your opponent, and he may walk away. If your opponent makes an extreme first offer, either break off or quickly counter. Remember that the point midway between the bids often becomes a focal point. The pattern of concessions usually gets smaller as parties approach their aspirations or their reservation price. The smaller the zone of agreement in distributive negotiations, as in other games, the longer it takes to come to an agreement.

What are some of the normative ethics of negotiation? Disputants often fare poorly if they act greedily and deceptively. In many negotiations, there are joint gains. Those that often fare best are those who seek to enlarge the pie and then negotiate to get a fair share of gains. Others may use less-than-ethical tactics.

Summary

Natural gas, although an environmentally desirable fuel, has the drawback of high transportation costs relative to oil. Because of these higher costs, natural gas markets tend to be more regional, and large suppliers and consumers may have market power. Where gas is far from its customers or across deep oceans, it may be liquefied by chilling to $-161.5°C$, shipped to customers, and then regasified, as is the case in the Asia-Pacific LNG market.

LNG projects typically require cheap natural gas supplies and large upfront capital investment. For banks to supply such funds, they will require a variety of information about the company involved. Such information could include the company's mission and management, environmental philosophy and records, investment strategies, financial information and performance, and risk management strategies.

Traditional LNG suppliers that have been in the market for more than two decades in order of their entrance to the market include Algeria, the United States, Libya, Brunei, the United Arab Emirates, Indonesia, Malaysia, and Australia. In the last decade and a half, more recent entrants in order of their appearance are Qatar, Nigeria, Trinidad and Tobago, Oman, Egypt, Norway, Equatorial Guinea, Russia, Yemen, and Peru. The Asia-Pacific region is still the largest market for LNG, but its dominance has been decreasing with the continuing entrance of new buyers and sellers.

Given the large capital costs, projects are often owned by a consortium of companies, and contracts have traditionally been for 20 years or longer. However, with recent new entrants and a more liquid market, contracts have become more flexible, and about one-fourth of supplies were traded on the spot market in 2011. Prices have more often been indexed to crude oil or products, but indexing to pipeline gas at hubs is becoming more common than in the past.

Japan is the largest importer of LNG, and because it negotiates contracts through METI, it has some buyer market power. Prior to 1986, Japan was the only buyer from its suppliers. As a monopsonist, Japan faced a supply curve rather than a given world price. With such monopsony power, Japan could minimize its costs for LNG by buying where MFC is equal to MRP. Compared to a competitive solution, Japan would pay a lower price for LNG and buy less. If Japan is able to price discriminate and pay each supplier its reservation price, Japan could reduce its costs even more. Although Japan is no longer the only purchaser for LNG into the Asia-Pacific market, it has been the only purchaser of US LNG exports, all from Alaska. This situation will change as new export capacity comes online in the United States.

If sellers are able to cooperate with each other or are few in number, they will have some monopoly power on the selling side of the market. The result will be a bilateral monopoly, with a monopolist selling to a monopsonist. Although we may be able to determine the reservation price of each player, the negotiating skills and bargaining power of the two players in the market will determine the exact price.

10

Game Theory and the European Natural Gas Market

The problem faced by each participant is to lay his plans so as to take account of the actions of his opponents, each of whom, of course, is laying his own plans so as to take account of the first participant's actions.

—Dorfman, Samuelson, and Solow (1958)

Introduction

Since the end of World War II, Europe has seen shifting market shares for primary energy consumption (electricity generation from fossil fuels and nonfossil fuels, such as nuclear, hydropower, geothermal, wind, and solar). These patterns for consumption and production by major fuel source are shown in figures 10–1 and 10–2 for three major geopolitical groupings that include parts of Europe. These groupings are chosen because of their historical roots in European energy markets and the availability of such grouped data.

Since the focus of this chapter is on the European natural gas market, consumption has been converted to their commonly used units for the gas market, billions of cubic meters (1 bcm = 10^9 m^3). My conversions assume that 1 m^3 of natural gas equals 36,800 Btus.

Western Europe includes countries that have been capitalist and democratic in the post–World War II era, and most were part of the OECD in the 1960s. These countries are: Austria, Belgium, Denmark, Finland, France, West Germany, Greece, Iceland, Ireland, Italy, Luxembourg, Netherlands, Norway, Portugal, Spain, Sweden, Switzerland, Turkey, and the United Kingdom, with Cypress, the Faroe Islands, Gibraltar, and Malta also included in the Western European total. Since separate data is not available, the combined Germany has been included in West European data after the 1990 reunification.

The other two country groupings for Europe are former communist countries that were behind the Iron Curtain, a term popularized by Churchill in a speech in 1946. Eastern Europe includes the former communist countries of Albania, Bulgaria, Hungary, Poland, Romania, and Czechoslovakia, which now consists of the Czech Republic and Slovakia. It also includes the former Yugoslavia, which now consists of Bosnia and Herzegovina, Croatia, Kosovo, Macedonia, Montenegro, Serbia, and Slovenia. The former East Germany (GDR) is included in Eastern Europe prior to 1990. Together, Eastern and Western Europe will be referred to as *Europe*, as in US EIA (n.d.d).

The second grouping, the former Soviet Union (FSU), now consists of Armenia, Azerbaijan, Belarus, Estonia, Georgia, Kazakhstan, Kyrgyzstan, Latvia, Lithuania, Moldova, Russia, Tajikistan, Turkmenistan, Ukraine, and Uzbekistan. Since the FSU contains a European and an Asian portion, the US EIA database refers to this grouping of countries as *Eurasia*. For those countries that have broken apart, no data has been found prior to the division, so those years are not included in the separate country breakdowns in table 10–1, while the later data is presented separately but also aggregated for comparison across time. Although the FSU, prior to 1991, would have been referred to as the Union of Soviet Socialist Republics or the USSR or Soviet Union for short, for consistency's sake I will tend to refer to it as the FSU in the short form, even for the earlier period.

Between 1947 and 1956, energy shortages were the chief energy policy concern in Western Europe. In 1951, Belgium, France, Italy, Luxembourg, the Netherlands, and West Germany (FRG) signed the Treaty of Paris, which created the European Coal and Steel Community (ECSC) to cooperate on energy supply. In this first step toward economic integration, members agreed to promote free trade in coal and steel within the community (EU, n.d.c).

Coal and Oil Consumption

Coal was king in the major groupings in 1953. It had more than a 70% share of primary energy consumption in Western Europe. Coal had more than an 85% share in Eastern Europe, and even in the now gas-rich FSU, coal's share topped 60%. Although coal did not dominate in all the countries in this chapter, its share was more than 50% in 18 of the 32 countries with available data.

These 18 countries included the largest energy consumers and together accounted for about 90% of total energy consumption in Europe and the FSU. In the 14 countries where coal was not king, oil had more than 50% of the market in one-half of them. Oil's share topped coal on all islands in the sample (Cyprus, the Faroe Islands, Iceland, and Malta) and in Gibraltar, Greece, and Albania. Of these seven, only in Iceland is oil no longer the dominating fuel. Rather, Iceland is unique in getting more than three-fourths of its primary energy from hydro and geothermal sources.

Of the remaining seven countries, Romania was also unique. Natural gas was more than 60% of the market in 1953, with coal and oil sharing about equally in 40% of the market. Its gas fields, with first discoveries in what was then the Austro-Hungarian empire beginning in 1909, had been producing for decades. Although gas production and share have fallen, gas is still the dominant source, accounting for about 30% of Romania's primary energy consumption.

Hydropower dominated in Norway and Switzerland, with more than one-half of primary energy consumption in 1953, and it remains their dominant source. The four remaining countries had more diversified sources of energy in 1953. All had significant hydropower backing out enough other fuels so that no fuel had more than 50% of the market. Hydro dominated in Italy but was soon surpassed by oil, which remains the dominant energy source. Coal dominated in Finland but was also surpassed by oil, with hydro gaining ascendancy by 2011. Oil dominated in Sweden, but hydro surpassed oil in Sweden by the early 1980s. Oil dominated in Portugal and remains the dominant source today.

These exceptions pointed to a coming tide, for coal was about to be toppled from a throne it had held for many decades. Coal's share of consumption generally trended down decade by decade. By 2011, its share had fallen by more than 45 percentage points in the FSU and Eastern Europe and almost 60 percentage points in Western Europe. Total coal consumption also generally fell as well in Western Europe. In Eastern Europe and the FSU, where energy was generally cheaper, hard currency scarcer, and the environment less of an issue, coal consumption trended up until their economies collapsed with the transition from communism in the early 1990s. In Eastern Europe, coal consumption continued its downward trend after 1990, whereas in the FSU, it rebounded somewhat starting in the late 1990s. See table 10–1 for a little more regional detail with information by country (available at http://dahl.mines.edu/t1001country.pdf).

From 1957 to 1972, energy was no longer an issue for Western Europe. The coal shortages of the early 1950s turned into a coal surplus by 1959, as cheap Middle Eastern oil increasingly displaced coal. Large oil finds in the Middle East, North Africa, and the FSU both before and after World War II kept oil cheap and abundant until 1973. In all three regions, oil consumption generally increased from 1950 to 1979, with average annual growth rates of about 8.5% in Western Europe, 10.5% in Eastern Europe, and 7.5% in the FSU (see fig. 10–1).

Through the Council for Mutual Economic Assistance (Comecon, which consisted of Bulgaria, Czechoslovakia, Hungary, Poland, Romania, and the Soviet Union from 1949 to 1991) and other special trade agreements that also included Yugoslavia, the FSU traded oil and natural gas. The Druzhba (Friendship) oil pipeline began deliveries to Eastern Europe from the FSU in 1962. The FSU, which held 90% of Comecon energy resources, supplied most of Eastern Europe's oil imports through 1990. With Soviet export oil prices typically negotiated and fixed for five years and later based on a moving five-year average, Eastern Europe generally enjoyed prices below world levels until the early 1980s (Aleksandrova 2009; Curtis 1992).

Between 1973 and 1985, after the Arab oil embargo and Iranian Revolution, the focus turned to security of supplies and prices. There was a diversification away from OPEC oil. Whereas OPEC supplied more than 93% of Western European consumption in 1973, by 1983, that share had fallen to 63%, and total oil imports had fallen by 45%. The largest beneficiaries from this substitution were the North Sea, Mexico, and the Soviet Union (IEA, n.d.n).

Table 10–1. Population and energy consumption across time, region, and source (Eastern Europe, Western Europe, and the former Soviet Union)

Years	Pop 10^6	E/Pop m^3 10^3	E bcm	%Coal	%Oil	%Ngas	%ElecP
Western Europe							
1953	314.1	2.2	678.6	72.0	19.7	0.7	7.6
1963	349.4	2.1	744.6	48.2	40.1	2.1	9.5
1973	379.9	3.3	1,263.3	20.0	59.9	11.0	9.1
1983	401.9	3.4	1,372.6	19.8	48.7	15.1	16.4
1993	443.2	3.9	1,710.0	16.5	44.9	18.0	20.6
2003	467.6	4.2	1,960.3	13.6	42.3	23.6	20.4
2012	494.7	3.8	1,887.3	13.5	38.3	24.4	23.8
Eastern Europe							
1953	109.4	1.3	139.0	88.0	7.0	4.1	0.8
1963	118.9	2.1	251.0	77.8	14.0	6.7	1.5
1979	132.9	3.7	490.7	51.9	29.6	14.8	3.7
1983	135.7	3.7	499.3	53.1	24.6	17.3	5.0
1993	120.7	2.6	312.7	48.0	20.8	21.6	9.6
2003	119.5	2.6	307.2	39.2	25.4	23.5	11.9
2012	116.3	2.7	307.8	38.1	26.3	20.0	15.6
Former Soviet Union							
1953	189.6	1.2	226.8	62.8	31.4	3.4	2.3
1963	224.5	2.5	551.2	39.2	38.5	18.5	3.8
1973	250.0	3.7	936.2	28.8	41.9	25.2	4.0
1983	272.8	5.0	1,370.8	23.5	37.1	33.3	6.1
1993	291.2	4.4	1,292.3	18.5	21.2	50.7	9.6
2003	286.4	3.9	1,129.7	17.3	19.4	51.9	11.4
2012	288.9	3.8	1,097.4	21.2	22.9	55.9	0.0

Sources: Data between1980 and 2012: US EIA (n.d.d); energy data between 1950 and 1979 is extrapolated back using UN (n.d.a); and population data between 1950 and 1979 is extrapolated back using Maddison (2010). *Note:* m^3 = cubic meters converted from Btus at a rate of 36,800 Btu/m^3; Pop = population; E = energy consumption; Ngas = natural gas; and ElecP = primary electricity. EIA computes primary electricity as a gross heat rate including the waste heat loss with an efficiency rate of around 1/3. This computation is made even for renewable generation, such as hydro, where no heat is lost in the generation process. The exact gross heat content are available in US EIA (n.d.d).

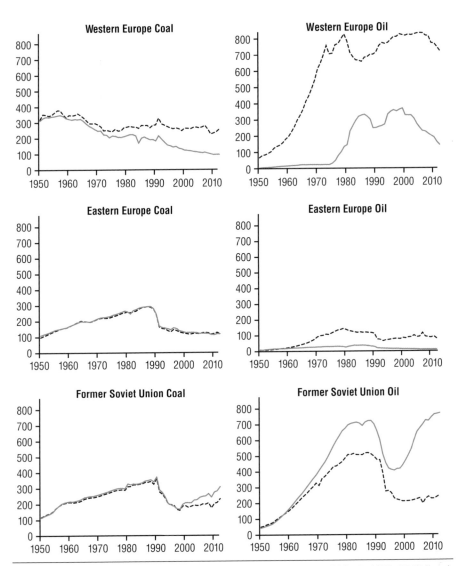

Fig. 10–1. Coal and oil consumption (dotted line) and production (solid line), 1950–2012 (bcm)
Source: Constructed from data in US EIA (n.d.d), UN (n.d.a), and Maddison (2010), as noted in sources for table 10–1.

When world oil prices fell by one-half from late 1985 to 1986, Western European oil consumption began a slow rebound to pass 1979 consumption by 1998. It leveled off and then trended down in 2007 to 2012 as the result of high prices, recession, and a push into biofuels. Patterns varied somewhat across countries, with very little rebound after 1985 in Denmark, Finland, France, Germany, Italy, Norway, Sweden, Switzerland, and the United Kingdom, but with a more pronounced rebound in the rest of Western Europe. Typically, those countries with policies to reduce transport fuel use or change fuel choice toward diesel and biofuels saw the least rebound. In Eastern Europe, Soviet oil export prices lagged behind world prices, and the downward trend after 1979 continued, with a rebound only beginning as their economies started to recover in 1993 and continuing to trend upward until 2008. Oil consumption generally trended down from 2008 to 2012.

In the FSU, oil consumption peaked in 1987 and then fell about 60% by 1999. Surely the precipitous fall in income during the transition contributed to this reduction. For the FSU as a whole, income as measured in international 2011 dollars converted with purchasing power parity (PPP) indexes, fell by 41% from 1990 to 1995 (IMF 2012). (PPP indices are international price indices created by dividing the cost of a typical basket of consumer goods in local currency by the cost of the same basket of goods in international dollars. (For more information on using PPP indices, see dahl.mines.edu/st10/st10.pdf, question 34.) Although the severity and timing of recovery varied from country to country within the FSU, most countries saw GDP declines of 25% or more, with recoveries beginning between 1994 and 1996. However, an interesting point is that even after income began to rebound, oil consumption continued to fall through 1999 and then rebounded only 19% through 2012, despite real income more than tripling. Thus, FSU oil consumption in 2010 was less than half of its pretransition peak in 1987. No fossil fuel had regained its pretransition high by 2012, with the largest drop in oil, followed by coal.

Country breakouts for the FSU are only available since 1992. Although a few of the smaller countries consumed more oil in 2010 than in 1992, the majority consumed far less, especially the largest consumers, including Russia, Kazakhstan, and Ukraine.

Coal and Oil Production

The rich and changing diversity in energy use across European countries has been influenced by a variety of factors, including market conditions, government policy, and energy endowments. Known endowments for coal and oil are indicated by production for the three regions shown above in figure 10–1, with more detail by region shown in table 10–2. For more detail and updates on energy use by country, see http://dahl.mines.edu/t1002country.pdf.

The column in table 10–2, with consumption divided by production (Cons/Prod), is indicative of whether a region is an exporter or an importer of total primary energy. For a self-sufficient country with no net imports, consumption equals production minus transportation and distribution losses, minus any stock drawdowns for stockable fuels. Such losses are above and beyond the fuel used for transporting and distributing energy and include oil spills, gas and heat loss from leaks, and frictional losses in electricity, as well as any theft. IEA (n.d.f) suggests that such losses are miniscule for coal, oil, and oil products, averaging less than 1% for

European natural gas and between 1% and 2% for FSU gas. They are more significant for electricity and average around 6% of generation for Europe and around 11% in the FSU. Thus, for a self-sufficient country with no exports, production divided by consumption for total energy use should be a few percentage points above 1. For a net exporter, it should be more than a few percentage points above 1, and for a net importer, it should be more than a few percentage points below 1.

Table 10–2. Primary energy production and relative share by energy source in Europe and Eurasia

Years	Prod/Cons	Total E	%Coal	%Oil	%Ngas	%ElecP
Western Europe						
1953	0.809	386.62	87.2	2.0	0.8	10.0
1963	0.580	432.14	74.6	4.6	3.3	17.5
1973	0.398	502.39	44.2	4.7	25.8	25.3
1983	0.599	821.97	26.5	24.3	21.2	28.0
1993	0.610	1,043.90	16.8	27.2	20.9	35.0
2003	0.595	1,167.10	9.9	30.2	25.1	34.8
2012	0.263	495.50	19.1	28.9	52.0	0.0
Eastern Europe						
1953	1.073	149.22	86.4	9.0	3.7	0.8
1963	0.934	234.49	82.0	9.3	7.0	1.7
1973	0.802	308.87	73.9	9.5	13.6	3.0
1983	0.770	384.53	71.6	7.3	14.8	6.3
1993	0.744	232.49	66.9	6.6	14.0	12.6
2003	0.652	200.20	63.1	7.1	10.7	19.1
2012	0.477	146.80	79.8	6.8	13.3	0.0
Former Soviet Union						
1953	0.947	214.88	65.0	28.8	3.7	2.5
1963	1.065	586.92	38.4	40.6	17.4	3.7
1973	1.118	1,046.56	26.6	47.3	22.4	3.7
1983	1.194	1,637.26	19.9	43.5	31.4	5.2
1993	1.212	1,565.99	15.7	29.0	47.2	8.1
2003	1.476	1,667.65	13.7	35.8	42.6	7.9
2012	1.822	2,000.10	15.4	38.5	39.1	7.0

Sources: Data between 1980 and 2012: US EIA n.d.d; energy data between 1950 and 1979 is extrapolated back using UN (n.d.a); population data between 1950 and 1979 is extrapolated back using Maddison (2010).
Note: Prod = production; Cons = consumption; Prod/Cons = production divided by consumption; and Total E = total primary energy production.

Figure 10–1 shows coal production followed similar patterns to coal consumption, and coal was the dominant fuel produced in all three regions in 1953. However, production trended down faster than consumption in Western Europe with increasing imports. Consumption and production closely tracked each other in Eastern Europe over the whole post-war period. With poor endowments of other fuels, coal still dominates production there. Coal production rebounded faster than consumption after the transition in the FSU, and it is now a net exporter. Coal production in the combined three regions was about 15% lower in 2012 than in 1953. The large dip in Western Europe combined with a small dip from Eastern Europe was not quite made up for by the significant increase in the FSU. Note that the coal production blip downward in 1984 in Western Europe resulted from the UK coal strike. Also recall the discontinuity in 1991, since East Germany is included in Western Europe after reunification in all figures.

Around one-half of the countries in the three regions had recorded oil production in 1953, with oil making up less than 10% of energy production for all but five countries: Austria, Albania, Hungary, Romania, and the FSU. Of these producers, the FSU accounted for about three-fourths of total oil production for the whole region, and Romania added another 14%.

Oil production held a very small share of Western European energy production until the North Sea finds (http://dahl.mines.edu/T1002country.pdf). Norway's Ekofisk was the first North Sea oil field to be discovered. Since 1975, Ekofisk oil has landed in Teesside, England. Statfjord, the largest oil field in the North Sea, with considerable gas reserves, was discovered in 1974, and is expected to produce through 2020. It is owned 15% by Britain, with the remainder owned by Norway. Oil is transported from the field by shuttle tankers. The bulk of Western European oil is produced by Norway and the United Kingdom from the North Sea, with small amounts from the Netherlands, Denmark, and Germany (Kemp 2011).

From figure 10–1, we can see the heavy Western European dependence on oil imports over the post-war period. The share of imports declined with the North Sea oil buildup from 1980–1986 but has generally trended up since then. Production also generally trended up, peaking in 2007. Since then, production and consumption have both fallen with consumption falling slightly more. This Western European production is still very much a North Sea story, with slightly more than one-half coming from Norway and another one-third coming from the United Kingdom.

Eastern Europe has been an importer over most of this period as well. Production has been dominated by Romania, an oil producer since 1857 (Museum of Romanian Oil Industry, n.d.). It produced 85% of Eastern European oil in 1953. Its production rose and peaked in 1976, while Eastern European oil production topped out a couple of years later. In 1983, Romania's share of Eastern European production had fallen to about 51%, with Albania and Hungary adding about another 30%. By 2010, oil production in Eastern Europe was still dominated by Romania (45%), with Hungary, Poland, and Albania adding another 30%. Eastern

European oil imports, although increasing since the transition, are not as large as they were through most of the 1970s and 1980s, when they were getting cheap Soviet oil (Curtis 1992).

The FSU, always the largest oil producer of the three regions, was largely an isolated market through the late 1960s. However, with double-digit annual production growth rates from 1953 to 1963, it soon had enough oil and infrastructure to begin significant exports through the Druzhba pipeline. The oil production was dominated by Russia, first in the Volga-Ural area until the 1970s but moving east, with western Siberia now producing about two-thirds of Russian oil. About 78% of total Russian exports are now to Eastern and Western Europe through the Druzhba, as well as some by pipeline to ports on the Black and Baltic Seas. The Eastern Siberia–Pacific Ocean (ESPO) pipeline began oil deliveries to the Russian Pacific port of Kozmino in 2009, with a spur to China beginning deliveries in 2011 (US EIA, n.d.a; Vatansever 2010).

In 1993, Russia produced about 88.4% of FSU oil, trailed by Kazakhstan (5.2%) and Azerbaijan (2.6%). However, with independence and Western investment, Kazakhstan and Azerbaijan leaped forward. By 2010, they accounted for 12.1% and 8% of FSU output, with the Russian share falling to about 77%. With falling consumption and rising production in these three oil-rich countries, the share of exports continued to rise as oil and infrastructure increased.

Oil leaves Kazakhstan by pipeline with small rail and barge shipments. The Caspian Pipeline Consortium (CPC) commissioned a pipeline in 2001 that connects to the Russian Black Sea port of Novorossiysk. A Soviet legacy pipeline connects to the Russian network at Samara, destined for the Russian Baltic port of Primorsk. Another pipeline to China began deliveries in 2009 (Paris 2006; US EIA, n.d.a).

Azerbaijan has also seen increasing export infrastructure with three oil export pipelines. The Baku–Supsa Pipeline commenced oil deliveries in 1999 through Georgia to the Black Sea; the Baku–Novorossiysk Pipeline commenced oil deliveries in 1997 through Russia to the Black Sea; the Batumi–Tbilisi–Ceyhan (BTC) pipeline commenced oil deliveries in 1999 through Georgia and Turkey to the Mediterranean (US EIA, n.d.a).

See Guoyi (2011) for an excellent geological description of oil and gas provinces worldwide and US EIA's *Country Analysis Briefs* (US EIA, n.d.a) for further information on oil markets in many oil-rich countries.

Natural Gas Markets

Europe first used gas made from coal (coal gas or town gas) in two basic chemical reactions:

$$C + H_2O + energy \rightarrow CO + H_2$$

$$C + 2H_2O + energy \rightarrow CO2 + 2H_2$$

Typical output was 12% hydrogen, 25% CO, 7% CO_2, and 56% nitrogen. However, this mix was potentially dangerous because of the high amount of toxic carbon monoxide (CO). The Lurgi process invented in the 1930s improved the process. It occurs under higher pressure, and two additional reactions rid us of the CO and create methane, which is less toxic:

$$C + 2H_2 \rightarrow CH_4 + energy$$

$$CO + 3H_2 + energy \rightarrow CH_4 + H_2O$$

The Lurgi process is used by Sasol in South Africa and Great Plains Synfuel Plant in North Dakota. Gas from Lurgi or other sources can be further changed to liquids (gas-to-liquids, GTL), based on the Fischer Tropsch (FT) process. FT may also be used for natural gas in remote areas where it is too expensive to transport in the gaseous form, or as LNG where liquid is more profitable than the gaseous fuel. FT is represented in following reaction with $n > 1$:

$$(2n + 1)H_2 + nCO \rightarrow C_nH_{(2n+2)} + nH_2O + energy$$

FT is used by Sasol and in the new Pearl GTL plant in Qatar.

Less economical town gas was typically replaced whenever natural gas became available. Equipment to burn town gas had to be modified to accommodate the higher heat content of natural gas. In 1953, natural gas use was less than 5% of energy consumption in all three major regions. Only four countries received more than 5% of their energy consumption from natural gas, all from local production: Austria (9.9%), Italy (8.7%), Hungary (6.4%), and Romania (62.3%). However, from this very small base, gas became the fastest growing fuel, averaging double-digit growth in all three regions from 1953 to 1963. This growth was heavily dependent on gas discoveries and infrastructure development. See figure 10–2 for natural gas production and consumption by major region, with more detail in tables 10–1 and 10–2.

In 1957, Western European economic integration and gas flow were further enhanced when the Treaty of Rome joined the countries in the ECSC into the European Economic Community (EEC), also called the common market. The EEC's goals were to abolish all barriers to the flow of goods, services, people, and capital between members and to establish common customs barriers to all nonmembers. The EEC then added Great Britain, Ireland, and Denmark (1971); Greece (1981); Spain and Portugal (1986); and Turkey (1996).

The most astonishing change to the natural gas market was in the FSU. Some gas was produced prior to World War II and used locally in the Baku and Caspian region, but no pipelines existed to move gas to markets, and surplus gas was typically flared. Before and during World War II, some significant gas finds were made in the Northern Caucasus and Volga region, and a 540-mile gas pipeline to Moscow had been completed by 1947 (Krylov et al. 1998).

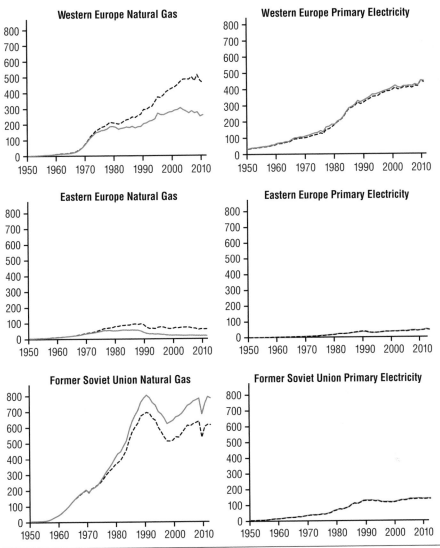

Fig. 10–2. Energy consumption (dotted line) and production (solid line) for natural gas and primary electricity in Eastern Europe, Western Europe, and the former Soviet Union

Sources: Constructed from data from US EIA (n.d.d); UN (n.d.a); and Maddison (2010), as detailed in table 10–1 sources.

By 1953, the FSU produced almost one-half of the natural gas in the three regions, and that dominance has only increased over time. By 1963, more than 10,000 miles of gas transmission lines had been completed, and domestic gas consumption had grown at an estimated average annual rate of almost 26% for the previous decade. Natural gas had increased from less than 4% to more than 17% of FSU primary energy consumption in a decade (Owen 1975; UN, n.d.a).

Transmission lines continued to increase in the FSU, more than quadrupling from 1960 to 1970. The Central Asian pipeline in 1967 allowed gas production areas in Turkmenistan, Uzbekistan, and Kazakhstan to connect to Russian industrial centers (Krylov et al. 1998). By 1973, natural gas had risen to provide around 22% of primary energy consumption in the FSU.

Despite rapid growth, gas still constituted a small share in 1963 in Eastern Europe (7%) and Western Europe (3.3%). However, this was about to change. When a huge natural gas field was discovered in Groningen, the Netherlands, in 1961, this gas was fed into existing networks, displacing the more expensive town gas. Rapid development of the gas grid in Europe followed, supplying Dutch gas to West Germany (1963), Belgium (1966), France (1970), and Italy (1971). These early international contracts set the generally accepted precedent for future international sales, with long-term contracts indexed to oil or oil products. (For more on contract details, see Melling [2010].)

The first contracted LNG shipments reached the Canvey terminal in the United Kingdom from Algeria in 1964, and imports to France commenced in 1970. Spain began importing LNG from Libya in 1971 (Högselius et al. 2010). As a result, natural gas consumption grew on average more than 20% per year in Western Europe from 1963 to 1973, with gas share climbing to 11%.

Although minor natural gas exports from the FSU had been made to Poland since 1944, significant exports did not commence until the late 1960s. FSU began exporting gas to Comecon countries through the Bratstvo (Brotherhood) natural gas pipeline, which began deliveries from the Shebelinka field in Ukraine to Uzhgorod and Czechoslovakia in 1967, with an extension to Austria in 1968 (Högselius et al. 2010). Average natural gas consumption remained in the double-digit range (10.2%) in Eastern Europe from 1963 to 1973, and its share had also climbed to more than 12%.

The 1973 Arab oil embargo, along with the Iranian Revolution of 1979 and the accompanying dramatic oil price increases, considerably altered consumption patterns. The Brotherhood pipeline allowed connections to numerous Eastern and Western European countries. Gas exports began in 1973 to East and West Germany, in 1974 to Italy and Bulgaria, in 1975 to Hungary, and in 1976 to France. The Soyuz pipeline was linked to the large Orenburg gas field in the Volga-Ural region and began deliveries to the border crossing at Uzhgorod, Ukraine, in 1978, with connections to deliver exports to the former Yugoslavia (Högselius et al. 2010).

Sales to Eastern Europe were typically at lower prices than those to the west. Again, natural gas was the fastest growing fossil fuel in Eastern Europe, and its share rose to more than 17% by 1983. The natural gas pipeline network continued to increase in the FSU, with mileage more than doubling between 1970 and 1983. Substituting natural gas for oil domestically allowed more profitable oil exports to garner much-needed hard currency, and FSU gas share jumped to more than 30% by 1983.

In 1977, Norway began to export pipeline gas from Ekofisk through the Stat/Norpipe system to the continent and from Frigg to St. Fergus, Scotland. Norwegian gas from Statfjord came through Emden, Germany, while British gas was delivered to Scotland by pipeline through the Brent field. Spain began importing Algerian LNG in 1974. Italy began importing LNG from Libya in 1975. Deliveries of Algerian gas through the 667-mile Trans-Mediterranean pipeline to Italy began in 1983 (Högselius et al. 2010; Stern 1984). As a result, natural gas's share increased to 15% in Western Europe by 1983.

Oil and gas discoveries in the FSU continued to move east with gas finds in the Timano-Pechoro Basin and huge finds in western Siberia in the late 1960s and 1970s (AAPG 1975). The most controversial pipeline to move these supplies west was to have originated in the Yamal Peninsula in Siberia to Uzhgorod. When contract volumes were less than expected, it was built from Urengoy, which was closer and had easier access. US opposition to this project took the form of trade sanctions in effect from December 1981 to November 1982. Since the Soviets proved much better at laying pipe than building compressor stations, the US embargo also included parts for those compressor stations, particularly rotor shafts and gas turbine blades. John Brown of Edinburgh, AEG Kanis of West Germany, and Nuovo Pignone of Italy normally imported these parts from General Electric (based in the United States) (Curtis 1992; Davis 1984; Zickel 1991).

Initially, five countries made contracts to buy Soviet gas: Austria, France, West Germany, Italy, and Switzerland. Belgium, Spain, and the Netherlands had also considered buying Soviet gas but decided against it. No pipeline business went to Belgium or the Netherlands, which may have influenced their decisions to cancel their contracts. This agreement for Soviet gas involved building a 2,760-mile pipeline from the Urengoy gas field, just north of the Arctic Circle, to the Czechoslovakian border through Ukraine, tying into the Brotherhood pipeline system. Gazprom now refers to this pipeline (Urengoy–Pomary–Uzhgorod) as the "Brotherhood (Bratstvo)" pipeline (Gazprom Export 2013).

Part of the Brotherhood pipeline was an extension of the huge Northern Lights pipeline system built from the 1960s through the 1980s to bring Siberian gas into Moscow and Leningrad, and eventually on to Belarus. It allowed tie-ins to the Baltic states and deliveries to Finland beginning in 1974. This pipeline, a 1.42-meter (56-inch) high-pressure pipeline, formed the fourth of seven pipelines from the Urengoy gas field, which was discovered in 1966 and came onstream in 1978. The line was completed ahead of schedule in 1983, and deliveries began in 1984 to

Austria, West Germany, and France. This engineering feat crosses permafrost, mountains, dense forests, and some 700 rivers. It is estimated to have cost $15 billion in the currency of the day, with 40 compressor stations at intervals of 100 to 120 km. (Krylov et al. 1998; Zickel 1991).

Agreements to sell Norwegian gas from Statfjord, Gullfaks, and Heimdal were signed in 1981, with deliveries of 15 billion m^3 a year in 1985 to Belgium, West Germany, France, and the Netherlands. By 1993, natural gas's share in Western Europe had risen to 18% (Stern 1984, Stern 1995).

Pipeline expansion continued in the FSU, with transmission mileage increasing by almost 50% from 1983 to 1992 (Owen 1975; Korchemkin 1993). Six 56-inch pipelines from Yamburg, the second-largest gas field in western Siberia, to European Russia were completed by 1993. By then more than 90% of Russian gas consumption came from western Siberia (Itoh 2010). The Central Asian Pipeline System was expanded. Natural gas jumped to first place, with 51% of primary energy consumption.

Meanwhile, natural gas growth in Eastern Europe slowed, and with the transition, it plunged by more than 20% from 1990 to 1992, leaving gas consumption considerably lower in 1993 than in 1983. Nevertheless, natural gas's share jumped to 21% because the other fossil fuels showed even larger drops. Consumption then cycled from 1993, ending up more than 10% lower by 2010. Similarly, the FSU saw a 25% decline in natural gas consumption from a peak in 1991 to a trough in 1999. It had recovered around 60% of the decline by 2009 but consumption again fell in 2010 as Russia went into a recession, along with many of its richer energy clients but recovered in the following year.

The 1990s began an exciting time in Europe, leading to closer political and economic union. With the fall of the Berlin Wall in 1989, Germany reunified. The European Bank for Reconstruction and Development (EBRD) was founded in 1990 for the financial support of the transition economies of Central and Eastern Europe. The European Union was created in 1992 with the 12 EEC members; membership rose to 27 by 2013. It committed to a European Monetary Union (EMU) that commenced with the Euro in 1999 with 11 members, rising to 17 by 2013. In 1993, the Single Market Act reduced nontariff barriers to trade within the EU and created a single EU market. Closer political union within the EU is based on adoption of a common foreign and security policy and cooperation on justice and home affairs (EU, n.d.c).

Western European energy policy (after the oil price decreases in 1985) focused on three areas: environment, competition, and Eastern Bloc reforms. Environmental concerns favored gas, and natural gas consumption continued its high growth as infrastructure came online. Natural gas was the fastest growing energy source from 1993 to 2003. Imports from Norway jumped in 1996 as natural gas from the huge Troll field (discovered in 1979) and related fields began deliveries through pipelines to Germany, Belgium, and subsequently France, with trans-shipments to the Netherlands as well. The Troll contract, signed in the late 1980s, brought a doubling of Norwegian exports. The contract was unique because volumes were contracted without requirements that the gas come from specific fields (as in dedicated reserve contract clauses). Although much of the gas would come from Troll, other fields could be used to fulfill the contract. The contract required three new pipelines to the continent: Zeepipe to Belgium, which began making deliveries in 1993; Europipe, which began making deliveries to Germany in 1995; and the Frapipe pipeline, which began making deliveries to France in 1996, also allowing deliveries to Spain and Italy (Stern 1995, 2005).

The Maghreb-Europe submarine pipeline from Algeria to Spain began transporting gas to Spain in 1996 and to Portugal in 1997. The Interconnector from Bacton, England, to Zeebrugge, Belgium, completed in 1998, allows flows of gas in either direction, and it increases flexibility and arbitrage possibilities in the European natural gas market. Zeebrugge has developed as a gas hub for trading, as well as an LNG terminal and Zeepipe terminal for Norwegian gas.

The Yamal–Europe pipeline will eventually connect the Yamal Peninsula with the Western European gas grid through Belarus and Poland. Initial portions began deliveries to Germany in 1997. The Russians were keen to finish this pipeline to bypass Ukraine, which had previously accounted for more than 90% of Russian exports to the west. Ukraine feels entitled to the preferential prices they received while part of the Soviet Union. Gazprom, the Russian gas monopoly, does not concur. Disputes over lack of payment, gas prices, transit payments, and gas theft have erupted sporadically since 1992. Such disputes have resulted in Russian curtailment of gas to Ukrainians and the subsequent Ukrainian diversion of transit gas bound for Europe. The dispute in 2009 left both the Russians and their European customers with a desire to diversify transport links. Price and payment disputes also erupted in 2007 with Belarus.

These desires for diversification have led to two new pipelines from Russia. Blue Stream, which includes a submarine pipeline under the Black Sea from Russia to Turkey, began deliveries in 2006. Nord Stream includes two submarine pipelines under the Baltic Sea from Russia to Germany. The first of two pipes began deliveries in 2011, and the second began deliveries in 2012. A third pipeline, South Stream, would connect Russia with Bulgaria, with the potential to also make deliveries to Greece, Italy, and Austria. This third pipeline project was canceled in late 2014. The accumulated export capacity for all the Russian pipeline connections westward are shown in table 10–3.

Trouble erupted again in late 2013 when the pro-Russian president of Ukraine signed a cooperation deal with Russia in exchange for Russian loans and lower gas prices. These loans allowed Ukraine to pay their natural gas debt for 2013. However, riots and unrest erupted and the president fled to Russia in February of 2013. Fighting by pro-Russians in the Crimea and a dubious referendum led to Russia annexing the Crimea in March amid much ongoing protest from the West. Gazprom raised gas prices and threatened a gas cut off for unpaid gas bills in April.

By April, the violence by pro-Russian speakers had spread to other eastern areas of the country. The violence continued into June, with peace talks being brokered in Europe with Russia threatening and then cutting off of natural gas supplies (BBC Europe 2014; CSIS 2013/14; *OGJ Newsletter* 2014). Gas supplies recommenced in December, and Ukraine paid its Russian gas bill and prepaid for additional natural gas in early 2015. However, sporadic fighting continues in 2015.

Table 10–3. Outlet capacity of export pipelines at the FSU border (bcm/year)

Pipeline	Capacity (bcm)	Destination of Exports
Via Ukraine:		
Orenburg–Western border (Uzhgorod)	26	Slovakia, Czech, Austria, Germany, France, Switzerland, Slovenia, and Italy
Urengoy–Uzhgorod	28	Slovakia, Czech, Austria, Germany, France, Switzerland, Slovenia, and Italy
Yamburg–Western border (Uzhgorod)	26	Slovakia, Czech, Austria, Germany, France, Switzerland, Slovenia, and Italy
Dolina–Uzhgorod; 2 lines	17	Slovakia, Czech, Austria, Germany, France, Switzerland, Slovenia, and Italy
Komarno–Drozdowichi; 2 lines	5	Poland
Uzhgorod–Beregovo; 2 lines	13	Hungary, Serbia, and Bosnia
Hust–Satu Mare	2	Romania
Ananyev–Tiraspol–Izmail and Shebelinka–Izmail; 3 lines	27	Romania, Bulgaria, Greece, Turkey, and Macedonia
Total via Ukraine:	**143**	
Via Belarus:		
Yamal–Europe (Torzhok–Kondratzki–Frankfurt/Oder)	31	Poland, Germany, Netherlands, Belgium, and United Kingdom
Kobrin–Brest	5	Poland
Total via Belarus:	35	
St. Petersburg–Finland; 2 lines	7	Finland
Blue Stream (design capacity)	16	Turkey (possibly to Greece, Macedonia)
Total Existing Export Capacity:	**201**	
New Pipelines:		
Nord Stream	55	Germany, France, Czech, and other
South Stream	63	Bulgaria, Serbia, Greece, Italy, and other
Subtotal new capacity:	**118**	

Source: EEGA (2013).
Note: bcm = billion cubic meters.

Russian gas potential for export remains impressive. It is estimated to have more than one-third of the world's explored reserves, which may exceed the energy equivalent of Saudi oil reserves. It has 9 of the world's 15 largest gas fields and more than an 80-year supply of gas at its 2011 production rate (BP 2012). Three-quarters of these reserves are in western Siberia. Urengoy, the largest gas field in the world, had initial reserves of 7.76 trillion cubic meters (10^{12} m^3). Other super fields in the area are Yamburg, 4.75×10^{12} m^3; Bovanenkovskoye, 4.15×10^{12} m^3; Zapolyarnoye, 2.67×10^{12} m^3; Medvezhye, 1.55×10^{12} m^3; and Kharasavei, 1.27×10^{12} m^3.

Europe's early concern with Soviet contracts was their degree of dependence on Soviet energy sources. However, the Soviets, and subsequently the Russians, have proven to be fairly reliable trade partners, although weather and transit disputes with Ukraine have briefly interrupted supplies. Dependence on Russian gas supplies, as well as the degree of concentration of energy from other suppliers, is reported in table 10–4, which shows recent imports to Europe and FSU countries by origin.

Some of these reserves are now slated for a new market as Russia and China signed a 30-year contract for the largest gas deal ever in May of 2014. The deal, 10 years in the making, is for 38 bcm per year at $350/thousand cubic meters expected to begin deliveries from East Siberia through a new pipeline in 2018. The gas is indexed to oil with take or pay clauses in the contract (OGJ Editors 2014).

Although Russia dominates gas production and exports in the FSU, accounting for more than three-fourths of both in 2013, it is not the only gas producer and exporter. According to BP (2014), Turkmenistan comes in second. With large gas discoveries in the late 1950s, the Turkmen gas industry took off. The Central Asia–Center (CAC) gas pipeline system was built from 1960 to 1988 to bring Turkmen gas to Uzbekistan, Kazakhstan, and on to Russia.

Since gaining independence, two additional export links have been constructed to bypass the Russian gas system: the Central Asia–China (CAC) gas pipeline, which began deliveries in 2009, and the Dauletabad–Salyp Yar pipeline, which began deliveries to Iran in 2010. The former pipeline also allows exports from Kazakhstan and Uzbekistan. In 2013, Turkmenistan accounted for about 8% of FSU natural gas production and 13.5% of FSU exports to other FSU countries. If we include its exports to China (24.4 bcm) and Iran (4.7 bcm), that share edges up 14.3% (BP 2014). Turkmenistan also promises more to come. Recent huge natural gas discoveries have moved Turkmenistan to rank fourth in natural gas reserves, and a new pipeline to Pakistan and India through Afghanistan may see the light of day when political conditions permit.

Table 10–4. Natural gas imports into Europe and Eurasia (LNG and pipeline), 2013 (bcm)

To	Neth†	Norw*	UK!	Ot Eur**	Kazak†	Russia!	Turkmn†	Ot FSU	Iran†	Qat#	Alg*	Nig#	Ot MENA*	S&C Amr#	Tot Im
Austria	–	1.2	–	0.5	–	5.1	–	–	–	0.0	–	0.0	0.0	0.0	6.8
Belgium	5.4	9.5	2.5	–	–	12.3	–	–	–	3.2	–	0.0	0.0	0.0	32.8
Czech R	–	3.8	–	–	–	7.2	–	–	–	0.0	–	0.0	0.0	0.0	11.0
France	6.5	15.8	–	0.6	–	8.1	–	–	–	1.8	5.3	1.2	0.1	0.0	39.2
Germany	22.4	33.5	–	–	–	39.8	–	–	–	0.0	–	0.0	0.0	0.0	95.8
Hungary	–	–	–	–	–	5.9	–	–	–	0.0	–	0.0	0.0	0.0	5.9
Italy	8.6	1.1	–	0.6	–	24.9	–	–	–	5.2	11.5	0.0	5.2	0.0	57.1
Neth.	–	4.8	1.6	13.0	–	2.1	–	–	–	0.0	–	0.0	0.0	0.0	21.5
Poland	–	–	–	1.8	–	9.6	–	–	–	0.0	–	0.0	0.0	0.0	11.4
Slovakia	–	–	–	–	–	5.3	–	–	–	0.0	–	0.0	0.0	0.0	5.3
Spain	–	3.8	–	1.6	–	–	–	–	–	3.5	14.6	3.1	0.2	3.5	30.3
Turkey	–	0.2	–	0.1	–	26.2	–	3.3	8.7	0.4	3.8	1.3	0.3	0.0	44.2
UK	9.5	29.2	–	3.3	–	–	–	–	–	8.6	0.4	0.0	0.1	0.1	51.2
Ot Eur	0.8	1.8	4.9	7.6	–	15.9	–	–	–	0.8	2.7	1.3	0.1	0.1	35.8
Tot Eur	**53.2**	**104.7**	**8.9**	**29.1**	**–**	**162.4**	**–**	**3.3**	**8.7**	**23.4**	**38.3**	**6.9**	**6.0**	**3.6**	**448.5**
Belarus	–	–	–	–	–	18.1	–	–	–	0.0	–	0.0	0.0	0.0	18.1
Russia	–	–	–	–	11.5	–	9.9	6.4	–	0.0	–	0.0	0.0	0.0	27.8
Ukraine	–	–	–	1.8	–	25.1	–	–	–	0.0	–	0.0	0.0	0.0	26.9
Ot FSU	–	–	–	–	0.2	5.6	1.1	3.8	0.7	0.0	–	0.0	0.0	0.0	11.4
Tot FSU	–	–	–	1.8	11.7	48.9	11.0	10.1	0.7	0.0	–	0.0	0.0	0.0	84.2
Tot Ex → **Eur & FSU**	**53.2**	**104.7**	**8.9**	**30.9**	**11.7**	**211.3**	**11.0**	**13.4**	**9.4**	**23.4**	**38.3**	**6.9**	**6.0**	**3.6**	**532.7**

Source: Compiled from information in BP (2014).

Note: † indicates all shipments in the column are by pipeline. * indicates shipments in column include both pipeline and lng. # indicates that all shipments in column are by LNG. ** indicates shipments in column include pipeline, LNG and transhipments. Neth. = Netherlands, Norw = Norway, UK = United Kingdom, Ot = other, Eur = Europe, Kazak = Kazakhstan, Turkmn, FSU = former Soviet Union, Qat = Qatar, Alg = Algeria, Nig = Nigeria, MENA = Middle East North Africa, S&C Amr = South and Central America, Tot Im = total imports, Tot Ex = total exports.

Although Uzbekistan currently has significantly fewer known reserves than Turkmenistan, in 2013 it was the FSU country ranked third in production (7.1%) (BP 2014) and in exports to Europe and FSU countries (4.4%) (BP 2012). Although natural gas was first discovered in Uzbekistan in 1953, it tended to be a net importer from other Soviet republics until the mid-1980s. It has been tied into the CAC pipeline system since the 1970s, allowing imports from Turkmenistan and exports to Kazakhstan, and it began exports to China in 2012. Other producers and exporters of lesser note are Kazakhstan and Azerbaijan. However, the stature in the gas market of all these non-Russian countries could change if infrastructure allows more access to outside markets.

Meanwhile, policy changes in the European Union have changed the market structure within Eastern and Western Europe. The push for increased competition in energy markets with the Single Market Act had major ramifications in the energy sector. Downstream oil was already reasonably competitive in many countries, and the European Commission pushed for liberalization in places where it did not exist, such as Spain, Portugal, and Greece. For coal, there was continued pressure to enforce the subsidy reduction policies. Gas and electricity, which were highly concentrated, came under even more pressure. The price transparency directive for gas and electricity required pricing and consumption information be supplied to the EEC Statistical Office on a regular basis. Third-party transit rights were implemented for electricity in 1991 and for natural gas in 1992.

The European Energy Charter Treaty (EECT), signed in 1994, was aimed at supporting energy reform in the former Eastern Bloc countries and assuring safe energy supplies from those areas. The treaty has the goal of encouraging free and nondiscriminatory trade and investment. Other provisions encourage transit at nondiscriminatory and reasonable rates. The investment criteria include provisions for outside investment under the same conditions as prevail for nationals. Each signatory is obligated to advance competition and is encouraged to promote technology transfer and environmental goals. The environmental provisions, although nonbinding, request that signatories prevent energy-related pollution that affects other states through a polluter-pays principle. Although the EECT provisions that govern competition law are not as encompassing as those of the EU, they do promote lower barriers to competition, the free movement of goods within internal markets, and price transparency within the gas and electricity markets (Wälde and Lubbers 1996). More recently, the Electricity Directives, discussed in chapter 6, and the Natural Gas Directives have sought to reduce monopoly power by opening up the EU markets.

The desire for more competition, lower carbon emissions, and natural gas import security led to diversifying supplies by also importing more LNG. From 2001 to 2011, LNG's share of total imports to Europe increased from about 11% to about 20%. Qatar provides the largest amount of LNG with regular exports having commenced to Spain in 1997, Belgium in 2006, and France, Italy, and the United Kingdom in 2011. Algeria was the earliest supplier of LNG and added new customers, such as Slovenia in 1992, Turkey in 1994, Portugal in 1997, and Greece in 2000. However, it fell to second in exports for Europe by 2011, having been passed by Qatar in 2010. Nigeria followed in third place. It commenced exports to Italy and Spain in 1999, Portugal and Turkey in 2000, and France in 2002. Together these three suppliers accounted for more than 80% of LNG imports into Europe. With high prices from 2011 to 2013, gas consumption and imports fell and the pendulum shifted back, leaving pipeline's share back at 2001 levels. The largest reductions in imports came from Qatar, where LNG increased and was diverted to the high-priced Asia-Pacific market (BP 2014).

Russia is also beginning diversification into LNG exports, allowing it access to a much broader array of markets. The first exports from Sakhalin commenced in 2009, with the largest shipments going to Japan. The first exports from Yamal are expected in 2016 (Shiryaevskaya 2013).

Primary Electricity

When measuring total energy consumption, we need to avoid double counting. For example, we would be double counting if we counted crude oil consumption and oil products consumption, or if we counted fossil fuels for electricity generation and the electricity generated by the fossil fuels. In energy balances, typically the fossil fuels are counted as primary energy sources and oil products, and electricity generated from them as secondary energy. Primary electricity is then all electricity generated from sources other than fossil fuels (i.e., nuclear, hydro, and other renewables). In the 1950s, primary electricity was less than 10% of primary energy consumption in our three major regions and consisted almost exclusively of hydropower, except for a small amount of geothermal energy in Italy. However, in a few Western European countries, its share was more than one-fourth of primary energy consumption. It was close to one-third in Finland, Italy, and Sweden and was nearly two-thirds (60%) for Norway and Switzerland. Primary electricity has not always been the fastest growing energy source; it has vied for that honor with natural gas. However, for decades its share has generally trended upward in all regions.

Nuclear power was introduced in the late 1950s and 1960s in many countries, and early Western European adopters included the United Kingdom (1957) and France (1958). France is still by far the largest Western European producer, with almost one-half of Western European generation. Other adopters of nuclear power included West Germany (1961), Italy (1963), which phased out nuclear power in 1987, and Sweden (1964), which reversed a phaseout policy in 2009. The other two regions followed, with production commencing in the FSU and East Germany in 1966 and in the former Czechoslovakia in 1972. The former Czechoslovakia is still the largest Eastern European producer, with 43% of generation. Since reunification, Germany has pledged to phase out nuclear power by 2022.

The buildup of nuclear power was most pronounced in Western Europe in the 1970s and early 1980s, with a decade lag for the other two regions. Nuclear passed hydro in Western Europe in 1985 and in Eastern Europe in 1987, but it did not catch up in the FSU until 2003, where nuclear and hydro have been neck-and-neck ever since. During most of the 1980s and 1990s, primary electricity's share of Western European energy consumption was slightly higher than that of natural gas in Western Europe. Natural gas regained the lead in 1999, and by 2011, natural gas's share was 25% to primary electricity's 23%. Although it gained on gas in Eastern Europe, primary electricity never caught natural gas, but by 2011, its share was about 14% to natural gas's 19%. The FSU's primary electricity was only a percentage point smaller than natural gas in 1953, but that spread widened over time. By 2011, FSU natural gas had about one-half of the total market, whereas primary electricity was only around 11%.

The other interesting development in primary electricity is other nonhydro renewables, which include geothermal, solar, tide, and wave (Geo&STW) and bioenergy and waste (BioE&W). They had not gained much prominence in the FSU as of 2011, when they had still less than 2% of primary electricity consumption, which in turn is only about 11% of primary energy consumption. They were more significant in Eastern Europe, where they constituted about 13.5% of primary electricity, which in turn is about 16% of primary energy. Most of this other renewable capacity in Eastern Europe has developed since 2003. Around 60% is generated from BioE&W. About one-half of this generation is in Poland, with much of the remainder split between the Czech Republic and Hungary. Another 28% of generation is wind. This capacity is a little more diversely distributed, but Poland is still the largest wind generator, with about 45% of Eastern European capacity. Bulgaria and Hungary jointly contribute another 40%. There is a small amount of solar generation, mostly in the Czech Republic.

Western Europeans are, of course, the champions when it comes to other renewable generation. With the European Union's target of 20% of their total primary energy consumption from renewables by 2020, policies to promote renewables had yielded 22% of their primary electricity from other renewables by 2011, almost 30% from hydro and the remainder from nuclear. The 2011 breakdown of electricity generation by source, including primary electricity as well as thermal generation in all three regions, is shown in figure 10–3.

This relatively higher share for other renewables in Western Europe is relatively recent. In 1980, about 2% of primary electricity was from other renewables, of which about three-fourths came from BioE&W, with most of the rest from geothermal.

Wind generation began in Western Europe in the 1980s, and it is now approaching 50% of other renewable generation, as shown in figure 10–3. Germany is now the largest wind generator, with Spain a close second. Together they generate about 49% of the wind power in Western Europe. France, Italy, and the United Kingdom have passed Denmark, and each generate between 5.5% and 8.5% of the Western European total.

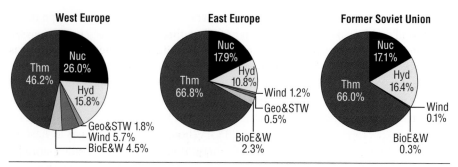

Fig. 10–3. Regional non-hydro electricity generation by source, 2011
Source: US EIA (n.d.d).
Note: Nuc = nuclear; Hyd = hydro; Geo&STW = geothermal, solar, tide, and wave; BioE&W = biofuels and waste; and Thm = thermal.

Tidal power developed in France in 1966 is still producing. It is thought to have accounted for most of the generation in the solar, wave, and tide category through about 1990. At that time, solar power started slowly growing, with its share of other renewables hovering around 1% until it took off in 2005. Solar power had jumped to 12% of other renewables by 2011. About 40% this new solar capacity is in Germany, 23% is in Italy and another 20% is in Spain. This category, however, remains a very small share of total electricity generation.

Western Europe's geothermal share has trended down since the 1980s. Italy has consistently dominated Western European production, but it has lost share, mostly to Iceland. By 2011, Italy's share had fallen to 50%, and Iceland's share had risen to almost 42%.

European Market Structure

The above statistical survey, largely from the data set compiled from data sources referenced in table 10–1 notes and from BP (2014), shows that the European and Eurasian energy mix has undergone a dramatic transition, particularly since 1973. This period has demonstrated a flexibility and speed of adjustment unexpected at the time of the oil embargo in 1973. Diversification has been out of oil and coal and into natural gas and primary electricity. The biggest adjustment out of oil came at the expense of heavy fuel oil, and refineries adapted to produce more of the lighter fuels.

With the transition from communism and European policy changes, there has been a significant movement to diversify from the old Soviet-era gas network. This is evidenced by dramatic changes in ownership and in market structure. Nevertheless, energy production and distribution in Europe and the FSU still tends to be characterized by very large firms.

Through the early 1980s, gas distribution and transmission in Western Europe was usually under government control or ownership, while production was a mix of public and private ownership. In Germany and Switzerland, the gas industry was mostly privately owned by large oil, gas, coal, and utility companies, which were responsible for the production, transmission, importation, and distribution of gas. Private companies produced Dutch gas, but they sold the gas to Gasunie, with the government owning one-half.

The Belgian transmission company, Distrigas, was one-third government owned. Wholly government-owned companies such as Österreichische Mineralölverwaltung (OMV) (Austria), Neste (Finland), Gaz de France (France), Snam (Italy), Bord Gáis Eireann (Ireland), Statoil (Norway), Instituto Nacional de Hidrocarburos (INH) (Spain), and British Gas (United Kingdom) were all heavily involved in their domestic gas industries. Some newer markets for gas (Greece, Denmark, and Turkey) also have serious state involvement in the industry. Under communism, the FSU and Eastern Europe had total state control over the gas industry.

Thus, these markets have historically been dominated by a few large players. These huge exporters negotiated with large transmission and distribution companies in the importing countries, usually with natural monopoly status. The three major exporters into the continental European market had state companies that heavily or exclusively directed the export of natural gas: Norway (Statoil), Russia (Gazprom), and Algeria (Sonatrach).

Thus, no European economy had an open market for gas. International gas contracts required government approval and were very complex because of the large infrastructure involved. A typical contract was 20 years in length and often had provisions for contract revision, with prices typically tied to some grade of fuel oil.

Germany had been a hub with gas connections from the Netherlands, Norway, and the Soviet Union through Belarus and Poland. It still operates a sophisticated grid through which gas from several sources can be distributed by means of gas blending stations. Austria was another hub where Soviet gas could be routed northwest or southwest. Its gas field at Baumgarten and increasing gas storage (see table 10–5) still provide flexibility and supply security against minor disruption.

However, times have been changing. Many of the state companies in both Eastern and Western Europe have been privatized or are being privatized, as shown in table 10–6. This is less the case further east.

The FSU gas-producing countries, including Russia, still have heavy state ownership in the industry. With slightly more than one-half ownership of Gazprom, the world's largest gas company, the Russian government owns a controlling share and, under Putin, certainly seems in control.

The UK gas market is the most restructured and functions more like the North American market, with a wholesale market based on the National Balance Point

Hub. Northern Europe, which is connected to the United Kingdom, is more open than the more isolated southern European regions. FSU countries outside the EU do not seem to be aspiring very strongly to more competitive gas markets.

EU gas market directives have been influential in opening up EU markets, and there are now more entrants from the private sector. Directives in the early 1990s called for more transparency in gas pricing and more interconnection, with nondiscriminatory access to transmission connection across borders. EU Gas Directive 1998/30/EC (EU 1998) was more forceful and resembles model 3 from Hunt and Shuttleworth (1996), discussed in chapter 6. The directive addressed several key points:

- Governments were required not to discriminate in authorizing companies in member countries to build and operate gas facilities.
- Grid system transmission, storage, LNG, and distribution facilities were required to be nondiscriminatory.
- Numerical targets and deadlines were established for opening up the transmission grid to large customers that should include electricity generators using natural gas.
- Accounts for transmission, distribution, and storage were required to be separated from other natural gas activities.

Table 10–5. Gas storage capacity in the European Union, 2012

Country	Capacity Working Gas (10^6 m³)	Capacity as % Annual Consumption	Daily Maximum Capacity Withdrawal (10^6 m³)	Injection (10^6 m³)
Austria	7,451	29.1%	85	69
Belgium	700	4.4%	15	8
Bulgaria	450	15.4%	3	3
Czech Republic	3,432	40.8%	55	39
Denmark	1,025	24.5%	18	8
France	12,700	31.5%	337	160
Germany	20,455	28.2%	491	258
Hungary	6,130	60.3%	79	45
Italy	16,487	23.1%	284	136
Netherlands	5,258	13.8%	216	59
Poland	2,052	13.4%	39	22
Portugal	171	3.3%	7	2
Romania	2,684	19.4%	2	2
Slovakia	2,905	46.5%	38	30
Spain	4,620	14.4%	179	9
United Kingdom	4,319	5.4%	113	53
Ukraine	32,965	61.4%	301	197

Source: Gas Infrastructure Europe (2013).
Note: m³ = cubic meters. Working gas capacity = total gas storage minus base gas. *Base gas* is a permanent inventory to maintain enough pressure to make withdrawal possible.

Table 10–6. Major gas companies in Europe

Country		Major Gas Companies	Internet Homepage	Business	Major Stock Holders
EU	Austria	Österreichische Mineralölverwaltung (ÖMV)	www.omv.com/	P,T,D	State (31.5%), Abu Dhabi International Petroleum Investment (24.9%)
EU	Belgium	Distrigaz	www.distrigas.be/	T	ENI* (99%)
EU	Denmark	Danish Oil & Natural Gas (Dong)	www.dong.dk/	P, T	State (76%), SEAS-NVE (11%), Syd Energy (7%)
EU	Finland	Gasum	www.gasum.fi/frindex_eng.htm	T, D	Fortum** (31%), Gazprom (25%), Ruhrgas (20%), State (24%)
EU	France	Gaz de France SUEZ	www.gdfsuez.com/	P,T,D	State (36%), GBL (5.2%), and Employees (2.9%)
EU	Germany	Ruhrgas	www.ruhrgas.de/	P, T	BEB (25%), Schubert KG (15%), Bergemann (58.76%)
		Thyssengas	www.thyssengas.de/	T	Shell(25%), PWR Power (75%)
		Wingaz	www/wingas.de	T	Wintershall Ag. (50%), Gazprom (50%)
		BEB	www.beb.de	P,T,D	Esso (50%), Shell (50%)
EU	Greece	DEPA	www.depa.gr/	T,D	State (85%)
EU	Ireland	Bord Gais Eireann	www.bge.ie/	T,D	State (100%)
EU	Italy	Snam	www.snam.it/	P,T,D	Cassa Depositi e Prestiti # (30%) ENI (8.54%)
EU	Netherlands	Gasunie	www.gasunie.nl/	T,D	State (100%)
		Nederlandse Aardolie Maatschappij (NAM)	www.nam.nl/	P	Shell (50%), Exxon Mobil (50%)
	Norway	Statoil	www.statoil.com/	P,T	State (67%)
		Norsk Hydro	www.hydro.com/	P,T	State (34.26%)
EU	Portugal	Petroleos e Gas de Portugal (GALP)	www.galp.pt	P,T,D	Amorin (33.34%), ENI* (33.34%), Parapublica (7.00%)
EU	Spain	Enagas	www.enagas.es/	T	Gas Naturall (5%), SEPI (%%), and others (90%)
	Switzerland	Swissgas	www.swissgas.ch	T	Erdgas(26%), Gaznat(26%)Gasverbund(26%), Swissgas(16%)
	Turkey	Botas	www.botas.gov.tr	T	State (100%)
EU	UK	British Gas Group	www.bgplc.com/	P,T,D	Black Rock (7.06%)

Country		Major Gas Companies	Internet Homepage	Business	Major Stock Holders
EU		Centrica	www.centrica.co.uk/	P,T,D	None >5%
		National Grid	www.transco.uk.com/	T	
EU	Poland	PGNiG S.A.	www.pgnig.pl	P,T,D	State (72.43%)
		EuRoPol GAZ s.a. Joint venture	www.europolgaz.com.pl/	T	PGNiG (48%), Gazprom (48%), Gas-Trading (4%)
EU	Czech Rep.	RWE transgas	www.rwe.cz/en/index/	P,T,D	100% RWE Group
		Vemex	www.vemex.cz/en/	T,D	Gazprom Germania (50.14%), Centrex Europe Energy & Gas(33%), KKCG Oil & Gas (16.86%)
EU	Hungary	MOL Hungarian Oil and Gas Company	www.mol.hu/en/	P, T	State (24.6%)
		Panrusgaz	www.panrusgaz.hu/	D	Gazprom export, EON Ruhrgas International, Centrex Hungaria
EU	Romania	Petrom		P	OMV (100%)
EU	Romania	WIEE JV with Gazprom	www.wiee.ch/	T,D	50% Wintershall, 50% Gazprom
EU	Bulgaria	Bulgargaz	www.bulgargaz.bg/en/	T,D	State (100%)
		Overgas	oldsite.overgas.bg/	D	DDI Holdings (50%), Gazprom Export (49.51%), Gazprom (0.49%)
	Turkey	Türkiye Petrolleri Anonim Ortaklığı (TPAO)	www.tpao.gov.tr/tp2/	P, T	State (100%)
		Botas	www.botas.gov.tr/	T	State (100%)
	Azerbaijan	SOCAR	new.socar.az/socar/en	P,T	State (100%)
	Kazakhstan	KazMunaiGas	www.kmgep.kz/eng/	P,T	State (100%)
	Russia	Gazprom	www.gazprom.com	P,T,D	State (50.002%)
	Turkmenistan	Türkmengaz	www.oilgas.gov.tm/	P,T	State (100%)
	Uzbekistan	Uzbekneftegaz	www.ung.uz/	P,T,D	State (100%)
	Ukraine	Naftogaz Ukrayiny	www.naftogaz.com/	P,T,D	State (100%)

Note: P = production; T = transmission; D = distribution; and JV = joint venture. *ENI is 30% owned by the Italian government. **Fortum Oyj is 75.5% owned by the Finnish government. EU in the first column indicates the country is a member of the European Union.

Dissatisfied with the pace of opening up, EU Directive 2003/55/EC set a goal of full retail competition, like model 4 from Hunt and Shuttleworth (1996), with a much faster time table than the 1998 directive. It mandated full access to the grid for large customers in 2004 and also mandated similar access for small customers in 2007. It reiterated the call for open, nondiscriminatory access to transmission and distribution systems for member companies and required their accounts to be separate from the rest of the gas supply chain. It also required member states to have regulators for the gas grid and allowed for tariff regulation for both transmission and distribution.

Still dissatisfied with the degree of competition and open access, they passed EU Regulation (EC) No. 715/2009 and EU Directive 2009/73/EC, which strove for a fully functional integrated single EU gas market by 2014, with independent system operators. Owners of the transmission and distribution grids could be operators provided they were not also involved in other aspects of the business, such as gas supply. It was this provision that caused the EU to prevent Gazprom from buying part of a gas trading platform in Austria in 2011.

To provide for more regional cross-border transmission cooperation, the European Network of Transmission Operators for Gas (ENTSOG) was developed. It is comprised of 40 transmissions system operator (TSO) members. Their challenge is to develop the rules for the integrated market, including issues of capacity allocation, balancing, congestion management, and tariff guidelines (ENTSOG, n.d.). The desire for more coordination between regulators in member states led to the development of the Council of European Energy Regulators (CEER, n.d.). In addition, the European Agency for the Cooperation of Energy Regulators was created as an EU regulatory agency for natural gas (ACER, n.d.).

These evolving EU regulations and institutions, along with national regulations, have increasingly opened up the European gas markets. So far, the UK market is the most liberalized, resembling that of North America. Gas hubs for trading and transit have developed, beginning with the United Kingdom's National Balance Point (NBP) in 1996, followed by Zeebrugge in 2000. Heather (2012) lists 12 major hubs; the most active are NBP and the Title Transfer Facility (TTF) in the Netherlands. Organized exchanges for regulated, anonymous trading of natural gas, other energy products, and natural gas futures and options have further helped to open the market. (See chapters 18 and 19 for more on futures and options contracts.) Natural gas futures and options can be traded at NBP and TTF on the ICE (ICE 2013a). Two continental exchanges, APX-Endex and European Energy Exchange (EEX), also allow for gas spot trading.

This liberalization of the European market has provided opportunities for Gazprom. Although it is almost a monopoly in Russia, it has striven to move heavily into the European downstream. Some of their downstream ownership is detailed in table 10–6; a complete list of subsidiaries is available from Gazprom (2013a).

European contracting is also changing. Long-term contracts indexed to oil still exist and are in the majority, but it is estimated that about 42% of European gas was acquired on the spot market in 2011 (Heather 2012). Lower European spot prices arose from a weak European economy. Further, some LNG destined for the United States was backed out by shale gas and found its way into the higher priced European market. This spot gas was cheaper than the contracted gas indexed to oil price, causing consumers to challenge the higher prices. This led Gazprom to issue some rebates in 2012. Another sign of possible times to come was a long-term contract in 2012 for Norwegian gas that was linked not to oil but to European hub gas prices (Statoil 2012) and more recently an Azeri natural gas contract was pegged to European natural gas prices (Patel 2014). See the interactive IEA (n.d.i) map for gas trade flows.

Cournot Duopoly

Although the EU gas market is more open than it once was and is striving to be competitive where possible, it is still characterized by a few non-EU countries exporting to it. To analyze such a market, we turn to game theory models, beginning with duopoly theory, in which two sellers face competitive buyers. We will consider three models in this framework.

1. *Cournot model.* Each firm chooses the quantity to maximize profits based on the other firm's output.
2. *Bertrand model.* The two firms set price instead of quantity.
3. *Stackelberg model.* One firm is a leader in the market and sets quantity, knowing how the other firm will react.

We begin with a Cournot model with two players. Suppose that Norway and Russia are the only two gas exporters to Germany. Assume German inverse import demand of

$$P = 660 - 2(q_N + q_R)$$

where P, q_N, and q_R are price and outputs for Norway (N) and Russia (R).

The costs for Norway and Russia are

$$C_N = q_N^2$$
$$C_R = 2q_R^2$$

Profit functions for the two countries, given the other country's output, are price times quantity minus cost, as follows:

$$\pi_N = (660 - 2(q_N + q_R))q_N - q_N^2$$
$$\pi_R = (660 - 2(q_N + q_R))q_R - q_R^2$$

First-order conditions for profit maximization for Norway are as follows:

$$\partial \pi_N \big/ \partial q_N = 660 - 2(q_N + q_R) - 2q_N - 2q_N = 0$$

This equation can be rearranged to the following:

$$\partial \pi_N \big/ \partial q_N = 660 - 2q_R - 6q_N = 0$$

Solving for q_N gives us Norway's reaction function. This function shows the profit maximizing quantity for Norway, given Russia's output:

$$q_N = 660 \big/ 6 - \left[2 \big/ 6 \right] q_R = 110 - \left(1 \big/ 3 \right) q_R$$

First-order conditions for profit maximization for Russia are as follows:

$$\partial \pi_R \big/ \partial q_R = (660 - 2(q_N + q_R)) - 2q_R - 4q_R = 0$$

This equation can be rearranged to a Russian reaction function as follows:

$$\partial \pi_R \big/ \partial q_R = 660 - 2q_N - 8q_R = 0 \longrightarrow q_R = 82.5 - 0.25q_N$$

These two reaction functions are shown in figure 10–4, labeled N for Norway and R for Russia.

For this model to be in equilibrium, each firm has to be on its reaction function. For example, suppose Norway is producing at q_{1N} and Russia at q_{1R}. Notice that Russia is on its reaction function but Norway is not. Norway was expecting Russian production to be q_{2R}. When Norway finds Russian production to be q_{1R}, it would increase its production to q_{2N}. The market is now closer to the equilibrium, where both are on their reaction function, but Russia is no longer on its reaction function and would want to adjust its production downward. Such adjustments should continue until the market reaches equilibrium.

The equilibrium is where both firms are on their reaction function or at the solution to the following system of equations:

$$q_N = 110 - (1/3)q_R$$
$$q_R = 82.5 - 0.25q_N$$

One way of solving this equation is by substituting the first equation into the second:

$$q_R = 82.5 - 0.25(110 - (1/3)q_R) = 55 + (1/12)q_R \rightarrow q_R = 60$$

We find q_N by substituting q_R back into the q_N reaction function:

$$q_N = 110 - (1/3)60 = 90$$

Thus Norway, the lower cost producer, gets a larger share of the market. We find the price in the market by substituting q_N and q_R back into the demand equation, as follows:

$$P = 660 - 2(90 + 60) = 360$$

We can find profits for the players by substituting their respective prices and quantities back into their profit functions:

$$\pi_N = Pq_N - C_N = 360 \times 90 - 90^2 = 24{,}300$$
$$\pi_R = Pq_R - C_R = 360 \times 60 - 2 \times 60^2 = 14{,}400$$

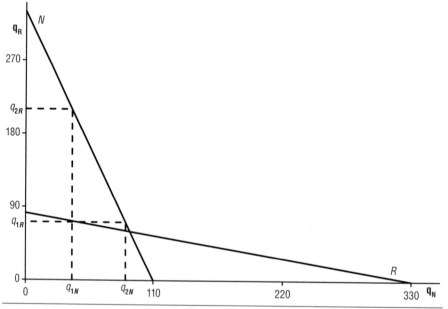

Fig. 10–4. Reaction functions for a duopoly

We know that we need marginal cost to be less than price to continue production, so we check whether or not the firms should shut down.

$$MC_N = \partial C \big/ \partial q_N = \partial q_N^2 \big/ \partial q_N = 2 \times q_N = 180$$

$$MC_R = \partial C_R \big/ \partial q_R = \partial (q_R^2) \big/ \partial q_2 \quad 4 \times 60 = 240$$

MC is smaller than price in both cases, so the firms should produce, and the low-cost producer makes more profits than the higher cost producer.

If this model is extended to more suppliers, then each player would have a reaction function that depends on the production of the other players. For a simple numerical example with three players (*triopoly*), see http://dahl.mines.edu/st10/st10.pdf, question 23.

For a more complicated example, the Norwegian Central Bureau of Statistics modeled the European gas market as a dynamic triopoly that optimized across time with Norway, Russia, and Algeria as the major players in order to work out optimal gas development trajectories (Bjerkholt et al. 1990; Hoel et al. 1990).

Before going on to two other game theory models, let's compare these results to earlier models in the next two sections.

Duopoly Compared to Competitive Market

In the previous example, the firms are taking advantage of the fact that they have some pricing power and are facing a downward sloping demand instead of taking price as given. This will be compared to the case in which each player behaves in a competitive manner and operates where price is equal to marginal cost.

In a competitive market, the supply curve equals the horizontal sum of the marginal cost curves. For the above example, the cost and marginal cost curves are the following:

$$C_N = q_N^2, \; C_R = 2q_R^2, \; MC_N = 2q_N, \; MC_R = 4q_R$$

Recall from chapter 7 how to get marginal cost and supply for the whole market. Set $P = MC$ for both suppliers and invert, as follows:

$$P = MC_N = 2q_N \longrightarrow q_N \frac{P}{2} \text{ and } P = MC_R = 4q_R \longrightarrow q_N = \frac{P}{4}$$

Then horizontally sum both markets to get supply or marginal cost as in the following:

$$Q = q_N + q_R = \frac{P}{2} + \frac{P}{4} = \frac{3}{4} P$$

Or solve for market marginal cost as follows:

$$P = \frac{4}{3} Q = MC$$

The country marginal cost curves are represented in figure 10–5, panels (a) and (b), and their inversion and horizontal sum gives the market marginal cost curve or the competitive market supply curve in panel (c).

Solve for equilibrium by setting the above inverse demand equal to market marginal cost as follows:

$$P = 660 - 2(q_N + q_R) = 660 - 2Q = MC = \frac{4}{3}Q \longrightarrow Q = 198$$

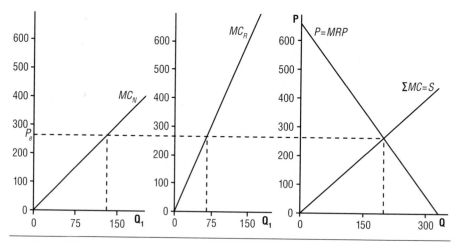

Fig. 10–5. Competitive market with two suppliers

Solving for market price results in the following:

$$P = 660 - 2 \times 198 = 264$$

Notice the competitive market has a lower price and higher quantity. Thus, duopolists can increase their profits by restricting production.

Marginal cost must be equal to price for both suppliers, allowing us to solve for each q:

$$MC_N = P = 2q_N = 264 \rightarrow q_N = 132$$

$$MC_R = 4q_R = 264 \rightarrow q_R = 66$$

Norway has economic profits as follows:

$$\pi_N = 264 \times 132 - (132^2) = 17{,}424$$

Russia has economic profits as follows:

$$\pi_R = 264 \times 66 - 2(66^2) = 8{,}712$$

Both countries make lower profits than in the duopoly case. These profits, called rents, are the result of having some resources that are lower cost than the marginal or last cubic meter sold. Norway makes more rents in this case because it has lower costs.

Monopoly Compared to Competitive and Duopoly Market

A third case for these two gas producers would be to have the two producers get together and monopolize the market. In that case, the producers should operate where marginal revenue equals marginal cost, as in figure 10–6.

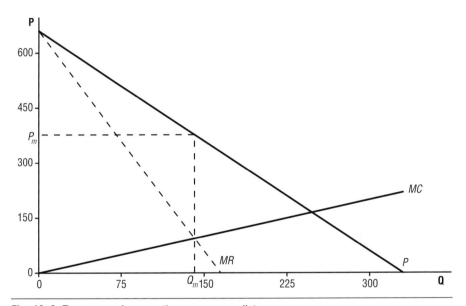

Fig. 10–6. Two gas producers acting as a monopolist

Setting marginal revenue equal to marginal cost and solving yields the following:

$$MR = 660 - 4(q_N + q_R) = MC = \frac{4}{3}(q_N + q_R)$$

$$q_N + q_R = 123.75 = Q_m$$

Now they produce less than what they would have produced together in a competitive market. To maximize combined profits, they should behave as a multiplant monopoly and each should produce where MR equals MC. MR for the market is $MR = 660 - 4(123.75) = 165$.

Norway should produce the following:

$$MC_N = 165 = \frac{q_N}{2} \longrightarrow q_N = 82.5$$

Russia should produce the following:

$$MC_R = 165 = 4q_R \rightarrow q_R = 41.25$$

Price in this market and profits are substantially higher than in either the competitive or duopoly case, as follows:

$$P_m = 660 - 2(123.75) = 412.5$$

Profits in the two markets are also considerably higher:

$$\pi_N = 412.5 \times 82.5 - (82.5^2) = 27,225$$
$$\pi_R = 412.5 \times 41.25 - 2(41.25^2) = 13,612$$

Other Game Theory Models: Bertrand and Stackelberg

In the competitive model and the monopoly under profit maximization, the results are predictable. However, when we get to small number bargaining situations, outcome depends on the strategy of the opponents. We have considered the Cournot duopoly in which firms pick quantity given the quantity of an opponent. Let's consider two other strategies. In the Bertrand duopoly, it is assumed that one firm sets price assuming the other firm holds price constant. To see what might happen in this case, suppose that Norway sets the price at P_N. If $P_N > P_R$, then Norway has no market, and if $P_N < P_R$, then Norway has the whole market. Each has an incentive to lower price to get the whole market until the competitive solution is reached.

In the Stackelberg model, one of the players is more powerful than another, perhaps because of asymmetric information or lower costs. In this model, the dominant player knows the other player's reaction function and uses that knowledge to maximize its profits. The dominant player will make more profit than in a Cournot model, but the nondominant player will make less. For a numerical example, see http://dahl.mines.edu/st10/st10.pdf, question 21. However, if both players try to dominate, there will be no market equilibrium.

Limit Pricing Model

A last model suggests the effect of potential entrants and is called the limit pricing model. Demand in figure 10–7 is for an energy service that can be satisfied by a fossil fuel or some more expensive unlimited backstop fuel. The backstop is a source yet to be developed on any large scale, such as photovoltaic, microhydro, or hydrogen fuel cells. Let's assume that the low-cost fossil fuel supplier is a monopolist in the fossil fuel market. The higher cost backstop (with average costs as shown by AC_b in figure 10–7) is a potential competitor. Without the backstop, the monopolist should operate where marginal revenue (MR_m) equals marginal cost (MC_m) charging P_m and producing Q_m.

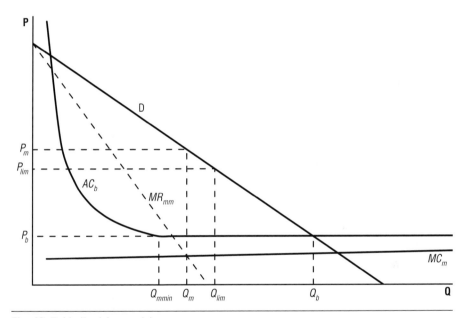

Fig. 10–7. Limit pricing model

However, at this price the backstop would have an incentive to enter. An alternative would be to charge a price that is equal to the minimum of the average cost curve for the backstop or slightly below this backstop price (P_b) to discourage entry.

However, the monopolist may be able to do better than price at the long run minimum of the backstop. To see how, consider the following argument. Note that Q_{min} is the smallest output that gives the backstop maximum economies of scale. Let the monopolist pick a quantity Q_{lim} such that if the backstop entered at Q_{min}, then $Q_{min} + Q_{lim}$ would drive price below the long run minimum average cost. This threat would stop the backstop from entering the market. The limit quantity would be $Q_{lim} = Q_b - Q_{min}$, and the price charged would be P_{lim}.

Thus, a monopolist threatened with a lower cost backstop may still be able to charge a higher-than-competitive price. The threat of increased production lowering price should discourage the discerning entrant. However, the monopolist needs to be careful that the price charged is not so high as to encourage entrants that may subsequently be protected under antitrust or other legislation.

Summary

Energy use has changed substantially in Western Europe in the decades since World War II. Coal was king in 1950 with postwar shortages, but cheap Middle Eastern and North African oil, and then Dutch gas and North Sea oil, turned coal shortages into a coal glut. With the oil price run-ups in the 1970s and the political uncertainties of secure oil supplies, Europe diversified into nuclear and natural gas. Long-distance pipelines from Norway, Algeria, and the former Soviet Union have continued to expand and bring in gas from further and further away to a hungry European market. Since the 1990s, the economic and monetary unions are knitting Western Europe ever closer together. In addition, the transition to communism has broken Eastern Europe loose from the former Soviet grip, weaving it ever closer to Western Europe as well. Meanwhile, FSU countries are moving further apart and trying to diversify away from Soviet-era trade ties. Sellers want to break the monopsony hold of Gazprom and the Russian pipeline network; Russia wants to break the monopoly power of Ukrainian and Belarusian transit pipelines.

With strong policy initiatives, other renewables for electricity generation have been growing in Western Europe and now account for more than 20% of primary electricity generation. Wind generation is close to one-half of total renewable generation, with Germany and Spain now in the lead. Solar energy had accounted for only about 1%, but its contribution increased between 2005 and 2011, reaching 12%. The largest producers are Germany, Italy, and Spain. Eastern Europe is following suit, but other renewables for primary electricity generation have not yet made as large inroads, and they are dominated by biocombustible fuels.

Large companies have dominated the Western European gas industry, often with monopoly control over parts of the industry in their respective countries, particularly for transmission and distribution. In such an oligopoly market, large companies and governments negotiate for long-term international contracts. Game theory is a useful technique for studying strategic interaction in these types of small-number bargaining situations. We have considered four different game theory models for insight into the operation of such markets.

In the Cournot noncooperative game model, each player maximizes profits, taking the other player's output as given. Players are competing with quantities, and the optimization problem gives each player a reaction function. Market equilibrium is the solution to the system of reaction functions. Cournot players will make more profits than competitive players but will make fewer profits jointly than if they collude and act as a monopolist. As we add Cournot players to the market, we will approach a competitive price and quantity.

In the Bertrand noncooperative game, market participants compete on price. An important application of this model is auctions. Interestingly, the Bertrand solution is the same as the competitive solution. When firms compete based on price, they bid away monopoly profits.

In the Stackelberg game, one of the players is more powerful than another. The dominant player knows the other player's reaction function and uses that knowledge to maximize its profits. The dominant player will make more profits than in a Cournot model, but the nondominant player will make fewer. However, if both players try to dominate, there will be no market equilibrium.

Earlier chapters have considered simple competitive markets that can maximize welfare when there are no externalities, market participants are fully informed but do not have market power, and there are no natural monopolies. With natural monopolies in some segments of the market, some segments may be regulated, as in gas transmission and distribution, and others may be required to compete, as in marketing and trading natural gas. North America has already gone down this path, and Europe is following. Whereas a Cournot model may have better represented the European market of decades ago, more players, more competition, market hubs, and government directives are transitioning to more market-based pricing. This includes a growing spot market, as well as traditional long-term contracts. Price indexing to a market hub gas price is a fait accompli in North America and the United Kingdom and is starting to happen on the European continent as well.

11

Externalities and Energy Pollution

An external cost exists when . . . an activity by one agent causes a loss of welfare to another agent and the loss of welfare is uncompensated.

—David W. Pearce and R. Kerry Turner

Introduction

The extraction, conversion, and consumption of energy may pollute the air and water, create solid wastes that are toxic and nontoxic, emit unusable heat into the air and water, degrade the land, and contribute to global climate change.

Burning fossil fuels creates a number of emissions, including carbon dioxide (CO_2), nitrogen oxides (NO_x), sulfur dioxide (SO_2), carbon monoxide (CO), and particulate matter (PM). For example, when we burn methane (CH_4), the most common component of natural gas, which is the cleanest burning fossil fuel, we have the following reaction:

$$CH_4 + 2O_2 \rightarrow CO_2 + 2H_2O + energy$$

Thus, we add oxygen and methane and get carbon dioxide, water (H_2O), and energy. None of the products of this combustion is toxic. However, the CO_2 buildup is thought to cause global climate change. It causes the Earth's atmosphere to absorb more heat and reflect less, much the same as a greenhouse traps heat.

A molecule with more carbon is heavier and emits more CO_2 when burned. For example, when the heavier natural gas molecule ethane (C_2H_6) is burned, it emits twice as much CO_2 as methane. This chemical reaction would be as follows:

$$C_2H_4 + 3O_2 \rightarrow 2CO_2 + 2H_2O + energy$$

The lightest hydrocarbons are gaseous. With more carbon, the hydrocarbons move to liquid and finally to solid states. Thus crude oil, which is a mixture of liquid and gaseous hydrocarbon compounds, emits more carbon than natural gas. Coal, which is a solid, emits more carbon than oil. Heavier hydrocarbons contain more energy, but their energy content does not increase in the same proportion as their CO_2 emissions. Thus, heavier hydrocarbons emit more CO_2 per unit of energy. For example, burning hard coal emits around 93.28 metric tonnes of CO_2/billion Btus; burning oil emits around 74.54 tonnes of CO_2/billion Btus; and burning natural gas emits around 52.91 tonnes of CO_2/billion Btu (US EIA 2011a). (Divide by 1.055 to approximate emissions per billion kilojoules.)

Energy is not the only source of greenhouse gases such as carbon dioxide, but it is the predominant source. Other sources include methane from natural leakages, animal flatulence, and decaying plants, along with nitrous oxide (N_2O), ozone (O_3), and chlorofluorocarbons (CFCs) from aerosols and refrigeration units. CFCs have also been shown to contribute to the deterioration of stratospheric O_3 and are being phased out of use.

With incomplete combustion, results are somewhat different. For example, suppose methane is burned in the following chemical reaction:

$$5CH_4 + 9O_2 \rightarrow 2CO + 3CO_2 + 10H_2O + energy$$

In this case, the ratio of carbon to oxygen is lower than in the first case, and we end up with some toxic CO, which bonds with hemoglobin in the blood. It can cause dizziness and headaches, and in high enough concentrations, it can be fatal. Many countries regulate the emissions of CO, including the United States, India, and China. The European Union also regulates CO emissions. For a global sample of such regulations for automotive emissions, see Delphi (2012/2013).

Such incomplete combustion is the reason that oxygenated fuels are required in some air pollution nonattainment areas of the United States at certain times. Leaner mixtures, since they contain more oxygen, burn hotter and more completely. If combustion temperatures are very hot (>1,000°C or >1,832°F), it is less likely that there is incomplete combustion. However, the heat causes the nitrogen in the air to bond with the oxygen to create nitrogen oxides as follows:

$$N_2 + O_2 \rightarrow 2NO$$
$$2NO + O_2 \rightarrow 2NO_2$$

Nitric oxide (NO) is clean, colorless, and nontoxic, but it is unstable and soon reacts and turns into other compounds. Nitrogen dioxide (NO_2) is a reddish brown irritating gas. If you add water to these nitrogen oxides (NO_x), you form nitrous and nitric acids in the following reactions:

$$4NO + 2H_2O + O_2 \rightarrow 4HNO_2$$
$$4NO_2 + 2H_2O + O_2 \rightarrow 4HNO_3$$

Hydrocarbons often also contain impurities. For example, suppose you burn sour gas, which includes some sulfur. This can be represented by the following chemical reaction:

$$4CH_4 + 5S + 13O_2 \rightarrow 5SO_2 + 4CO_2 + 8H_2O + energy$$

In addition to CO_2 and H_2O, you also get SO_2. When SO_2 interacts with the atmosphere, it reacts with H_2O and oxygen to produce sulfurous and sulfuric acid as follows:

$$SO_2 + H_2O \rightarrow H_2SO_3$$
$$2SO_2 + 2H_2O + O_2 \rightarrow 2H_2SO_4$$

These acids mix with rain to create acid rain. Such rain can damage flora and fauna, buildings, and equipment. It can increase acidity in water bodies, damaging fish, and can dissolve heavy metals, such as lead, out of the soil. Acid rain also reduces organic decomposition, washes nutrients out of the soil, and damages plant leaves and needles.

Heavier hydrocarbons produce additional pollutants, including particulates, unburned hydrocarbons, and ash. Volatile organic compounds (VOCs) are produced from incomplete combustion of hydrocarbons and from evaporation of organic-based liquids, such as gasoline, solvents, paint thinners, and cleaning solutions. VOCs react with NO_x when in sunlight to produce O_3. The majority of urban smog consists of this ground-level (tropospheric) O_3.

Examples of solid wastes from energy production include overburden from coal mining, radioactive solid waste from nuclear electricity generation, and drilling muds from oil exploration. Drilling muds are typically not harmful to the environment, since they are mainly made up of water and bentonite, which is a product of pure clay. Nuclear power plant thermal emissions can change the ecosystem of a body of water. Coal mine runoff and crude oil spills are also problematic. For some of the more dramatic crude oil spills see http://dahl.mines.edu/b1101.pdf.

Pollution as a Negative Externality

To understand how energy pollution affects energy economic policies, we begin our analysis with figure 11–1. In this figure, D equals the demand or the marginal benefits in the oil market; MC_{pv} represents the private marginal costs or the private supply curve. We assume that the production and transportation of oil creates oil spills and other pollution. With the included external costs of such pollution, the supply curve representing private and external costs or all social costs would be MC_{sc}. The private market allocation in this market would be at P_{pv} and Q_{pv}. At the private market solution, Q_{pv}, the true social costs are greater than the benefits, with the area abc representing the social losses.

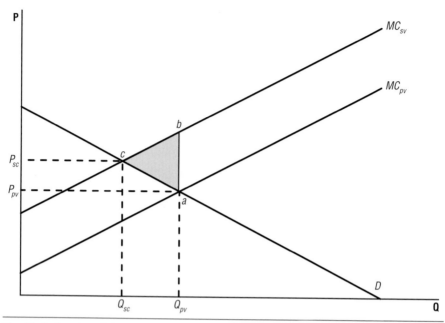

Fig. 11–1. Supply and demand in a market with negative externalities

If we could internalize the externality, then the social costs would equal the private costs and the market price and quantity would be P_{sc} and Q_{sc}. If we knew the social costs, we could internalize this cost with a tax equal to ab in this market. That would put the market price equal to the social optimum and would put us at the price and quantity that maximized social welfare (i.e., producer plus consumer surplus). In chapter 4, we noted that taxes can distort the market and cause social losses. However, in this case, the tax corrects a social loss and makes the market more efficient. If a tax is removed from a product where it is causing inefficiencies, and is instead levied on energy to correct inefficiency, we get an even more beneficial result called a *double dividend*. We remove social losses in two markets at the same time.

Optimal Level of Pollution

Another interesting way to look at externalities focuses on the polluting emission itself and considers its costs and benefits. Costs include health costs, property damage, and aesthetic costs. The benefit of an emission is the savings from not having to clean up the environment, which has been used as a waste repository free of charge. Suppose the costs and benefits for water pollution are as shown in figure 11–2.

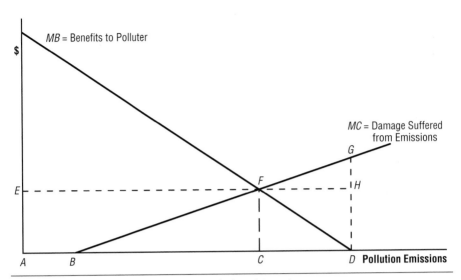

Fig. 11–2. Costs and benefits of pollution emissions into water

A small amount of pollution (less than B) causes no damage, as the natural environment is able to absorb it. However, as pollution increases, we exceed the environment's carrying capacity, and marginal damage (or the damage of the last unit of pollution) gets larger as the pollution increases. Thus, the environment becomes less able to cope as pollution increases. On the benefit side, we note that marginal benefits are high for the first units of pollution, suggesting it is very difficult to emit almost no pollution. However, as the amount of pollution increases, it becomes increasingly easier to not emit or clean up the last unit of pollution. For example, suppose you have a stack gas scrubber at an electric power plant that removes 96% of the SO_2. Further reductions would require another scrubber to catch the exhaust from the first scrubber to take out 96% of the last 4%, or 3.84%. This next 3.84% is then very expensive to abate.

In figure 11–2, if those who suffer the damages of pollution have rights to the water, pollution will be at B. With less pollution than B, they will suffer no damage. When there is more pollution than B, they will suffer damages. If polluters have rights to the water, then the pollution level will be at D. Since they do not benefit from pollution beyond D, they will not pollute more than D. Since there are benefits for all pollution less than D, they will pollute up to point D.

Alternatively, if property rights are not well-defined, the pollution level would likely be at D as well, as those having no recourse will not be able to stop the pollution. From an economic point of view, the optimal level of pollution, or what economists would call an *efficient* level of pollution, is at point C. At pollution levels less than C, the benefits of pollution are greater than costs, so society benefits from pollution. After C, costs are greater than the benefits, so society loses if we pollute more than C.

If the polluter is large and the group suffering damages is also large, the two are likely to get together and negotiate an optimal solution. For example, suppose the polluter is a coal mine with runoff that makes a local river more acidic. A refinery downstream uses the water in its processes. Acidic water has to be cleaned up or it corrodes the refinery equipment. Suppose that initially the coal mine has the property rights and pollution is at D. For the last unit of pollution, the refinery would be better off if it paid anything less than GD to get the coal mine not to pollute. Similarly, for the next-to-last unit of pollution, there is some payment from the refinery to the coal mine to not pollute that would make both the refinery and the coal mine better off. Through negotiation, the firms could devise some payment schedule that would make them both better off until we reach point C. For less pollution, there is no payment between them that would make them both better off. We could start at B and make the same arguments why the firms should end up at C.

This result was first suggested by Coase (1960). He noted that given well-defined property rights and no transaction costs, an optimal level of pollution can be derived by bargaining between polluter and sufferer, no matter who holds the original property rights. Transaction costs include money costs as well as the time and effort required for the transaction.

However, transaction costs are often very high. Take the example of a refinery polluting a low-income neighborhood. Although the residents as a whole may suffer more damages than the refinery gains by polluting, it may be difficult and costly to organize and negotiate a better solution with the refinery. Further, they may not have access to any market where they can convert the health damages they will suffer into the cash to make payments to the refinery. In such a case, economists believe markets will fail, leaving room for government intervention.

Tietenberg and Lewis (2011) discuss the following four government policies relating to pollution:

1. Set a pollution standard permitting pollution of AC.
2. Set a tax on pollution of AE.
3. Sell pollution permits equal to AC.
4. Subsidize cleanup or abatement of AE.

Setting standards and enforcing them tends to be how US environmental policy historically has been implemented. It is called command and control. If the regulation is obeyed, the firm would pollute C and abate or no longer pollute amount CD. The polluter's losses would be the cost of abatement or area CFD in figure 11–2.

The last three, more market-oriented policies are often called *incentive-based policies*. They include a tax on pollution, selling pollution permits, or offering a subsidy to help abate pollution. In figure 11–2, if a tax equal to AE per unit were set, then the optimal level of pollution would also occur. For pollution before C, the benefits are greater than the tax; it is beneficial to pollute and pay the tax. After

C, the benefits of pollution are less than the tax, and the facility is better off not polluting. The distributional effects of the policy indicate the gains and losses from the policy. The total tax the polluter would pay would equal *AEFC*, and the cost of abatement would be *CFD*. If pollution had been at *D* before the policy, the benefits to society would be equal to *FGD*. The government would have revenues of *AEFC*, which would be more than enough to compensate for the losses from pollution of *BFC*.

The government would have to give out or sell pollution permits equal to *AC*. If the government auctioned off permits, firms would have to buy them in order to pollute. If the price of a permit was less than *AE*, the firms would want to pollute more than *AC*. There would not be enough permits, and polluters would bid up the price of the permits. Similarly, if the price of the permits was higher than *AE*, there would be forces pushing the price down. The cost of the permits to the polluters should be the same as in the tax case.

A subsidy on abatement would have to be equal to *AE*. Then for any pollution after *C*, the benefits to pollution would be less than the subsidy. It would be better to take the subsidy and abate. However, before *C*, the benefits to pollution are higher than the subsidy. The polluter is better off polluting than taking the subsidy and abating. In this case, the polluter gets a total subsidy of *CFHD*, but abatement only costs *CFD* for a net profit of *FHD*.

From the above discussion, in a perfect world where we know the optimal level of pollution, a number of policies could get us to the optimum. However, who pays for the reduction in pollution or the distribution effects may vary. With taxes, permits, and standards, the polluter and those who purchase the polluter's products pay; with a subsidy, the victims or the government (i.e., taxpayers) pay. From a strict efficiency or socially optimal point of view, we would choose the policy with the lowest cost, including transaction costs, which can include contracting, enforcement, measurement, and unintended side effects.

Economists generally favor the *polluter-pays principle* as likely to be more efficient for a number of reasons. With a subsidy the government must pick a baseline, and the subsidy is paid on any pollution reduction below this baseline. One can imagine a lot of gaming, lobbying, and other gyrations by polluters to get the baseline set high.

Victims are often a more diverse group than polluters and incentives exist for each victim to free ride. Alternatively, if the government pays, it must somehow raise taxes elsewhere. Such taxes are likely to distort other markets, causing deadweight losses. Both situations increase transaction costs.

Economists also note another potential distortion. There may be the political temptation to use such subsidies as a trade policy to favor domestic industry and contravene free trade agreements.

Baumol and Oates (1988, 215–228) argue that a subsidy, by lowering average costs for firms, could even encourage entry into the industry. This could further increase levels of pollution and subsidy costs to the taxpayers. In a competitive market with entry, instead of seeing a price signal that includes the cost of pollution, consumers could see a lower price and consequently overconsume the product.

Putting efficiency aside, from a distributional point of view, it seems fairer to make the responsible polluting parties pay. For all these reasons, the OECD adopted the polluter-pays principle in 1972 (OECD 1992).

Regional Differences in Optimal Pollution Levels

When it comes to standards, the government often picks a technology and applies it uniformly across regions and firms. Economists are skeptical of standards for three reasons. First, it is unclear whether policymakers are better at picking cleanup technology than the more decentralized market. Second, it is not true that pollution damages or costs are uniform across regions. Third, the benefits of pollution are also hardly uniform across firms.

To explore further, take the model in figure 11–2 and let pollution costs vary across regions. For example, automobile emissions in the middle of Wyoming may cause far less damage than those in the middle of Manhattan. In figure 11–3, suppose that MC_1 equals marginal cost or marginal damage in Wyoming, while MC_2 equals marginal cost or damage from emissions in New York City. From an economic efficiency point of view, the emissions should be E_1 in Wyoming and E_2 in New York City.

This difference in damages is also reflected in different environmental restrictions across regions. In the next section, we will see such differences in the evolution of vehicle emission restrictions across countries.

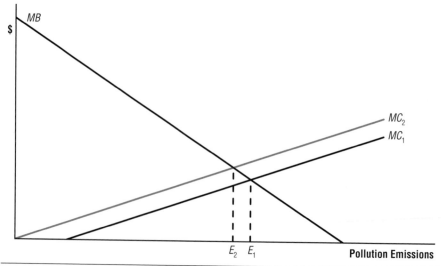

Fig. 11–3. Varying marginal costs by area

Evolution and International Comparison of Vehicle Emission Standards

Poor air quality in California, particularly in Los Angeles, caused it to be a US leader in air pollution control legislation. The state passed its first Air Pollution Control Law in 1947, seven years before the first US federal law. In 1952, scientists traced California's infamous smog to vehicle emissions. The first California statewide tailpipe vehicle emission standards were set in 1966 on hydrocarbons (HC) and CO. In 1967, the California Air Resources Board (CARB) was established to regulate air pollution. It set new air standards for PM, SO_2, NO_2, and CO in 1969 and specific NO_x standards for vehicles in 1971. In 1975, CARB required catalytic converters in new vehicles sold in California. A year later, it limited lead in gasoline, which fouled catalytic converters. In the early 1990s, California also started requiring cleaner gasoline and diesel fuels. US federal regulations tended to follow California's lead, but with a lag and typically less stringent standards. For more detail on the evolution of California and subsequent US standards, see CARB (2012). However, by 2004, the US Tier 2 regulations on vehicle emissions had caught up to California standards. Some of the milestones in US federal standards are shown in table 11–1, beginning in 1973.

The actual regulations typically vary by age of vehicle and vehicle category. Thus, the table contains representative regulations. The regulations continued to become more stringent and encompassing as the decades rolled by. The European Union started its vehicle emission restrictions in 1992, and they have evolved also. By 2005, EU CO restrictions were more stringent than in the United States. By 2009, the EU restrictions for particulate matter matched those for the United States. However, NO_x emission restrictions are less stringent than in the United States, making it easier for light duty diesel vehicles to enter European markets. Lower European taxes on diesel fuel also have encouraged a much higher penetration of light duty diesel trucks in Europe. In the United States, diesel's share still remains well below 5% of light duty vehicle registrations (US EIA 2009a).

The United States does not have specific regulations for hydrocarbon emissions, but their fuel efficiency standards called Corporate Automotive Fuel Efficiency (CAFE) standards indirectly restrict hydrocarbon emissions. The US standards have changed little from the 1990s to date. However, standards are set to increase fuel efficiency significantly in the coming decades. Auto efficiency measured by miles per gallon (MPG) is to increase from an average of 27.5 MPG (8.554 liters/100 km) to 35 MPG in 2020 and 54.7 MPG in 2025. Light duty truck standards are to see an even more dramatic increase to the same levels starting from 22.2 MPG in 2007.

Table 11–1. Milestones in US and EU vehicle emissions restrictions

Stage Units	Date	CO	HC	HC + NO$_x$	NO$_x$ grams/kilometer	PM	PN #/km
US Light Duty Vehicles							
1st US Federal 1973	1973	24.38	2.13	–	1.88	–	–
Tier 0 1981	1981	2.13	0.26	–	0.63	–	–
Tier 1 1994	1994	2.13	0.26	–	0.25	0.05	–
Tier II (Bin 5)	2004	2.13	0.26	–	0.03	0.01	–
US Light Duty Truck							
1st US Federal	1973	24.38	2.13	–	1.88	–	–
Tier 0 1988	1981	6.25	0.50	–	0.75	0.16	–
Tier 1 1994*	1994	2.13	0.50	–	0.25	0.05	–
Tier II (Bin 5)	2004	2.13	0.50	–	0.03	0.01	–
Positive Ignition (Gasoline) Vehicle Class M1							
Euro 1	1992	2.72	–	0.97	–	–	–
Euro 2	1996	2.20	–	0.50	–	–	–
Euro 3	2000	2.30	0.20	–	0.15	–	–
Euro 4	2005	1.00	0.10	–	0.08	–	–
Euro 5	2009	1.00	0.10	–	0.06	0.01	–
Euro 6	2014	1.00	0.10	–	0.06	0.01	6.0 × 1,012
European Compression Ignition (Diesel) Vehicle Class M1							
Euro 1	1992	2.72	–	0.97	–	0.14	–
Euro 2	1996	1.00	–	0.70–0.90	–	0.08–0.10	–
Euro 3	2000	0.64	–	0.56	0.50	0.05	–
Euro 4	2005	0.50	–	0.30	0.25	0.03	–
Euro 5	2009	0.50	–	0.23	0.18	0.01	–
Euro 6	2014	0.50	–	0.17	0.08	0.01	6.0 × 1,011

Sources: DieselNet (2013); US EPA (2012a); US EPA (2012b); and US EPA (2007).
Note: PN = number of particles; – indicates no limit in effect; * = light duty diesel trucks have a higher NO$_x$ emission limit. See original sources for more notes and details. Category M1 is passenger vehicles with no more than eight seats. US Tier 2 also has regulations on formaldehyde and nonmethane organic compounds.

Other countries are phasing in vehicle standards as well, with varying time schedules (An et al. 2011). The Euro standards seem to be the more popular, and the phase-in schedules for a number of countries are shown in http://dahl.mines.edu/b1102.pdf.

Abatement across Firms

In the previous section, the focus was on cost and benefits of emissions. In this section, we will look at control from a different angle and focus on the costs of abatement. The benefits of pollution are the benefits of not having to clean up pollution or the avoidance of abatement costs. For example, let the benefits of pollution emissions above be expressed as $MB_E = 10 - 2E$. The optimal amount of emissions (E) from the firm's point of view would be where the marginal benefits of emissions would be $0 = 10 - 2E$ or $E = 5$.

If the firm abates (A), the amount of polluting emissions would be $E = 5 - A$. Since the cost of abatement is the benefit from polluting, $MB_A = 10 - 2E = 10 - 2(5 - A) = 2A$.

Just as damages may vary across regions, costs of abatement may vary across polluters. Let's consider two such polluters and see how economic incentives can accomplish cleanup at the lowest cost. Let the amount that needs to be abated be the distance CD from figure 11–2. Suppose figure 11–4 represents the costs of SO_2 abatement for two electric power producers. The costs of abatement are the costs of stack gas scrubbers and other pollution control devices and any foregone output that results from abatement. In the figure, read the amount abated for firm one from the left-hand axis and the amount abated for firm two from the right-hand axis. The horizontal axis represents the amount that the law requires to be abated and is the optimal amount of abatement or the amount CD from figure 11–2.

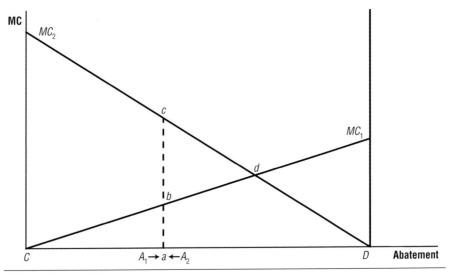

Fig. 11–4. Marginal abatement costs for two firms

Suppose the allocation of abatement is at point a, with firm one abating A_1 and firm two abating A_2. Notice that at this allocation, the cost for firm one to abate is ab, but for firm two it is ac. It would be cheaper for society if firm one abated more, and firm two abated less. This would be true until we arrived at the point where MC_1 crossed MC_2. Beyond that point it would be more expensive for firm one to abate than firm two. Thus, from a social point of view, if the government set the same standard for both firms at a, the losses would be cbd.

The government could set each firm's abatement rate, with firm one required to abate more than firm two, or if it knew cost, it could pick a tax rate to achieve the optimal level of abatement for each. This would, however, require that the government know the cost situation for each firm. Since such information is proprietary, it is unlikely that the government would get the targets or the taxes right. Each firm has an incentive to exaggerate abatement costs to be required to abate less. In addition, it would seem politically difficult to impose different standards on different firms.

For these reasons, economists tend to favor marketable permits from among various polluter-pays policies. Such permits require that the government make a decision on the optimal level of pollution. Once that is decided, pollution permits are auctioned. If the price of permits is too low, firms will want to pollute too much, resulting in a shortage of permits, and their price will be bid up. Similarly, too high a price will result in a surplus of permits. The market will determine the price of the permits, and then firms with low abatement costs will abate more, and firms with high abatement costs will abate less, as is socially efficient.

However, we know that markets can also fail. Big players in a permit market may be able to game the system and manipulate prices to their advantage. In an international context, market coordination may be a problem. With no central allocation, the sum of countries' domestic allocations may be greater than the targeted level of pollution. (See, for example, the growing pains for the European carbon dioxide market [Ellerman and Joskow 2008].)

Early US pollution regulations favored pollution standards or the command-and-control approach. However, with the high cost of this approach and at the urging of economists, market incentives increasingly have been introduced into the regulations. A convincing argument for using marketable permits (also called *cap and trade*) comes from the SO_2 regulations in the Clean Air Act Amendments of 1990. These regulations, championed by the Republicans, set up a market in SO_2 emissions permits beginning in 1995. Before the regulations were passed, estimates of compliance cost, which would determine permit price, were between $170 and $1,000 dollars per short ton of SO_2, with a few even higher. Once implemented in 1995, permits averaged around $80 per ton (US CEA 1997). Thus, by allowing firms to choose their own abatement strategy, the costs of the program were much lower than expected. This result is illustrated in figure 11–5. The original regulations were on the 263 most-polluting generating units, mostly east of the Mississippi. I was able to find data for 249 of these generating units, and they are ranked in order of their SO_2 emissions in pounds per million British thermal units (lbs/MMBtu) as of 2000. (Multiply by 0.426 to get kilograms per million kilojoules.) The rising gray line measures their emissions in 2000. The solid vertical black lines are their emissions in 1985. Note that there is no pattern to which polluters reduced emissions most. Some highly emitting firms are still highly emitting and buying permits, while other firms that were highly emitting before have found it cheaper to abate. These striking results are one of the reasons that economists have been strongly in favor of tradable CO_2 permits.

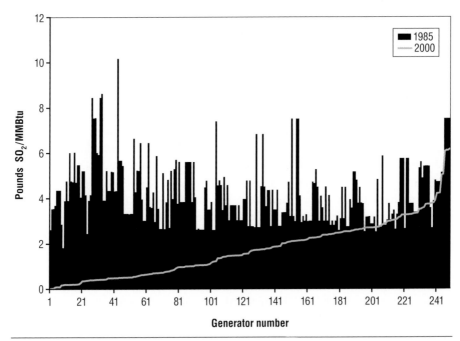

Fig. 11–5. SO_2 emissions for 249 regulated generating units, 1985 and 2000
Sources: US EPA (1999); US EPA (2002); and US EIA (1997).

Economists widely view this program as a success. It established SO_2 emissions reductions faster than expected and at a lower cost. It reached its target of 8.95 tons of SO_2 in 2007, which is 50% of the 1980 emissions, despite considerably more electricity being generated. In addition to allowing trading, earlier railroad deregulation and falling rail rates made low-sulfur western coal cheaper, which further contributed to lower compliance costs.

Nevertheless, problems with the program began when SO_2 restrictions were to be increased. Challenges to the rules led the courts to rule against unlimited interstate trading in 2008, essentially gutting the program. Republicans, once the champions of cap and trade for the right reasons, demonized cap and trade in the US House of Representative's Waxman Markey Bill in 2009. This bill sought to require permits to emit CO_2 and to allocate permits by trading. The bill died in the Senate in 2010. In the next chapter, we will investigate CO_2 permit trading in Europe.

Difficulties Measuring Costs and Benefits of Pollution

In theory, we can derive the optimal amount of pollution from energy sources. In practice, however, we need accurate estimates of both costs and benefits. Determining the marginal benefit of pollution is probably easier than determining the marginal cost. The marginal benefit is the opportunity cost of pollution, or the cheapest way not to produce that pollution, whether it comes from reducing production of final product, buying abatement equipment, or using different production materials. If the abatement is a standard process, one can use market prices to estimate costs. For new processes, the costs will be less reliable and have to be derived from engineering estimates of equipment and production costs.

Some damages are easier to measure than others. For economic damages, we can go to markets. If we want to measure the damage of acid rain on timber, we can go to biologists and get measures of reduced growth and lost harvest and multiply this lost harvest by market prices. If particulate matter makes our clothes dirtier, we may be able to monetize the change by measuring extra cleaning bills.

Other damages are harder to measure because there are no markets for them. For example, CO has health effects of increasing morbidity (sickness) and mortality. Although we may be able to use medical evidence to quantify statistical days of illness and death, it is not easy to change these into monetary measures. A number of ways have been devised to try to measure such nonmarket costs. One direct approach is *contingent valuation*, in which surveys determine either willingness to pay to avoid something unpleasant or willingness to accept a payment to put up with an unpleasantness.

An indirect approach would include *hedonic pricing*. This is pricing that would fit a function on empirical data (e.g., the price of a house as a function of the house's characteristics such as area, number of bathrooms, or local amenities, along with air quality or other pollution factors). The decrease in housing price from each negative externality would be a measure of how the market values the negative externality.

The value of a life (V) can be measured by the value people seem to place on their lives as measured in the market place. For example, suppose that workers are willing to increase the risk of death in a given year by 1 in 10,000 for a $500 higher salary. If they die, their loss is V, and if they live, their loss is zero. Their expected loss (L) is as follows:

$$L = 0.0001V + 0 \times 0.9999$$

Remember the expected value of a random variable is $\sum_{i=1}^{n} X_i P(X_i)$, where X_i is the value assigned to an event, $P(X_i)$ is the probability of the event occurring, n is the number of events that can occur, and $\sum_{i=1}^{n} X_i P(X_i)$. If workers are willing to take the above expected loss for \$500, they are setting their expected loss equal to \$500, as follows:

$$0.0001V + 0 \times 0.9999 = \$500 \rightarrow V = \$500/0.0001 = \$5,000,000$$

The solution for V gives the implicit value of a life.

Summary

Energy production, transportation, conversion, and consumption are all significant sources of pollution. Since pollution is a negative externality, it creates costs that consumers and producers do not directly account for and respond to in their production and consumption decisions. This failure to directly observe negative externalities leads to overconsumption of energy services and social losses.

From an economic point of view, the socially optimal level of pollution is not a zero level of pollution but one that maximizes the net benefits of pollution. The net benefits of pollution are equal to the total benefits minus the total costs. Total benefits are the savings from not having to use scarce resources to clean up the mess, and total costs include the damage to the ecosystem, health, property, and recreation resulting from pollution. With decreasing marginal benefits of pollution and increasing marginal costs of pollution, the optimal level of pollution is where marginal benefits are equal to marginal costs. The optimal level may vary across regions, and the policies passed reflect political preferences and systems. Regulations on vehicle emissions demonstrate how policies vary and evolve. Since a cleaner environment is typically a luxury good, restrictions typically start in richer countries. Laws also have tended to become stricter, better enforced, and disseminated across borders as knowledge, technology, regulatory bodies, and preferences evolve.

There are various ways to arrive at a socially optimal level of pollution. If transaction costs are low, no party has disproportionate market power, and the parties are well informed, the Coase (1960) theorem suggests that private parties might negotiate an optimal level of pollution regardless of who has the property rights to the environment. However, when there are many parties and transaction costs are large, the government needs to step in. If the government is well informed, it can determine the optimal level of pollution and policies to obtain that level. The government could use command and control by setting pollution standards, and it may even dictate the technology. Alternatively, the government could use more market-based economic approaches and subsidize pollution abatement, tax or charge for pollution, or issue marketable pollution permits.

Economists favor the polluter-pays principle and economic incentives such as taxes or marketable pollution permits. Subsidizing abatement, rather than charging for pollution, may encourage firms to exaggerate the amount of pollution they would produce without the subsidy and inflate the cost of abatement. With a well-designed marketable permit or cap and trade program, those firms with cheaper abatement will abate more and those with higher abatement costs will abate less. The market will set the costs of the permit to be the marginal cost of the last unit abated, which should be the same across all polluters. Although in theory cap and trade should deliver abatement at the least cost, poorly designed programs may fail to deliver, or political wrangling may fail to enact such programs.

Determining the optimal level of pollution requires knowledge of the benefits and damages from pollution. The benefits of pollution are measured by abatement costs. These costs can be determined by market prices for well-developed abatement technologies but must be estimated from engineering estimates for new technologies. Damages include those that have market measures and those that do not. For nonmarket costs, contingent valuation and hedonic pricing may be used, and life may be valued using the amount of compensation needed for people to work in riskier occupations.

12

Public Goods and Global Climate Change

With climate change, those who know the most are the most frightened.
With nuclear power, those who know the most are the least frightened.

—Unknown

Introduction

Greenhouse gases, along with water vapor, trap heat at the earth's surface, decreasing the amount radiated back into space. Such gases include not only CO_2, but also methane (CH_4), nitrous oxide (N_2O), and chlorofluorocarbons (CFCs). Together, greenhouses gases, particularly carbon dioxide (CO_2), have made earth habitable and kept humans from freezing to death for millennia. However, too much of a good thing may be putting us in the hot seat.

CO_2, which is currently building up in the atmosphere, has increased from around 290 parts per million (ppm) in 1860 to 400 ppm in May 2013 (Encyclopedia of Earth 2012; NOAA, n.d.). This buildup is indisputably coming from human-made sources. About 10% to 20% results from deforestation, and much of the rest from burning fossil fuels (Harris et al. 2012). If current rates continue, this concentration is expected to exceed 600 ppm by 2050 (IEA 2012b, 247).

Other gases contribute to the buildup as well. A molecule of methane has more than 20 times the heat-trapping capacity of a molecule of CO_2, while a molecule of nitrous oxide has more than 300 times the heat-trapping capacity of CO_2. Although a molecule of carbon dioxide has less trapping capacity, it is thought to contribute between 60% and 70% to global climate change, because of its sheer volume (US EPA 2013a).

The Intergovernmental Panel on Climate Change (IPCC) has conducted five large multi-year studies on climate change published in 1990, 1995, 2001, 2007, and 2013/14. Links to these and other IPCC reports can be found at IPCC (n.d). IPCC's fourth report, published in 2007, won the Nobel Peace Prize. As the evidence has mounted, each successive report has increased the scientific consensus and concern that the massive increases of CO_2 and other greenhouse gases (GHG) will increase overall average global temperatures and cause climate change. Melting ice caps could cause coastal cities to be submerged. Reduced ocean currents could cause Europe to become colder, with tropical storms of increased intensity. Shifts in agricultural patterns might cause the current major grain growing regions to become drier with grain belts moving north. Such a shift in climate could also benefit some areas of the world.

Precisely what, when, and where is harder to pinpoint. However, the more we know, the more likely it seems that the day of reckoning is approaching, giving us less time to mitigate and adapt. For example, locking in fossil-fuel-intensive electricity generation now that will last many decades leaves us less maneuverability to incorporate other technologies in the future.

Public Goods

Buildup of CO_2 is considered an externality, since emissions affect the climate and other factors outside the emitting market. But abatement in this case has the special characteristics of a public good. With a pure public good, one person's consumption of the good does not influence or reduce another person's consumption of it (nonrivalry in consumption). You also are not able to exclude someone from using such goods (nonexcludability). For example, in the often-cited classic case of a pure public good—a lighthouse—one person looking at the lighthouse does not diminish the possibility of another looking at the lighthouse, and you cannot stop passersby from looking at the lighthouse. In this case, cost-of-production is independent of the scale of use. Once produced, marginal cost of an extra unit of consumption of a pure public good is zero.

In the environmental context, examples of public goods include control of global pollutants, such as CO_2, to prevent global warming, maintenance of biodiversity, and control of chemicals that damage stratospheric O_3 levels. If CO_2 levels are decreased, many benefit because the climate is not disrupted. If biodiversity is preserved, everyone benefits from any new medicines developed or any new applications for genetic material preserved. If we phase out CFCs and stop destroying the stratospheric ozone that protects us from harmful ultraviolet radiation, we all benefit from lessened risk of blindness and skin cancer. In these cases, we expect the market to misallocate and produce too little of the public good from an economically efficient point of view. Therefore, the private market would be likely to produce too little biodiversity and too little control of CFCs and CO_2.

To see why, consider an ecosystem with two consumers: Coastal Clarence, a coastal dweller, and Inlander Ingrid, an inland dweller. Their demands for CO_2 control, which are their marginal benefits of abatement (*MB*), are shown in figure 12–1. Assume that the benefits of abatement are higher for the coastal dweller, because global warming would raise the sea level and flood the coast. For the sake of simplicity, assume that CO_2 control is a constant-cost industry and that the marginal cost of removing or abating an additional ton of CO_2 from current emissions is constant at MC. Let abatement be the number of tons of CO_2 removed.

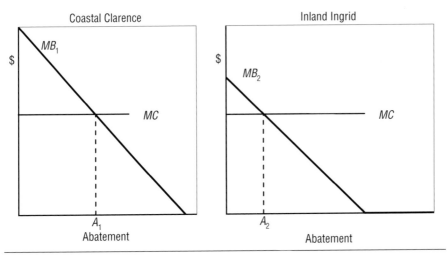

Fig. 12–1. Coastal and inland demand for CO_2 abatement

If the world consisted only of Clarence, he would maximize the net benefits of CO_2 abatement by abating A_1. If the world consisted only of Ingrid, she would abate A_2. With both persons, Ingrid would like Clarence to abate A_1, and she would like to have a free ride and spend nothing on abatement. Clarence would prefer that Ingrid abates A_2, so he could have a free ride and only abate $A_1 - A_2$. The fact that you cannot exclude anyone makes both individuals want a free ride. Thus, we would expect to get less of the public good than either person's optimum.

A second problem with the market allocation of public goods is that there is a positive externality for the other group for each ton of CO_2 abated. Thus, the total social benefit of a ton of CO_2 abated is the sum of the benefits from the inland and the coastal person's MBs. We can represent the total social benefit of abatement as the vertical sum of the two benefit curves shown in figure 12–2.

Figure 12–2 shows the optimal level of abatement is A_s, which is higher than the private optimal level for either the coastal or inland person. However, due to externalities and the effects of free riding, the market is unlikely to produce A_s. The social losses in a private market depend on how much public good is produced. For example, if the private market produced A_p, as in figure 12–3, then the social losses would be the area abc.

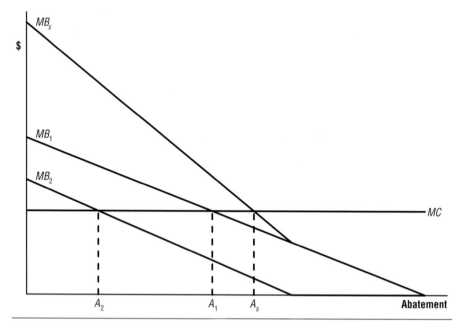

Fig. 12–2. Social optimum for CO_2 abatement

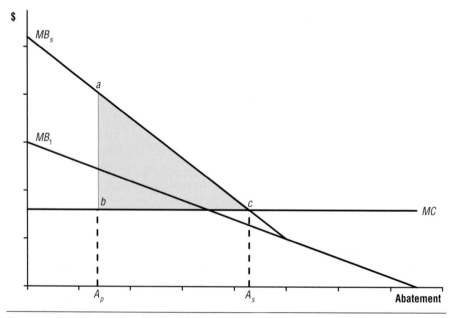

Fig. 12–3. Social losses for private market production of public goods

It may be helpful to review a numerical example of this concept. Let the marginal benefit for CO_2 abatement for our two consumers be expressed as follows:

$$MB_1 = 150 - A \text{ and } MB_2v = 110 - 1.1A_2$$

If the marginal cost of abatement is 80, then Coastal Clarence's (1) optimum would be the following:

$$MB_1 = 110 - 1.1A_1 = 80 \rightarrow A_1 = 70$$

Inlander Ingrid's (2) optimum would be as follows:

$$MB_2 = 110 - 1.1A_2 = 80 \rightarrow A_2 = 27.273$$

See these solutions in figure 12–4.

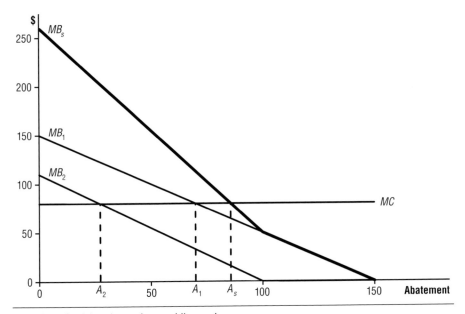

Fig. 12–4. Social optimum for a public good

However, Ingrid would want Clarence to abate 70, leaving her to abate nothing, whereas Clarence would want Ingrid to abate 27.273, leaving Clarence to abate only $70 - 27.273 = 42.727$. Free riding creates pressure to produce less than the individual optimums of either 27.273 or 70. In addition, the social optimum is greater than either individual's optimum. To see this, compute the social optimum. First, vertically sum the two marginal benefit curves. For each level of abatement, vertically sum the marginal benefits. Since abatement is a public good at any level of abatement (A), both will be able to enjoy the benefits so $A = A_1 = A_2$. This vertical summation is as follows:

$$MB_1 + MB_2 = 150 - A + 110 - 1.1A = 260 - 2.1A$$

Notice that this is the marginal benefit curve for $0 < A < 100$. After that, the marginal benefit for Ingrid becomes zero, and total benefits are only the benefits accruing to Clarence. Thus, the kink in the social marginal benefit curve is at $A = 100$.

Now compute the social optimum from the following:

$$260 - 2.1A_s = 80 \rightarrow A_s = 85.714$$

Since $A_s = 85.714$, we are to the left of the kink and this is the social optimum. If A_s had been to the right of the social optimum, we would have to recalculate the social optimum using the marginal benefit to Clarence, who is the only beneficiary after the kink. From a social point of view, we should abate 85.714. Since the private sector is unlikely to provide this, we need the government to step in to accomplish abatement.

Since abatement costs money, members of local industry are often worried that environmental protection will make them noncompetitive in global markets. Porter (1990) argues that strong environmental regulations that require abatement may actually produce net savings by stimulating new processes that save resources. He offers anecdotal evidence from the nonferrous metals industry. Economists are a bit skeptical of his arguments. If all these cost savings are readily available, wouldn't profit-maximizing firms snatch them up in the absence of regulation?

Who pays for the abatement is an issue that will have to be solved politically. For even if we can decide the value of A_s, we still need to figure out how to allocate abatement costs. All of the policies discussed in chapter 11 have possibilities: command and control, taxes, cap and trade, and subsidies.

Our first policy criterion has been to set marginal social benefits equal to marginal costs. However, these cases all assume that costs and benefits are known with certainty, which is not the case; there are large uncertainties in measuring the benefits of CO_2 abatement because we do not totally understand the process of global warming (IPCC 2013). So how should we proceed? We could do nothing and just wait and see, but it might prove to be our downfall. CO_2 resides in the air for more than 100 years, CH_4 for about 10 years, N_2O for about 170 years, and CFCs from 60 to 100 years.

If we can quantify the uncertainties, we might proceed as above. For example, suppose scientific evidence suggests two possible scenarios. The first has a 75%

probability and the second has a 25% probability. The two scenarios have the following global marginal benefit functions for CO_2 abatement:

$$MB_1 = 100 - A \text{ and } MB_2 = 50 - 0.2A$$

Let marginal abatement cost = 27.5. With such information, we can maximize the expected value of abatement. Then, the expected marginal benefit of abatement is the following:

$$E(MB) = \sum_{i=1}^{n} MB_i \times P(MB_i) = MB_1 \times 0.75 + MB_2 \times 0.25$$

$$= (100 - A) \times 0.75 + (50 - 0.2A)^* = 87.5 - 0.8A$$

To maximize the net benefits of abatement, we set the expected marginal benefit of abatement equal to the marginal cost as follows:

$$MC = E(MB) = 27.5 = 87.5 - 0.8A \rightarrow A = 75$$

Two Other Abatement Policies

In the last section, we considered the economically efficient level of abatement, where marginal benefits equal marginal costs. Two other criteria are the no-regrets policy and the policy of minimaxing regrets.

The no-regrets policy has us undertake activities that help remove existing market distortions and work to mitigate global warming as well. For example, current subsidies to fossil fuel industries cause market distortions. Removing these subsidies would raise costs in these industries and help reduce CO_2 emissions. Automobiles cause local air pollution problems as well as traffic congestion. Policies to reduce the amount of driving and amount of fuel used would help solve these problems plus contribute to CO_2 reductions. Taxing parking subsidies given to employees by businesses would reduce a market distortion, increase the cost of driving to work, and cause some to switch to mass transit. Encouraging bicycling and walking to work could improve health in countries where obesity is an increasing problem. These policies would all help solve existing problems, so we would have no regrets if we implemented them, even in the unlikely event that climate change turned out to not be a problem.

Another approach to abatement policy aims at minimizing maximum regrets (called the *minimax regret* criterion). This option can be represented in the following game theoretic framework, which is a slight adaptation from an example from Pearce and Turner (1990). In this example, we need to quantify the possibilities: there is climate change (CC) or there is no climate change (NCC). We can respond in two ways: we can do nothing (N) or we can mitigate (M). Represent this choice set as follows:

	Climate Change	No Climate Change
Do Nothing	RCC-N	RNCC-N
Mitigate	RCC-M	RNN-M

The R-values inside the matrix indicate the regrets attached to a particular choice set, which are the costs of each strategy. For example, RCC-N would be the cost or regret of doing nothing, if climate change occurs. The minimax regret strategy would include choosing the option that minimizes the maximum loss. Thus, we would look at the maximum loss for the climate change scenario and for the non–climate change scenario, and then take the option that minimizes the maximum loss.

This is most easily seen with a numerical example. Suppose the following:

	Climate Change	No Climate Change	Maximum
Do Nothing	700	0	700
Mitigate	300	200	300

In the above scenario, if we do nothing, our maximum losses are 700; if we mitigate, our maximum losses are 300. If there is no climate change, it costs us the mitigating cost of 200. If there is climate change, it costs us the 200 for mitigating plus any extra costs the mitigation cannot prevent. The policy that would minimax our regrets or losses is to mitigate and pay the 200. We can think of the mitigating cost of 200 as an insurance policy to forestall the possibility of the higher loss of 700.

Another option we could add to the above example would be to adapt rather than mitigate. With adaptation, carbon emissions would be allowed to increase, but we would adapt to the consequences. However, such adaptation, discussed in a later section, may take more time than we have, given the current rate of increase in GHGs.

Now, if we decide by one criterion or another that we want to mitigate or abate CO_2, we know that in the case of a public good such as CO_2 reduction, we do not expect the market to produce the economically efficient amount of abatement. Countries contribute different amounts to the problem, and each country would prefer that other countries bear the burden of the reductions, while they receive the benefits. We can see the different contributions to the problem in table 12–1, which shows historical global carbon emissions from 1980 through 2010 by region. The table also shows income and income per capita. Incomes are converted to US dollars by exchange rates (e.g., if there are 3.75 Saudi riyals per US dollar, and Saudi GDP is 1,347 billion riyals, then Saudi GDP per capita in dollars = 1,347/3.75 = \$359.2 billion). For updates and statistics for more counties, see http://dahl.mines.edu/t1201country.pdf.

Table 12–1. Carbon dioxide, GDP, and population for regions

Region/ Country	CO_2 1980 (10^6 t)	CO_2 1990 (10^6 t)	CO_2 2000 (10^6 t)	CO_2 2010 (10^6 t)	Pop 2010 (10^6)	GDP 2010 (10^9 \$)	GDP/Pop 2010 (\$)	CE/Pop 2010 (10^6 t)
N. America	5,475.3	5,814.3	6,818.8	6,605.7	456.6	15,383.1	33,691	14.5
United States	4,776.8	5,040.6	5,861.3	5,610.1	310.2	13,247.0	42,700	18.1
C&S America	627.4	716.3	991.3	1,257.7	480.0	2,424.7	5,051	2.6
Brazil	185.7	237.3	344.4	453.9	201.1	1,097.3	5,456	2.3
Europe	4,680.3	4,545.6	4,457.8	4,370.3	606.0	15,707.5	25,920	7.2
France	488.9	367.7	401.7	395.2	63.3	2,226.1	35,145	6.2
Germany	1,056.0	990.6	854.7	793.7	82.3	2,943.3	35,770	9.6
UK	613.6	601.8	560.3	532.4	62.6	2,324.7	37,128	8.5
Eurasia	3,081.9	3,820.8	2,261.1	2,454.1	282.9	1,261.8	4,460	8.7
Russia	—	—	1,498.8	1,633.8	139.4	899.7	6,454	11.7
Middle East	490.7	729.9	1,095.0	1,785.9	212.3	1,370.4	6,454	8.4
Saudi Arabia	176.9	208.0	290.5	478.4	25.7	359.2	13,961	18.6
Africa	537.1	725.7	887.2	1,145.2	1,015.5	1,256.0	1,237	1.1
Nigeria	69.1	82.5	80.8	80.5	152.2	153.3	1,007	0.5
South Africa	235.0	298.0	386.0	465.1	49.1	288.2	5,868	9.5
Asia & Oceania	3,541.5	5,262.9	7,227.2	14,161.4	3,799.7	13,799.9	3,632	3.7
Australia	198.8	267.6	356.3	405.3	21.5	842.7	39,169	18.8
China	1,448.5	2,269.7	2,849.7	8,321.0	1,330.1	3,825.7	2,876	6.3
India	291.2	578.6	1,003.0	1,695.6	1,173.1	1,227.8	1,047	1.4
Japan	947.0	1,047.0	1,201.4	1,164.5	126.8	4,598.3	36,263	9.2
World	**18,434.2**	**21,615.5**	**23,738.4**	**31,780.4**	**6,853.0**	**51,196.7**	**7,471**	**4.6**

Source: US EIA (n.d.d).
Note: 10^6 t = millions of metric tonnes; Pop = population; CE = carbon emissions; N. America = North America; C&S America = South and Central America; UK = United Kingdom.

China is the largest emitter, with about 8.3 billion metric tonnes (gigatonnes [Gt]) of CO_2, followed by the United States with 5.6 Gt. Together they emit more than 40% of the global carbon dioxide. Add the next three largest emitters, India, Russia, and Japan, and we account for nearly 55% of carbon emissions. Of these countries, only Japan made a binding commitment to reduce GHG emissions by 2012, which it did not meet. Their CO_2 emissions are estimated to be about 6% higher in 2012 than in 1990. Russia had agreed to a zero increase in emissions, but this commitment was not binding, as their actual emissions had fallen by more than 25% by 2012 (BP 2014).

If we look at per capita emissions, the story is different. India is far below the global average of 4.6 tonnes per person. Chinese per capita emissions are about one-third of those in the United States, Japanese about one-half, and Russian emissions about 60%. Other high emitters per capita, which top even the carbon-hungry United States, include Australia, which generates much of its power from coal, along with Qatar, Trinidad and Tobago, Saudi Arabia, Singapore, and the United Arab Emirates, which are large exporters of petroleum-based products such as crude oil, LNG, marine bunkers, and petrochemicals.

Energy Conservation and Its Cost

As we seek to reduce our future carbon footprint, we will need to master new technologies. Switching to lower carbon fuels is an option, as is conserving and using fewer fossil inputs for any given energy use. Numerous opportunities for cost-effective conservation are thought to exist. For examples, see IEA (2011a) and McKinsey & Co. (2009).

Let's take lighting as an example. Lighting accounts for a nontrivial portion of energy consumption, particularly in wealthier countries. In the United States, residential and commercial lighting accounts for about 13% of total energy use (US EIA 2013a). Street and industrial lighting put this percent even higher. Estimates suggest that if a household that gets power generated from coal switches from incandescent bulbs to compact fluorescent lightbulbs (CFLs), one-half tonne of coal or more could be saved over the lifetime of the CFLs (Audubon 2009).

First, let's determine whether the savings from the CFL technology could make the switch cost-effective. To determine the answer to this question, we need to compare capital and operating costs per unit of energy service for the traditional versus an energy-conserving technology. For example, a 75-watt (75/1,000 = 0.075 kW) incandescent bulb lasts about 600 hours. However, an incandescent light is very inefficient, with about 90% of the energy lost to heat. A 20-watt (0.020 kW) compact fluorescent bulb will produce the same amount of light and will last around 8,400 hours. Suppose the lights will run 1,200 hours per year and electricity costs \$0.11/kWh. The interest rate (r) is 10%, and interest is compounded monthly.

Assuming the consumer finds the light from the two technologies equally pleasing, let's determine the cost of an hour of light from each. The hourly operating cost for the incandescent bulb (o_i) is the number of kilowatts per bulb times the cost per kilowatt hour = (kW/bulb) × (cost/kWh) = cost/(bulb × h) = (0.075) × 0.11 = \$0.00825 per hour. Similarly, the hourly operating cost for the fluorescent (o_f) is (0.020) × 0.11 = \$0.0022 per hour.

The capital cost per unit of output, called the *levelized cost* for each bulb, is a bit harder to compute. Since we have to pay for the bulb up front but get the services spread out over time, we must allocate the capital cost across output and across time. To see how, let X be the monthly output of light (1,200/12) that begins immediately and lasts for n years. K is the initial capital cost. You pay your bill monthly. We assume output is spread equally over the year. Let Lc_k be the unit cost per kilowatt hour—the levelized cost—that would make the net present value of the future flow of service equal to the purchase price:

$$K = \frac{Lc_K X}{1 + \frac{r}{12}} + \frac{Lc_K X}{(1 + \frac{r}{12})^2} + K + \frac{Lc_K X}{(1 + \frac{r}{12})^{n \times 12}}$$

Then:

$$K = Lc_K X \sum_{i=1}^{12n} \frac{1}{(1 + \frac{r}{12})^i}$$

Solving for Lc_K gives the following:

$$Lc_K = \frac{K}{X} \Bigg/ \sum_{i=1}^{12n} \frac{1}{(1 + \frac{r}{12})^i}$$

A package of incandescent bulbs costing $K = \$1.20$ would last $n = 1$ year and would run $X = 100$ hours per month. Applying the above formula to the life of incandescent bulbs, we get the following:

$$Lc_{Ki} = \frac{1.20}{100} \Bigg/ \sum_{i=1}^{12 \times 1} \frac{1}{(1 + \frac{0.10}{12})^i} = \frac{1.20}{100} \Bigg/ 11.375 = \$0.00105$$

Capital cost per unit of light is lower than operating cost for the incandescent lighting.

A compact fluorescent costing $K = \$4.00$ would last $n = 7$ years and would run $X = 100$ hours per month. Applying the above formula, we find the following:

$$Lc_{Kf} = \frac{4.00}{100} \Bigg/ \sum_{i=1}^{12 \times 7} \frac{1}{(1 + \frac{0.10}{12})^i} = \frac{0.04}{60.237} = \$0.00066$$

For the compact fluorescent, operating costs are lower than capital costs. Adding capital and operating costs can be shown as follows:

Total incandescent costs $= Lc_{Ki} + o_i = \$0.00825 + \$0.00105 = \$0.00930/\text{kWh}$.
Total compact fluorescent costs $= Lc_{Kf} + o_f = \$0.0022 + \$0.00066 = \$0.00286/\text{kWh}$.

Thus, the compact fluorescent lighting costs are less than one-half the cost of the incandescent. However, as the interest rate goes up or electricity price goes down, the fluorescents have less of a cost advantage over the incandescent bulbs.

(See http://dahl.mines.edu/ch12m.xlsx for a program to calculate costs at different interest rates and prices for electricity and bulbs.)

A second way to compare the two options would be to look at the present value of costs providing a similar amount of electricity. Thus, the present value of seven years of electricity for the fluorescent lighting would be the following:

$$PV_f = 4.00 + 100 \times 0.0022 \times \sum_{i=1}^{12 \times 7} \frac{1}{(1 + 0.1/12)^i}$$

$$PV_f = 4.00 + 100 \times 0.0022 \times 60.237 = \$17.252$$

The present value of seven years of electricity for the incandescent would be as follows:

$$PV_f = 1.20 \left[\sum_{i=0}^{6} \frac{1}{(1 + 0.1/12)^{i \times 12}} \right] + 100 \times 0.00825 \times \sum_{i=1}^{12 \times 7} \frac{1}{(1 + 0.1/12)^i}$$

$$PV_f = 6.355 + 100 \times 0.00825 \times 60.237 = \$56.050$$

Again, we can see the much higher cost of the incandescent lighting.

Although the compact fluorescents are clearly cheaper, and this cost differential has widened over time, economists have puzzled over why rational consumers still buy incandescent lights. Other examples abound where engineering estimates show a clear cost advantage of a cheaper, more efficient technology, yet the cheaper technology may not be adopted. We will come back to why this might be the case in the next section after considering a few other technology options.

An even newer technology, light emitting diodes (LEDs), promises greater savings in cost and energy. Long-lasting LEDs have higher upfront costs. However, they last more than five times as long as CFLs, reducing maintenance costs, and they use less energy. Recently, their costs have been rapidly decreasing, and they are now competitive for high-intensity use, such as street lighting in the United States, and for off-grid battery-powered applications in India. The quality of their light and their flexibility, with the promise of further cost reductions in the near term, indicate that this technology will likely come to dominate lighting markets (Kar, n.d.).

Buildings are an important area for potential energy savings. For example, in the United States they account for 39% of total energy use, 68% of electricity consumption, and 38% of carbon dioxide emissions. Green buildings are designed for a minimal carbon footprint. With a few percentage points increase in building costs, you may be able to cut energy and electricity use in half and make a building a healthier place to live or work. See US Environmental Protection Agency (US EPA 2013c) for numerous links and resources on green building and energy certification programs, such as LEED and Energy Star.

Passive solar, which requires positioning new buildings to take advantage of solar heat, may not even add to cost. In the northern temperate zone, windows on the south side with eaves provide shade in summer but let in light and heat in winter when the sun is lower on the horizon. Trees used as a windbreak can reduce heat loss in the winter and provide shade in the summer.

Insulation is another area where it may be cost-effective to conserve. The amount of heat flow (Q), measured in British thermal units per hour through a wall or window, can be represented as the following:

$$Q = \frac{A \times \Delta T}{R} \qquad (12\text{--}1)$$

where
A is the area, ft^2,
ΔT is the temperature difference on the two sides, °F, and
R is the resistivity of the material the heat passes through, ft^2-hour-ΔT (°F)/Btu.

Resistivity can vary considerably across materials. For example, hard wood and flat glass have Rs nearer 1, while 6 inches of fiberglass insulation provides a whopping R of 19. The R values for some common building and insulation materials can be found at http://dahl.mines.edu/b1201.pdf.

Industry is another area where conservation and process changes may be cost-effective. Examples can be found from the three most energy-intensive industries: steel, cement, and aluminum. In the steel industry, coke and energy use can be reduced through better fuel preparation and injection, continuous casting and rolling, and using hot metal directly from the blast furnace, requiring no reheating. Mini-mills relying on electric furnaces also conserve energy. In the cement industry, replacing a wet process with a dry process has conserved energy. In the aluminum industry, electricity use has been reduced through the use of improved electrodes and their spacing, improved anodes, the addition of lithium to the alloy, no reheating for fabrication, and better insulation. Again, however, the higher the interest rates are and the lower the energy costs are, the less likely it is that investments to change processes will be economical.

Energy Efficiency Gap and Policy Options

So if all this potential exists, why is it not implemented? Gillingham et al. (2009) provide a nice survey of the issues and research relating to adoption of energy efficient technology. They point out that such investment requires a higher initial cost for potentially lower future energy operating costs. To compute expected future savings requires energy consumers to form expectations about future energy prices and other related operating costs, any related environmental regulations or policies, maintenance requirements, and equipment life.

Studies that form such expectations and compute lifetime costs, as in the examples above, often find an efficiency gap, since the adoption of new energy-efficient technology is slower than would be warranted by economic costs alone. A substantial amount of literature has been devoted to investigating whether this gap exists, and if so, what should be done about it.

If consumers are rational and markets are efficient, we would not expect such a gap. Using this line of thought, the apparent gap is not really a gap. Gillingham et al. (2009) present various hypotheses that researchers have considered in this line of thought. They include the following:

- Perhaps the products are not the same. Some consumers prefer the light from incandescents and are willing to pay more. Some consumers prefer larger, safer, and more comfortable automobiles and are willing to pay more.
- The studies are based on some average assumed usage, but consumers are heterogeneous. Low-intensity users will not save as much as high-intensity users and may not switch. Further, researchers with a vested interest in a technology may form biased expectations that are more favorable to the new technology.
- There may be high transaction or other hidden costs that the researchers do not take into account.
- With uncertain future energy savings, consumers may use a high discount rate. Indeed, studies that compute a discount rate based on actual technology adoption suggest discount rates between 25% and 100%.
- Coupling irreversibility of such investments with future uncertainty, there may be some option value in waiting to have more information before making an investment.

If no gaps exist, there is no problem to fix. Although results are not conclusive, the studies surveyed by Gillingham et al. (2009) did not find strong evidence for any of the above hypotheses.

Alternatively, Gillingham et al. (2009) cite a number of market or behavioral failures with policy options depending on the specific failure. Examples of market failures with potential policies are as follows:

- If energy prices are too low because environmental externalities are not included, we can employ cap and trade or energy taxes to internalize these costs.
- If utilities use average cost pricing so electricity is overused during peak periods, we could move to real-time prices.
- If oil markets and prices do not take into account energy security issues, we could use strategic petroleum reserves paid for through oil taxation.
- If capital markets fail and consumers are unable to get loans for such projects, we could implement financing guarantees and loan programs.

Markets fail with positive externalities as well. R&D may have positive externalities in the form of learning by doing and other spillovers that an individual firm will not take into account when investing in R&D. This may lead governments to give R&D tax credits, public funding, or other incentives for early adoption.

One of the assumptions of perfect competition is perfect information, which we know is not often the case. Providing information on cost-effective conservation for households could conserve energy and save on consumer energy bills. Consumers may have bounded rationality, but if informed of the cost advantages of CFLs, they may switch. Such information has some aspects of a public good. There is nonrivalry in consumption; my consumption of it does not preclude you from also consuming it. If I teach you, then we both know it. The marginal cost of an extra user may be minimal, particularly with new information technologies. These public good attributes suggest that the market may underproduce information, and there may be room for the government to produce and disseminate information about effective conservation measures or even regulate what measures are implemented.

The higher energy use of incandescent lighting has caused a number of countries to begin phasing them out, including Brazil and Venezuela (2005), the European Union and Australia (2009), Argentina, Russia, and Canada (2012), and the United States (2014) (Lighting Science 2012). Economists might be skeptical of such a ban for a couple of reasons. Economists are typically hesitant about governments picking technologies and deciding how others should spend their money. Since the pocketbook is one of the most sensitive parts of the body, who has a greater incentive to spend money wisely than the individual consumer? It might be better to provide information. If the misallocation arises from energy or carbon being underpriced, wouldn't it be better to send the proper price signal through appropriate policies?

Sometimes the misallocation occurs because there is asymmetric information. For example, a home builder may care more about the upfront costs of the building than the operating costs. In a perfect world, the operating costs would be factored into the sale price of the home. More efficient homes would cost more to build and would sell for a higher price. However, the buyer may not have information on how efficient the home is. For this reason, building codes are often implemented requiring certain insulation and other standards. However, if the problem is only one of missing information on one side of the market, an argument can be made to require the information be provided and let consumers decide on the level of energy efficiency they want to purchase.

A last set of failures noted by Gillingham et al. (2009) are behavioral and relate to decision making under uncertainty. First, although our *homo economicus* of economic theory is assumed to be rational when life is certain, rationality may be bounded in a real world full of uncertainty and cognitive constraints for our *homo sapiens*. Second, prospect theory suggests that consumers are anchored to some reference point, often the status quo, and they value losses and gains differently. They are risk averse with respect to new technology and require a higher expected return from an uncertain technology than from a sure bet. They are risk seeking with respect to losses and prefer a higher expected loss to a sure loss. The result is that consumers weigh gains less and losses more, biasing them away from new technologies.

The third bias mentioned is a heuristic bias. In complex situations, consumers may develop rules of thumb to make their lives easier and minimize the transaction cost of analyzing every decision. These rules are formed in various ways and may or may not minimize expected cost (McFadden 1999). (For fascinating accounts of many types of nonrational behavior, see Kahneman [2011].) Since behavioral biases are thought to be cognitive rather than market failures, the recommended policy responses are not market based and include information programs, education, and product standards.

Integrated resource planning (IRP) might also be considered a no-regrets policy. IRP is designed to aid in providing minimum-cost energy services. With IRP, all environmental and social costs and benefits must be considered with equal emphasis on demand- and supply-side alternatives. Thus, if it costs less to improve the efficiency of energy production using capital than to produce more energy, the efficiency option would be undertaken. Cost savings can be accomplished through demand-side management (DSM) or through more traditional, least-cost supply-side alternatives. DSM cost savings come from reducing resource consumption through increased efficiency.

Kar (2010) provides an innovative and inspiring example of IRP in JABA village in India. Although he does not refer to his methodology as IRP, he applies IRP principles and finds that conservation and solar power are a better solution for providing electricity for the world's poorest rural consumers than putting in more fossil fuel grid-based electricity.

IRP was developed in North America in the mid-1970s by state and provincial regulators of private electric utilities with monopoly franchises. Regulatory commissions pressed electricity utilities to adopt IRP to improve electricity efficiency on the demand side. These policies were designed to reduce energy use and electricity service costs. By 1991, 31 US states required utilities to implement IRP. Furthermore, the US National Energy Policy Act of 1992 required all state PUCs to design, implement, and evaluate IRP programs for all electric utilities.

So, why did regulators encourage IRP rather than just let the market operate? If energy efficiency was cheaper than energy supply, why didn't consumers invest in energy efficiency rather than buying electricity? In an increasing-cost, competitive world with marginal-cost pricing, no externalities, perfect information, and perfect capital markets, we believe that markets would allocate energy efficiently. If DSM is more efficient, consumers will reduce demand rather than increase supply. However, we live far from such an ideal world.

For example, electricity generation often uses average-cost pricing. During peak periods, marginal cost is typically much higher than average cost. During off-peak times, marginal costs are less than average. Thus, average-cost pricing leads consumers to overconsume electricity during peak periods and underconsume electricity in off-peak periods. Thus, capital is idle during off-peak periods. If load could be shifted from peak to off-peak, even if total consumption of electricity remained the same, it would still promote efficiency. Off-peak generating capacity would be used more intensively. Less generating capacity would be needed during the peak load period. Load shifting can be accomplished by three means: changing the electricity tariff structure, direct load control, and changing technologies of energy-using equipment. As electricity becomes more deregulated and competitive with marginal cost prices charged by time of use, mandated IRP should become less necessary.

Government Failure

We have noted that with public goods, markets may fail to allocate costs and resources properly, and we need governments to step in. However, governments can also fail. In regulating pollution, governments have imperfect information. For climate change issues, the timing or exact outcomes cannot yet be determined with precision. Nor can the economists measure the exact climate change costs. The available evidence suggests that under the current emission trajectories, the costs could be huge, especially for the poor. Further, the government may not have the in-house expertise and resources for cost abatement and may have to rely on the industry being regulated for abatement costs (Goodstein 1999).

For example, suppose you are in a polluting industry and know you are going to have to buy permits to pollute. If the control authority is going to choose the number of issued permits by computing where marginal benefits of pollution (abatement costs) are equal to the marginal damages of pollution, you will have an incentive to overstate your costs or the marginal benefits of pollution. If the true control costs are MB_1, but you can convince the control authorities the costs are MB_2, then more permits will be issued (E_2 instead of E_1). The permits will then cost less to buy (C_3 instead of C_2), and the firm can profitably emit more pollutants (fig. 12–5).

Alternatively, if you know a tax is going to be levied, you would want to understate your compliance cost. In the above figure, if your compliance costs were really MB_2, but you could convince the control authorities that they were MB_1, then your tax would be lower (C_1 instead C_2). You could also profitably pollute more (E_3 instead of E_2).

On the other side of the fence, environmental groups have an incentive to overstate the damages from pollution and understate the cost of abatement. The regulatory costs of reconciling extreme positions can be quite high. Small but well-focused groups can influence the decision process through lobbying, votes, publicity campaigns, and funding. Typically, environmentalists are better at getting votes and mustering popular opinion, but industries have more dollars for campaign contributions, lobbyists, lawyers, and expensive studies. Goodstein (1999) argues these pressures can lead to tough but poorly enforced environmental laws.

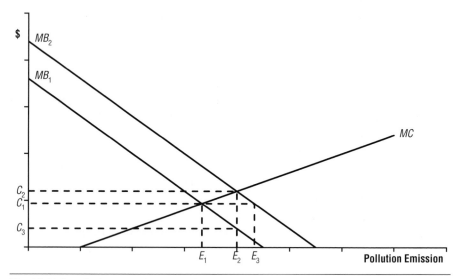

Fig. 12–5. Permits issued under different abatement cost scenarios

Global Carbon Policy

Policy toward public goods in a global setting becomes even more challenging. In this section, we will briefly consider the evolution of climate policy to demonstrate these challenges. In the 1970s, scientists became able to accurately measure current atmospheric concentrations of CO_2 as well as thousands of years of historical concentrations from Antarctic ice cores. With mounting concern over increasing atmospheric CO_2, the IPCC, noted above, was formed in 1989. In 1990, global carbon emissions were estimated at 22.6 Gt. In 1992, the UN Framework Convention on Climate Change (UNFCCC) was presented for signatures in Rio de Janeiro. (More information on this meeting, also called the First Earth Summit, and subsequent meetings can be found at UNFCCC [2013e].)

UNFCCC was an international nonbinding agreement to limit GHG emissions to levels that would not endanger the climate. No quantitative targets were set, but the treaty recognized that reductions were to be unequally distributed and based on equity, responsibility, and capability. It also called for signatories to provide inventories of GHG emissions. A good snapshot of GHG inventories available past and present for numerous countries can be found at Emission Database for Global Atmospheric Research (EDGAR, n.d.). There are currently 193 signatory countries to the agreement, called *parties*.

Although CO_2 emissions fell with the fall of communism and the economic collapse in the former Soviet Union and Eastern Europe, for the most part, they continued to increase elsewhere from 1992 to 1995. The parties to the conventions then began meeting annually to map out strategies for cutting greenhouse gases. These conventions of the parties, abbreviated COP with the number of the meeting added, have met in various cities in the world. The latest as of this writing was COP20 (Lima, Peru, 2014).

In 1997, COP3 met in Kyoto and adopted one of the most pathbreaking of the COP agreements, the Kyoto Protocol. It required mandatory GHG emission target reductions for some Annex I countries, mostly countries in the OECD and former Soviet Union. Annex I countries and their targets, which varied across countries, can be seen at UNFCCC (2013d). For other Annex I countries, restricted increases were lower than would be expected under business as usual (BAU). Most were relative to the base year 1990 and required collective reductions below 1990 levels of 4.2% to be achieved between 2008 and 2012. The six gases targeted in the agreement are: CO_2, CH_4, N_2O, perfluorocarbons (PFCs), hydrofluorocarbons (HFCs), and sulfur hexafluoride (SF_6). Together they are collectively referred to as GHGs. Their cumulative warming effects are combined into a unit called *carbon dioxide equivalents* (CO_{2e}). (For a CO_{2e} calculator, see US EPA [2013b].)

Subsequent COP meetings were used to refine and develop implementation strategies. COP6 (Bonn, Germany, 2001) agreed to three flexibility mechanisms: Emissions Trading (ET), discussed in the last chapter, the Clean Development Mechanism (CDM), which allows approved projects in developing countries to fulfill a portion of an Annex I country's targets, and Joint Implementation (JI), which allows approved projects in another Annex I country to fulfill a portion of an Annex I country's targets. For more information on these methods and other flexibility methods, see UNFCCC (2008).

In addition, carbon sinks or land-use changes that absorb or reduce emissions could be allowed to fulfill a portion of an Annex I country's targets. Changing land use patterns referred to as land use, land-use change, and forestry (LULUCF) are currently being inventoried and can decrease as well as increase emissions. Three funds were agreed upon for the following purposes: (1) to support climate change programs; (2) to provide assistance to the least developed countries for adaptation to climate change; and (3) to support adaptation to climate change.

Under the auspices of these flexibility mechanisms, the European Union began an emissions trading system (ETS) in 2005. Under their cap and trade law, a limited number of tonnes of CO_2 are allowed to be emitted—an assigned amount (AA)—with each tonne designated as an assigned amount unit (AAU). Large legally covered emitters are required to surrender a certificate, called an emission allowance unit (EAU), for each tonne of CO_2 emitted. Emitters without enough certificates to cover all emissions must purchase units on the market. Excess units may be banked for later use during a given trading period, or they can be sold. Trading periods designated so far are: (1) 2005–2007; (2) 2008–2012; and (3) 2013–2020.

As with all complex government programs, there seem to be lots of acronyms to keep track of. A certified emission reduction (CER) unit is a credit for 1 tonne CO_{2e} from activities undertaken under the clean development mechanism. An emission reduction unit (ERU) is a credit for 1 tonne CO_{2e} from activities undertaken under joint implementation. A removal unit (RMU) is a credit for 1 tonne CO_{2e} from activities relating to land use, LULUCF. EUAs, ERUs, and CERs, along with their futures and options, trade on the European Climate Exchange, owned by the ICE. Although the European Union has met its targets overall, the ETS has had its growing pains. Permits have been overallocated, and prices have been volatile and are currently very low, with weak economies. Much of the reduction in emissions has come from transitional economies. These economies may have reduced emissions anyway as they cleaned up and modernized their communist-era infrastructure and have had to pay market prices for fossil fuels.

Issues addressed in subsequent COP meetings included technology transfer, adaptation, forestry protection, financing and post-2012 strategies. With Russia's ratification, the Kyoto Protocol had enough signatories to go into effect in 2005. At that point, meeting of the Kyoto parties (MOP) was added to subsequent COP meetings. The 2005 meeting in Montreal, Canada became COP11/MOP1.

COP13/MOP3 (Bali, 2007) adopted the Bali Action Plan that provided a timeline for a post-2012 agreement to be signed in 2009, which was postponed in COP15/ Mop5 (Copenhagen, Denmark, 2009). However, in the Copenhagen Accord, parties did agree to a goal of keeping global temperatures from rising more than 2°C by 2050. Attainment of this goal is expected to require CO_{2e} concentrations to stabilize at 450 ppm by 2050. Relating to the accord, voluntary pledges have also been made by dozens of countries that affect 2020 emissions. (See UNFCCC [2013a] and UNFCCC [2013b] for current information on these pledges.) Studies suggest that if pledges are honored, CO_2 emissions would fall slightly below 2005 levels but would not put us on track for the 2050 goal of 450 ppm (Glomsrød et al. 2012).

By 2011, COP17/MOP7 (Durban, South Africa) negotiated a treaty called the *Durban Platform* to further postpone a legally binding treaty, with terms defined by 2015 and implementation in 2020. By August 2011, 191 countries plus the European Union, the Cook Islands, and Niue had signed and ratified the Kyoto Protocol. The signatories, date of signing and ratifying, and any 2012 targets can be seen at UNFCCC (2013a). The United States signed the Protocol, but the US Congress never ratified its commitment to reduce GHG emissions 7% below 1990 levels by 2012. However, with new shale gas developments, US GHG emissions were well below original expectations at only 5% above 1990 levels in 2012 (US EIA 2013d). Canada withdrew from the Protocol in December 2011, when it became quite clear they would not come close to meeting their 6% reduction target (Jull 2012). By 2010, the European Union met its Kyoto targets, transitional economies more than met their mark, and Japan was close but since has been stymied by the unfortunate events at Fukushima. Australia and New Zealand did not meet their targets (BP 2014).

With the Doha amendment from COP18/MOP8 in 2012, a number of countries have made Quantified Emission Limitation or Reduction Commitments (QELRCs) for 2013 to 2020. Countries in the European Union committed to a 20% reduction from 1990 by 2020, while Australia committed to a very modest 0.5% reduction below 2000 levels by 2020. Commitments have also been received from Belarus (12%) and Ukraine (24%), while the United States, Russia, Canada, and New Zealand have made no commitments (UNFCCC 2013c).

Despite the fact that the majority of Americans believe global warming is happening (Leiserowitz, et al. 2013), sadly, the US Congress has passed no comprehensive greenhouse gas reduction legislation. However, more than 20 states have implemented renewable energy portfolios requiring increasing power to be generated from renewable energy sources, and the US EPA has proposed a rule to give state-level guidelines for CO_2 emissions under the auspices of the 1990 Clean Air Act (U.S. Federal Register 2014).

Adaptation

An alternative to mitigating is to let carbon dioxide increase and adapt to the consequences. Dikes could be built or inhabitants relocated in response to sea-level rises. More resilient crop strains could be developed to withstand reduced water availability or more variation in precipitation. Seed and gene banks could be developed to minimize extinction from ecosystem loss. More trees could be planted to absorb some of the emissions. Stronger buildings could be built to withstand stronger storms. Better emergency response systems could be developed for increasingly extreme climate events as well as risk-sharing programs to support those most negatively impacted by climate change (Wilbanks et al. 2007).

Economists would argue that if adaptation is the cheaper alternative, everything else being equal, it should be undertaken instead of mitigation. However, many of the same uncertainties and policy issues exist with adaptation as with mitigation. Alternatively, given the increasing buildup of GHG from emerging markets, the length of time these gases remain in the atmosphere, and the slow pace of government response, some adaptation is more than likely to be a necessity. Costs of adaptation have received far less attention than costs of climate change and mitigation, but work has been increasing. See, for example, National Research Council (2010), and you can follow ongoing work on adaptation at http://www.unfccc.int.

Summary

Pure public goods provide a classic example of poorly defined property rights due to nonrivalry and nonexcludability. In such a case, one person's consumption of the good does not influence or reduce another person's consumption. Nor can we exclude someone from consuming a pure public good, such as mitigation of CO_2 to prevent climate change. If CO_{2e} is reduced, many benefit, whether they contribute to mitigation or not. Free riding becomes a strong temptation, and a less-than-efficient reduction of CO_{2e} may be attained. Therefore, governments need to step in to determine the optimal level of CO_{2e} reduction and to initiate policies to attain this optimal level. The optimal economic level is where marginal benefits of reduction equal marginal costs. Due to nonrivalry of consumption, the total benefits of a unit of mitigation are the sum of benefits to all who are affected by mitigation, while the costs should include the direct costs as well as any opportunity costs, such as foregone output.

Although there is still some disagreement regarding the extent and timing of climate change, the scientific consensus is that the buildup of CO_2 and other greenhouse gases traps heat in the lower atmosphere and warms the Earth's surface, thus affecting the climate. The economic solution from the last chapter would be to maximize net benefits by setting marginal benefit equal to marginal cost to find the optimal level of abatement. The policies suggested there to accomplish the GHG reductions also apply here (tax, cap and trade, command and control, and subsidies), as well as others. Economists prefer the polluter-pays principal and endorse the first two of these policies. However, some worry that cap and trade, which should in theory get us to our- goal at the least cost, may be too prone to tampering to be effective in an international setting or in local settings with weak institutions. In addition, carbon costs would be more volatile than with a carbon tax.

With high levels of uncertainty over exact costs and benefits of GHG abatement, an alternative is to maximize the expected value of net benefits. Clearly no-regrets policies should be implemented. They correct other market distortions while limiting GHG buildup. Abatement policies that minimax regrets can be thought of as insurance policies to ensure that the most horrific outcomes do not occur.

One way to reduce emissions is to reduce energy use. To evaluate new technologies that conserve energy, both operating and capital costs need to be distributed over the life of the project. Compact fluorescent and LED lighting, passive solar, trees for windbreaks, insulation, and continuous processes are all possibilities for economically efficient conservation techniques. High interest rates, uncertainty, unfamiliarity, and high transaction costs are all barriers to efficient conservation. Markets might fail from positive and negative externalities not reflected in the price of the energy. Public good aspects of information and new technology may cause markets to invest too little, leaving a role for governments to encourage and support the development and transfer of technology. Cognitive failures may also favor the status quo, and again, governments might step in with information, education, and standards to help consumers make more efficient choices.

Where governments own an industry, conservation options and demand management should be given equal weight for increasing energy services. All environmental and social costs and benefits should be considered, with equal emphasis on demand- and supply-side alternatives.

Whether or not conservation projects are economically efficient depends on the amount of energy saved, the cost of capital, and the interest rate. When comparing different conservation projects, the most valid comparison is to a similar production profile of services. With similar services, levelized costs per unit or the present value of costs are both valid comparisons.

We can also see government regulatory failure from the lack of information, misinformation, and pressure from interest groups. Firms and environmentalists may misrepresent costs to the government. Small but well-focused groups may influence the decision-making process through lobbying, votes, publicity campaigns, and funding.

Industrial countries contribute about one-half of the world's CO_2 emissions. The transitional economies emit around 14%, and developing countries are responsible for the remainder. It is relatively straightforward to compare carbon emissions across countries. More uncertainty prevails in the measurement of other greenhouse gases, as well as emissions from land-use changes.

From the discussion of public goods and our observations on policy over the last two decades, it is clear that GHG policy is not an easy issue. In addition to free riding and external benefits, there is also the challenge of allocating abatement in the international arena while reducing the scientific uncertainty and promoting sustainable development. Markets and governments seem to both be failing us. There has been some progress on policy, and emissions are lower than they would be without policy. European countries are trying to lead the way, and some have impressive results so far. However, not enough of the rest of us are following. The US government, in particular, has been a disappointing failure. So we each must do our part. I am going to plant a tree before going on to the next chapter, and you can sequester this book in the attic or deep under the garden rather than burning it when you have had your fill.

Doing nothing, adapting, and mitigating are all possible actions we can take. In a certain world, we would want to pick the option or portfolio of options with the highest net benefit. In an uncertain world with measures of how likely different future scenarios would occur, we would want to pick the option that has the highest expected net benefit. Events and likelihoods are shrouded in uncertainty, but the potential outcomes could be catastrophic. Thus, we can think of mitigation and adaptation plans as insurance policies as we hope for the best but prepare for the worst.

13

Safety and Security

If interactive complexity and tight coupling—system characteristics—inevitably will produce an accident, I believe we are justified in calling it a "normal accident," or a "system accident"... given the system characteristics, multiple and unexpected interactions of failures are inevitable.

—Perrow (1999)

Introduction

We all want and expect safe and secure energy. When we flip the switch, we expect the lights to come on. When we pull up to the pump, we expect the gasoline to flow. When we turn up the thermostat, we expect the furnace to light. Further, we do not want destruction of life, limb, or property anywhere along the energy supply chain. It takes massive amounts of capital and an impressive array of technology to deliver these valuable capital-intensive energy products to our vehicles, homes, and industries. However, ever since humans have designed equipment that can leak, implode, explode, collide, crash, and burn, it inevitably has. Adding more complexity, tighter coupling, more deadly substances, more hostile environments, greater speed, and greater volumes in complex networked systems has only increased such risks (Perrow 1999).

Accidents and failures happen for a variety of reasons, and the results can sometimes be quite spectacular:

- Equipment can fail from normal wear and tear, poor design, or random flaws.
- Humans can cause failure through deliberate commission, as in terrorist attacks.
- Humans can cause failure through omission, as in failing to implement proper maintenance, failing to take proper precautions and implement safe operating procedures, or by not reacting appropriately in threatening situations.

315

- Nature can throw unexpected curve balls in the form of tsunamis, volcanoes, asteroids, hurricanes, tornadoes, fires, floods, and more.
- One type of failure can cascade into others in tightly coupled systems, with complicated interactions.

Reason (1990) notes that in a study of 180 significant potentially dangerous nuclear events around the mid-1980s, 52% resulted from human failures, while 40% were caused by design and manufacturing problems. Poor procedures and training were responsible for more than one-half of the human errors. The most serious event prior to the study date occurred at Three Mile Island, Pennsylvania, in 1979. It was caused by design, equipment, and human error.

Although designs vary, there are typically four main safety features of reactors:

1. Ceramic fuel pellets create a slight delay in the nuclear reaction to allow time for insertion of control rods to shut down the reaction.
2. Cladding around the fuel rods keeps radioactive materials from contaminating the cooling water.
3. The thick steel plates of reactor vessels contain the reaction.
4. A containment building of steel or reinforced concrete is designed to trap radiation in the event of a reactor vessel breach. (IAEA, n.d.)

In the event of a problem in which the reaction could run out of control, the first line of defense is to shut down the reactor. The shutdown is accomplished by dropping neutron absorbing control rods or materials into the reactor core. This shutdown process is referred to as a *scram* or *trip*, depending on the reactor design. However, if the reactor is overheated, warping may make it impossible to insert the control rods. Then one hopes the reactor vessel is strong enough to contain the reaction, or if all else fails, that the containment building will provide a final safeguard.

At Three Mile Island, a stuck valve resulted in a loss of some cooling water. Ambiguities on the control panel caused operators to mistakenly dump even more cooling water. The overheated reactor prevented control rods from being inserted to trip the reactor. Although a partial meltdown occurred, only small amounts of radioactivity were released into the atmosphere when steam was vented to reduce pressure in the containment building. Luckily, the containment building held, preventing any further contamination. However, it took 14 years and around $1 billion to clean up the environmental mess (*NY Times* 1993).

The design and operator errors were much more serious for the Chernobyl reactor in Ukraine in 1986. During a power-down test, improper cautionary procedures and design flaws resulted in reactor instability. Power surges and an explosion blew the lid off the reactor vessel. The graphite, which was intended for absorbing neutrons and shutting down the reaction, caught fire. With no proper containment building, reactive materials were thrown into the atmosphere, spreading across parts of Russia and Europe in the ensuing days and weeks. Although estimates of the total long-term damage vary considerably, the direct death toll for emergency workers during the first year after the accident was 134. Premature deaths from cancer in the decades since are presumed to be in the thousands. Direct costs such as cleanup, resettlement, and lost agricultural and forest output are estimated to be in the hundreds of billions of dollars. (For more information, see the summary of an extensive study by international organizations on the accident and its aftermath at IAEA [2006].)

The most recent nuclear catastrophe occurred in 2011, when an earthquake struck off the northeast coast of Japan. Three of the six nuclear reactors at the Fukushima Daiichi Nuclear Power Plant were offline at the time. The remaining three were automatically shut down, and emergency diesel backup generators came online to maintain cooling and other essential operations. The ensuing tsunami knocked out the backup generators and their fuel tanks, causing the reactors to overheat and melt down, releasing significant amounts of radioactive particles. In this accident, the emergency equipment worked properly, but the tsunami wave was higher than the design capabilities of the surrounding protective wall. There was no direct loss of life from the accident, and currently the health effects are thought to be small (Canadian Nuclear Safety Commission 2012). However, it is estimated that cleanup and compensation costs will exceed $200 billion in the next 10 years (Japan Center for Economic Research 2011). See the US Nuclear Regulatory Commission (US NRC, n.d.) and the International Atomic Energy Agency (IAEA, n.d.) for more information regarding all aspects of the nuclear power industry.

Pathak and Pathak (2011) find that human and management failure also make a significant contribution to coal mine accidents. Such accidents, along with the above nuclear and other accidents along the energy supply chain, happen with some regularity. Occasionally these accidents result in large damages, both in monetary and human terms. Sovacool (2008) documents 279 high-profile energy accidents from 26 countries across the globe from 1907 to 2007. Since only accident reports published in English were sought out, almost 80% are from English-speaking countries. To be included, the accident must have been unintentional, caused at least $57,850 of property damage (in real 2013 dollars) or one death, and been along the energy supply chain from production through to distribution.

Although hydropower is often considered benign, the deadliest accident in Sovacool's inventory was the failure of the Shimantan Dam in China in 1975, causing 171,000 deaths and billions of dollars in property damages. As with Fukushima, nature presented a situation that far exceeded the design capacity of the facility. The dam broke when heavy rains filled its reservoirs to twice their design capacity. In Sovacool's survey, nuclear accidents caused the most property damage, followed by oil. Given the large damages estimated for Fukushima, nuclear likely still ranks first. Many of the oil-related accidents involve oil spills. Damages can include the lost oil, the cost of cleaning up the fugitive barrels, and the damage the oil can inflict on ecosystems. For some of the more dramatic oil spills see http://dahl.mines.edu/b1101.pdf.

The most cataclysmic oil spill accident happened in 2010 in the Gulf of Mexico at the deep water exploratory Macondo well. The well was being drilled by the *Deepwater Horizon* semisubmersible rig owned and leased to BP by Transocean. BP and Transocean employees were conducting the operation with a subcontract to Halliburton through Sperry Rand for the cement job and a subcontract to Cameron for the blowout preventers. Current drilling procedures require the circulation of mud during well drilling to keep oil, gas, and water from flowing into and up the wellbore, known as a *kick*. If a kick occurs, proper drilling mud pressures need to be reestablished. If efforts to regain control fail, a piece of equipment called a *blowout preventer* (BOP) can be used to shut in the well to prevent the liquids and gas from blowing out of the wellhead. Once the well is completed, drilling fluids are replaced with cement to stabilize and seal off various zones from the wellbore. In the case of the *Deepwater Horizon* accident, operators failed to recognize a kick. Gas migration to the surface from the ensuing blowout caused an explosion and fire that sank the rig after two days. Eleven men died on that fateful night, and the uncontrolled well continued to spew crude oil in the Gulf waters for almost three months. A faulty cement job was credited as the most prominent cause of the accident by an official investigation team (Deepwater Horizon Study Group 2011; JIT 2011). In addition to the loss of life, BP estimated the total cost for the accident to be more than $40 billion (Ramseur 2012).

Sovacool found natural gas accidents to be the most common in sheer numbers and also the most likely to include an explosion. Although fatalities were often involved, natural gas accidents trailed behind nuclear and oil in property damage. Explosions were also prominent in coal mine accidents, which were ranked fourth in both property damage and number of fatalities. Such explosions often involved methane, the lightest and main component of natural gas. If the gas released during the mining process ignites, it will cause an explosion. Canaries were the first methane detectors in coal mines. A dead canary signaled the potential presence of methane, which could result in the deaths of miners from suffocation or explosion. Coal dust can also ignite and cause explosions.

Sovacool found no high-profile accidents for other renewable sources. However, they are not without their casualties. Paul Gipe's worldwide database catalogs 76 deaths in 14 countries related to wind electricity from 1980 to 2012 (Gipe 2013). Geothermal energy has been known to suffocate with sulfur dioxide poisoning and to cause explosions. Solar photovoltaics most often have fatalities from roof falls during installation or maintenance. For more information on dangers of renewable energy sources, see the US Department of Labor (US DOL, n.d.).

Disruptions and damage can also be the results of deliberate actions. Figure 13–1 shows a sample of oil disruptions going back more than a century, all of which were set off by political events. The disruptions are estimated as the maximum barrel per day reduction during the course of the incident. The earliest disruptions noted in the figure occurred in Russia. Historically, the Russian Empire (subsequently the Soviet Union and then the Russian Federation) has always had a prominent role in oil markets. In 2013, Russia was the largest oil producer, with about 13% of the world's output of 76 million barrels per day (Mbbl/d). It falls to third after Saudi Arabia and the United States if we add in NGLs (EIA, n.d.d). In 1900, the Russian Empire accounted for about one-half of world oil production. Although Russia has never regained such an overriding role, its two revolutions caused more than a 5% shortfall in world oil production (API 1971).

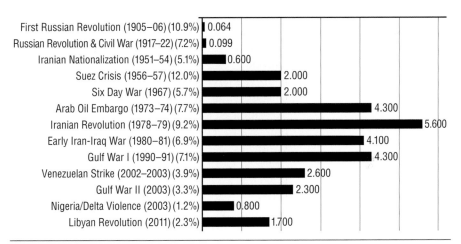

Fig. 13–1. Significant oil disruptions
Sources: Yergin (1991); API (1971); US EIA (n.d.d); Crude Oil Peak (2012).
Note: 10^6 bbl/d indicated on the right, and percent of world oil and lease condensate production is in parentheses on the left after the date.

Egyptian President Gamal Abdel Nasser's seizure of the Suez Canal caused the largest percentage disruption. The rest of the shortfalls involve countries that were or came to be members of OPEC. The Iranian nationalization in 1951 was eventually followed by a US CIA–backed coup, and the multinationals again regained control. The largest barrel-per-day disruption occurred during the Iranian Revolution. Iraq was involved in three of the disruptions, twice in wars it triggered and the most recent war orchestrated by the Bush administration in the United States.

The damages from such disruptions vary significantly. With existing spare capacity, other producers may step in and the results may be minimal. If spare capacity is low, immediate price spikes serve to allocate the existing production. Longer term, the higher prices may feed back to the macro economy. (See Hamilton [2011] for a survey of studies on the effects of oil price shocks on the US macro economy.)

One of the above incidents involved the Suez Canal. The US EIA refers to this canal and contiguous pipelines as well as other narrow global points where significant oil or petroleum products pass through confined areas known as *chokepoints.* They note seven such points, shown on the map in figure 13–2.

Blockage of any of these points by saboteurs, pirates, or accidents could cause significant disruption to world oil markets. To understand the effect such a blockage might have, take the following example. The West, fearing Iran is developing nuclear weapons, began an embargo of Iranian oil in mid-2012. In retaliation, Iran threatened to block off the Straits of Hormuz. In the event of such a blockage, let's consider what might happen.

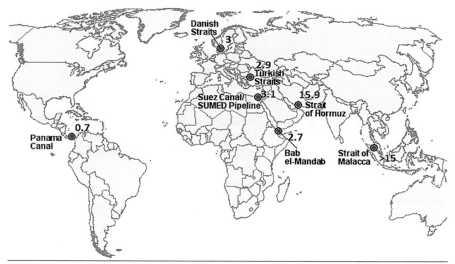

Fig. 13–2. Oil and product world chokepoints
Source: US EIA (2012c).

World oil production plus lease condensate was about 76 Mbbl/d in 2012. It is estimated that about 17 Mbbl/d pass through the Straits of Hormuz (US EIA 2012c). Brent crude oil averaged about $111 in 2012 (US EIA, n.d.b). Historical data suggests that the very short run elasticity of oil demand might be –0.05 (Dahl 1993). Remembering our analysis from chapter 4, this disruption suggests that the proportionate price change would be as follows:

$$\frac{\Delta P}{P} = \frac{\Delta Q}{Q} \Big/ \varepsilon_p = \frac{-17}{76} \Big/ {-0.05} = 4.47$$

The price elasticity result suggests that with such a large disruption, the price of oil could more than quadruple, spiking to more than $500 per barrel in the very short run.

Given such potential costs, how should we respond to potential accidents and disruptions? In the coming sections, we will consider how both markets and governments might respond.

Market Responses to Uncertainty and Disruption

Humans typically tend to be risk averse. That is, they often prefer a sure thing to a gamble with an uncertain but higher expected outcome. Thus, an oil seller might prefer a certain price of $100 a barrel to a 50% chance of receiving $50 per barrel (/bbl) and a 50% change of receiving $175/bbl. In the gamble, the expected price is the following:

$$E(X) = \sum_{i=0}^{n} X_i P(X_i) = 75 \times 0.5 + 175 \times 0.5 = \$125$$

Similarly, for two gambles with the same expected outcome, the seller is likely to prefer the one with lower uncertainty. Uncertainty is often measured as the variance of the gamble (σ^2) equal to the expected squared deviation from the mean or by its standard deviation (σ):

$$\sigma^2 = E(X - E(X))^2 = \sum_{i=1}^{n} (X_i - E(X))^2 P(X_i) = (75 - 125)^2 \times 0.5 + (175 - 125)^2 \times 0.5 = \$2500$$

Or $\sigma = (2,500)^{0.5} = \$50$. The seller is likely to prefer the above gamble to one with a 50% chance of receiving $50/bbl and a 50% chance of receiving $200/bbl. This latter gamble would have the same expected value as the former or $E(X) = 0.5 \times 50 + 0.5 \times 200 = \$125$, but its variance and standard deviation are considerably higher: $\sigma^2 = 0.5 \times (50 - 125)^2 + 0.5 \times (200 - 125)^2 = \$5,625$, and $\sigma = \$75$.

One way to reduce the variance in the market is to diversify. To illustrate, suppose there are two buyers whose markets are independent of each other. Let the expected gambles be the same for the two buyers and the same as the first gamble above: a 50% chance the price will be $75/barrel and a 50% chance the price will be $175. We call such values with probabilities attached *random variables*. The function that describes the random variables is called the *probability distribution*. Now suppose the buyer splits purchases equally between the two markets. The following shows the possible price pairs in the two markets: (X_1, X_2) = (75,75), (75,175),(175,75),(175,175). (See http://dahl.mines.edu/courses/dahl/ps/st1/ps_st1.htm for a review and references for basic probability theory.) If prices in the two markets are independent, the probability that each possible X_1 and X_2 choice occurs is the probability of their intersection = $P(X_1 \cap X_2) = P(X_1) \times P(X_2)$. Thus, the probability of each possibility for (X_1, X_2) is as follows:

    a. $P(75,75) = P(X_1=75) \times P(X_2=75)=(1/2)(1/2) = (1/4)(1/4)$
    b. $P(75,175) = P(X_1=75) \times P(X_2=175)=(1/2)(1/2) = (1/4)(1/4)$
    c. $P(175,75) = P(X_1=175) \times P(X_2=75)=(1/2)(1/2) = (1/4)(1/4)$
    d. $P(175,175) = P(X_1=175) \times P(X_2=175)=(1/2)(1/2) = (1/4)(1/4)$

Now create a new random variable that is the average price paid for oil in each case, a to d, and the likelihood of getting that price is the following:

a. $(75 + 75)/2 = 75$, $P(75) = 1/4$

b. and c. $(75 + 175)/2$ or $(175 + 75)/2 = 125$, $P(125) = P(b) + P(c) = 1/4 + 1/4 = 1/2$

d. $(175 + 175)/2 = 175$, $P(175) = 1/4$

Computing the mean and variance of this new random variable $\overline{X}$, you will find $E(\overline{X}) = \$125$, and $\sigma_{\overline{x}}^2 = \$1,250$. Notice by diversifying into two markets, the mean is the same but the variance has fallen by half.

In the above case, we assumed that prices in the two markets are completely independent. However, let's relax that assumption and assume they are negatively correlated so that when price in one market is low the other is always high, and vice versa. In that case there are only two possibilities, and their probabilities are: $P(X_1 = 75, X_2 = 175) = 1/2$ and $P(X_1 = 175, X_2 = 75) = 1/2$. Notice we have removed all uncertainty in this market, and the average price is always $\$125$. Thus, the more the two markets move in the opposite direction, the more we can reduce risk and uncertainty.

If we take the opposite extreme, where the two prices always move together, we get the two cases: $P(X_1 = 75, X_2 = 75) = 1/2$ and $P(X_1 = 175, X_2 = 175) = 1/2$. In this case, there is no advantage to diversifying, for the average price has the same probability distribution as the original random variables with the same mean and variance. However, provided the prices are not perfectly correlated as in the last case, we can reduce uncertainty by diversifying or not "putting all our eggs in one basket."

The results can be generalized as follows for any two independent random variables X_1 and X_2, with identical mean μ and variance σ^2. Then the expected value and variance for the average of the two random variables $(\frac{X_1 + X_2}{2})$ are $E\left(\frac{X_1 + X_2}{2}\right) = \mu$ and $\sigma_{\overline{x}}^2$ and $\frac{\sigma^2}{2}$. (See http://dahl.mines.edu/st13/st13.pdf question 1 for a derivation of this general case). Thus, by diversifying into two independent markets with the same mean and variance, we can reduce by one-half the variance of our price for the general case.

With more complicated computations, this result can be generalized even further to the variance of average price, if we buy different shares (α_i) from n independent markets with the mean and variance in the ith market equal to μ_i and σ_i^2. Then the average price of the formula is the following:

$$\overline{X} = \alpha_1 X_1 + \alpha_n X_n$$

The mean and variance of $\overline{X}$ are as follows:

$$E(\overline{X}) = \sum_{i=1}^{n} \alpha_i^2 \mu_i \text{ and } \sum_{i=1}^{n} \alpha_i^2 \sigma_i^2 = \sigma_{\overline{X}}^2 = \sum_{i=1}^{n} \alpha_i^2 \sigma_i^2$$

If prices are positively correlated in all markets, the variance will be larger, and if the prices are negatively correlated, the variance will be even smaller. However, unless the prices are perfectly positively correlated, the variance is lowered by diversifying.

The same analysis could be done on an X that represents the quantity of product available or the quantity of product purchased. If we are worried about disruptions, embargos, business cycles or accidents, which would disrupt our sales or purchases, we may lower our risk by diversifying. We can see this desire for diversification in table 13–1, which shows oil trade with OPEC by major region in 2012. Although North Africa is closest to Europe, and it sends the largest quantity there, it also sells about a third of its oil exports to other markets. The Middle East is closest to the Asia-Pacific market via supertanker, yet it distributes around one-fourth of its exports to other markets. Similarly, buyers diversify as well. In 2012, North America got one-third of its oil from the Middle East and Africa, yet it is closer to Latin America. Similarly, Europe got 20% of its oil from the Middle East, yet it is closer to North and West Africa.

Table 13–1. OPEC flows of crude oil, 2012 (1,000 bbl/d)

	Europe	North America	Asia & Pacific	Latin America	Africa	Middle East	Total World
Middle East	1,805	2,309	12,368	175	448	292	17,397
Iran	162	–	1,839	–	101	–	2,102
Iraq	547	559	1,205	105	–	7	2,423
Kuwait	99	220	1,699	–	52	–	2,070
Qatar	–	–	586	2	–	–	588
Saudi Arabia	991	1,423	4,577	68	216	281	7,557
UAE	7	106	2,461	–	79	4	2,657
North Africa	1,139	298	303	29	2	–	1,771
Algeria	424	263	92	29	–	–	809
Libya	715	34	211	–	2	–	962
West Africa	1,097	1,310	1,192	253	179	–	4,031
Angola	353	86	1,101	47	76	–	1,663
Nigeria	744	1,224	91	206	103	–	2,368
Latin America	69	708	570	736	–	–	2,082
Ecuador	–	229	21	108	–	–	358
Venezuela	69	479	550	628	–	–	1,725
Total	**4,110**	**4,625**	**14,433**	**1,193**	**629**	**292**	**25,281**

Source: OPEC (2013).
Note: UAE = United Arab Emirates. Totals may not exactly equal the sum of components because of rounding errors.

Just as countries may diversify across suppliers, they may also diversify across fuels. In 1973, hydro and nuclear together produced about one-fourth of the world's electricity, about the same amount of electricity as oil. However, with the price run-ups and oil insecurities of the 1970s, there has been both a relative and absolute shift out of oil for electricity generation. Natural gas gained the largest relative share, followed by nuclear and coal. Although hydropower generation has more than doubled in capacity, it has not kept pace with the previous three, and it lost share. Other renewables for electricity generation, especially wind and solar, have grown quite spectacularly recently, averaging more than 20% annually for the past decade. However, they started from such a small base that they have not yet even caught up to oil. Wind and biofuels make up the largest two categories of the other renewables, each with more than 40%. The largest wind generation is in the United States, China, and Spain. The largest biofuel and waste capacity is in the United States, Germany, and Brazil. Solar has a few champions, such as Germany, Spain, and Japan, but it is still less than 1% of world generation and still less than 4% of capacity in these top producers (see fig. 13–3 for recent global shares of all major electricity sources).

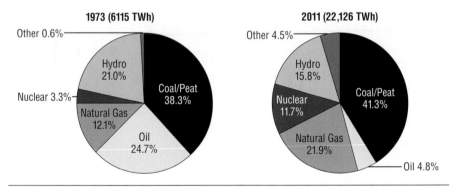

Fig. 13–3. Share of global electricity consumption by fuel
Source: IEA (2013a).

Markets will also help in the case of disruption. Shortfalls will be followed by higher prices that will help allocate existing supplies. The higher prices will signal buyers to purchase less and suppliers to produce more.

Importing companies may hold inventories to help tide them over in the event that weather or other problems cause short-term disruptions. With shortages and higher prices, they will be able to profit from drawing down inventories when prices are high. So how much inventory should they hold? First, they need to evaluate the likelihood of a disruption times the profit or loss from a disruption. They would want to hold inventory up to the point where the marginal expected benefits of the inventory equal the marginal expected costs. The expected benefits could be a reduction in product losses, or it could be additional revenue gained from the sale of inventory at higher prices during disruptions.

Producing countries may hold spare production capacity to be able to take advantage of price spikes and keep prices more stable. For example, from 2001 to first quarter 2012, US EIA (2012a) estimated that OPEC spare capacity averaged about 2.6 million barrels a day (fig. 13–4). The figure shows higher reserves during economic weakness, such as in the early 2000s, when oil price averaged around $40 per barrel. Spare capacity was much lower with the price run-up from 2004 to 2008. It rose dramatically with the recession in 2009 but then trended down as oil prices increased.

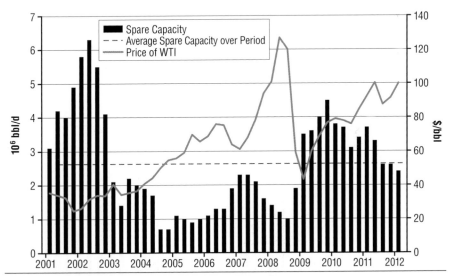

Fig. 13–4. OPEC spare capacity in millions of barrels a day
Source: US EIA (2012a).

Governments and Energy Security

So, do governments need to step in and make contingencies for disruptions? As economists, we are always suspicious of governments interfering in markets. Thus, we need to be convinced that markets would fail and governments would fail less. For example, we could consider government intervention in the market in the following situations:

- Companies are unable to correctly evaluate risks.
- There are externalities to a disruption, with higher energy prices having negative impacts on inflation, unemployment, and economic growth.
- Governments can exploit greater economies of scale than companies.
- Governments can reduce risk at lower cost through diplomacy or other means.

(See Bohi and Toman [1996] for a nice discussion of the role of government in energy security.)

Although it is debatable whether governments will handle disruptions better than the market, numerous countries now have strategic stockpiles of oil. It is argued that such stocks can provide a deterrent to a deliberate reduction in oil supplies and can be tapped to temper price spikes during accidental disruptions. Since 2001, the IEA has required that importing members hold stocks equal to 90 days worth of the previous year's net imports (IEA, n.d.a). Figure 13–5 shows historical total petroleum stocks and those held by the private sector for countries for the IEA as reported by US EIA (n.d.d). The difference between these is due to government-held stocks. These strategic IEA government reserves began in the late 1970s and reached nearly 40% of petroleum stocks by 2011. China, India, and Russia, as well as some other countries, have also started or are planning strategic petroleum reserves (Clayton 2012; Russia & India Report 2012).

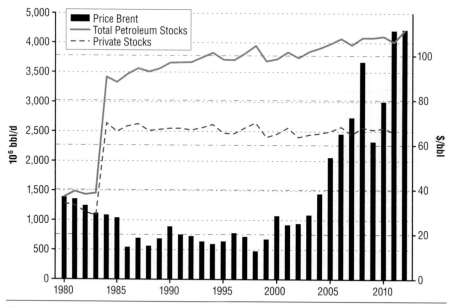

Fig. 13–5. Petroleum stocks in IEA countries (millions of barrels)
Source: US EIA (n.d.d).

Critics of strategic government stockpiles argue that such stocks are expensive to maintain and that the government is taking on a task that should rightly be done by the private sector. With government stocks, there will be a tendency for private companies to free ride and hold less. We can see some evidence for these arguments in figure 13–5. Despite IEA oil product consumption rising by about 30% from 1984 to 2005 before backing off somewhat from high prices and a weak world economy, private stocks remained relatively flat during the whole period. All the substantive growth in reserves was in those controlled by governments.

Second, governments may pay more for stocks than the private sector as they respond to political motivation and election cycles. Thus, when oil prices are high and oil is an issue, they may be more likely to stockpile. When prices are low and oil is not an issue, they may be less likely to stockpile. Between 1984 and 2011, a simple correlation shows some evidence of such behavior, with a positive relation between government-controlled stocks and oil prices.

A third argument is that governments, being quite risk averse, are less likely to release stocks during a disruption in case the situation worsens. However, the IEA has released strategic reserves on three occasions as of this writing: during the Gulf War in 1991, after Hurricane Katrina in 2005, and during the Libyan disruption in 2011 (Herron and Favole 2011). There was speculation that there might be a release in response to the embargo of Iranian oil if markets got tight enough, but that never happened.

Additionally, governments may spend money on R&D to reduce dependency on insecure fuels. Since information has public good aspects, the case for government intervention may be stronger, and we do see governments investing. Figure 13–6 shows the R&D investment into energy efficiency.

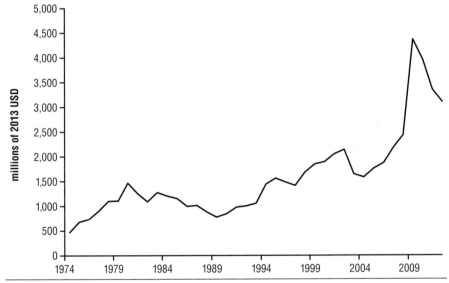

Fig. 13–6. IEA countries' investment in energy efficiency
Source: IEA (n.d.o).

The figure shows a tripling of investment in energy efficiency after the supply shocks and oil price run-ups of 1974 and 1979. There was also a doubling of such investments with the demand shocks and corresponding price run-ups from 2004 to 2008. It trended down during the 1980s when markets were awash in oil and price run-ups were eventually halted. However, it then resumed a generally upward trajectory, with decreases after bouts of economic weakness.

Energy Accidents

How might the market respond to the risk of accidents? Suppose the loss from an accident can be represented by $L(A)$ and the probability of an accident is $P(A)$. The expected loss (EL) from an accident is then the following:

$$EL = L(A) \times P(A) + 0(1 - P(A))$$

Further, a firm can spend X dollars on safety standards and preventative measures. This X might reduce the losses if an accident occurs or reduce the probability of an accident. Thus, we can write the expected loss from an accident as follows:

$$EL = L(A,X) \times P(A,X)$$

Assume that $\frac{\partial EL}{\partial X} < 0$, $\frac{\partial^2 EL}{\partial X^2} > 0$. Thus, increasing X reduces losses. However, the slope of this marginal reduction in expected losses ($\frac{\partial EL}{X}$) gets larger as X increases. Since the slope of $\frac{\partial EL}{\partial X}$ is negative, an increase makes it less negative or closer to zero, and the incremental reduction in losses gets smaller in absolute value as X gets bigger. Let's consider the easiest and least expensive safety precautions first. However, we know it is impossible to forestall all accidents and make any activity totally safe. Thus, each additional dollar of precaution has a smaller incremental effect than the last. Our total expected loss from an accident is any expected accident losses plus our expenditure on safety ($EL(X) + X$). Suppose the functions are those shown in figure 13–7. Then the optimal level of precaution that minimizes losses is at X^*.

Now let's develop the general principle for picking X^*. Minimize our total expected loss $EL(X) + X$ with respect to X as follows:

$$\frac{\partial EL(X)}{\partial X} + \frac{\partial X}{\partial X} = 0$$

The first-order conditions are as follows:

$$-\frac{\partial EL(X)}{\partial X} = 1 \qquad\qquad (13\text{–}1)$$

Equation (13–1) tells us to increase expenditure on precaution up to the point where the last dollar spent is just offset by the marginal reduction in losses from the expenditure. Second-order conditions to determine when this condition is a maximum are the following:

$$\frac{\partial^2 EL(X)}{\partial X^2} > 0$$

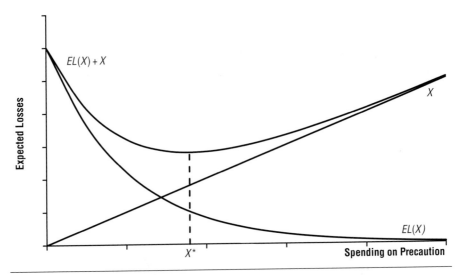

Fig. 13–7. Optimal spending on safety precaution (X^*)

Our second-order conditions hold from the properties of EL as discussed above.

To make a risk assessment and apply the principles developed above, managers will need to know the following:

- the probability of an accident, such as a well blowout
- the hazards and damages caused by an accident, such as loss of life, injuries, loss of property, and environmental damage
- the economic value of the damages
- available safety systems, their costs, and their effectiveness

So the market should provide incentives to invest in risk reduction using the above principles. Do we think the market will fail to provide the right amount of safety investment, leaving room for regulation? Accidents typically have externalities, with innocent bystanders suffering damages. The market is unlikely to consider these externalities. Thus, the firm may take higher risks because it generally does not bear all the costs of taking such risks. Such a situation is called a *moral hazard* and may arise from asymmetrical information. The parties that may bear some of the costs of the accident may not even be aware of the risk. Thus, governments step in with liability laws so that external losses will be internalized. However, even with liability laws, a firm can always go bankrupt, which truncates the maximum damages it can incur. Further, losses may have to be recouped through the courts, which can be an expensive process.

If there are information failures, private firms may not be able to correctly judge probabilities, losses, or safety options. This is more likely the case for small firms than for larger firms. Since information has aspects of a public good, the market may underinvest in safety information. Governments may choose to step in. They may perform ex ante activities to support research and development in safety measures, pass safety standards, spend money on safety inspections, and implement fines for safety violations. Governments may perform ex post activities as well. They can require firms to pay liabilities and fines, and they could even inflict jail sentences if accidents are found to be the result of negligence or failure to follow required safety standards.

With safety regulation, governments have to decide not only on the regulations themselves but also the means of enforcement. We can think of the regulator as the agent, whose objective is to achieve the optimal level of safety for society. The firm is the agent responsible for carrying out this task. Typically, this process includes a moral hazard, as the regulated agent has more information than the regulatory agent. Here Becker (1968), in a classic article on crime and punishment, gives us some insight. The government has limited resources to inspect and discover whether the firm is complying. Suppose it wants firms to invest X in safety and it fines them X if they do not comply. If we catch one-half of the violations, the expected loss to the firm from fines would be $E(X) = 0.5 \times X + 0.5 \times 0 = 0.5X$. To raise their expected losses from noncompliance, we could allocate more resources to catching violations, thus raising the probability of noncompliance from 0.5, or we could raise the fine. Therefore, the government's nontrivial challenge is to pick an optimal combination of inspections and fines to arrive at the optimal safety target using the minimal amount of resources. We think of the regulator as the agent for society. However, there is also the risk of regulatory capture, in which the regulator turns a blind eye to infractions in return for bribes or the prospect of later joining the regulated firm.

Another principal/agent problem in this scenario is that management is the agent for stockholders, who are the principals. If an accident occurs and management can simply move on, stockholders may unwittingly bear more of the risk for underinvestment in safety. Hence, jail time or fines for management may also help internalize accident risks.

As a result, there are arguments in favor of regulation relating to safety. See Vasquez Cordano (2012) for a more complete discussion and modeling of safety regulation in energy and mineral industries. In the next section, we will consider how nuclear power was encouraged in the United States and how the promotion might have influenced investment in safety practices.

US Government Promotion of Nuclear Power

In the early years of US nuclear energy, the government was a strong proponent. Private industry was more risk averse and concerned about liabilities and, thus, did not move into nuclear energy as quickly as the government would have preferred. The US federal government passed the Price-Anderson Act in 1957, which set a nuclear liability limit for 1957 to 1967 at $560 million. In the event of an accident, the utility would have been responsible for the first $560 million, and the government would cover the rest of the damages. The act has been extended a number of times and is still in place. As of 2012, the liability had been raised to $11.6 billion and the law has been extended to 2025 (US NRC 2012).

Let us consider two possible effects of this act. We will do the first analysis with our standard supply and demand diagram before the act was passed. In figure 13–8, suppose that D is the demand for nuclear energy and S is the private market supply. Under this scenario, the amount of nuclear energy would be Q_1 and the price would be P_1.

Reducing the risk of loss reduces cost to S'. Thus, we have more nuclear power than without Price-Anderson. We could take a more extreme case, as in figure 13–9, where S represents private supply and D represents private demand. In such a case, the risk costs are so high that the private market would not have developed nuclear power without the insurance. Consequently, the Act clearly should have had the effect of increasing the amount of nuclear energy and may even have spurred the creation of the industry, when the private market might not have.

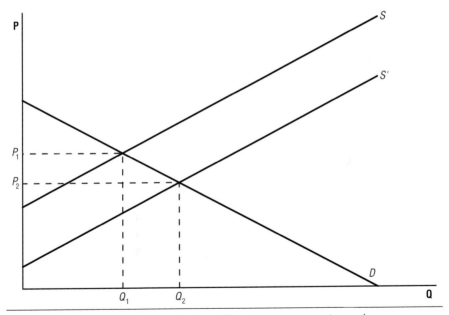

Fig. 13–8. Nuclear power with (S) and without (S') government support, case 1

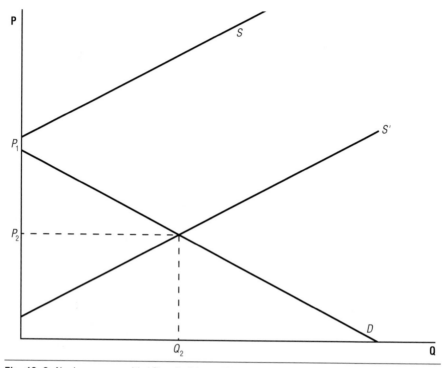

Fig. 13–9. Nuclear power with (*S*) and without (*S'*) government support, case 2

A second effect that such a bill may have is on the level of precaution in building and operating a nuclear power plant. Nuclear power plant owners have numerous options for making their plants safer, including more shielding in the form of lead and concrete, backup safety systems, and personnel training. The more precautions that are taken, the more expensive the last precaution is likely to be on the margin. This marginal cost of precaution is represented as *MC* in figure 13–10. The marginal benefit of extra precaution is the reduction in expected liability, which is represented by *MB*.

From the figure, it is quite easy to predict the expected outcome. With no Price-Anderson Act, the profit maximizing utility should invest in precaution up to the point where the marginal benefit of precaution equals the marginal cost of precaution at A_n. With the $11.6 billion limit, the profit maximizing amount of precaution is A_{pa}. The economic incentives of the bill may result in a nuclear power industry that is larger and less safe than would have existed in the private market.

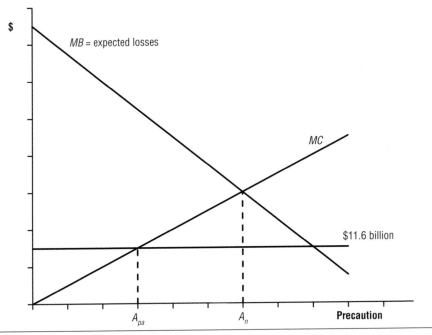

Fig. 13–10. Optimal nuclear safety precaution with and without the Price-Anderson Act

Summary

Energy disruptions occur with some regularity. They may be deliberate acts, such as the sabotage of pipelines or workers' strikes, with political or economic motives. Or disruptions may be due to accidental events, such as nondeliberate fires and explosions. In such cases, we do not know whether the event will occur ahead of time. The best we can hope for is to have an idea about the likelihood of such an event occurring.

In this chapter, we considered numerous oil disruptions and three high-profile accidents at Three Mile Island, Chernobyl, and the Macondo well. Markets can be expected to handle such events in various ways. During the disruption, ideally, higher prices will signal to producers to increase production and signal to consumers to decrease consumption. Existing production will be allocated to its highest value use. Without government interference, we can expect markets to clear, albeit at higher prices. If the market is reasonably competitive, government price restrictions are likely to result in shortages, with the potential for black market activity. With expectations that markets may be disrupted, firms are likely to hold some inventories of storable products that are bought at lower prices and sold at higher prices. Producing countries may hold excess capacity that can be used to calm excited markets. Both consumers and producers may diversify sales and purchases to lower the risk of disruption.

If there are negative externalities associated with the disruption, governments may hold strategic stockpiles. Such stockpiles act as insurance and may provide deterrence for deliberate disruption. However, political motivations for creating such stockpiles may mean they are not an economic bargain. Governments may buy high and sell low or may be hesitant to utilize the stockpiles, and firms may free ride on these public goods. Alternatively, governments may spend money on R&D for energy efficiency and to reduce dependency on insecure fuels.

Firms will respond to the threat of accidents by investing in safety measures. Producers should invest up to the point where the marginal investment in safety just offsets the incremental decrease in expected damages. Lack of information and externalities may contribute to markets making incorrect investment in safety precautions. Markets may assume more than the optimal level of risk because of the existing moral hazard, and some of the potential damages are external to the firm. Even with liability laws, the damages can be truncated by the firm going bankrupt and may be hard to collect through the courts. Moral hazard may also exist within the firm. Stockholders may bear more of the risk than their agents, the managers, who may benefit from lower safety investment while avoiding financial responsibility in the event of an accident.

Thus, governments often intervene with regulatory activity both ex ante and ex post of any accident. They may subsidize research on safety, pass regulatory standards, make inspections, and fine safety infractions. Again, there may be a moral hazard if the regulated firms are tempted to invest less than the optimal amount in safety, as they may not get caught. Small resources for inspection may be offset with larger fines to raise the expected losses of violation. In the event of an accident, governments can require firms to pay liabilities, as well as give additional punishments in the form of fines and jail sentences. Their challenge is to provide the appropriate mix of ex ante and ex post incentives to attain the proper level of security.

Likewise, government oversight or regulation can fail. Governments may also lack information necessary to identify the appropriate levels of stockpiles or precautions. They will typically not be able to monitor every player and may not be able to choose policies that promote socially optimal behavior. Regulators may serve their own and not the public's interest and be subject to regulatory capture. Governments should also recognize that liability limits on risky ventures may provide us with more of the risky ventures at lower levels of safety. As both the markets and the government may fail to provide the optimal level of safety, at the end of the day, we need to utilize the institution that fails us least (Joskow 2010).

14

Allocating Fossil Fuel Production over Time and Oil Leasing

The world has been exhausting its exhaustible resources since the first cave-man chipped a flint, and I imagine the process will go on for a long, long time.

—Robert M. Solow, Nobel Laureate in Economics

Introduction

Many energy decisions are dynamic because what we decide today influences energy markets tomorrow. If we sign a long-term contract (such as an oil lease) today, it binds our activities over the life of the contract. If we invest in long-lived capital that uses or produces energy, it influences energy markets for years to come. Industrial capital often lasts more than 20 years, while many consumer durables last more than 10. Table 14–1 lists the typical lifetimes of a variety of energy-using capital purchases.

With renewable energy resources (such as tree plantations) and nonrenewable energy resources (such as fossil fuels), the timing of usage matters. The timing of harvesting trees for lumber influences total yield; use of a nonrenewable resource today negates its use tomorrow. Economic decisions today often influence what happens tomorrow, and thus, the economic decision maker must consider these intertemporal effects. Economic decision making should be dynamic, optimizing across time rather than optimizing for one point in time (static). In this chapter, we focus on dynamic analysis with applications to fossil fuel production patterns and oil leasing. We begin our analysis with fossil fuel reserves.

Table 14–1. Typical lifetime of energy-using plant, equipment, and appliances

Power Plants	Years	Appliances	Years
Coal/nuclear	40–50	Air conditioner	9
Combined cycle natural gas	30	Clothes dryer	13
Wind turbine	20	Color TV	8
Photovoltaic	20	Dishwasher	9
Infrastructure		Electric range	13
Buildings	80	Furnace electric	15
Transmission and distribution networks	40	Garbage disposal	12
Transport Equipment		Gas range	15
Pipeline	30	Microwave oven	9
Aircraft	30	Refrigerator	13
Automobile	15	Washing machine	10

Sources: Consumer Reports (2009); IEA (2011b, 68).

Reserves and Reserves-to-Production Ratios (R/P)

Table 14–2 illustrates current proven crude oil reserves by region and by major reserve holders in the world. Reserves are fairly concentrated, with roughly half of them found in the Middle East and another 20% in South America (mostly Venezuela). North America adds another 13%, much of which is in the oil sands of Canada. Other major areas contain between 3% and 8% of global reserves, with the exception of Europe, which is rather poorly endowed with oil.

Conventional natural gas reserves are also concentrated but perhaps not as much as for oil. The Middle East has around 40% of the reserves, followed by the former Soviet Union with almost another third. New shale gas reserves will certainly change the picture, but it is not yet clear by how much and where.

The reserves-to-production ratio (R/P) indicates how many years of production these reserves would provide if production remains constant. Global R/P for oil is pegged at near 60 years, conventional natural gas at a few years less, and prolific coal at more than a century. These numbers suggest that at current production and reserve rates, oil and gas are scarcer fossil fuels with coal being most abundant.

Table 14–2. Conventional coal, oil, and gas proven reserves for selected countries

Product/Year Region/Country	Production O 2013 10³ bbl/d	Production Ng 2012 10⁹ cf	Production C 2011 10⁶ ston	Reserves O 2013 10⁹ bbl	Reserves Ng 2012 10¹² cf	Reserves C 2011 10⁹ ston	Reserves/Production* O 2013 years	Reserves/Production* Ng 2012 years	Reserves/Production* C 2011 years
North America	13,366	30,812	1,187	214	412	267	44	13	225
Canada	3,350	5,070	74	173	61	7	142	12	98
Mexico	2,562	1,684	17	10	17	1	11	10	77
US	7,455	24,058	1,096	31	334	259	11	14	236
C & S America	6,670	5,738	104	326	270	16	134	47	155
Brazil	2,024	598	6	13	15	7	18	25	1,204
Colombia	1,003	421	95	2	5	7	6	11	79
Venezuela	2,300	803	3	298	195	1	354	243	211
Europe	2,948	10,183	779	12	147	91	11	14	116
Germany	56	434	208	0	6	45	12	14	215
Netherlands	22	2,840	0	0	46	NA	30	16	NA
Norway	1,530	4,155	2	5	71	0	10	17	4
Poland	19	219	153	0	3	6	22	15	39
UK	810	1,323	20	3	9	0	11	7	13
Eurasia	12,836	28,125	558	119	2,165	251	25	77	451
Kazakhstan	1,568	416	128	30	85	37	52	204	289
Russia	10,019	21,685	355	80	1,680	173	22	77	488
Middle East	24,132	19,292	1	802	2,800	1	91	145	1,083
Iran	3,200	5,649	1	155	1,168	1	132	207	1,083
Iraq	3,054	23	0	141	112	NA	127	4,888	NA
Kuwait	2,650	548	0	104	64	NA	108	116	NA
Qatar	1,553	5,523	0	25	890	NA	45	161	NA
Saudi Arabia	9,685	3,585	0	268	284	NA	76	79	NA
UAE	2,820	1,854	0	98	215	NA	95	116	NA
Africa	8,529	7,489	285	128	510	35	41	68	123
Algeria	1,484	3,053	0	12	159	0	23	52	NA
Angola	1,784	27	0	10	11	NA	16	408	NA
Egypt	539	2,141	0	4	77	0	22	36	NA
Nigeria	2,369	1,190	0	37	180	0	43	152	5,938
South Africa	4	42	279	0	1	33	11	13	119
Asia & Oceania	7,573	17,227	5,529	45	505	318	16	29	57
Australia	332	1,902	443	1	28	84	12	15	190
China	4,164	3,811	3,878	24	107	126	16	28	33
India	772	1,460	634	5	41	67	19	28	105
Indonesia	825	2,559	397	4	141	31	13	55	78
Malaysia	516	2,176	3	4	83	0	21	38	1
World	**76,054**	**118,866**	**8,444**	**1,646**	**6,809**	**980**	**59**	**57**	**116**

Source: US EIA (n.d.d).
Note: Regional totals include countries not listed. UAE = United Arab Emirates. NA indicates no production (P), reserves (R), or values indicated. O = crude oil; G = natural gas; and C = coal, including hard coal and lignite; bbl/d = barrels per day; cf = cubic feet; ston = short ton. P and R are rounded to the nearest whole number in the table; R/P computed from original nonrounded numbers.

R/Ps for US oil and gas are around 11 and 14 years. Does this mean that in 15 years the United States will be out of oil and gas and will be importing all its supplies? Probably not, and to see why, consider figure 14–1. In 1900, the United States had about 40 years of oil reserves, and by 1940, it still had more than 13 years of oil left. This was despite the fact that production in 1940 was considerably higher than production in 1900. By 1980, the United States still had more than 9 years remaining, which was even a bit higher by 2000. Thus, an R/P of 10 years in the United States really means that the oil industry has a developed inventory of 10 years of oil reserves at current production levels. In a mature industry, 9 to 10 years is considered an efficient target. As reserves deplete, the industry will find and develop new reserves. However, the fact that the R/P has fallen from 40 years in 1900 to 10 years in 2000 suggests that oil is getting relatively scarcer in the United States. Further, as reserves and production have fallen, consumption and imports have increased. However, with changes in technology, new reserves of formerly uneconomic unconventional shale oil and gas have increased US reserves of both hydrocarbons. Although it is too early to tell just how much reserves we can extract, optimism is currently running high in the US oil and gas patch.

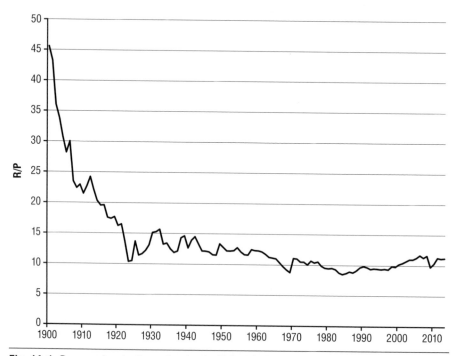

Fig. 14–1. Reserves/production ratios for the United States
Source: US EIA (n.d.b).

Dynamic Two-Period Competitive Optimization Models without Costs

To gain some intuition on how reserves (R) should best be produced over time, we begin with a simple two-period model with competitive markets and no production costs and proceed to add complications (Griffin and Steele 1986). The goal of the producer is to choose the production over the two periods (Q_0, Q_1) that maximizes present value of reserves from production. P_0 is equal to the price in the current period and P_1 is the price in the future period. The present value of production over the two years is the following:

$$PV = P_0 Q_0 + \frac{P_1 Q_1}{(1+r)}$$

We maximize this result subject to our resource constraint as follows:

$$R = Q_0 + Q_1$$

In this model, we neglect future discoveries of reserves and maximize the value of existing ones. To maximize, we use the Lagrange multiplier technique, which gives our optimizing function as follows:

$$\Im = P_0 Q_0 + \frac{P_1 Q_1}{(1+r)} + \lambda(R - Q_0 - Q_1)$$

The Lagrange multiplier in front of the constraint incorporates the constraint into the optimization. The interpretation of the multiplier is interesting: mathematically, it is $\lambda = \partial \Im / \partial R$. Along the constraint, $\Im$ is equal to the present value of our reserves. Thus, $\partial \Im / \partial R$ is the change in the present value of reserves for a change in reserves and is a measure of the present value of an additional unit of reserves. In a general Lagrangian problem, λ is called the *shadow value of the constraint*, and it represents the change in the value of our objective function for a change in the constraint. Here it is the present value of an additional barrel of reserves.

In such a problem, we treat λ as a choice variable, just as we do with our other choice variables. Then first-order conditions for a maximum are the following:

$$\Im_\lambda = R - Q_0 - Q_1 = 0 \qquad \text{(14–1)}$$

$$\Im_{Q_0} = P_0 - \lambda = 0 \qquad \text{(14–2)}$$

$$\Im_{Q_1} = \frac{P_1}{(1+r)} - \lambda = 0 \qquad \text{(14–3)}$$

Note that $\Im_\lambda = 0$ cleverly forces us to be on the constraint. From equations (14–2) and (14–3), we solve for the following:

$$P_0 = \lambda \tag{14-4}$$

$$\frac{P_1}{(1+r)} = \lambda \tag{14-5}$$

Setting equation (14–4) equal to equation (14–5) yields our solution:

$$P_0 = \frac{P_1}{(1+r)} \quad \longrightarrow \quad P_1 = (1+r)P_0 \tag{14-6}$$

Equation (14–6) requires the price to increase at the interest rate. Why does this condition maximize the present value of reserves? Suppose that price increases more slowly than the interest rate, as follows:

$$P_1 < P_0\,(1 + r)$$

In this case, money in the bank is worth more than oil in the ground. It is better to produce more oil now, sell it for P_0, and put the money in the bank to collect interest.

We can easily demonstrate this with a numerical example. Suppose the interest rate and prices are $r = 0.1$, $P_0 = 100$, and $P_1 = 105$. Then $P_1 < (1 + r)P_0$ or $105 < 1.1 \times 100 = 110$. If you leave oil in the ground, it will be worth $105 in a year. But if you sell the oil for $100 and put the money in the bank, you will have $1.1 \times 100 = \$110$ in a year. Thus, you should produce more oil now and put the money in the bank. However, if everyone produces more now and less next period, price P_0 will fall and P_1 will increase until the price goes up at the rate of interest. Alternatively, if the price goes up faster than the interest rate, oil in the ground is worth more than money in the bank. Producers should produce less oil now and more in the next period. Such a production change will drive oil prices up now and down in the next period. Production should be reallocated until price increases by the interest rate. Thus, in our stylized world, if producers dynamically optimize in this simple model, market forces should cause price to increase at the interest rate.

Next we will use some simple two-period examples to better illustrate the intuition of dynamic modeling.

Model One (No Costs, No Income Growth)

Suppose demand is as follows:

$$Q = 140 - 2.5P + 1Y$$

Reserves (R) = 120 and income (Y) = 90. To begin simply, marginal cost (MC) = 0. Demand and inverse demand at the above income in the current period are the following:

$$Q_0 = 140 - 2.5P_0 + 1(90) = 230 - 2.5P_0 \rightarrow P_0 = 92 - 0.4Q_0$$

See this demand in figure 14–2, with Q_0 varying from 0 to our total reserves of 120.

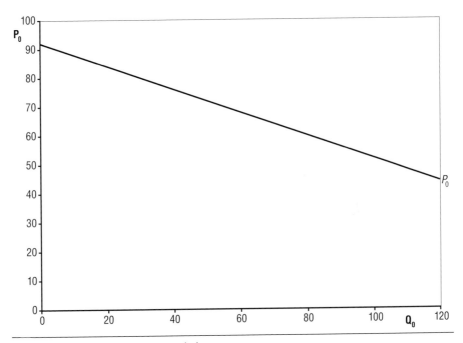

Fig. 14–2. Demand in the current period

If income does not grow, inverse demand next period is the following:

$$P_1 = 92 - 0.4Q_1$$

Flip this function over and graph it from right to left on figure 14–2 and see the result in figure 14–3. On the horizontal axis, the top labels show consumption in the current period, Q_0. The bottom labels show consumption during the next period, Q_1. For example, if we consume all reserves in these two periods and Q_0 is 0, then Q_1 is 120; when $Q_0 = 30$, then $Q_1 = 90$, etc. The graph allows us to easily compare the price received in each period. For example, when $Q_0 = 30$ and $Q_1 = 90$, P_0 is greater than P_1. When $Q_0 = 60$ and $Q_1 = 60$, the price is equal in the two periods. If Q_0 is greater than 60, and Q_1 is less than 60, then P_0 is less than P_1.

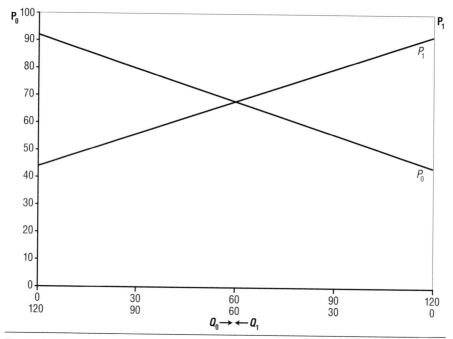

Fig. 14–3. Demand for oil now and in the next period

We are now almost ready to make our economic choice. As economists, we always want to consider our return in each period and sell where we get the most profit. But since a dollar in the second period is not equal to a dollar in the first period, we first have to discount price in the second period by dividing by $(1 + r)$, assuming $r = 20\%$, in figure 14–4. This is a fairly high interest rate but is easier to visualize on the diagram.

Note that discounted price in period 1 is not parallel to the nondiscounted price in period 1 and moves further from the nondiscounted price as price increases. The graph shows that Q_0^* and Q_1^* are the optimal allocation. Before Q_0^*, selling more in the current period yields a higher return, since $P_0 > P_1/(1 + r)$. After Q_0^*, however, we are better off selling in the next period, since $P_0 < P_1/(1 + r)$. At the optimal allocation Q_0^* and Q_1^*, prices in the two periods are P_0^* and P_1^*.

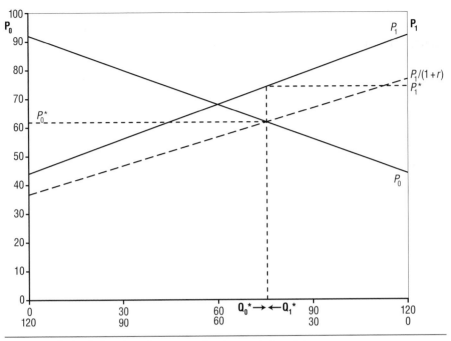

Fig. 14–4. Optimal allocation of a resource in a two-period model

We can also solve this model numerically to get a more precise solution. At the optimal allocation Q_0^* and Q_1^*, prices in the two periods are P_0^* and P_1^*. The first-order condition requires the following:

$$P_0 = \frac{P_1}{(1+r)} \quad \longrightarrow \quad 92 - 0.4Q_0 = \frac{(92 - 0.4Q_1)}{1.2} \qquad \textbf{(14–7)}$$

The reserve constraint tells us the following:

$$Q_0 + Q_1 = 120 \qquad \textbf{(14–8)}$$

Solving equation (14–8) for Q_1 and substituting into equation (14–7) yields the following:

$$92 - 0.4Q_0 = \frac{92 - 0.4(120 - Q_0)}{1.2}$$

Solving for quantities in the two periods then results in the following:

$$Q_0 = 75.455 \text{ and } Q_1 = 120 - Q_0 = 120 - 75.455 = 44.545$$

Substituting quantities into the inverse demand equation yields prices in the two periods:

$$P_0 = 92 - 0.4 \times 75.455 = 61.818 \text{ and } P_1 = 92 - 0.4 \times (44.545) = 74.182$$

The price that the producer receives, minus his marginal economic costs, was called the producer surplus in our earlier competitive static models. This surplus (often referred to as *Hicksian rent*) occurs because some units have lower costs than others, whereas all units receive the price of the marginal unit. In a dynamic problem, marginal cost also includes the foregone opportunity of selling the resource in the future. This scarcity-induced surplus or opportunity cost, referred to as *Hotelling rent*, is the difference between current price and marginal costs in the current period for the marginal producer. In this contrived problem with no economic costs, the whole price is rent. The present value of these rents is easily computed as follows:

$$PV_r = P_0 Q_0 + \frac{P_1 Q_1}{1.2} = 75.455 \times 61.818 + \frac{44.545 \times 74.182}{1.2} = 7,418.175$$

We can also represent the present value of consumer surplus in this example as the area below demand in period zero plus the area below the discounted demand in period one minus discounted rents (PV_r).

$$PV_{cs} = \int_0^{75.455} P_0 dQ_0 + \int_0^{44.545} \frac{P_1}{1.2} dQ_1 - PV_r$$

$$= \left(92Q_0 - \frac{0.4}{2} Q_0^2\right)\Big|_0^{75.455} + \frac{1}{1.2}\left(92Q_1 - \frac{0.4}{2} Q_1^2\right)\Big|_0^{44.545} - 7,418.175 = 1,469.401$$

This is the shaded area *abcde* in figure 14–5. With linear demand, consumer surplus can be even more easily computed by using triangles *abc* + *cde* from figure 14–5 as follows:

$$\left(abc = 0.5(92 - 61.818)75.455 = 1,138.691; cde = 0.5\left(\frac{92 - 61.818}{1.2}\right)44.545 = 330.717\right)$$

With no externalities in this market, this competitive solution maximizes the present value of social welfare, which is the present value of consumer surplus plus the present value of producer surplus or Hotelling rents. We can see this in figure 14–6. At the competitive optimum, the present value of social welfare (producer plus consumer surplus) is the area above zero and below *aeh*. Notice if we move to another allocation, the social welfare is reduced. For example, if the allocation is at point *d*, then the present value of social welfare is the area above zero and below *abch*, and we lose the area *bce*.

In the next section, we discuss the more interesting case that occurs when income grows.

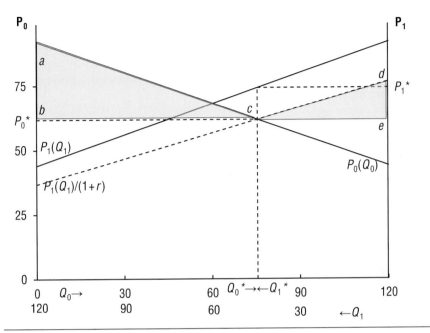

Fig. 14–5. Consumer surplus in a two-period model

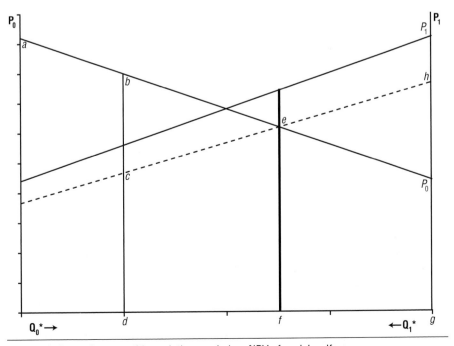

Fig. 14–6. Dynamic competitive solution maximizes NPV of social welfare

Model Two (No Costs, Income Growth)

In model two, let income grow at 10%, increasing demand in the later period. The income change is exaggerated in figure 14–7 to more clearly show what happens. Discounted future price crosses the present demand at a lower current consumption, m_2, instead of m_1.

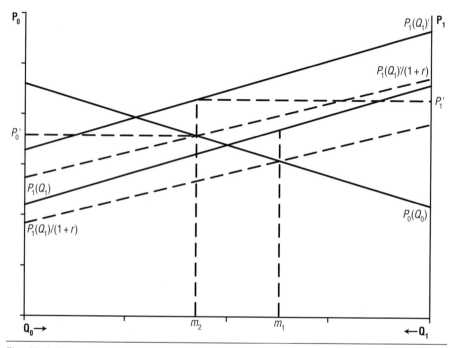

Fig. 14–7. Change in resource allocation over time with income growth

The model tells us that less is consumed now and more is consumed in the future. With higher income in the future, future demand $P_1(Q_1)$ is higher, bidding up price and moving consumption to the next period, where it has become more valuable. Note that the price is higher in both periods in this scenario than with no income growth.

Now imagine what will happen if the interest is lowered or the future is discounted less.

Model Three (No Costs, No Income Growth, Lower Interest Rate)

Suppose in model one (no income growth, no cost model), technical change falls, inducing a reduction in the interest rate as shown in figure 14–8.

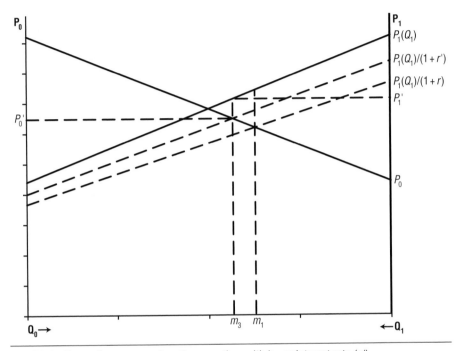

Fig. 14–8. Change in resource allocation over time with lower interest rate (r')

Figure 14–8 suggests that we should consume at m_3 instead of m_1, less now and more in the future. With a lower interest rate, producers discount the future less. They face poorer investment opportunities, making it more attractive to shift production to the future. Lower production now raises current price; increased production next period lowers the future price.

Next let's see what happens with increased reserves.

Model Four (No Costs, No Income Growth, Increased Reserves)

As we search for new reserves and as technology improves, total known reserves may increase. For example, US proven oil reserves jumped from 29.6 billion barrels in 1970 to 39.6 billion barrels in 1971 with finds in Alaska. Western European reserves jumped from 3.7 billion barrels in 1971 to 14.2 billion barrels in 1972 with finds in the North Sea. Canadian reserves jumped from about 50 billion barrels in 1998 to about 180 billion barrels as technology evolved, making its oil sands competitive with conventional reserves. As of 2014, Brazilian and US reserves are expected to increase sharply if the presalt and shale oil reserves are as prolific as expected. If we increase reserves, stretching out the above graph, we expect prices to fall in both periods. However, if reserves increased to 500 with only two periods, we could even attain the interesting situation shown in figure 14–9.

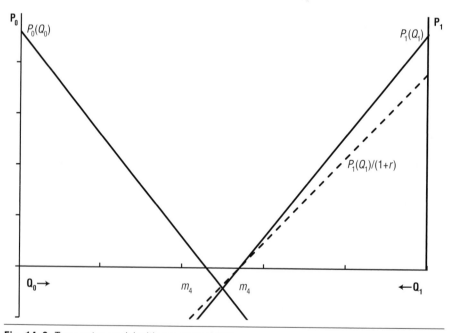

Fig. 14–9. Two-sector model with reserves of 500

Now there are so many reserves that to consume them all we need to drive the price down to negative amounts. In such a market, there is no scarcity across time and no need to dynamically optimize. We can consume all we want now and in the future period and still have reserves left. We should operate in each period where price equals marginal cost, bringing us back to the static optimum. In this simple case with no costs, P would be zero and the consumption (m_4), identical in both periods, would be as follows:

$$P = 0 = 92 - 0.4 = 0 \rightarrow Q = 230$$

Consumption is 230 in each period, and the reserves left in the ground after the next period would be $500 - 2 \times 230 = 40$. In this case, there is no producer rent, since price is zero, but consumer surplus is quite high, since consumers get the oil for nothing, as follows:

$$CS = (0.5 \times 92 \times 230) + (0.5 \times 76.67 \times 230) = 19{,}396.67$$

Next let's add a bit more realism by including costs in the model.

Model Five (No Income Growth, with Costs)

In model five, oil production is a constant cost industry in each period, with marginal cost MC_0 and MC_1. Now the net benefits to producers in each period are Hotelling rents ($P - MC$). The new optimization condition is then given in eq. (14–9):

$$P_0 - MC_0 = \frac{P_1 - MC_1}{(1+r)} \qquad \textbf{(14–9)}$$

The condition tells us to set the present value of Hotelling rent (or user cost) equal on the margin across time. This Hotelling rent is not required in order to entice a supplier to produce, since marginal costs include a normal rate of return. However, if Hotelling rent is maximized, it helps the market allocate scarce resources efficiently across time. To see why, suppose the user cost in the current period is greater than the present value of user cost in the next period:

$$P_0 - MC_0 > \frac{(P_1 - MC_1)}{(1+r)} \qquad \textbf{(14–10)}$$

Here, money in the bank is more valuable than oil in the ground. It is better to produce the oil now, sell it, and put the money in the bank. Producing more oil now and less in the next period would lower the price now and raise it in the next period. We should keep producing more until oil in the ground and money in the bank are equal in value.

Mathematical optimization, or the same arguments as for figure 14–6, will show that the condition for a private market in equation (14–9) provides the socially optimal allocation provided that the following are true:

- There are no externalities in the market (demand and cost curves represent true social benefits and costs).
- Property rights are well-defined, so the producer is assured of having property rights in both periods.
- The interest rate represents the social discount rate.

In the simpler model (with no costs), price rose at the interest rate. To see how price increases in this case, rearrange equation (14–9) as follows:

$$P_1 - MC_1 = (1 + r)(P_0 - MC_0) \qquad \text{(14–11)}$$

Equation (14–11) indicates that user cost goes up at the interest rate. Solving for P_1, we have the following:

$$P_1 = (1 + r)(P_0 - MC_0) + MC_1 = (1 + r)(P_0) + MC_1 - (1 + r)MC_0$$

If marginal cost rises faster than the interest rate ($MC_1 - (1 + r)MC_0 > 0$), then price rises faster than the interest rate. If marginal cost rises exactly at the interest rate ($MC_1 - (1 + r)MC_0 = 0$), then the price rises at the interest rate. If marginal cost rises slower than the interest rate, price rises slower than the interest rate. If marginal cost falls enough, prices might even decrease in this model.

Adelman (1993) argues that petroleum resource costs are determined by a race between depletion and technology. Depletion tends to increase costs, while technology decreases them. Falling prices would be a case of technology winning. Important technical changes have reduced oil exploration and production costs. Recent increases in materials and engineering expertise in an overheated world economy have raised costs. The price plunge at the end of 2014 can be expected to lower them.

To continue our cost analysis, we model marginal cost as constant across production and time, which implies the race between depletion and technology is a tie, and the economy is not overheated. Assume model one (constant income and interest rate) with constant marginal costs of 20, as seen in figure 14–10. Our optimization conditions tell us to first subtract a constant marginal cost from each demand to find user costs and then discount the future period user cost to find the optimal allocation. These new darker lines using price minus cost now help us make the allocation at m_5.

By including costs relative to model one, you decrease current consumption from m1 to m5 and increase future consumption. Future discounted costs do not reduce future revenue as much as today's cost reduces today's revenue, making the future relatively more desirable. (A similar numerical example is worked out at http://dahl.mines.edu/st14/st14.pdf, question 12.)

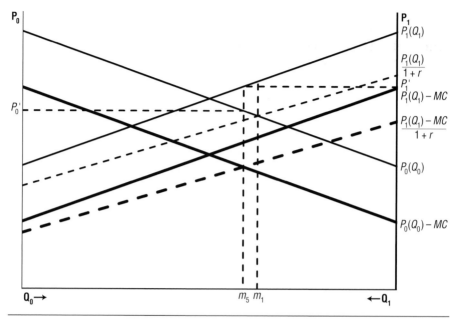

Fig. 14–10. Allocation in a two-period dynamic model with constant marginal cost

In the earlier models (without costs), we saw that as interest rates rise, the future becomes less valuable, and we consume more today. Let's see what happens in the model when we add costs and the interest rate goes to infinity. A model with an interest rate of infinity may be expressed as follows:

$$P_0 - MC_0 = \lim_{r \to \infty} \left[\frac{P_1 - MC_1}{1 + r} \right] \longrightarrow P_0 - MC_0 = 0$$

Thus, we want to operate where price equals marginal cost. In this case, the future is totally discounted; it is worth nothing to us. With an infinite discount rate, we are back to the static solution. Such a case may happen in the following situations:

- An oil firm in danger of being nationalized in an oil-producing country might totally discount the future and produce where price equals marginal cost.
- A government with high deficits and a threatened political future might also produce where price equals marginal cost for revenues spent on social programs today to ensure that there will be a political tomorrow.

Technical change has been an important factor reducing costs in the more mature oil-producing areas in the world, such as the United States. For example, horizontal and directional drilling, 3-D seismic, and improved fracturing techniques are unlocking massive shale gas and oil reserves, which were previously noncommercial. Improvement in deepwater drilling, including dual gradient drilling, is opening up new frontiers (Paganie 2012). Presalt oil, often in very deep water, had been hard to find as seismic waves were scattered and became difficult to interpret in their trip through the low-density salt. However, better algorithms and more powerful computers are making greater sense of these images, and related oil exploration efforts in the Gulf of Mexico, offshore Brazil, and offshore West Africa are currently hot plays (Durham 2010).

Model Six (No Income Growth, No Costs, with Backstop Technology)

As oil depletes and prices increase over time, we will gradually switch away from oil to other cheaper abundant products (commonly called *backstops*) and leave the remaining high-cost reserves in the ground. The view that we will never totally deplete our oil resources is most strongly expressed by Adelman (1993): "Minerals are inexhaustible and will never be depleted. . . . How much was in the ground at the start and how much will be left at the end are unknown and irrelevant."

Jaramillo et al. (2008) estimated per barrel costs of natural gas to liquids of around $50 to $60, and coal to liquids of around $60 to $70 per barrel. Given the abundance of these reserves, they might be considered backstop fuels. However, the existence of such backstops does not mean we should completely ignore time in our decision process. If we use up low-cost reserves now, we will not have them in the future. Time is still a factor in the process of deciding when to shift to backstop technologies.

In our model, the backstop is more expensive to produce than our low-cost reserves and is abundant; we assume it can be produced at a constant cost (P_{bk} = $70). Since any reserves at prices higher than $70 per unit will not be purchased, the backstop price puts an upper limit on the price we can charge for our product. It replaces the existing demand for all prices greater than $70 (fig. 14–11).

Demand in the current period ($P_0(Q_0)$) becomes the curve *abc*, and demand in the next period ($P_1(Q_1)$) becomes *def*. Using these new truncated demands we find $P_1 / (1 + r) = P_0$. P_1 = $70 is the price that we would charge in the second period. If the current price is P_0 = $70, we will want to produce all the reserves now and put the money in the bank to produce interest instead of waiting until next period to get the $70. This will drive reserve prices down in the current period until the following applies:

$$P_0 = 70/1.2 = \$58.333$$

At a price of $58.333:

$$Q_0 = m_{60} = 230 - 2.5 \times 58.333 = 84.167$$

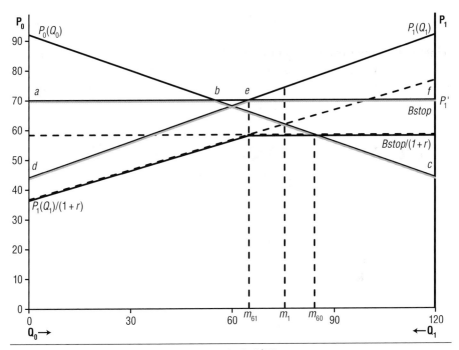

Fig. 14–11. Two-period model with a backstop fuel of $70

This leaves reserves of the following:

$$120 - 84.167 = 35.833$$

At a price of $70, demand in the next period is expressed as follows:

$$Q_1 = 230 - 2.5 \times 70 = 55$$

Or Q_1 = the distance from 120 back to m_{61}. Then Q_1 is satisfied with the existing reserves of 35.833, and the remaining consumption of $55 - 35.833 = 19.167$ is filled in from the backstop. This amount is the quantity between m_{61} and m_{60} in figure 14–11. Notice that under dynamic optimization, the fixed resource gradually approaches the backstop price.

The same analysis can be done for a price control as for a backstop. In the price-controlled case, the ultimate price is the controlled price, and the excess demand in the future period is a shortage instead of consumption from the backstop.

Dynamic Multiperiod Models

All of the above examples are for only two periods: a current and a future period. Although a simple model, it gives us the basic intuition about how to optimize over time. We always need to compare the price today with the discounted opportunity cost in other periods; if we produce today, we give up the opportunity to produce in another period. More realistically, our reserves will last many years through many periods.

Suppose we have n periods. By the same reasoning as above, the discounted present value of all prices should be equal across time.

$$P_0 = \frac{P_1}{(1+r)} = \frac{P_2}{(1+r)^2} = \ldots = \frac{P_n}{(1+r)^n}$$

All our reserves will be gone after n periods, expressed as follows:

$$Q_0 + Q_1 + Q_2 + \ldots + Q_n = R$$

Since we do not know n, typically such models will be solved iteratively (by computer) to also choose the optimal n.

Dynamic Models with Market Imperfections

So far we have also assumed that markets behave well. However, we also have market imperfections in dynamic optimization, just as we have them in static optimization. We will briefly consider four such problems:

- negative externalities
- private interest rate unequal to social interest rate
- property rights not well-defined, as in the United States under the law of capture
- firms with monopoly power

You can use the previous examples to understand the effect of negative externalities on optimal allocation. Suppose that energy production creates pollution so that private costs are higher than social costs. Comparing model one and model five shows how the market would misallocate resources. When we added costs to the model, we decreased current output and shifted consumption to the future. Thus, ignoring costs or using costs that are too low would cause us to consume too much today.

Earlier in this chapter, we argued that if the private market is competitive and there are no externalities, property rights are well-defined, and producers dynamically optimize, then social welfare is maximized as well. However, all of these criteria must be met for this to be true. This result requires producers to respond to the correct discount rate, which is the one that represents society's rather than the producer's opportunity cost. If the private rate is higher, our earlier modeling demonstrates that the resources would be exploited too rapidly. If the private rate is lower, the resources would be exploited too slowly.

In our discussion, we have used a discount rate that reflects the rate of return on an asset put into the bank. In reality, the discount rate should be the opportunity cost of capital. This could be money in the bank or another investment if it were the next best alternative. To determine when private and social rates diverge, we note that the cost of capital for a firm is composed of a risk-free rate (often represented by a safe government bond rate) plus a risk premium. This risk premium compensates capital owners for the fact that actual returns may differ from expected returns; the riskier the industry, the higher the required risk premium. If risks are higher for individual companies than for society as a whole, then private decision makers would require a higher risk premium, and social and private discount rates would diverge. This divergence would cause a misallocation of resources, with too much oil produced in the present and too little saved for the future.

Alternatively, in a society that is very dependent on oil sales, the risks of dependency on one product may be higher for society than for an integrated oil company. An integrated company's oil-producing division may do well when prices are high, but the product sales division may do better when oil prices are low, demonstrating the cyclical nature of the industry. In the event that the social discount rate is higher than the private discount rate, the private market will produce too little oil today and save too much for the future when compared to the social optimum.

Absence of well-defined property rights can cause problems as well. A well-defined property right typically entitles the owner to the use and proceeds of an asset over time. If it is exclusive and enforceable, the owner can effectively exclude others from the use and proceeds. Further, if it is divisible and transferable, the owner can divide and sell it whole or piecemeal. Poorly defined property rights can cause a misallocation of resources. For example, in the early years of the US oil industry, the law of capture prevailed. If A drilled into a pool of oil, and the oil came out of A's well, it was A's. If B drilled into the same pool, and the oil came out of B's well, it was B's. In such a case, property rights were poorly defined. If you did not produce the oil and another did, you lost the rights to the oil. You did not have exclusive use of the oil, even though it was under your property. In such a situation, producers had incentive to produce as much as possible for fear they would lose access to the oil. Price was driven down until it equaled marginal cost because producers infinitely discounted the future. As a society, we lost the Hotelling rents, which well-defined property rights would be more likely to preserve.

Two solutions to this problem have been utilized. In the United States, state commissions regulate oil production and well spacing. Elsewhere, fields are more often government-owned and leasers are required to *unitize* fields. A unitized field is operated as a unit, and each owner or leasee shares pro rata in the costs and revenue. Such a policy encourages producers to maximize the overall value of the field over time, since they will then maximize the present value of their share. An exception is the giant Middle Eastern offshore gas field known as the North Field for Qatar's share and as the South Pars field for Iran's share. In reality, these two fields actually consist of one large field, which is the largest known gas field on earth. With substantial foreign capital and investment profits, Qatar has been able to develop and produce more from its portion than has Iran, and the law of capture has thus far prevailed.

The last market imperfection that we will consider in a dynamic context is a monopoly producer. The monopoly has pricing power and would consider the whole demand curve $P(Q)$, not just the price. Since the monopoly looks at marginal revenue instead of price, the optimizing condition for a two-period model with no costs is the following:

$$MR_0 = \frac{MR_1}{(1+r)}$$

For the monopolist, marginal revenue, not price, goes up at the interest rate (figure 14–12). The competitive allocation is, as usual, m_1. The darker lines are the new marginal revenue and discounted marginal revenue curves that determine the monopoly allocation. The optimal level of production for the monopolist is at m_7. Prices in the two periods are P_0' and P_1', respectively.

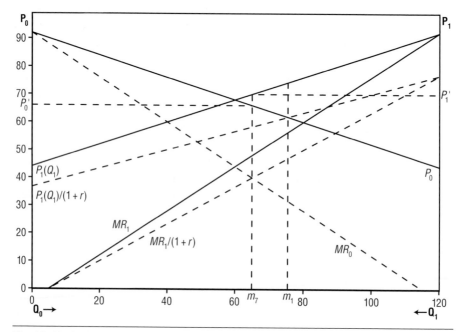

Fig. 14–12. A monopoly producer in a two-period model

How does m_7 compare to a competitive allocation? If you discount a higher price, the discounted price curve shifts down in a steeper manner than when you discount the lower marginal revenue. Thus, it is sometimes said that the monopoly is the conservationist's best friend, since the monopolist produces less now and puts off production to the future, permitting the resource to last longer. Adding costs to the monopoly model brings expected results. The monopolist would produce where marginal revenue minus marginal cost in the current period is equal to discounted marginal revenue minus marginal cost in the next period.

So, what does the analysis in this book tell us so far about how oil prices will evolve? It tells us that evolution depends on a shifting mélange of forces in the market, including market structure. It is affected by the dynamic and myopic behavior of firms and political decisions of producing and consuming governments. Oil price evolution further depends on costs, available reserves, short-run capacity, and interest rates.

Taxing and Bidding Decisions

In most areas of the world, the government owns the mineral wealth. Government oil companies develop and produce the minerals, though the private sector is often involved (table 14–3). Various types of agreements exist between companies and governments (Duval et al. 2009a). Private companies often bid for the right to search for oil and gas reserves and may pay taxes and royalties on the oil they produce from their leases. Private companies may be hired to develop and produce the governments' reserves. Our modeling thus far has implications for taxing and bidding systems for these governments' leasing, development, and production decisions for their petroleum properties.

Typically, a national oil company, a government ministry, or both, represent a government. Their goal should be to provide private-sector companies with the minimum rate of return required to enable them to develop and produce the mineral and to garnish the scarcity or Hotelling rents for the local inhabitants. Two types of agreements are entered into: concessions and production sharing. If the government grants a concession (as in the United Kingdom), it is granting an ownership right (often called a lease) to the company, which in turn pays taxes and royalties.

Table 14-3. Companies with significant oil or gas production or refinery capacity, 2011

Company	Prod. Liq. (10⁶ bbl)	Res. Liq. (10⁶ bbl)	Prod. Gas (bcf)	Res. Gas (bcf)	Ref. Cap. (10³ bbl/cd)	% Gov't. Owned	Homepage
OPEC - National Oil Companies							
PDVSA (Venezuela)	912	211,170	713	195,100	1,016	100	http://www.pdvsa.com
NIOC (Iran)	1,306	151,170	6,040	1,168,000	1,772	100	http://en.nioc.ir
Sonatrach (Algeria)	466	12,200	2,730	159,000	652	100	http://www.sonatrach-dz.com
NNPC (Nigeria)	795	37,200	920	180,458	445	100	http://www.nnpcgroup.com
ADNOC=Abu Dhabi	858	92,200	NA	200,000	205	100	http://www.adnoc.ae
INOC (Iraq)	903	143,100	254	111,520	800	100	http://www.oil.gov.iq
KPC (Kuwait)	914	101,500	422	63,000	936	100	http://www.kpc.com.kw
Saudi Aramco	3,395	264,520	2,720	275,200	2,107	100	http://www.saudiaramco.com
QPC (Qatar)	299	25,380	4,135	890,000	80	100	http://www.qp.com.qa/
Other National Oil Companies							
PetroChina Co. Ltd.	886	11,128	2,396	66,653	2,675	86	http://www.petrochina.com.cn/ptr
Sinopec (China)	322	2,848	517	6,709	3,917	76	http://www.sinopec.com
EGPC (Egypt)	244	4,400	1,330	77,200	NA	100	http://www.egpc.com.eg
ONGC (India)	203	2,594	897	8,825	3	69	http://www.ongcindia.com
Petronas (Malaysia)*	164	2,080	1,353	22,848	310	100	http://www.petronas.com.my
ENI (Italy)	308	3,134	1,491	16,198	767	32	http://www.eni.it/en_IT
PEMEX (Mexico)	949	10,264	2,407	17,224	1,703	100	http://www.pemex.com
Petrobras (Brazil)	741	10,783	4	12,381	2,044	54	http://www2.petrobras.com.br/ingles/index.asp
OAO Rosneft (Russia)	869	18,351	452	30,017	1,293	70	http://www.rosneft.com
OAO Gazprom (Russia)	236	5,284	18,123	140,248	843	50	http://www.gazprom.com
Statoil (Norway)	1,343	2,276	1,434	17,681	304	67	http://www.statoil.com

Large Private Oil Companies

Royal Dutch Shell (Netherlands/UK)	561	4,313	2,462	47,662	4,194	—	http://www.shell.com
Total SA (France)	448	5,784	2,226	30,717	2,314	—	http://www.total.com
BP (UK)	787	10,565	2,744	45,130	3,322	—	http://www.bp.com
ExxonMobil (US)	662	10,113	3,303	45,375	5,788	—	http://www.exxonmobil.com
Chevron Texaco (US)	676	1,805	6,455	28,683	2,559	—	http://www.chevron.com
Conoco Phillips (US)	274	3,817	1,632	17,600	2,568	—	http://www.conocophillips.com
OAO Lukoil (Russia)	649	13,123	754	23	551	—	http://www.lukoil.com
Total These Companies	20,167	1,161,102	67,915	3,873,453	43,169		
Total World	**26,484**	**1,523,225**	**116,514**	**6,746,751**	**88,054**		

Sources: True and Koottungal (2011, table 1); *OGJ* (2012b); *OGJ* (2012c); OPEC (2013); US EIA (n.d.d); and company Web sites.
Note: Companies with significant oil production or refinery capacity; *some values for 2010. CNPC = China National Petroleum Corp.; EGPC = Egyptian General Petroleum Corp.; ENI = Ente Nazionale Idrocarburi; KPC = Kuwait Petroleum Corp.; NIOC = National Iranian Oil Co.; NNPC = Nigerian National Petroleum Corp.; ONGC = Oil and Natural Gas Corp. Ltd. (India); PDVSA = Petróleos de Venezuela; Pemex = Petróleos Mexicanos; Petrobras = Petróleo Brasileiro SA; Petronas = Petroliam Nasional Bhd; QPC = Qatar Petroleum Corp.; Saudi Aramco = Saudi Arabian Oil Co.; Sinopec = China Petroleum & Chemical Corp.; Sonatrach = Société Nationale pour la Recherche, la Production, le Transport, la Transformation, et la Commercialisation des Hydrocarbures; and Statoil = Den Norske Stats Oljeselskap ASA. Statistics for a few additional companies can be found at http://dahl.mines.edu/t1403.pdf.

If the government enters into a production sharing agreement (PSA), it retains ownership. The company receives a share of production for providing services to the government (or a cash payment) rather than a portion of production, as seen in a service contract (Johnston 1994, 2003, 2008). PSAs began in 1966 with Pertamina of Indonesia (Machmud 2000), and samples of some production-sharing contracts can be found in Bindemann (1999). Despite philosophical distinctions regarding ownership between concessions and production sharing agreements, they can be designed to have identical economic impacts.

Leasing and taxation systems vary considerably across countries, across time, and even within countries. We will consider five policies in the leasing taxation context and their effect on production profiles and economic efficiency:

- unit tax (t_u)
- royalty on price (ad valorem) (t_a)
- user cost or rent tax (t_r)
- bonus bidding (T)
- work bidding (W)

A unit tax would affect production in the same way that costs affect production (as in model five). That is, the tax would decrease current consumption and push production off to the future. However, since taxes are not true costs (they are considered transfers by economists), a unit tax would distort the true costs of production.

Under the concessionary system, governments usually charge a royalty on the price of oil, equivalent to an ad valorem tax (t_a). The objective function for a resource under such a tax would be to maximize the after present value of after tax profit, as follows:

$$\text{Maximize: } PV = (1 - t_a)P_0Q_0 - TC(Q_0) + \frac{[(1 - t_a)P_1Q_1 - TC(Q_1)]}{(1 + r)}$$

Subject to the following resource constraint:

$$R = Q_0 + Q_1$$

The Lagrangian function is the following:

$$\Im = (1 - t_a)P_0Q_0 - TC(Q_0) + \frac{[(1 - t_a)P_1Q_1 - TC(Q_1)]}{(1 + r)} \lambda(R - Q_0 - Q_1)$$

First-order conditions are the following:

$$\Im_\lambda = E - Q_0\, Q_1 = 0$$

$$\Im_{Q_0} = (1 - t_a)P_0 - MC_0 - \lambda = 0$$

$$\Im_{Q_1} = \frac{[(1-t_a)P_1 - MC_1]}{(1+r)} - \lambda = 0$$

Since these are dissimilar from first-order conditions for the socially optimal production derived previously, a royalty levied on price alone distorts the market and results in deadweight losses.

The easiest way to track the direction of the distortion is to understand that the tax shifts down the demand curve, with the shift wider at higher prices. Deducting marginal cost from this lower price and discounting the lower $[(1 - t_a)P_1 - MC]$ causes a smaller shift in the future period than in the case of no tax and calls for less consumption in the current period and more in the future period. However, a percentage tax (t_r) on user costs or rents does not distort the market.

This can be demonstrated as follows. The Lagrangian function is:

$$\Im = (1 - t_r)[P_0 Q_0 - TC(Q_0)] + \frac{(1-t_r)[P_1 Q_1 - TC_1(Q_1)]}{(1+r)} + \lambda(R - Q_0 - Q_1)$$

where TC is the total cost, including any opportunity costs.

First-order conditions are as follows:

$$\Im_l = R - Q_0 - Q_1 = 0$$

$$\Im_{Q_0} = (1 - t_r)(P_0 - MC_0) - \lambda = 0$$

$$\Im_{Q_1} = (1 - t_r)\frac{(P_1 - MC_1) - \lambda}{(1+r)} = 0$$

These conditions can be rearranged as follows:

$$(1 - t_r)(P_0 - MC_0)\frac{(1-t_r)(P_1 - MC_1)}{(1+r)}$$

Cancel $(1 - t_r)$ from both sides, and you are left with the following:

$$P_0 - MC_0 = 0$$

This is the same allocation as without a tax. Thus, a rent tax would not distort the allocation of the resource over time. However, rent taxes are usually levied on accounting profits rather than on economic profits or Hotelling rents. Businesses are not allowed to deduct their opportunity costs, which we would need to do to compute rents. Thus, in a perfect world, the tax should be on $(P_0 Q_0 - $ accounting $TC(Q_0) - $ opportunity cost).

Bonus bidding requires firms to pay money up front for the right to search for oil. Since this is a fixed, upfront payment, it would not distort the production profile. If the bidding is competitive, we expect companies to bid away excess profits or rents and only make a normal rate of return. Thus, economists typically favor bonus bidding so that the government or society receives the rents rather than the producers.

Critics of this policy argue that large upfront costs may prevent some firms from bidding, thus making the market less competitive. This has led some governments to allow smaller companies to make joint bids. Mead (1994) argues that lowering US restrictions on foreign firms bidding on acreage would also increase competition.

Another argument against this type of bidding is called the *winner's curse*. When bidding for acreage, bidders do not know the actual hydrocarbon potential. Capen et al. (1971) argue that under bonus bidding, firms that consistently overestimate resource potential usually win the bids. But by overestimating potential, they make less than a normal rate of return and pay the resource owner more than the rents for the property. Although popular with engineers and geologists, economists discount this argument. Firms that consistently make incorrect bids and reap lower-than-expected profits would not remain in business. They would have to learn from incorrect bids and curb their "wild-eyed" optimism in order to survive.

Some governments use work bidding, which is typically measured in such things as kilometers of seismic run, wells drilled, and production developed. Work bidding may also be done in conjunction with other types of bidding. In strict work bidding, the company that promises to do the most drilling or develop the most resources the most quickly wins the bid. Often, this type of bidding is used by a government with a political goal of quickly increasing employment or gaining revenue, which will be satisfied if the oil is developed sooner. With competitive work bidding, the winning company will bid away the entire rent, whereas in an optimal production profile, producers should equalize the present value of the rents over the life of the project.

Of the five types of policies outlined, economists prefer competitive bonus bidding as the most likely to promote economic efficiency while garnishing the maximum rents for the government. Mead (1994) sums up empirical research on bonus bidding that supports this conjecture.

A Foray into the Real World

In all the previous cases in this chapter, the modeling was for discrete decisions. However, oil is produced continuously, and we can employ somewhat fancier mathematical techniques to solve for a continuous production profile across time. See, for example, http://dahl.mines.edu/b1401.pdf.

All of the above models assume some fixed ultimate reserves with perhaps exogenous shifts in reserves. In reality, as Adelman (1990) indicates, total reserves are unknown and probably unknowable. However, producers are making upstream decisions every day as to how many leases to acquire, how much seismic survey to shoot or buy, and how many wells to drill. Current and expected future prices and costs provide an incentive to invest or disinvest in exploration and development. With economic depletion, reserves are not gone. Rather, the current and expected prices and costs of exploration and development do not warrant continued development and production.

Some dynamic modelers have specifically included investment decisions in their models. The classic example is in the widely cited paper by Robert Pindyck (1978). His dynamic model with endogenous exploration results in more interesting price and production trajectories than in a simpler Hotelling model. However, we still do not get the volatile real-world prices as in figures 3–3, 7–7, and 8–4.

In the real world, producers make lumpy investment for an unknown future. Capital costs are high and operating costs are low. Long-term marginal costs are high, and short-term marginal costs are low. Much of OPEC's market power derives from restricting investment. When demand increases unexpectedly, as in the years 2004 to 2008, spare capacity falls, and prices can shoot up. As consumption falls and non-OPEC producers increase production, spare capacity again resurfaces, and OPEC must cut production if it wants to shore up the price. Alternatively, when demand suddenly dropped, as with the Asian crisis, prices dropped like a rocket, only to rebound as consumption recovered and production fell in and out of OPEC. Thus, with lumpy investment and the lag between exploration and production, producers are often responding to short-term cost and price levels. As a result, cycles of shortage and surplus can develop.

Summary

About one-half of the world's proven oil reserves are located in the Middle East, and more than two-thirds of the world's gas reserves are found in the Middle East and the former Soviet Union. R/P ratios for these countries are often many decades, and when used in conjunction with reserves figures, R/P can indicate the relative abundance of the resource in particular countries. Since production today influences production tomorrow for nonrenewable resources, economic efficiency requires dynamic optimization of resources.

We considered two-period and multi-period models to illustrate the basic principle of allocating these resources efficiently across time. Theory suggests that in a competitive discrete model with fixed resources, well-defined property rights, and no costs, the price of the resource would go up at the interest rate. Higher income growth in the future would raise prices and shift production to the future.

If we add costs to the model, the user cost (or price minus marginal cost) should rise at the interest rate and production should shifts toward the future. This user cost (or Hotelling rent) helps allocate resources efficiently across time. To the extent possible, opportunity costs should be considered both across projects and across time for better resource allocation.

Raising the interest rate discounts the future and shifts production to the present. However, if the interest rate goes to infinity, we are back to the familiar competitive case of price equal to marginal cost. Marginal costs may be a function of production if there are economies or diseconomies of scale. They may rise with depletion and fall with technical change. In the model with costs, prices can fall over time if costs fall fast enough.

Adelman argues that although oil and gas reserves are technically finite, we will continue to use them until they become too expensive, at which point we will switch to alternative energy sources. Often we call such alternatives backstops or backstop technologies. Modeling suggests that if we do not interfere in markets and producers dynamically optimize, oil and gas prices should gradually approach the price of these backstops.

We can use discrete or continuous mathematical modeling to show that under ideal conditions a competitive market model maximizes social welfare. However, market failures can cause markets to allocate resources inefficiently in dynamic models just as they do in static models. Negative externalities will cause the market to produce too much in the current period, such as when the private interest rate is larger than the social interest rate. Poorly defined property rights that do not enable a producer to exclude another from its reserves are also likely to cause too much production today. Unitized production and proration have both been used to alleviate this problem.

Our last market failure example had the opposite effect. In a monopoly case, marginal revenue minus marginal cost increases at the interest rate. Monopoly firms tend to restrict output now and shift output to the future. By allocating production over a longer time horizon, they are able to keep prices higher and extract more of the rents for themselves.

We have considered both competitive and monopoly models in earlier chapters where static optimization was accomplished. The first-order conditions for dynamic models compared to static models are summarized as follows:

Static Perfectly Competitive

$P = MC$

Dynamic Perfectly Competitive

$$P_0 - MC_0 = \frac{(P_1 - MC_1)}{(1+r)}$$

Static Monopoly

$MR = MC$

Dynamic Monopoly

$$MR_0 - MC_0 = \frac{MR_1 - MC_1}{(1+r)}$$

Most of the world's mineral resources are owned by governments, which allow development and production by national oil companies or through production sharing or concessionary systems. A government's goal should be to allow companies a competitive rate of return while reserving the rents for its economy. In considering five leasing taxation systems for their efficiency effects, we saw that unit taxes, royalties on price, taxes on accounting profit, and work bidding all distort the optimal production profile. A tax on Hotelling rents would not distort the profile but may be difficult to implement, since user costs or rents are hard to measure.

Conventional profit tax schemes typically do not allow companies to deduct opportunity costs, while companies have an incentive to inflate cost numbers to reduce their tax liability. Theory and studies suggest that competitive bonus bidding would not distort the production profile and would be most likely to transfer the rent to the government. However, if high-cost, high-risk acreage is so expensive that only very large firms can enter, there may be some danger that the bidding may not be competitive. In such a case, allowing joint bids from smaller firms and international bidding can provide a more competitive playing field.

Another potential problem is the winner's curse. Firms that more often overestimate reserves and overbid are more likely to win the bid, yielding the government more than the Hotelling rents. However, such firms cannot stay in business in the long run. Therefore, as with other loss-making activities, economists expect the market to correct this behavior in the long run.

The models considered here assume a fixed amount of reserves that will be depleted over time. The known demand and cost curves and the resulting price and production profiles are smooth and predictable. However, in the real world, reserves are developed. We will never completely exhaust oil and other resources, nor will we ever know how much of the resource truly exists. Further lumpy investment with long investment/production cycles implies that costs in the short term diverge widely from costs in the long term. Thus, short-term production decisions resulting from shocks to the system can cause the non-Hotelling swings in inventory, excess capacity, and price that we see daily in the market.

15

Supply and Costs Curves

Some Common Pitfalls in Decision Making . . . Ignoring Opportunity Costs . . . Failing to Ignore Sunk Costs.

—Robert H. Frank, Professor at Cornell University

Introduction

Primary energy comes from a variety of sources. The majority comes from fossil fuels, which are commonly measured by volume or weight. For comparison across countries and energy sources, these weights and volumes are typically converted to energy content. The most common units of such measurements are British thermal units (Btus), joules (J), or million tons of oil equivalent (Mtoe). Since electricity has no weight or volume, it is typically measured using energy content by kilowatt hours (kWh).

Primary energy is measured at the point of first consumption. Thus, commercial primary energy, which excludes traditional biomass, includes all fossil fuels and all nonfossil/biomass generated electricity, such as nuclear, hydro, wind, and solar sources. Electricity generated from fossil fuels has already been counted as primary energy consumption. If we counted both the fossil fuels and the electricity generated by fossil fuels, we would be double counting energy. Thus, fossil-generated electricity is designated as *secondary energy*, as are oil products. Noncommercial traditional biomass fuels also would be considered primary energy supply, with any power produced by biomass being secondary energy.

A country's *total primary energy supply* (TPES) includes domestic primary energy production plus imports less exports and an adjustment for bunkers and inventory change. To compare TPES across sources or across countries, conversions across fossil fuels are fairly straightforward. Comparisons are based on the heat content of the fuels, which varies by fuel and country (US EIA, n.d.d.).

Converting primary electricity or electricity from nonfossil sources to gross energy supplied is more problematic, for the following reason. If one country

367

generates electricity from 1,000 Btus of coal at a 33% efficiency rate, it generates 333 Btus of electricity output but measures energy consumption at 1,000 Btus. The difference is the unusable heat lost in the process. Electricity generated from renewable sources and nuclear power may be measured by output. Thus, if a second country generates 333 Btus of electricity by nuclear power and power is measured as output, the end use of electricity would be the same as the first country, but it would appear that the first country consumed three times more energy than the second. For this reason, when comparing total energy consumption across sources and countries, gross electricity production may be computed as the energy content of an equivalent amount of fossil fuels. Thus, for the second country, net electricity consumption would be 333 Btus, but gross consumption would be 333/0.333 = 1,000 Btus. Different statistical sources that report electricity use or production in some heat content or calorific unit (British thermal units, kilojoules, or tonnes of oil equivalent) may convert primary electricity in different ways.

An example may be seen from table 15–1. Note the differences for four prominent highly cited sources of data. EIA adjusts one kilowatt hour of all primary electricity sources for heat loss, and IEA only adjusts sources that have heat in the generation process. The UN and BP convert no primary electricity sources for heat loss. However, when presenting consumption of electricity in kilowatt hours, the four sources use the same procedure by presenting the energy actually consumed as electricity.

Table 15–1. Example conversions of one kilowatt hour primary electricity to Btu energy

	Nuclear	Hydro	Geothermal	Wind	Solar (PV), Tide, Wave	Biomass
EIA: US 2012	10,452	9,516	9,516	9,516	9,516	9,516
IEA	10,339	3,412	10,339	3,412	3,412	NA
UN	3,412	3,412	3,412	3,412	3,412	3,412
BP	3,412	3,412	3,412	3,412	3,412	3,412

Sources: US EIA (n.d.d); IEA (2005); UN (2010); BP (2012).
Note: 1 kilowatt hour = 3,412 Btus. US EIA numbers are specifically for the United States, with all primary generation adjusted for some equivalent heat content. IEA suggests they take numbers by country when available, but I was unable to locate any specific country conversions. Numbers are based on their general conversions from kilowatt hours. They adjust for heat content for thermal processes (nuclear and geothermal) but not for other primary renewables. The UN and BP use the same conversions for all countries with no adjustment for heat loss.

Another distinction made for electricity is gross versus net generation. If measured in kilowatt hours, the distinction most often indicates gross as the kilowatt hours generated, and net as the gross minus the amount used by the energy sector, including transmission losses. If measured as heat content, gross may be kilowatt hours adjusted for heat content and net may be kilowatt hours not adjusted, as noted above.

Alternatively, when fossil fuels are burned, water vapor is produced. When this water vapor condenses, heat is lost. The IEA refers to gross heat content for fossil fuels as the amount of heat generated but net as gross minus the amount of heat lost to condense the water vapor. They indicate these losses as 5% to 6% for solid and liquid fuels and 10% for natural gas.

The evolution of primary and secondary energy consumption patterns is influenced by both the properties of the various energy sources and their costs. In this chapter, we first consider supplies from these sources, their characteristics, and how to "cost" these different energy sources.

Commercial energy is energy purchased in a market. Examples include oil, gas, and nuclear power but not wood, dung, or other biofuels that are gathered rather than purchased. The consumption of commercial primary energy to the world in 2011 was estimated at about 519 quadrillion Btus (548 exajoules). About 20% of this was consumed in China, 19% was consumed in the United States, and Russia, India, and Japan accounted for another 14% (US EIA, n.d.d).

Figure 15–1 shows the breakdown of global total primary energy supply (TPES) sources and the prominence of fossil fuels. This supply also includes estimates for noncommercial bioenergy and waste (bioE&W).

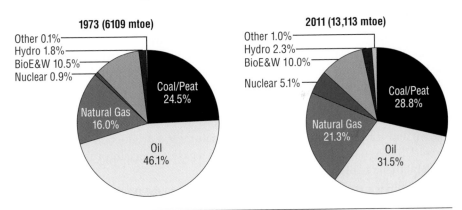

Fig. 15–1. Gross world primary energy consumption
Source: IEA (2013a).

Note: The IEA conversions for primary electricity sources are as shown in table 15–1. Conversions used by the EIA would give a larger share to primary electricity sources, and those used by the UN and BP would give a smaller share. Crude oil and other liquid hydrocarbons (labeled "oil" in fig. 15–1) provide the highest amount of energy of any source, at almost one-third of world primary energy supply. More than 60% of this use is for global transportation, primarily gasoline, diesel, jet fuel, and bunkers. In North America, transport use of oil is even higher, estimated at more than 75% (IEA, n.d.m).

Coal, discussed in chapter 3, constitutes more than one-fourth of world energy supply. Steam coal is used as an energy source, burned to create process heat and steam for electricity generation. Coking coal, which is about 10% of total coal and peat production by volume, is used in iron and steel production (IEA, n.d.b). It is not included in fig. 15–1.

Natural gas constitutes about one-fifth of global energy supply. *Nonassociated natural gas* is found alone, while *associated natural gas* is found with oil. In the United States, about 85% of gas reserves are nonassociated (US EIA, n.d.d). As natural gas is produced, pentanes (C_5H_{12}) and heavier hydrocarbons condense due to the pressure and temperature drop when arriving at the surface, and they are removed at the wellhead. These liquids are called *lease condensates.* They are lighter than WTI and can be blended into crude streams for refineries designed to run light crudes. They also can be used as a diluent for bitumen from oil sands or split into components including naphtha (typically C_6–C_{11}), which can be used to produce petrochemicals in plants designed to run naphtha. The remainder, called *wet natural gas,* is sent to gas processing plants, where NGLs are extracted, leaving *dry natural gas.*

In 2012, about 4% of US wet natural gas production (excluding lease condensate) by volume was NGLs; less than 1% was flared, and about 11% was reinjected to maintain well pressure. Worldwide, total wet gas production in 2012 was about 147 trillion cubic feet (4.2 trillion cubic meters), of which around 80% was dry, 5% was NGLs, 3% was flared or vented, and 11% was reinjected (US EIA, n.d.d). The share of gas flared or vented has continually declined over time, sometimes at the prodding of governments. However, where amounts of gas produced are small and nearby markets or infrastructure are absent, a higher share may be flared or vented. For example, in 2012, the US EIA estimated that in Angola, Cameroon, Gabon, and Iraq, more than 45% of gross natural gas production was vented or flared (US EIA, n.d.d). We know vented natural gas is especially problematic, if we recall methane's potency as a greenhouse gas.

The liquids extracted from natural gas (NGLs) make up a significant portion of global liquid hydrocarbon supply. In 2013, global oil production including lease condensate was about 76.1 million barrels a day, and NGLs added approximately another 9.3 million barrels per day. The NGLs, which are lighter, average around 70% of the energy content of crude oil. The largest producers of NGLs are those with significant natural gas production. Four countries contributed almost 60% of world supply in 2013: the United States, Saudi Arabia, Canada, and Russia (US EIA, n.d.d).

IEA (2013a) estimates that biofuels and waste constitute another 10% of primary energy supply. Such fuels have been recently promoted in the United States and Europe as transport fuels in the form of ethanol and biodiesel. However, most is still used in the residential sector of developing countries for cooking and heating. Around 80% of this fuel is estimated to be used in Latin America, Africa, and developing countries in Asia, three-fourths of which is used in the residential sector. However, in poorer countries, such fuels are typically gathered rather than acquired in the market, reducing our understanding of exactly how much and how such fuels are used. Figure 15–1 shows that biofuels have remained a roughly constant share at around 10% but there has been a relative shift from oil into gas and coal. Primary electricity in the form of nuclear (adjusted for heat rates) and hydro (not adjusted for heat rate) together accounted for another 8% of global energy supply in 2010. Nuclear is more than two-thirds of the combined portion.

Nuclear Fuels

US EIA (n.d.d) indicates that nuclear energy provides more than 5% of gross primary energy supply worldwide, more than 10% of total electricity production, and more than one-third of primary electricity supply. The world's largest nuclear generators in 2012 were the United States, France, and Russia. Together these four countries accounted for more than 45% of global production. Although Japanese production was small in 2012, it accounted for more than 10% of world nuclear production in 2010, prior to the Fukushima accident. China still produces less than 5% of global production but saw annual growth averaging more than 14.5% a year from 2000 to 2012. Other countries with double-digit annual growth during this same period were Pakistan and Brazil.

Prior to the nuclear accident at Fukushima-Daiichi, nuclear power was being increasingly considered as one of the bridges to a low carbon output world. Although some countries do not seem to have reduced their commitment to nuclear, such as China, others such as Germany and Japan have been reconsidering their options.

Uranium is the fuel source for the bulk of nuclear power (86% in 2012), with the remainder from recycled plutonium in spent fuels and nuclear weapons disposal (World Nuclear Association 2013a). More than two-thirds of known uranium resources are concentrated in five countries: Australia, Kazakhstan, Russia, Canada, and Niger (table 15–2). Production is even more concentrated, with the top four countries, Kazakhstan, Canada, Australia, and Niger, accounting for more than 70% of production.

The world uranium industry is also rather concentrated, with eight companies accounting for 85% of world production in 2011, as shown in table 15–3. Companies that are partially or totally owned by their respective governments account for 45% of production. Such companies include KazAtomProm, AREVA, and ARMZ.

Table 15–2. Known recoverable resources and mined production of uranium (tonnes)

	Resources 2011		Production 2012	
	Tonnes U	Percentage of World	Tonnes U	Percentage of World
Australia	1,661,000	31%	6,991	12%
Kazakhstan	629,000	12%	21,317	37%
Russia	487,200	9%	2,872	5%
Canada	468,700	9%	8,999	15%
Niger	421,000	8%	4,667	8%
South Africa	279,100	5%	465	1%
Brazil	276,700	5%	231	0%
Namibia	261,000	5%	4,495	8%
United States	207,400	4%	1,596	3%
China	166,100	3%	1,500	3%
Ukraine	119,600	2%	960	2%
Uzbekistan	96,200	2%	2,400	4%
Other	254,200	5%	1,851	3%
World Total	5,327,200		58,344	

Source: World Nuclear Organization (2012, 2013c).

Table 15–3. Largest uranium producers in the world, 2012

Company	Headquarter Country	U (t*)	%	Homepage	Government Holdings
KazAtomProm	Kazakhstan	8,863	15	http://www.kazatomprom.kz	100%
AREVA	France	8,641	15	http://www.areva.com	78.5% Gov't
Cameco	United States	8,437	14	http://www.cameco.com	Public
ARMZ	Russia	7,629	13	http://www.armz.ru/eng	100%
Rio Tinto	UK/Australia	5,435	9	http://riotinto.com	Public
BHP Billiton	Australia	3,386	6	http://bhpbilliton.com	Public
Paladin	Australia	3,056	5	http://www.paladinenergy.com.au	Public
Navoi	Uzbekistan	2,400	4	http://www.ngmk.uz/en	100%
Other		10,497	18		
Total		58,344	100		

Source: World Nuclear Association (2013c) and company homepages.
Note: *t = metric tonnes.

About 20% of world uranium production was from open pits in 2012. Underground mining accounted for much of the rest. With conventional underground methods (28%), ore is brought to the surface, crushed, ground, and treated with sulfuric acid to leach out the uranium oxides (U_3O_8). With underground in-situ methods (45%), an acid or basic solution is circulated underground and then pumped back to the surface, where the uranium oxides are recovered (World Nuclear Association 2013c). The ores are milled to a khaki-colored product called *yellow cake*, which is about 80% U_3O_8. Joint products such as nickel, cobalt, or molybdenum can also be extracted at this stage. In the next stage, yellow cake is converted to uranium hexafluoride (UF_6). Four companies convert more than 90% of the yellow cake to UF_6: Cameco, AREVA, Russia's Rosatom, and the US-based USEC (World Nuclear Association 2013d).

There are 495 commercial nuclear power plants operating or under construction worldwide, and most require enriched fuel. Exceptions are the Canadian CANDU and the old British Magnox reactors, but the latter are being phased out. (I only know of one Wylfa style in operation, http://www.magnoxsites.co.uk/site/wylfa/.) Natural uranium comes in two isotopes: U-235, which is fissionable but accounts for less than 1% of uranium, and U-238. For use in power plants, the uranium must be enriched or the share of U-235 must be raised to between 3% and 5%. The small difference in mass allows the UF6 gas to be enriched using gaseous diffusion (currently about 25% of low-level enrichment capacity) and by centrifuges (65% of low-level enrichment capacity). The older, more energy-intensive gaseous diffusion is projected to be phased out by 2017. A new laser enrichment technology, Global Laser Enrichment (GLE), is also under consideration by companies in the United States, Canada, and Japan. Another 10% of capacity for low-level enrichment takes highly enriched weapons-grade uranium and recycles it to low-level power plant grade (World Nuclear Association 2013).

The ability to enrich is politically sensitive, since such capacity has the potential to produce weapons-grade uranium with more than 90% U-235. Iran's enrichment program, claimed to be only for power generation, is the source of the current dispute and economic embargo of Iran implemented in July 2012 and partially lifted in January of 2014.

Enrichment is highly concentrated, with strong international controls through the IAEA. Four companies accounted for more than 95% of global capacity. Russia's Rosatom operated and owned much of the Russian capacity that accounted for 40% of global capacity in 2010, and its capacity is projected to increase. Kazakhstan, Armenia, and Ukraine have small equity shares in some of this Russian capacity. Rosatom was followed by URENCO's 23% share, owned by the United Kingdom and Netherland's governments, plus E.ON and RWE in Germany, with capacity in those three countries. US-based USEC has a 20% share, but it is old capacity and is set to shrink by about two-thirds by 2015. EURODIF, which closed down in 2012, had five owners: France (60%), Italy, Spain, Belgium, and Iran. It is being replaced by the new Georges Besse II plant. This plant is operated and mostly owned by a subsidiary of AREVA, with a total of 10% owned by GDF Suez, a Japanese partnership, and Korea Hydro and Nuclear Power (KHNP). It had about 15% of global share in 2010. Much of the remaining enrichment capacity is in China. It, too, is projected to increase and could approach 10% by 2020 (World Nuclear Association 2013f). The enriched UF_6, after being changed to uranium dioxide (UO_2), is typically encased in ceramic pellets for placement into zirconium fuel rods.

Around 30 countries produced nuclear power in approximately 430 plants in 2011. (See table 15–4 for those with more than 2% of generation or 2% of capacity, operating and under construction.) Nuclear waste disposal is an ongoing problem in most of these countries, and a number of countries have opted for reprocessing spent fuel, including France, the United Kingdom, Switzerland, Russia, Japan, China, and India. With reprocessing, plutonium and uranium are extracted and mixed with new fuel to create mixed-oxide fuel (MOX). This increases the energy extraction from the uranium by one-fourth and reduces the volume of waste to about one-fifth by volume (World Nuclear Association 2013e). The United States had originally intended to reprocess, but such activities have been banned since the late 1970s under fear of nuclear proliferation. All the final spent fuel in the United States and elsewhere is currently being stored at the reactor or off-site, pending decisions for final disposal. Deep geological internment is most generally considered to be the best option, but social protest at moving the fuels though populated areas, along with the storage option chosen, have slowed or stopped this process in many places (Alley and Alley 2013).

Table 15–4. Nuclear power reactors operating and under construction for selected countries

Country	2012 Nuclear Electricity Generation		2013 Operable Reactors		Reactors under Construction		Uranium Required
	billion kWh	% of total	No.	MWe net	No.	MWe gross	tonnes U
Brazil	15.2	3.1	2	1,901	1	1,405	325
Canada	89.1	15.3	19	13,553	0	0	1,906
Finland	22.1	32.6	4	2,741	1	1,700	728
France	407.4	74.8	58	63,130	1	1,720	9,254
Germany	94.1	16.1	9	12,003	0	0	1,934
India	29.7	3.6	20	4,385	7	5,300	1,261
Japan	17.2	2.1	50	44,396	3	3,036	4,425
South Korea	143.5	30.4	23	20,787	4	5,415	3,769
Russia	166.3	17.8	33	24,164	10	9,160	5,073
Spain	58.7	20.5	7	7,002	0	0	1,355
Sweden	61.5	38.1	10	9,399	0	0	1,469
Ukraine	84.9	43.6	15	13,168	0	0	2,356
UAE	0.0	0.0	0	0	1	1,400	0
United Kingdom	64.0	18.1	16	10,038	0	0	1,775
United States	770.7	19.0	103	101,570	3	3,618	18,983
World	**2,346.0**	**NA**	**435**	**374,524**	**66**	**68,309**	**66,512**

Source: World Nuclear Association (2013g).
Note: "Operable" indicates connected to grid and "under construction" indicates concrete has been poured or units are being refurbished. MWe = megawatts of electrical capacity; UAE = United Arab Emirates. See http://dahl.mines.edu/t1504.pdf for original source for information on additional countries. NA = not available.

For example, after the conclusion of a long study beginning in 1978, Yucca Mountain in Nevada was to be the final US repository. The US government was to have begun transporting spent nuclear fuel from power plants in 1998. However, with defunding by Congress and political opposition, the Yucca Mountain option was taken off the table and no replacement has been arrived at. France, Finland, and Sweden are credited with making the most progress toward permanent storage, but no waste has yet been injected. The first such storage is expected in 2020 (IAEA 2010; Feiveson et al. 2011).

In addition to currently operating generators, an additional 66 are under construction as of 2013 (see table 15–4 for countries with more than 2% of additional capacity). They will add about 18% to operating capacity. The majority are in Asia, with China accounting for about 44% of the new capacity and Russia adding another 13%. The World Nuclear Association indicates an additional 160 reactors are in the planning stages and have received a firm commitment, with expectations of being operational within the next 8 to 10 years. If all goes according to plan, these reactors would add almost another 50% of capacity compared to those operating in 2013.

Hydroelectricity

Hydroelectricity accounted for about 38% more power generated than nuclear worldwide in 2011 (IEA, n.d.d.). China, Brazil, Canada, and the United States (ordered by 2011 production) are the largest producers, with slightly more than one-half of total world production (US EIA, n.d.d). For more than 50 countries, the majority of their power generation is from hydro sources. Indeed, for 9 countries, hydro accounts for 99% or more of their electricity generation: Paraguay, Albania, Republic of Congo, Ethiopia, Lesotho, Mozambique, Zambia, Bhutan, and Nepal.

To generate hydroelectricity, falling water, rather than steam, turns a bladed wheel inside a turbine. The amount of electricity produced depends on the water flow per unit of time and the *head*, which is the turbine distance from the surface of the water. If the flow (F) is measured in cubic meters per second and the head (H) is measured in meters, then an approximation of the power created (P) in kW is the following:

$$P = 5.9HF$$

If the turbine is 10 meters from the top surface of the water and 20 m^3 flow through the turbine per second, then the capacity = 5.9 × 10 × 20 = 1,180 kW. If the turbine runs for an hour, it would produce 1,180 kWh. Since a higher head produces more power, one function of dams is to raise the water height (fig. 15–2). Dam storage also allows for better control of water flow for measured power output (releasing water in dry years or seasons and storing it in wet years or seasons).

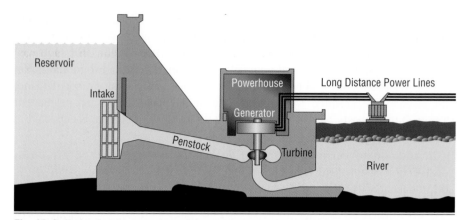

Fig. 15–2. Hydroelectric power from a dam
Source: TVA (n.d.).

Hydropower installations vary tremendously in size from run-of-river plants (no dam) with a capacity of less than 1 MW to multiple turbines at huge dams with capacities of thousands of megawatts (table 15–5). WEC (2010, 289) suggests there is considerable uncertainty over potential hydro resources, but we might realistically expect them to be twice the current capacity, with the largest potential in Asia.

Table 15–5. Sample of the world's largest hydro capacity dams

Name	River	Country	Capacity (MW)	Year Completion, Additions, Repairs
Three Gorges	Yangzte	China	22,500	2012
Ust-Ilimskaya	Angara	Russia	4,320	1980
Bratskaya	Angara	Russia	4,500	1984
Churchill Falls	Churchill	Canada	5,428	1961
Long Tam	Hongshui	China	6,300	2007
Sayano-Shushenskaya	Yenisei	Russia	6,400	1985/89/ 2009/14
Grand Coulee	Columbia	United States	6,809	1942/1980
Tucurui	Tocantins	Brazil	8,370	1984/1998
Guri	Caroni	Venezuela	10,200	1986
Itaipu	Parana	Brazil/Paraguay	14,000	1984/1991/2003
Millennium Project*	Blue Nile	Ethiopia	5,250	2014

Sources: Gulf Oil & Gas (2011); CIGB (c. 2012).
Note: *Grand Ethiopian Renaissance Dam, formerly known as the Millennium Dam.

Hydropower has few carbon emissions except initially when flooded vegetation decays. It also has no fuel costs, but power supply depends on rainfall. The dams provide flood control, the lakes behind them provide recreational boating and fishing and water for irrigation. However, changes in ecosystems are associated with large dams. When upstream areas are flooded to create the reservoir, downstream areas may suffer water reductions. Fish may not be able to migrate to spawning grounds over the dams. Populations are often displaced, and culture heritage may be destroyed. These ecosystem and cultural changes may generate considerable environmental opposition to new dams.

Other Renewable Energy Sources

Other renewable energy sources include biomass, wind, solar, and geothermal. They accounted for an estimated 10% to 11% of global energy consumption in 1973 and 2010 (fig. 15–1). Biomass, fuel derived from living organisms, constituted more than 90% of this other energy. It once heated huts and castles and is still an important source of energy in the developing world in the form of wood and dung. About 68% of the population in Africa and around 50% of the population in developing countries in Asia still used traditional biomass for cooking in 2010 (IEA 2012c).

Solar energy has provided heat and light from the dawn of human history and is also the indirect energy source for biomass and wind. Geothermal has heated baths for centuries, while the first use of geothermal for a power plant was in Italy in 1903. Even waves and tides created by the gravity between the earth, sun, and moon have energy potential. Tidal power is very limited worldwide, but a 240 MW plant in Rance, France has been in operation since 1966. Another somewhat larger project began operation in Sihwa, South Korea in 2011. A handful of other plants with extremely small capacity (less than 25 MW) are also operational around the world (WEC 2010, 544; Chanal 2012).

High use of biomass fuels in poorer countries is labor intensive, produces indoor air pollution, and may contribute to deforestation and erosion. Further, using dung for fuel instead of fertilizer may have repercussions on food production. Thus, as countries develop, they typically move away from renewable energy sources toward more concentrated, convenient, and efficient commercial fuels (petroleum products, natural gas, LPG or bottled gas, coal, and electricity). However, more recently, environmental concerns about global warming and technical advances have improved the long-term outlook for renewables. Barring some major technological breakthrough, renewables will undoubtedly need to play a key role in meeting global targets to reduce greenhouse gas emissions.

There is strong need to use traditional biomass more efficiently. For example, when cooking on an open fire, up to 95% of the energy is lost, whereas more efficient stoves could more than double the usable heat and reduce the pollution generated. Higher priced stove models see more impressive fuel savings, but with higher cost, they are often out of the economic reach of poorer households. For more information about these stoves for the developing world, see Global Alliance for Clean Cookstoves (2013) and Barnes et al. (2012).

In industrial countries, increases in solid wastes and concerns about global warming have spawned interest in generating power from garbage and waste from wood, agricultural crops, food processing, and animals. Worldwide, about 1.7% of electricity was generated from biomass and waste in 2011, about 20% of which was from waste (IEA, n.d.p). Although still a small share, biomass waste generation has grown at an average annual rate of more than 7% for the last decade through 2011, with the number of countries participating increasing from 67 to 85. With its strong commitment to carbon reduction, Europe accounted for around 45% of the global production in 2011 (US EIA, n.d.d).

Wind power is really a form of solar power, as the uneven heating of the earth's surface causes the winds to blow, while the earth's rotation, water bodies, vegetation, and terrain modify wind patterns. Windmills, long popular in the American plains to pump water for cattle, were first used to generate electricity in Ohio in 1887. Small turbines were put to use in the 1920s (Nixon 2008). Now, modern wind turbines are often the most competitive new renewable source for electricity generation. About 2.1% of the world's electricity was generated from wind in 2011, with an average annual growth rate of more than 24% for the preceding decade. The United States was the largest wind generator in 2011, with about 27% of global generation. Five states contributed about one-half of this generation: Texas with about 20% of US production, followed by California, Iowa, Minnesota, and Oregon. China and Germany came next and their combined share had slightly less than one-half of US generation. Other countries with more than 10% of global capacity are Germany and India (US EIA, n.d.d).

Unit or Levelized Costs of Wind Electricity

Currently, a typical wind turbine has three fiberglass blades, a standard gearbox, and a generator to turn mechanical energy into electrical power. It lasts about 20 years. Installed costs in 2010 for onshore wind farms were typically from $1,800/kW to $2,200/kW, with costs lower in China and India. The turbines account for 64% to 84% of installed cost, with installation accounting for the remainder. Offshore installed costs were typically double those onshore, with installation accounting for one-half or more of the costs. If it is known how much power such a turbine will produce in a year, these upfront capital costs can be converted into a per unit electricity cost. Since electricity is produced over time and capital costs are expended up front, levelized capital costs need to be calculated by using discounting to distribute the capital costs over the production profile for electricity (IRENA 2012a).

Suppose that a 600 kW turbine is installed. The wind allows the turbine to run at full capacity 25% of the time. Assume that the turbine is totally idle the rest of the time. The turbine then generates $600 \times 24 \times 365 \times 0.25 = 1,314,000$ kWh per year. Assume the turbine is paid for up front and takes one year to install, after which electricity generation commences. The generator lasts 20 years and stops producing at the beginning of year 22 with no down time. Power is paid for at the beginning of the year, starting one year after construction until year 21, and the real discount (or interest) rate is 10%. Take the initial cost per kilowatt to be the midpoint of the above estimates or ($1,800 + $2,200)/2 = $2,000. The total cost of the generator installed is $2,000 × 600 = $1.2 million. Decommissioning at the end of the turbine's life is just paid for by the salvage value of the unit. The levelized capital cost per kilowatt hour (Lc_k) is the amount that should be charged per kilowatt hour over time to recoup the capital costs. Lc_k can be computed from the following equation:

$$\$1,200,000 = \sum_{i=1}^{21} \frac{Lc_k \, 1,314,000}{(1+0.10)^i}$$

Solving for Lc_k results in the following:

$$Lc_k = \frac{\dfrac{\$1,200,000}{1,314,000}}{\displaystyle\sum_{i=1}^{21}\frac{1}{(1+0.10)^i}} = \frac{0.913}{8.649} = \$0.106$$

Total levelized costs would include the levelized capital cost plus operating and maintenance (O&M) cost. If operating costs are paid and recouped during production, they will not need to be levelized. Economic levelized costs do not consider taxes. Taxes are a transfer payment and not a true cost of production. However, a utility will see them as a cost. Thus, a utility's levelized cost would include O&M, taxes, and related costs. Such computations are often the first step in an integrated resource plan (IRP), usually required by public utility commissions (PUCs).

Solar Energy

Solar power can be used to generate heat passively or through the use of solar panels, which can be either flat or concentrating. Passive solar water heaters and cookers can be built quite inexpensively; the cookers may cost as little as $10, using an insulated box and a reflector (Teach a Man to Fish 2010).

Concentrated solar heat can be used to generate steam to run a power plant. Early 10 MW experimental models were built in Barstow, California by the US government and a consortium of companies. Problems included designing mirrors to track the sun and a medium for power storage. Solar One, which operated from 1982 to 1988, used water as a receiver, storage, and transmission mechanism. Solar Two, which operated from 1996 to 1999, used molten nitrate salt for the same purpose, which stored heat better and allowed more continuous power generation

(National Renewable Energy Laboratory 2001). The 11 MW Planta Solar 10 (PS10) in Spain is the first commercial grid-connected concentrated solar power plant in the world. It began operating in 2007 and receives a feed-in tariff of 27.1188 Euro cents per kilowatt hour (around $0.35 at 2013 exchange rates [US IRS, n.d.]) (National Renewable Energy Laboratory 2013a). By 2012, SolarPaces notes more than 50 concentrated solar power projects worldwide in 16 countries. Their database of existing and projects under construction is maintained at National Renewable Energy Laboratory (2013b).

There was an estimated 2 GW of concentrated solar capacity by 2012 (IEA-ETSAP and IRENA 2013), but costs per kilowatt hour are generally high, ranging from $0.14/kWh to $0.39/kWh (IRENA 2012a).

Alternatively, electricity can be generated by use of a photovoltaic cell (PV). PVs are typically made from two layers of silicon material or other semiconductors that have added impurities. When sunshine strikes the cell, it interacts with electrons to set up a direct current. For a summary of technologies and costs, see IRENA (2012b).

PVs were first developed in the 1950s in Bell Laboratories in the United States. Early US federal support was related to their use in the space program. By the late 1970s, they were being given US investment tax credits for terrestrial use. In the 1980s, some very small grid-connected PV plants were built in California, but until the early 1990s, three-fourths of PV applications were stand-alone, off-grid systems. In 1993, the first grid-supported distributed PV system was installed in California (US EIA 2003a).

Global grid-connected solar PV has expanded rapidly since 2000, with an average annual growth of more than 40% per year through 2011, often with strong policy encouragement. It reached an estimated global peak capacity of more than 63,000 MW by 2011. Germany has almost 40% of this capacity, followed by Italy, Japan, Spain, and the United States (IEA 2014). Although still quite expensive, with estimated utility-scale levelized costs in 2010 of $0.26/kWh to $0.59/kWh, its costs have dropped considerably in the last three decades. The PV learning rate is estimated to be about a 20% reduction in costs for each doubling of capacity (IRENA 2012b).

However, recent events may have changed the equation and driven these costs down even more. Although China is still trailing the Europeans in grid PV capacity, it produces about one-half of the world's solar panels. Their production, spurred by high sales and soft government loans, drove costs down even more. Higher production and a glutted world market have caused recent panel prices to plummet. As a result, a number of German firms have quit the business, beleaguered US manufacturers filed an antidumping complaint, and in May 2012, the US Commerce Department implemented a provisional tariff on some Chinese solar panels (*Economist* 2012a). In 2013, the large Chinese PV manufacturer, Suntech, went bankrupt (*Economist* 2013).

How things get sorted out and how much prices rebound remains to be seen. However, if past cost reductions and technical changes are indicative of future trends, solar is likely to eventually have a sunny future in many areas of the world. It is a low carbon option and is widely distributed around the world. It often matches population and energy-use patterns. However, it is intermittent, requires storage or alternative power sources, and is diffuse. A typical household in the United States would require about 11 square meters or 440 square feet of collecting surface to be self–sufficient (see chapter 21).

Geothermal Energy

Although geothermal energy has a longer history of contributing to grid electricity, it is the slowest growing of the other renewables, averaging only about 3.3% annual growth during the decade ending in 2012. Both wind and biofuel generated more than five times as much power globally in 2011. However, despite solar's dizzying gains in usage, it still had not caught up to geothermal by 2011 (US EIA, n.d.).

Geothermal energy, or heat from the earth's core, is tapped by drilling wells that are 1,000 to 10,000 feet deep. It is accessible near active tectonic plate margins (e.g., El Salvador, Italy, Iceland, Indonesia, Japan, Mexico, New Zealand, the Philippines, and the United States).

Geothermal sources with temperatures less than 300°F can be used for direct heat; if the source temperature is more than 300°F, it can be used for power generation. Figure 15–3 shows two technologies. If hot water is produced from the well, a binary-cycle plant is used. A heat exchanger raises the temperature of the water to steam, which runs the turbine. If steam is produced from the well, a flash power plant is cheaper, since the steam can be used to run the turbine directly.

Some estimates put geothermal potential at 8.3% of the world's electricity. About 30 countries generate power from geothermal. The United States, with around 23% of the world's generation in 2011, was followed by the Philippines, Indonesia, Mexico, Italy, and Iceland for a combined share of 86%. However, because new capacity requires exploration and drilling, its costs have risen rather dramatically, along with drilling for oil and gas, making it less competitive than in 2000. NREL estimated levelized costs in 2008 of $0.10/kWh (Schwabe 2011).

Updated estimates of plant costs in table 15–6 show the comparative expected total levelized costs for various power sources in the United States for new generation to be built in 2017. These costs suggest the $0.10/kWh for geothermal is expected to hold. Coal plants without carbon capture and sequestration include a cost increase equivalent to about a $15/tonne charge for CO_2.

With oil at more than $100/bbl, it is considered noncompetitive, with no cost estimates given. Their estimates suggest that conventional combined cycle gas baseload generation has the smallest levelized cost. Adding carbon capture and sequestration adds around 40% to the costs of fossil fuel plants. Gas turbines that run around one-third of the time during peak loads have approximately doubled the cost of combined cycle natural gas plants. Advanced coal and nuclear have similar costs, and biomass generation is only slightly higher. Geothermal and wind have similar costs, but since wind is nondispatchable, more backup generation is required. The solar options are considerably more expensive, especially concentrating solar thermal.

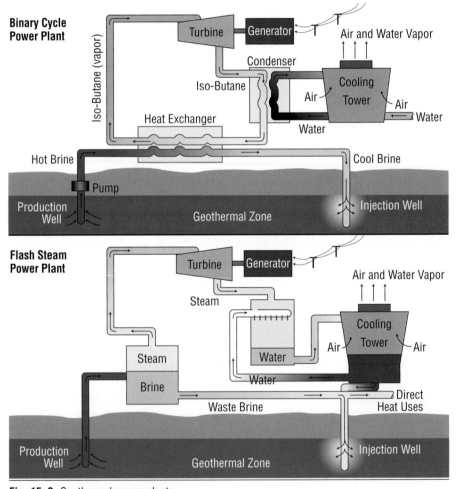

Fig. 15–3. Geothermal power plants
Source: Idaho National Engineering and Environmental Laboratory.

Table 15–6. Estimated levelized cost of new generation resources, 2017

Plant Type	Capacity Factor (%)	US Average Levelized Costs (2010 $/MWh) for Plants Entering Service in 2017				
		Levelized Capital Cost	Fixed O&M	Variable O&M (including fuel)	Transmission Investment	Total System Levelized Cost
Dispatchable Technologies						
Conv. Coal	85	64.9	4.0	27.5	1.2	97.7
Adv. Coal	85	74.1	6.6	29.1	1.2	110.9
Adv. Coal CCS	85	91.8	9.3	36.4	1.2	138.8
Natural Gas–Fired						
Conv. CCycle	87	17.2	1.9	45.8	1.2	66.1
Adv. CCycle	87	17.5	1.9	42.4	1.2	63.1
Adv. CCycle CCS	87	34.3	4.0	50.6	1.2	90.1
Conv. CTurbine	30	45.3	2.7	76.4	3.6	127.9
Adv. CTurbine	30	31.0	2.6	64.7	3.6	101.8
Adv. Nuclear	90	87.5	11.3	11.6	1.1	111.4
Geothermal	91	75.1	11.9	9.6	1.5	98.2
Biomass	83	56.0	13.8	44.3	1.3	115.4
Nondispatchable Technologies						
Wind	33	82.5	9.8	0.0	3.8	96.0
Solar PV	25	140.7	7.7	0.0	4.3	152.7
Solar Thermal	20	195.6	40.1	0.0	6.3	242.0
Hydro	53	76.9	4.0	6.0	2.1	88.9

Source: US EIA (2013c).
Notes: Adv. = advanced; Conv. = conventional; CCycle = combined cycle; CCS = carbon capture and sequestration; CTurbine = combustion turbine; PV = photovoltaic; and MWh = megawatt hours. Fuel costs are based on model runs in US EIA (2012b). In 2017 they are near $4/Mcf for natural gas, $44 per short ton for coal, and $110 per barrel for oil.

Inground and Aboveground Costs for Gas and Oil

Unit costs are computed for oil or gas in two ways. Unit inground capital costs are exploration and development costs divided by reserves found. Aboveground costs require us to take into account the production profile of the reserves. Often we can represent such a profile by an exponential decline curve. In such a curve for oil, production at time t is represented by equation (15–1):

$$Q_t = R_o e^{-\alpha t} \tag{15–1}$$

where
the decline rate is α per year and
e is the exponential function.

For example, if initial reserves R_0 equal 100, the decline rate is 10% (α equals 0.10), and production follows an exponential decline, then at the beginning of production (year zero) a well would produce as follows:

$$Q_0 = 0.1 \times 100e^{-0.1 \times 0} = 10$$

In a year, production would have declined to the following:

$$Q_1 = 0.1 \times 100^{-0.1 \times 1} = 9.05$$

After two years, the production would be as follows:

$$Q_2 = 0.1 \times 100e^{-0.1 \times 2} = 8.2$$

and so on.

Once we have the production profile, we also need to be able to discount the value of that production. In chapter 5, we reviewed how to discount with annual compounding. A dollars in n years, discounted with interest rate r, is equal to $A/(1 + r)^n$ today. If we compounded c times per year, the formula becomes $A/(1 + r/c)^{cn}$. For example, the net present value (NPV) of $20 in 10 years with quarterly compounding at 15% is the following:

$$NPV = \frac{\$20}{(1 + 0.15/4)^{4 \times 10}} = \$4.59$$

Recall compounding tells us how often we receive interest on the interest we have earned. In oil production, it is more convenient to be able to compound continuously, which tells us that we receive interest on our interest immediately. Mathematically, continuous compounding can be represented by letting the number of times we compound per period (c) go to infinity. The formula to discount in the continuous compounding case then becomes the following:

$$\lim_{r \to \infty} \frac{A}{(1 + r/c)^{cn}} = Ae^{-rn}$$

For example, the net present value today of $100 in 20 years with continuous discounting at a 15% interest rate is $100e^{-0.15 \times 20} = \$4.98$. Thus, if we have $4.98 today and hold it for 20 years with continuous compounding, we would have $100. The continuous compounding case is more realistic when we receive a flow of income with costs spread out over a year, which is often the case. However, if you find discrete computations easier or have discrete tables with the computations, you can easily convert the continuous rate (r_c) to a comparable discrete rate (r_d) as follows:

$$e^{-r_c t} = \frac{1}{(1 + r_d)^t}$$

Solve this equation for r_d: Since the natural log is the inverse of the exponential function, first take the natural log of both sides of the equation:

$$\ln(e^{-r_ct}) = \ln\left(\frac{1}{(1+r_d)^t}\right) \longrightarrow -r_ct = \ln(1) - t \times \ln(1+r_d)$$

Since $\ln(1)$ is zero, we can rewrite and then divide by $-t$ as follows:

$$-r_ct = -t\ln(1+r_d) \rightarrow \ln(1+r_d)$$

Take the exponential of each side and solve for r_d:

$$e^{r_c} = 1 + r_d \rightarrow r_d = e^{r_c} - 1$$

Thus, if you want to approximate a continuous rate of 0.08 with a discrete annual compounding rate, you would use the following:

$$r_d = e^{0.08} - 1 = 1.083 - 1 = 0.083$$

Suppose you produce oil for the next 20 years at an annual rate of $Q_1, Q_2, \ldots, Q_n$ that sells for $P_1, P_2, \ldots, P_n$. Q_1 is the amount you produce in the first year, Q_2 the amount in the second year, and so forth. You are paid P_iQ_i at the end of each year starting for $i = 1, \ldots, n$. Income is compounded at the rate of c times per year. To compute the net present value, you must determine when each payment is made and discount each payment back the correct number of periods and then find the sum. The net present value of this discrete stream of income is the following:

$$PV = \sum_{i=1}^{n} \frac{P_iQ_i}{(1+r/c)^{ic}}$$

If the income stream is a constant PQ every year, then the formula simplifies as follows:

$$PV = PQ \sum_{i=1}^{n} 1/(1+r/c)^{ic}$$

With continuous compounding, the present value of these annual payments can be computed by taking the limit of the above two expression as c approaches ∞. Then the present value of the above discrete income flow for n periods for the variable income case would be as given in equation (15–2):

$$PV = \int_1^n P_tQ_te^{-rt}dt \tag{15–2}$$

And for the constant income case, PV is as follows:

$$PV = PQ\int_1^n e^{-rt}dt$$

These tools can be used to distribute capital costs over the output of a project. First, distribute the costs of an oil well over the barrels produced. To compute such *levelized costs*, use the decline curve equation (15–1). To distribute costs over future production from these reserves, we want to compute a unit cost (Lc). Cost at time t will be production times unit cost or $Lc\alpha R_0 e^{-at}$. If we want the present value of cost incurred at time t, multiply by e^{-rt} to get $Lc\alpha R_0 e^{-at} e^{-rt}$. The present value of the future annual cost flows in a continuous discounting case is the integral of costs over the production profile or $\int_{t=i}^{n} Lc\alpha R_0 e^{-at} e^{-rt} dt$, where i is the period when production begins and n is the period when production ends. Unit costs will then be the amount Lc, which makes the discounted present value of total future costs equal to the initial capital costs for development as follows:

$$K = \int_i^n Lc\alpha R_0 e^{(-a-r)t} dt$$

Solving for Lc gives us the following:

$$Lc = \frac{K}{R_0 \alpha} \bigg/ \int_i^n e^{(-a-r)t} dt$$

The expression K/R_0 is the average cost of capital distributed over initial reserves and is referred to as *average inground costs*. We can integrate the denominator to be the following:

$$\left.\frac{e^{(-\alpha-r)t}}{-\alpha-r}\right|_i^n = \left[\frac{e^{(-\alpha-r)n}}{-\alpha-r} - \frac{e^{(-\alpha-r)i}}{-\alpha-r}\right]$$

By a few simplifying assumptions, we can develop a nice rule of thumb for levelized reserve costs, often called *aboveground costs*. First, if production begins immediately, then $i = 0$ and the last expression becomes $1/(\alpha + r)$. Note minus signs were cancelled in the last equation. Next, assume the constant decline rate lasts forever so that n approaches infinity. Although we know that no oil well produces forever, they can be online for many decades; further production eventually slows as we get far out in the production profile. (In the United States, wells that have daily production of less than 10 barrels of oil or less than 60,000 cubic feet of natural gas are called stripper wells.) Not only does production fall further out, it also gets discounted more. For example, at a commonly assumed decline rate of α = 0.10, a discount rate of r = 0.10 with a well life of 30 is shown in equation (15–3):

$$\frac{e^{(-\alpha-r)n}}{-\alpha-r} = \frac{e^{(-0.10-0.10)30}}{-0.10-0.10} = -0.012 \tag{15–3}$$

Eq. 15–3 indicates we have produced almost 99% of reserves by 30 years. When $n \to \infty$, the above expression approaches zero. Substituting in our assumptions of immediate production ($i = 0$) and an infinite production horizon ($n \to \infty$) yields aboveground costs of the following:

$$Lc = \frac{K}{R_0\alpha} \bigg/ \int_i^n e^{(-\alpha-r)t} dt = \frac{K}{R_0\alpha} \bigg/ \left[\frac{e^{(-\alpha-r)n}}{-\alpha-r} - \frac{e^{(-\alpha-r)i}}{-\alpha-r} \right] = \frac{K}{R_0\alpha} \bigg/ \left[\frac{1}{\alpha+r} \right] = \frac{K}{R_0} \frac{(\alpha+r)}{\alpha} \quad \textbf{(15–4)}$$

Thus, aboveground costs (Lc) approximately equal in-ground costs (K/R_0) adjusted by decline and interest rates. Let's apply the formula to a new field in the Norwegian North Sea. Suppose its decline rate is $\alpha = 0.13$ and the discount rate is $r = 0.10$. The field costs K = \$1 billion to develop and has reserves of R_0 = 100 million barrels. In-ground costs are K/R_0 = 1,000,000,000/100,000,000 = \$10.00. However, this in-ground cost does not take into account a required rate of return for holding oil and producing it over many years. Including these holding costs, the aboveground cost, which is also referred to as the *levelized cost*, is the following:

$$Lc = \frac{1,000,000,000}{200,000,000} \left(\frac{0.13+0.10}{0.13} \right) = 10 \times \left(\frac{0.23}{0.13} \right) = \$17.69$$

Alternatively, if αR_0 is initial production from the field designated by Q_0, $i = 0$, and $n \to \infty$, then we can rearrange the formulas in equation (15–4) as follows:

$$Lc = \frac{K}{Q_0} \bigg/ \int_0^n e^{(-\alpha-r)t} dt = \frac{K}{Q_0} \bigg/ \left[\frac{e^{(-\alpha-r)n}}{-\alpha-r} - \frac{e^{(-\alpha-r)i}}{-\alpha-r} \right] = \frac{K}{Q_0} \bigg/ \left[\frac{1}{\alpha+r} \right] = \frac{K}{Q_0}(\alpha+r) \quad \textbf{(15–5)}$$

The expression K/Q_0, referred to as *capacity cost*, is an important concept that we will return to in coming sections.

Unit Costs with No Decline Rate

The same technique applies to transporting the oil. Suppose that as production from your field decreases, you are able to find and develop nearby fields to keep the flow in the pipeline constant at Q_0. You pay for the pipeline immediately, it starts transporting oil in period i, and the pipeline lasts n periods. Again, distribute the capital cost over all barrels transported over the life of the project. The unit capital costs for pipeline transport are computed by finding Lc_p (the rate charged per unit of oil that will just cover the capital costs K). Compute Lc_p with the following formula:

$$K = \int_i^n Lc_p Q_o e^{-rt} dt$$

Solving for Lc_p is then possible, as shown in equation (15–6):

$$Lc_p = \frac{K}{Q_0} \bigg/ \int_i^n e^{-rt} dt = \frac{K}{Q_0} \bigg/ \left[\frac{e^{-ri}}{r} - \frac{e^{-rn}}{r} \right] \quad \textbf{(15–6)}$$

In this case, if production starts immediately ($i = 0$) and we approximate our problem with an infinite horizon, the denominator in square brackets simply becomes $1/r$. However, with continuous production, aggregate production after period n would not be trivial. In essence, we would be ignoring the last expression with n in it. Thus, we may not attain a valid approximation, if we integrate to ∞. For example, if we again assume the project life to be 30 years, the last expression with n in it is: $-\dfrac{e^{-rn}}{r} = -\dfrac{e^{-0.10\times30}}{0.10} = -0.498$. Letting n go to infinity in this case would not give us as good an approximation as for the -0.012 in the constant decline case of equation (15–3). However, notice that the higher the discount rate and the longer the life of the project, the better the approximation.

Now apply the formula to TransCanada's proposed controversial Keystone XL pipeline. The original proposal in July 2008 planned to begin construction in 2010 with deliveries in 2012 to the US Gulf. Its initial capacity was expected to be 510,000 bbl/d, with an ultimate capacity of 830,000 bbl/d. It included a 36-inch, 1,980-mile pipeline from Hardisty, Alberta to Port Arthur, Texas, with 41 pumping stations (Smith 2010). Since the pipeline originates in a foreign country, it requires a cross-border permit. In 2010, the US EPA evaluated TransCanada's draft as inadequate. Subsequent Environmental Impact Statements (EISs) were not accepted by the US State Department, with one of the sticking points being the crossing of the Sand Hills and the Ogallala aquifer. The Obama administration ultimately denied the cross-border permit in early 2012, citing that additional study was required before the permit would be issued (Parfomak et al. 2012). However, with elections over, many expected the permit would eventually be approved, with TransCanada moving the expected startup date to 2015 after the permit denial. As of June 2014, approval has not been forthcoming, although some members of Congress are threatening to pass a law to approve the crossing. However, so far this threat is more bluster than bite.

Let's compute the levelized costs of this pipeline without the most recent political delays. TransCanada estimates costs for the pipeline, including a 50-mile lateral to Houston, to be $7.6 billion, with $2.4 billion having been spent as of year-end 2011 (Smith 2012).

Now apply our tools to the proposed pipeline. First, take the simple case with maximum throughput of 830,000 bbl/d starting immediately and continuing for 30 years, with a total upfront cost of $7.6 billion. Keystone's annual capacity cost per barrel would be $K/Q_0 = (\$7.6 \times 10^9)/(365 \times 830{,}000) = \$25.086$ per barrel. Applying equation (15–3), with $r = 0.10$, $i = 0$, and $n = 30$, results in the following:

$$Lc_p = \frac{\$7.6 \times 10^9}{365 \times 830000} \Bigg/ \left[\frac{e^{-0.10\times0}}{0.10} - \frac{e^{-0.10\times30}}{0.10}\right] = \$2.64$$

Note that if you use the infinite approximation, the transport costs per barrel would be about 5% less, as follows:

$$Lc_t = \frac{\dfrac{\$7.4 \times 10^9}{365 \times 830 \times 10^3}}{\dfrac{1}{0.1}} = \$2.509$$

Rail shipment costs are thought to be more than \$10 per barrel, with a higher risk of spills. We can easily incorporate construction time and delays into the computation. For example, if the \$7.4 billion is spent equally in j payments, with the first payment now and each subsequent payment a year later, replace K by the present value of K using the appropriate compounding either at rate c times per year or continuously, as follows:

$$PVK_{discrete} = \sum_{k=0}^{j-1} \frac{K/j}{(1 + r/c)^{ck}} \quad \text{or} \quad PVK_{continuous} = \int_0^{j-1} \left(K/j\right) e^{-rk} dk \qquad \textbf{(15–7)}$$

Replace K in equation (15–6) with the appropriate PVK from equation (15–7). For example, if the pipeline began transporting oil at $j = 5$ with a life of 30 years, the cost of oil transportation would be considerably higher, as follows:

$$Lc_p = \frac{\dfrac{\sum_{k=0}^{5-1}\left(\$7.6 \times 10^9/5\right)e^{-0.10k}}{365 \times 830 \times 10^3}}{\left[\dfrac{e^{-0.10\times5}}{0.10} - \dfrac{e^{-0.10\times35}}{0.100}\right]} = \$3.600$$

All the same formulas can be developed for discrete decline and compounding. With a discrete decline rate ($Q_t = (1 - \alpha)^t Q_0$) and annual compounding, these formulas for production costs become the following:

$$Lc_p = \frac{\dfrac{K}{Q_0}}{\displaystyle\sum_{t=0}^{n}\left\{\dfrac{1-\alpha}{1+r}\right\}^t} = \frac{\dfrac{K}{Q_0}}{\dfrac{1 - \left\{\dfrac{1-\alpha}{1+r}\right\}^{n+1}}{\dfrac{r+\alpha}{1+r}}} = \frac{\dfrac{K}{Q_0}}{\left(1 - \left\{\dfrac{(1-\alpha)}{1+r}\right\}^{n+1}\right)\dfrac{1-r}{r+\alpha}}$$

With decline, we have $0 < \alpha < 1$, and without the decline rate, we have $\alpha = 1$ in the above equation. If $r > 0$ and $n \to \infty$, the denominator can be simplified to $(1 + r)/(r + \alpha)$.

Production and transportation costs include operating as well as capital costs. But operating costs (C_t) are not spent up front; they are distributed over the production profile (Q_t). We compute unit operating cost in year t as follows:

$$Lc_{op} = \frac{C_t}{Q_t}$$

Developing Cost Data

When we do not know the costs of investing in a large piece of capital, we may need to approximate them from similar projects. For example, Adelman and Shahi (1989) developed capital costs for finding and developing reserves (K) for a number of countries as follows. They had reasonably fair estimates for the number of oil wells drilled, the number of dry wells (W_d), and average well depth, but insufficient estimates on the costs of drilling these wells. Thus, they started with US well drilling *costs* (C_w), which are available by depth, and multiplied by 1.66, which is an estimate of the other costs over and above drilling costs. They multiplied this cost per well times the number of wells drilled that struck oil (W_o) plus the number of dry wells (W_d), yielding the following:

$$K = C_w(W_o + W_d) \times 1.66$$

Where they had information indicating well costs might be higher because of complex geology, such as in Iran or Nigeria, or for offshore areas, they added a correction factor. Since they were interested only in real physical economic costs, they did not include taxes in their computations. (A company would certainly include taxes as part of its costs, but from a social point of view, taxes are transfer payments that do not reflect real physical costs of production.)

Gas capital costs can be similarly computed. Since coal resources are generally known, exploration costs may be trivial, while coal development costs depend on the size of deposit, thickness of seam, depth, amount of overlying soil, amount of tectonic disturbance of the seam, angle of the seam, inflow of water and ethane, and heterogeneity of the deposit. Again, these capital costs can be distributed over the life of the project. For project evaluation and decision making, one can also rank projects by net present value and then undertake them. If capital is limited, you undertake the portfolio of possible projects within your budget constraint with the highest net present value. (See Stermole and Stermole [2012] for very detailed descriptions of project analysis.)

Estimating Total Energy Resources

How fossil fuel costs evolve depends upon the total resource base and the success of discovery efforts. Oil resources are often measured by one of two popular approaches represented by Hubbert (1962) and USGS (1995). Hubbert was a geologist who used the following logistics model to estimate US reserves:

$$Q_t = \frac{Q_\infty}{1 + \alpha e^{-\beta(t-t_0)}}$$

where

Q_t is cumulative oil production at time t,

Q_∞ is estimated total reserves,

α and β are parameters that are all estimated statistically from historical data, and

t_0 is the date of first production.

Such a curve for cumulative production has the shape shown in figure 15–4. The production profile would take a familiar bell-shaped curve.

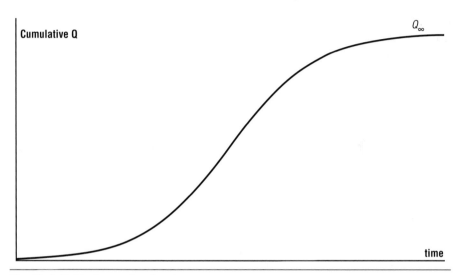

Fig. 15–4. Hubbert curve for oil and gas reserves

From his forecasted curve for oil, Hubbert concluded that US production would peak in 1970, which it did. His estimate of total US reserves was $Q_\infty = 170$ billion barrels. Since US cumulative production as of 2011 was 198.3 billion barrels, with another 27.7 billion barrels of estimated proven reserves and promises of even more from tight oil, his estimates of total reserves were clearly too low. Hubbert forecasted world oil would peak in 1995 (Grove 1974). However, 1995 came and went, and global oil production continued to increase; indeed, it was 18% higher in 2011 than in 1995, and 23% higher if natural gas liquids are included. Despite the failure of these predictions, others have taken up the peak oil hue and cry. However, these predictions by petro-pessimists, which have included Buz Ivanhoe, Colin Campbell, Jean Laherrere, and Matt Simmons, have not yet come to pass.

Economists, who are typically more optimistic than geologists, criticize Hubbert's model because it does not incorporate oil prices or technological change into reserve estimates. It also does not include undiscovered and nonconventional oil reserves (tight oil, tar sands, and heavy oil deposits), which are estimated to be at least equal to conventional reserves (IEA 2012c, 101). Despite these critiques, Hubbert retains his supporters, and papers that debate oil reserves, often with a Hubbert slant, can be found at the M. King Hubbert Center (http://hubbert.mines.edu/).

The United States Geological Survey (USGS), which publishes periodic assessments of reserves and resources, is more optimistic than Hubbert's followers in their estimates. Identified reserves (proven reserves) are typically those that have been located and are expected to be producible under current costs and operating conditions. The USGS estimates a probability distribution for the total remaining world resources by using geological data on sedimentary basins and probabilities from past discoveries, while taking into account minimum economic sizes. The mean, 95%, and 5% levels for their most recent probability distribution are shown in table 15–7 for oil, natural gas, and natural gas liquids.

Thus, for North America, proven reserves are about 204.5 billion barrels; the mean for additional resources is 83.4 billion barrels. The USGS estimated probability that additional resources are greater than 25.5 billion barrels is 95%, and estimated probability that additional resources are greater than 208 billion barrels is 5%. In all cases but for North America, USGS estimates identified reserves to be larger than cumulative production, suggesting that we have not yet produced one-half of world oil reserves.

Table 15–7. Oil, natural gas, and NGL reserves and resources, 2012

Reserves	Proven Oil Reserves (10^6 bbls)	Undiscovered Resources		
		F95	F5	Mean
Arctic Ocean/Former Soviet Union	98,886	15,984	177,175	66,211
Middle East and North Africa	865,936	43,316	212,678	111,201
Asia and Pacific	36,065	20,950	87,744	47,544
Europe	11,877	4,344	19,417	9,868
North America	204,468	25,500	208,032	83,386
South America and Caribbean	238,817	44,556	261,862	125,900
Sub-Saharan Africa	57,880	40,777	232,090	115,333
South Asia	9,294	3,323	9,339	5,855
Total	**1,523,224**	**198,750**	**1,208,337**	

Reserves	Proven Gas Reserves (bcf)	Undiscovered Resources		
		F95	F5	Mean
Arctic Ocean/Former Soviet Union	2,164,800	398,288	4,055,759	1,622,531
Middle East and North Africa	3,092,323	383,257	1,798,981	941,300
Asia and Pacific	419,146	322,102	1,364,079	738,328
Europe	146,942	60,677	300,138	148,625
North America	350,829	160,383	1,510,325	573,737
South America and the Caribbean	270,047	229,547	1,476,006	678,537
Sub-Saharan Africa	217,060	318,783	1,453,668	743,529
South Asia	85,604	77,055	275,231	159,039
Total	**6,746,751**	**1,950,092**	**12,234,187**	

Region	Proven NGL Reserves (10⁶ bbls)	Undiscovered Resources		
		F95	F5	Mean
Arctic Ocean/Former Soviet Union	NA	10,092	96,709	39,979
Middle East and North Africa	NA	11,941	60,994	30,676
Asia and Pacific	NA	8,690	40,082	21,042
Europe	NA	1,157	6,017	2,929
North America	NA	3,789	55,708	19,267
South America and the Caribbean	NA	6,850	46,580	21,001
Sub-Saharan Africa	NA	10,662	59,172	27,968
South Asia	NA	1,812	6,741	3,806
Total	**NA**	**54,993**	**372,003**	

Sources: USGS (2012); Proven reserves from *OGJ* (2012d).
Note: NA = not available, bbls = barrels, bcf = billion cubic feet. F5 indicates that 5% of the time, we would expect reserves higher than this amount. F95 indicates that 95% of the time, we would expect reserves higher than this amount.

Summary

Energy comes from a variety of sources, including fossil fuels, nuclear energy, and renewables. Using IEA conversions and data for 2011, about 31.52% of the world's energy comes from oil, another 28.8% from coal, 21.3% from gas, and the remainder from biofuels, nuclear, and hydro, with a dab of other renewables thrown in. All primary energy sources except traditional biomass gained relative share at the expense of oil between 1973 and 2011. Biomass has remained at around 10% of world consumption since 1973. But with government policy encouragement, waste has starting to be used to generate electricity. Bioenergy produces around 5% of renewable primary electricity generation with about 1% of the 5% coming from waste. About 2.4% of transport fuels come from biofuels, mostly for highway use (IEA, n.d.d., n.d.f., n.d.p.)

Hydropower, which generates about one-half of primary electricity, is used in many countries, ranging in size from small, run-of-the-river generators to huge dam complexes among the world's largest electrical facilities.

Nuclear power once represented most of the remainder of primary electricity. It now shares the field with rapidly growing other renewables, which currently produce around 13% of primary electricity. Nuclear fuel reserves and processing are fairly concentrated, while nuclear power generation is more widespread, with significant government involvement and oversight.

Other renewables (biomass, wind, solar, and geothermal) have a long history of use, with modern versions constituting a small but rapidly growing share of commercial energy. They are likely to continue to grow as environmental effects for nonrenewables come under increasing scrutiny, and if scarcity or taxes increase the cost of nonrenewables.

Fuel choices and how these choices will evolve in the future depend both on the properties of the fuels and their relative costs of production, transportation, and transformation. These costs include capital costs and operating costs per unit of output. Capital costs are usually concentrated at the beginning of a long-lived energy project, while the flow of product or service occurs over many years. To evaluate the profitability of energy projects, costs and benefits must be discounted to the same time period to determine whether project benefits outweigh costs and the project should be undertaken. To discount income at time t, (P_tQ_t) back to the present with discrete compounding c times per year, and also with continuous compounding, the following equations are used:

$$\text{discrete: } PV = \frac{P_tQ_t}{\left(1+\dfrac{r}{c}\right)^{ct}} \text{ and continuous: } PV = P_tQ_te^{-rt}$$

If you are more comfortable using annual discrete compounding than continuous compounding, convert a continuous rate (r_c) to a comparable discrete rate (r_d) using the following equation:

$$r_d = e^{r_c} - 1$$

These formulas can be extended to give the present value of an annual flow of income from now to n for discrete annual compounding and continuous annual discounting as follows:

$$PV = \sum_{i=0}^{n}\frac{P_iQ_i}{(1+r)^i} \text{ and } PV = \int_{t=0}^{n} P_tQ_te^{-rt}\,dt$$

Energy capacity cost equals upfront capital cost (K) divided by the capacity output in (Q). These capacity costs are often distributed over the life of the project into what are called levelized cost or the unit charge needed when sold that covers the initial capital costs at the required rate of return (r). For oil and gas, these are called aboveground costs. These levelized costs are computed by solving the following equations for Lc_k in the discrete and continuous cases as follows:

Discrete:

$$K = \sum_{t=0}^{n}\frac{Lc_kQ_t}{(1+r)^t} \longrightarrow \$_k = \frac{K}{\sum_{t=0}^{n}\dfrac{Lc_kQ_t}{(1+r)^t}}$$

Continuous:

$$K = \int_{t=0}^{n} \$_kQ_te^{-rt}\,dt \longrightarrow Lc_k = \frac{K}{\int_{t=0}^{n} Q_te^{-rt}\,dt}$$

For initial production of Q_0 and a constant decline rate of $\alpha > 0$, simplify the formulas to the following:

$$\text{Discrete: } Lc_k = \left. \frac{K}{Q_o} \middle/ \frac{1 - \left\{\frac{1-\alpha}{1+r}\right\}^{t+1}}{\frac{\alpha+r}{1+r}} \right. \qquad \text{Continuous: } Lc_k = \left. \frac{K}{Q_o} \middle/ \left(\frac{1 - e^{(-\alpha-r)\times n}}{\alpha+r}\right) \right.$$

When $\alpha = 1$ in the above equations, you have a zero decline rate. These computations are sometimes simplified by assuming an infinite life for the project. The formulas for $n \to \infty$ are the following:

$$\text{Discrete: } Lc_k = \left. \frac{K}{Q_o} \middle/ \frac{1}{\frac{r+\alpha}{1+r}} \right. = \frac{K}{Q_o}\frac{1+r}{r+\alpha} \text{ and Continuous: } Lc_k = \left. \frac{K}{Q_o} \middle/ \left(\frac{1}{\alpha+r}\right) \right. = \frac{K}{Q_o}(\alpha+r)$$

These approximations as $n \to \infty$ are better; the higher the interest rate, the longer the project life and the higher the decline rate.

For a constant decline rate for oil or gas, the above equations can also be rewritten in terms of initial reserves by substituting for $Q_0 = \alpha R_0$ as follows:

$$Lc_k = \frac{K}{R_o}\frac{(\alpha+r)}{\alpha}$$

Oil and gas costs also include the resources necessary to find them. The ultimate resource base influences these exploration costs. Measurements of ultimate resource base have followed two popular methodologies. Hubbert's technology fits a logistic curve to a cumulative production profile, while the USGS uses measurement of the world's sedimentary basins, along with probability distributions of known reserves, to estimate the probability of reserves in unexplored areas.

16

Energy Balances and Energy Demand

All societies require energy services to meet basic human needs (e.g., lighting, cooking, space comfort, mobility and communication) and to serve productive processes. . . . There are multiple options for . . . satisfying the global demand for energy services.

—IPCC, 2012

Introduction

Understanding energy demand and energy markets is important information for a whole host of stakeholders. Consumers would like to know price trends before they buy appliances and vehicles. International capital markets would like to know how and when emerging markets will provide modern energy services for their more affluent population. Electric utilities would like to know consumer electricity consumption patterns for investment into a capital-intensive infrastructure. Likewise, oil producers and refiners would like to know travel patterns and vehicle choice patterns before investing. Policy makers in a carbon-constrained world would like to know how consumers respond to policy initiatives designed to reduce fuel use and encourage fuel switching. Modelers would like to know the best equations to fit into their models for a whole host of modeling scenarios.

Modeling this demand requires understanding of who is using the energy and why. Energy is used by all sectors of the economy. Households want energy to fuel their vehicles, space condition their homes (heating, cooling, and lighting), and run their appliances. However, it is the transportation, space conditioning, and appliance services that consumers desire, not the gasoline, natural gas, or electricity. Likewise, industrial consumers and electric utilities want energy to produce process heat and run their vehicles and equipment, but it is the heat and equipment services they really desire. These desires drive the demand curves that appear in our energy models and help determine how responsive purchases are to price, economic activity, and other strategic variables. In short, there is a derived demand for energy. In this chapter, we will consider various energy uses, drivers of energy use, and models representing these underlying desires.

We begin with a brief description of global energy uses in the next section. This is followed by sections devoted to the optimization that energy users should undertake in making consumption decisions, as well as econometric issues in estimating energy demand.

Energy Balances

First we will consider aggregate energy balances, as shown in table 16–1 from the International Energy Agency's *Energy Balances*, measured in millions of metric tons of oil equivalent (Mtoe), where 1 toe = 41.868 gigajoules = 10 gigacalories = 39.68 million Btus (IEA, n.d.f). The top seven data rows of the table show supply information; the middle data rows (9–19) show energy transformations and intermediate use; and the bottom rows show energy consumption by sector. The final two rows show energy consumption for nonenergy use, including coking coal with a separate breakout of petrochemical use. Data for these and subsequent balance tables in this chapter are collected by IEA from questionnaires sent to countries. These questionnaires, along with a statistical manual describing data needed, data processes, and data relationships across questionnaires and conversions, along with data documentation and a glossary, can be found at IEA (2012a), IEA (2013a), IEA (2013b), IEA (n.d.d), and IEA (n.d.j).

Such statistics are collected for the 34 OECD countries, as well as more than 100 other countries and 15 subregions. The data is aggregated for the whole world in table 16–1, and into a number of other political and regional groupings in the online database. (See IEA [n.d.q] for energy balances by region and by country. Historical data going back to 1990 is available free online and data back to 1960 is available for some countries by subscription or purchase.)

The table columns represent the basic sources of energy. Petroleum products and electricity from fossil fuels are considered secondary sources. Coal/peat, crude oil, natural gas, nuclear, other renewables (geothermal electricity, solar photovoltaics, solar thermal, wind, and wave), biofuels and waste, and a small amount of heat consumed directly from geothermal energy are considered primary sources.

These balances give an overall snapshot of the energy situation at a given time for a given region, and as the name suggests, energy demand and supply flows must balance. To see how this balance is computed, start with the top rows of the table. The total primary energy supply (TPES) to a country or region is what the region produces plus what they buy from others (Imports) minus what they sell to others (Exports) minus international marine and aviation bunkers (Bunkers), which is fuel oil sold to ships and planes engaged in international transport, minus any stock or inventory changes (Stock Δ). When stock changes are positive, inventories are building up and less supply is available as would be indicated by a negative in the table. When stock changes are negative, inventories are being depleted, increasing the available supply as indicated by a positive stock change in the table. By convention, international marine and aviation bunkers or oil products that are used in international transport are not included in country and regional TPES, which represents domestic supply. However, they are included when all countries and regions are aggregated for the whole world, as in table 16–1. At the country or regional level we subtract bunkers as follows:

$$\text{TPES} = \text{production} + \text{imports} - \text{exports} - \text{bunkers} - \Delta \text{ stock}$$

For world coal in 2011, the amount of total primary energy supply was the following:

$$3{,}850.5 + 696.80 - 726.2 - 0 - 0 - 45.0 = 3{,}776.1$$

Since the world is a closed system and does not import energy from other planets or solar systems yet, we would expect imports to cancel out exports. However, they do not quite balance, largely the result of reporting lags and errors.

The stock changes in the table are generally insignificant but are a bit larger and negative for coal and natural gas. Since they reduce domestic supply, the negative sign indicates that their stocks increased in 2011.

Stock changes for biofuels and waste are reported and are quite small both absolutely and as a percent. The next two columns relating to electricity and heat do not record stock changes, since we cannot stockpile electricity and heat. However, we can store water behind dams and in pumped storage for future generation. A small amount of electricity is stored in batteries, but such data is not collected, nor is data on nuclear fuel storage reported.

Coal includes coking coal, while crude oil includes natural gas liquids and refinery feedstocks that are back flows from petrochemical plants. Nuclear is converted to a total heat equivalent by dividing electricity output by 0.33, while IEA measures hydro and other renewables by their electricity output.

Primary energy is transformed in a variety of ways, as shown in rows 8–19 in table 16–1. A negative value is energy in, and a positive value is energy out. For example, 2075.4 Mtoe coal are used as inputs into plants that produce only electricity. Another 180.8 Mtoe are used as inputs into plants that produce heat and electricity, called combined heat and power (CHP) or cogeneration plants. Some coal (109.5 Mtoe) is used in plants that produce heat alone and sell it under contract. Gas work plants convert coal into gas; most of this conversion to gas capacity is in China, South Africa, and the United States, while liquefaction plants change coal into crude oil and products. Most of this coal-to-liquids (CTL) capacity is held by Sasol in South Africa.

Table 16–1. World energy balances, 2011 (mtoe)

Supply & Cons.	Coal Peat	Crude Oil	Oil Products	Natural Gas	Nuclear	Hydro	Ot Renew	Bio & Waste	Elec	Heat	Total*
Prod	3,850.5	4,133.0	0.0	2,805.4	674.0	300.2	127.0	1,310.6	0.0	1.1	13,201.8
Imports	696.8	2,299.3	1,077.4	865.3	0.0	0.0	0.0	13.9	55.8	0.0	5,008.5
Exports (–1)	–726.2	–2,210.8	–1,164.0	–861.7	0.0	0.0	0.0	–11.6	–55.8	0.0	–5,030.2
M Bunk	0.0	0.0	0.0	0.0	0.0	0.0	0.0	0.0	0.0	0.0	0.0
Av Bunk	0.0	0.0	0.0	0.0	0.0	0.0	0.0	0.0	0.0	0.0	0.0
Stock Δ (–1)	–45.0	–1.9	3.1	–22.0	0.0	0.0	0.0	–0.7	0.0	0.0	–66.6
TPES	3,776.1	4,219.6	–83.6	2,787.0	674.0	300.2	127.0	1,312.2	0.0	1.1	13,113.4
Transfers	–0.3	–169.1	195.4	0.0	0.0	0.0	0.0	0.0	0.0	0.0	25.9
Stat Dif.	–143.7	6.8	3.5	10.3	0.0	0.0	0.0	–0.1	–1.7	–0.1	–124.9
Elec Plant	–2,075.4	–41.6	–203.8	–711.3	–670.4	–300.2	–99.8	–81.1	1,725.0	–0.4	–2,459.0
CHP Plant	–180.8	0.0	–25.5	–314.0	–3.6	0.0	–1.4	–42.8	177.8	153.1	–237.2
Heat Plant	–109.5	–0.8	–11.7	–92.9	0.0	0.0	–0.1	–10.8	–0.3	189.7	–36.5
Gas Works	–6.3	0.0	–3.8	3.2	0.0	0.0	0.0	0.0	0.0	0.0	–7.0
Oil Ref	0.0	–4,023.9	3,989.3	–0.8	0.0	0.0	0.0	0.0	0.0	0.0	–35.4
Coal Trans	–249.9	0.0	–3.2	–0.1	0.0	0.0	0.0	–0.1	0.0	0.0	–253.3
Liq Plants	–17.5	9.6	0.0	–10.7	0.0	0.0	0.0	0.0	0.0	0.0	–18.7
Ot Trans	–0.1	32.6	–34.6	–3.6	0.0	0.0	0.0	0.0	0.0	–0.4	–60.6
E Own	–85.4	–6.5	–206.7	–267.5	0.0	0.0	–0.2	–10.8	–164.4	–42.6	–784.1
Losses	–3.4	–7.9	–0.7	–19.0	0.0	0.0	–0.1	–0.2	–154.3	–19.4	–204.9
TFC	903.6	18.8	3,614.5	1,380.5	0.0	0.0	25.3	1,111.7	1,582.1	281.0	8,917.5
Industry	728.9	10.7	312.5	506.4	0.0	0.0	0.5	198.1	673.8	125.9	2,556.7
Transport	3.4	0.0	2,265.2	92.5	0.0	0.0	0.0	58.6	25.2	0.0	2,444.9

Other	132.0	0.5	435.6	610.2	0.0	24.8	855.0	883.2	155.1	3,096.4
Resident	79.1	0.3	205.7	415.4	0.0	10.0	826.2	428.5	107.6	2,072.8
Com&Pb	24.8	0.0	101.3	178.2	0.0	2.1	17.8	360.2	30.8	715.2
Ag&For	11.4	0.1	105.2	6.2	0.0	0.7	7.1	40.5	6.1	177.3
Fishing	0.0	0.0	6.1	0.1	0.0	0.1	0.0	0.4	0.0	6.6
Non-Spc	16.7	0.2	17.2	10.4	0.0	11.9	3.9	53.6	10.6	124.4
Non-EUse	**39.2**	**7.6**	**601.3**	**171.4**	**0.0**	**0.0**	**0.0**	**0.0**	**0.0**	**819.4**
PChmFsk	2.4	7.0	403.5	168.5	0.0	0.0	0.0	0.0	0.0	581.4

Source: IEA (n.d.f).

Note: M Bunk = international marine bunker fuel; Av Bunk = international aviation bunker fuel; Stock Δ = stock changes; TPES = total primary energy consumption; Stat Dif. = statistical differences; Elec = electricity; CHP = combined heat and power; Oil Ref = oil refineries; Coal Trans = coal transformation; Liq Plants = liquefaction plants; Ot Trans = other transformation; E Own = energy industry own use; TFC = total final consumption; Com&Pb = commercial and public sector; Ag&For = agriculture and forestry; Non-Spc = nonspecified; Non-EUse = nonenergy use; Losses = transportation and distribution losses; PChemFStk = petrochemical feedstocks; Ot Renew = geothermal, wind, solar, and wave; Bio & Waste = fuels and electricity from biological sources; and Oil Prod = oil products. Totals may include rounding errors. International marine and aviation bunkers are included in transport for world totals but are included separately for country and regional tables. Nuclear is measured as gross, not net, energy; (−1) indicates that although IEA labels these as positive values, the number recorded is actually minus their value.

Coal transformation is the loss in converting from coal to other products such as coke or the conversion of coke to blast furnace gas. *Own use* is coal used by the industry itself. For example, large draglines used to mine coal may run on self-generated electricity from coal. The coal lost in transport and distribution is designated as losses. The statistical difference is a fudge factor attributed to conversions and unexplained differences that are added in to make the balances really balance.

If you subtract all energy used in conversion processes and add back any energy outputs from the transformation, such as oil from coal liquefaction, you will get the total final consumption of energy (TFC). This final consumption in turn is divided into the following sectors:

- *Industrial.*
- *Transportation.*
- *Other.* This category is broken down into: a. residential; b. commercial and public services; c. agriculture, forestry, and fishing; and d. nonspecified.

In their extended balances, IEA breaks down transport use into road, rail, pipeline, air, and domestic shipping.

The last category is nonenergy use, with a separate category for petrochemical feedstocks. It also includes coking coal, bitumen for roads, lubricants, paraffin, and petroleum coke not used for energy. Most nonenergy coal use is coking coal, with a small amount used as petrochemical feedstocks. One-half of coal for petrochemicals is used in South Africa.

The column for crude oil contains many of the same categories. The table shows that more than one-half of crude oil produced enters into international trade, a considerably higher share than for oil products and natural gas, with their export share nearer 30%. Coal's share was less than 20%, and electricity's share was less than 5%. These shares reflect the distribution of resources as well as the relative ease of transporting a liquid. Since crude oil is too valuable to be used as a raw material, more than 95% is sent to refineries to separate out more valuable lighter products, including jet fuel, gasoline, and diesel fuel. As you will see later in the oil balances table, the crude oil in the transfer row is mostly NGLs that get blended into products. The crude oil used for the petrochemical row is also mostly NGLs.

A small amount of crude oil is burned to produce electricity. This is a balancing activity during peak periods when refinery capacity is not high enough to produce residual fuel oil or residual fuel stocks are not available. More than one-half of this crude is burned in Saudi Arabia and another quarter is burned in Japan; two other oil-rich Gulf countries, Iraq and Kuwait, account for the rest. A smaller amount of crude is used directly in the industry, about 40% of which is in Saudi Arabia.

Positive items under Transformations in the crude oil column include the crude coming from liquefying coal and natural gas. Around 85% of the liquid made from coal is from South Africa with much of the remainder liquefied in China. Some of the gas is also changed to liquids and its shown in the same row. Shell's giant Pearl plant in Qatar converts more 40% of the gas converted to liquid worldwide with a high percent of the remainder converted in South Africa and Canada. Other transformations include refinery feedstocks or processed oils that are blended back into the crude oil stream for reprocessing.

The columns for Crude Oil and Oil Products show the relationship between refineries and products. Crude oil includes NGLs that are stripped out of wet natural gas. In the transfer row, the gas liquids taken out for blending (−169.1) are shown as a negative under crude and a positive number under products (195.4). The fact that they do not exactly match could be a timing issue or statistical error. Note the negative value under crude oil (−4,023.9 Mtoe) going to refineries, which is transformed into the petroleum products that are output in the next column (3,989.3 Mtoe). If crude and products had been measured in volume, for example barrels or cubic meters, then the product volume would be larger than the crude input volume. This volume gain, typically around 3% to 5%, is called *refinery processing gain*. Such gain always occurs when the larger molecules in crude oil are broken up into lighter products with smaller molecules. These gains are included in oil balances when volume measures are used. US EIA (n.d.d) reports these gains by country.

In the Oil Products column and Transformation rows, you can see the oil products used to produce electricity and heat and those used in the energy sector itself. Later you will see that residual fuel oil is the product most often used for these purposes. The last rows in the table under TFC show that transportation accounts for more than 60% of end-use oil product demand. No other sector so clearly dominates oil product use on a global basis.

The next column shows natural gas balances. The electricity sector is the largest user of natural gas. Electricity and heat together consume about 40% of the world's natural gas, with the industry (18%), residential (15%), and commercial and public sectors (6.4%), accounting for an additional 39%. Most of the natural gas for heat plants is consumed in the former Soviet Union, and more than 60% of the gas used for CHP was consumed in the former Soviet Union. The sector shares are a bit different for the richer countries with more developed infrastructure, as represented by the OECD. For them, electricity takes about 36%, with the industry (20%), residential (19%), and commercial and public sectors (11%), accounting for an additional 50%.

In the next two columns, Nuclear and Hydro show their total production of electricity, with a small amount of heat in the case of nuclear. The electricity and heat output and uses are in the electricity and heat column. In the rows for Electricity, CHP, and Heat Plants, the inputs are negative, while the outputs are positive and are shown in the Electricity and/or Heat columns. Thus, electricity plants produce electricity and use a bit of heat as input. Combined heat and power plants output both electricity and heat, while heat plants output heat and use a bit of electricity. Inputs do not add up to outputs because coal, oil, gas, and nuclear are measured in gross inputs, while electricity is the net production before own use and distribution losses have been taken out. The negative values in the final column in the transformation rows show the net transformation losses. Combining transformation losses and gains with TPES gives us the total available product for end-use demand.

The Other Renewables column, which includes geothermal, solar, wind, and waves, amounts to less than 1% of total primary energy supply. Almost 80% goes to generate electricity and heat, with much of the remainder going to the residential, commercial, or nonspecified sectors.

Biofuels and waste include traditional fuels like rice, straw, wood, and dung, as well as burnable industrial waste, ethanol, and biodiesel. They are treated more like fossil fuels, with imports, exports, and stock changes listed in the table. The largest users in this category are residential users. They consume more than 60% of the global total, with about three-fourths of this residential use in developing countries in Asia and Africa. Industrial users come in second with another 15% of global consumption, with about two-thirds of this industrial use spread over countries not in the OECD. Plants that produce electricity, heat, or both, add another 10%. For this more modern use, about 80% occurs in the OECD. Transport fuels in the form of ethanol and biodiesel account for the 4.5% share, which is largely policy driven. More than 90% of this use (in order of consumption) is in the United States, Brazil, and the European Union. Other transformations, which include wood for charcoal production or peat formed into briquettes, are almost exclusively carried out in non-OECD countries. Almost 90% of such use is in Africa and the Asian developing countries, excluding China.

The next two columns show how the electricity and heat generated in the other columns are used. All of the electricity is accounted for as transformations from other columns. Total electricity generation is the sum of the positive numbers (1,725 + 177.8) in the electricity and CHP plant rows. Heat-only plants use a bit of electricity, as indicated with the −0.3. About 4% of electricity enters into international trade. Own use and losses claim close to almost another quarter roughly divided equally between them. Losses vary considerably across countries. They average more than 20% in India and China but less than 7% in the OECD.

For the world, about 43% of end-use electricity goes to industry, about 27% goes to the residential sector, and another 23% goes to the commercial sector. These shares tend to change as countries develop. Whereas these three sectors claim around one-third each in the richer OECD countries, in non-OECD countries, industry accounts for more than one-half of final electricity consumption. The residential and commercial sectors account for less than one-fourth each. For example, in China, which has chosen an energy-intensive industrial path of development, industry's share is close to two-thirds of end-use electricity demand. In India, which has chosen a softer, less energy-intensive, more service-oriented path, industry claims less than one-half. In both China and India, the residential sector claims less than 25% of electricity, and the commercial and public sector, which tends to develop later as countries become wealthier, claims less than one-half of that.

In the heat column, a small amount of heat (< 1%) not accounted for elsewhere, such as from heat pumps, is included in the production column. Heat is a smaller market and accounts for less than one-tenth of the energy delivered by electricity. Somewhat less than one-half of heat comes from CHP plants, with most of the rest from heat plants. Within the CHP plants, heat accounts for about 45% of the produced energy, and electricity the rest. All of the CHP production comes from countries in northern temperate climate regions. In order of heat production, these countries or regions include the former Soviet Union, Europe, and the United States. More than 90% of production from heat plants comes from the former Soviet Union and China, with most of the remainder from OECD Europe. Heat losses including own use for the whole world are slightly higher than for electricity. China sends more than one-half of its heat production to the industrial sector, whereas the richer but colder countries of the former Soviet Union send a higher proportion to the residential sector in the form of district heating.

The OECD data provides an even more disaggregate look at a number of industries in their extended energy balances. Figure 16–1 shows the IEA breakdown of global energy use for these end-use industries, along with the energy industries' own use, as reported previously in table 16–1. This same breakdown of energy use is also available by country, regional groupings, and by separate energy product.

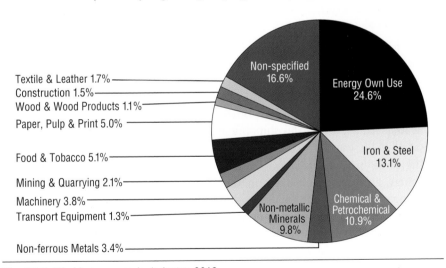

Fig. 16–1. World energy use by industry, 2010
Source: IEA (n.d.h).

The largest industrial user of energy is the energy industry itself, followed by iron and steel. The chemical and petrochemical industries come next. Their consumption does not include the energy used as feedstock. They are followed by nonmetallic minerals, which include glass, ceramics, bricks, and cement. The industries with the largest energy use are also among the US EIA group of industries categorized as energy intensive, with more than 4.8% of their operating costs coming from energy.

In industry, manufacturing takes the most energy. The seven most energy-intensive industries in the United States and some of the services they require include the following, as categorized by the US EIA (2009c):

- *Food and Kindred Products.* Process heating, process cooling, and machine drive.
- *Paper and Allied Products.* Wood preparation, pulping using mechanical or chemical means, and paper making including bleaching, drying, and packaging.
- *Bulk Chemicals.* Process heat, process cooling, machine drive, and electrochemical processes.
- *Glass and Glass Products.* Batch preparation and charging, melting, remelting, refining, and forming.
- *Cement.* Crushing, grinding, blending, and clinker production.
- *Iron and Steel Industry.* Coke production, melting, casting, and rolling.
- *Aluminum.* Smelter from ore and scrap, rolling, extruding to bars and wire, and fabricating to end products.
- *Metal-Based Durables.* Process heating, process cooling, machine drive, and electrochemical.

The IEA (n.d.h) also has a finer breakdown of transport fuel by type of use, as shown in figure 16–2. You can see the heavy dominance of road transport, with close to three-fourths of global use. World marine bunkers are used for international shipping, while world aviation bunkers are for airplane fuel used in international transport. If you add domestic and world use together, the total share going to air transport and to shipping are each between 10% and 11%. Further, almost all of the transport fuel use is from oil products, as shown in table 16–1.

The same breakdown shown in figure 16–2 can be seen for all countries, groupings, and energy products in the IEA database. For more detail on the breakdown between freight and passenger transport, as well as vehicle stock and trip characteristics, individual country survey data should be consulted. (See for example ORNL [2013] or Eurostat [2013].)

IEA does not have a breakdown of separate energy uses within the residential sector. For such breakdowns, again you need to turn to periodic individual country household surveys. For example, figure 16–3 shows US household energy use by service in 2012. More than 20% of household total energy use is for space heating, and almost one-half is used for heating, cooling, and water heating. Lighting adds almost another 10%, followed by the various appliance uses.

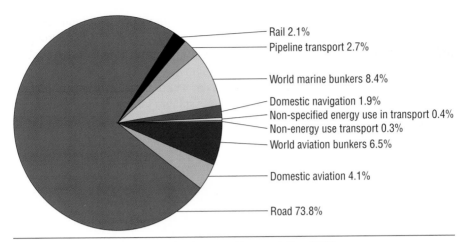

Fig. 16–2. World energy consumption by type of transportation, 2010
Source: IEA (n.d.h).

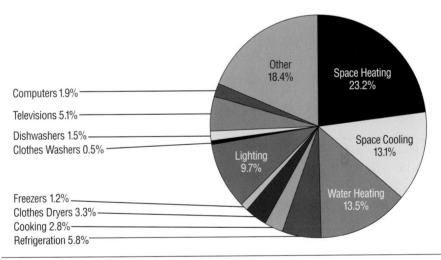

Fig. 16–3. Total US residential energy use by service, 2012
Source: US EIA (2014b, table A4).

For poorer countries, the basics take up a larger share. For example, in China, with similar latitude to the United States, about 66% of residential energy use was for space heating, water heating, and cooking in 2000, as opposed to the 41% shown above for the US (IPCC 2007, figure 6.3).

IEA combines the commercial sector with the government or public sector. Thus, the breakdown of energy consumption by service use for the commercial sector again requires going to country level surveys. Figure 16–4 gives the breakdown for the United States in 2012. With more heterogeneity in energy use, more than one-third of commercial use has no breakdown given. For uses where sectoral breakdowns are given, some interesting similarities and differences from the residential sector appear. Many of the same categories are included for both including heating, cooling, water heating, lighting, refrigeration, cooking, and personal computers (PCs). Both sectors consume more than 30% of their energy for space conditioning—heating, cooling, and ventilation (in the commercial sector), but the share is a few percentage points smaller for the commercial sector. However, the largest single share for the commercial sector is lighting, and they spend a somewhat larger share for refrigeration but a smaller share on water heating and cooking.

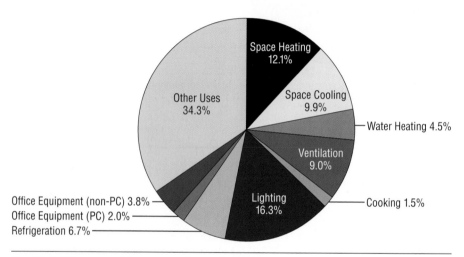

Fig. 16–4. US commercial energy use by service, 2012
Source: US EIA (2014b, table A5).

In poorer countries, the commercial sector is typically a smaller share of the economy, with more need for heat and water heating. For example, in the Chinese commercial sector in 2000, space heating comprised 45% of commercial sector energy use, with water heating adding another 22% (IPCC 2007, figure 6.3).

The IEA has energy balances by product for coal/peat, oil, natural gas, electricity/heat, and renewables, with 1990–2012 data currently available free online and additional historical statistics available for purchase in the energy database. Although space constraints preclude much discussion of these tables here, a sample for the world is included for each (see tables 16–2 through 16–6), with some salient features of the tables highlighted in the discussion.

The breakdown of coal use is shown in table 16–2. The unit for solids is weight in metric tonnes, not energy content as in table 16–1. Columns 2–6 contain primary solid coal sources ordered by quality with higher carbon, higher energy content coals coming first. Recall table 3–4 for some of these characteristics. Bituminous coal has the highest global consumption. Coking coals are high-quality coals for making coke for iron and steel making. They must be able to produce coke that is porous enough to allow hot air to travel up and the metal to travel down the furnace and strong enough not to collapse during the process. (For more on the steelmaking process, see http://dahl.mines.edu/st16/st16.pdf, question 1.)

Columns (7–11) are secondary source solid fuels, also measured in metric tonnes. Patent fuel consists of briquettes put together from coal particulates to make them usable. Briquettes are made from brown coal and peat. Coke is a product with high carbon content, no volatile compounds, and few impurities. It comes from coke ovens and gas works and is used for iron and steel making or as a fuel. Coal tar is a viscous by-product of coke or town gas manufacture.

The next four data columns (12–15) are by-product gases from coke ovens, gas works, and elsewhere. To roughly convert these gases into megatonnes of bituminous coal equivalent, divide the petajoule values by 27. These by-product gases are used in the energy industry and are also burned in power plants and other industries. In some cases, by-product gases are even used for heat in the residential, commercial, and public sectors. The last column contains balances for the small amount of peat consumed worldwide.

The breakdown of oil and natural gas liquids into refinery products and their uses are shown in table 16–3. Columns 2–3 in the table are primary products. Refineries run most of the crude oil and use about one-third of the NGLs.

NGLs consist of the liquids extracted from natural gas: ethane, propane, butane, isobutane, and condensate. Condensate, which consists of hydrocarbons from pentane (C_5H_{12}) and heavier, is a liquid at atmospheric pressure and is usually taken out at the gas wellhead and called lease condensate. It is called *plant condensate* if it is taken out at the gas plant. Condensate is sometimes called *naphtha* and *natural gasoline*. Condensate, naphtha, and natural gasoline are mixtures of hydrocarbons, similar to crude oil, and so no exact chemical formulas exist for them. For a nice discussion on the composition of lease condensates, see Braziel (2012).

NGLs are used for blending into oil and products, for petrochemical feedstocks, and for heating. The United States typically uses more liquid petroleum gases (propane and butane) taken from NGLs to produce petrochemicals. Europe and Asia use naphtha and small amounts of gas oil from crude oil.

Column 4 holds the secondary product, refinery feedstocks. These are products and backflows from NGLs and the petrochemical industry that are blended back into crude oil at the refinery.

The remaining table columns (5–12) show the secondary refined products. Naphtha is used mostly as a petrochemical feedstock. The majority of LPG is used in the residential sector, followed by petrochemicals. In richer countries, kerosene is used as a jet fuel, but in poorer countries, it is also used for heat and light. Gasoline is mostly used in the transport sector.

Table 16-2. Coal and peat use in the world in 2011 (solids in megatonnes (Mt), gases in petajoules (PJ))

Unit	Anthr	Coking	Ot Bitum	Sub-Bitu	Lig/Brwn	Pat Fuel	CokO Cok	G Cok	C Tar	Briq	GasWk Gas	CokO Gas	BlastF Gas	Ot Rec Gas	Peat
	Mt	Mt	Mt	Mt	Mt	Mt	Mt	Mt	Mt	Mt	Mt	PJ	PJ	PJ	Mt
Prod	91.9	970.0	4,749.5	717.5	911.0	13.7	636.1	2.1	12.6	11.2	354.8	3,248.1	5,268.0	137.8	16.5
OtSourc	3.7	0.0	163.7	0.4	0.0	0.0	0.3	0.0	0.0	0.0	2.4	2.7	0.0	0.0	0.0
Imports	53.7	261.7	731.8	81.8	5.0	0.7	23.0	0.0	2.3	0.5	0.0	0.0	0.0	0.0	0.6
Exports (−1)	−46.3	−282.9	−691.9	−119.3	−3.7	−0.2	−23.0	0.0	−0.9	−1.9	0.0	0.0	0.0	0.0	−0.2
Stock Δ (−1)	1.2	−12.6	−66.8	5.8	1.5	−0.1	−10.9	0.0	0.0	0.0	0.0	0.0	0.0	0.0	1.2
DomSup	104.3	936.2	4,886.3	686.2	913.8	14.0	625.4	2.1	14.1	9.9	357.2	3,250.8	5,268.0	137.8	18.1
Stat Dif	−4.3	−9.1	−172.2	5.3	−1.8	−1.3	−15.7	0.0	−0.4	0.0	−1.1	−24.3	−4.8	0.0	−0.1
Transform	36.1	879.7	3,546.4	647.7	862.4	0.0	447.2	0.0	0.3	2.0	77.3	717.8	1,060.0	44.6	15.6
Elec Plant	23.9	42.2	3,037.6	642.3	603.7	0.0	0.0	0.0	0.0	0.6	3.8	370.7	581.4	17.7	4.6
CHP Plant	4.6	1.1	199.1	4.4	215.2	0.0	0.0	0.0	0.1	1.1	19.4	201.7	400.8	23.6	6.4
Heat Plant	0.2	0.5	193.0	0.2	12.0	0.0	0.0	0.0	0.0	0.3	1.4	145.4	77.8	3.3	1.4
Ot Trans	7.4	835.9	116.8	0.9	31.4	0.0	447.2	0.0	0.2	0.0	52.7	0.0	0.0	0.0	3.3
E Own	0.0	5.0	108.0	0.0	1.8	0.0	1.6	0.0	1.4	0.0	8.1	705.4	364.7	12.9	0.3
Losses	0.0	0.0	3.5	0.0	0.1	0.0	0.1	0.0	0.0	0.0	0.0	12.8	48.7	9.5	0.2
TFC	63.8	42.4	1,056.2	43.7	47.7	12.7	160.7	2.1	11.6	7.8	270.7	1,790.6	3,789.7	70.8	1.9
Industry	59.4	40.6	791.3	39.3	30.9	4.1	142.5	2.1	8.2	3.5	131.1	1,701.3	3,789.5	70.8	1.0
Transport	0.0	0.0	6.4	0.3	0.1	0.0	0.0	0.0	0.0	0.0	0.0	0.0	0.0	0.0	0.0
Resident	3.1	0.1	120.0	0.3	12.7	8.6	0.7	0.0	0.0	3.2	118.3	70.5	0.0	0.0	0.6
Com&Pub	0.8	0.0	43.2	1.0	1.7	0.0	0.4	0.0	0.0	0.3	21.2	7.9	0.0	0.0	0.1
AgForFsh	0.0	0.0	20.8	0.1	0.3	0.0	0.6	0.0	0.0	0.0	0.0	0.0	0.0	0.0	0.2
Non-Spec	0.1	0.2	28.1	2.5	2.1	0.0	0.0	0.0	0.0	0.1	0.0	0.0	0.0	0.0	0.0
Non-EUse	0.3	1.5	46.5	0.2	0.0	0.0	16.4	0.0	3.4	0.6	0.1	10.9	0.2	0.0	0.0
PChFsk	0.1	0.0	2.4	0.0	0.0	0.0	0.0	0.0	0.9	0.0	0.1	0.1	0.0	0.0	0.0

Source: IEA (n.d.b).

Note: Mt = megatonne; Stock Δ = stock changes; Stat Dif = statistical differences; Transform = energy transformations; Elec = electricity; CHP = combined heat and power; Ot Trans = other transformation; E Own = energy industry own use; Losses = transportation and distribution losses; TFC = total final consumption; Resident = residential sector; Com&Pub = commercial and public sector; AgForFsh = agriculture, forestry, and fishing; Non-Spec = non-specified; Non-EUse = nonenergy use; PChFsk = petrochemical feedstocks; Anthr = anthracite coal; Coking = coking coal; Ot Bitum = other bituminous coal; Sub-Bitu = subbituminous coal; Lig/Brwn = lignite/brown coal; Pat Fuel = patent fuel; CokO Cok = coke oven coke; GCok = gas coke; C Tar = coal tar; Briq = brown coal (BKB) peat briquettes; GasWkGas = gas works gas; CokO Gas = coke oven gas; BlastF Gas = blast furnace gas; Ot Rec Gas = other recovered gases; Prod = production; OtSourc = other sources; (−1) indicates that although IEA labels these as positive values, the number recorded is actually minus their value.

Table 16–3. World petroleum statistics, 2011 (million metric tonnes).

	Crude Oil	NGL	Ref FStks	Naphtha	LPG	Motor Gasol	Av Gasol	Jet Kero	Ot Kero	Gas/Diesel	Fuel Oil
Production	3,637.3	342.6	0.0	254.9	113.0	898.6	1.7	248.5	72.2	1,279.9	515.3
Ot Sources	0.0	0.0	30.4	0.0	0.0	0.0	0.0	0.0	0.0	0.0	0.0
Imports	2,153.0	26.5	73.4	100.3	73.0	163.7	0.8	61.4	12.7	300.3	246.2
Exports (−1)	−2,050.4	−70.3	−10.4	−99.9	−80.2	−164.1	−0.2	−65.4	−23.0	−311.7	−290.7
Stock Δ (−1)	−3.1	−0.2	1.2	0.0	−1.1	2.7	0.0	0.1	−0.8	4.4	2.0
DomSupply	3,736.8	298.6	94.6	255.3	104.6	901.0	2.3	244.6	61.1	1,273.0	472.8
Transfers	8.0	−193.9	35.3	12.6	123.9	13.0	0.0	0.7	−7.6	−5.3	−16.0
Stat Dif.	5.4	0.1	3.5	−2.9	1.3	5.5	−0.1	−2.0	−1.0	2.8	−3.2
Transform	3,724.7	98.0	133.5	25.5	7.0	0.1	0.0	0.1	0.5	77.1	140.1
Elec Plant	41.0	0.0	0.0	0.5	1.2	0.1	0.0	0.1	0.2	73.8	116.5
CHP Plant	0.0	0.0	0.0	1.5	0.1	0.0	0.0	0.0	0.0	1.4	13.4
Heat Plant	0.8	0.0	0.0	0.0	0.1	0.0	0.0	0.0	0.0	1.0	8.4
Oil Ref	3,682.9	98.0	133.5	0.0	0.0	0.0	0.0	0.0	0.0	0.0	0.0
OtTrans	0.0	0.0	0.0	23.5	5.6	0.1	0.0	0.0	0.3	1.0	1.8
E Own Use	6.2	0.2	0.0	0.2	3.6	0.4	0.0	0.0	0.1	14.8	25.8
Losses	7.8	0.0	0.0	0.0	0.1	0.1	0.0	0.1	0.0	0.0	0.1
Final C	11.5	6.6	0.0	239.3	219.2	918.9	2.2	243.2	51.8	1,178.6	287.5
Industry	8.4	2.0	0.0	6.1	18.9	4.5	0.0	0.4	3.9	128.7	77.8
Transport	0.0	0.0	0.0	0.0	19.3	906.3	2.2	241.4	0.3	821.6	191.3
Residential	0.0	0.2	0.0	0.0	96.8	0.4	0.0	0.0	34.9	56.9	1.1
Com&Pub	0.0	0.0	0.0	0.0	16.5	1.4	0.0	0.2	8.0	65.3	6.3
AgForFsh	0.0	0.0	0.0	0.0	2.8	4.8	0.0	0.0	1.0	96.0	4.2
Non-Spec	0.2	0.0	0.0	0.0	2.3	0.2	0.0	1.2	2.0	6.8	3.2
Non-EUse	2.9	4.3	0.0	233.1	62.6	1.2	0.0	0.0	1.6	3.3	3.7
PChmFStk	2.4	4.3	0.0	233.1	61.2	1.2	0.0	0.0	1.4	3.3	3.6

Source: IEA (n.d.m).

Note: In the database, international marine bunker fuel and international aviation bunker fuel are reported in the transport total for the world as in this table, but separately for tables for individual countries and regions. Abbreviations from the first column are as follows: Ot Sources = other sources; Stock Δ = stock changes; DomSupply = total domestic energy supply; Stat Dif. = statistical differences; Transform = energy transformations; Oil Ref = oil refineries; OtTrans = other transformations; E Own Use = energy industry's own use; Losses = transportation, distribution, and other losses; Final C = total final consumption; Com&Pub = commercial and public sector; AgForFsh = agriculture, forestry, and fishing; Non-Spec = nonspecified; Non-EUse = nonenergy use; PChmFStk = petrochemical feedstocks. Abbreviations from the first row are as follows: NGL = natural gas liquids; Ref FStks = refinery feedstocks; LPG = liquid petroleum gases; Motor Gasol = gasoline; Av Gasol = aviation gasoline; Jet Kero = jet fuel from kerosene; Ot Kero = other kerosene; and Gas/Diesel = gas oil and diesel oil; (−1) indicates that although IEA labels these as positive values, the number recorded is actually minus their value.

Gas oil (called *number 2 fuel oil* or *distillate* in the United States) and diesel fuel are essentially the same product except that the specifications are stricter for diesel to comply with motor needs and environmental regulations.

More than two-thirds of all final consumption of liquid fossil fuels goes to the transport sector. Around two-thirds of the final demand shown in the Gas/ Diesel column goes to the transport sector in the form of diesel. Although not shown separately in the world table, around one-third of residual fuel oil is used for international marine bunkers. A small portion is used in domestic marine and rail transport, and most of the rest goes to industry and for electricity generation. Worldwide gasoline has the largest share of the transport market at 42%, followed by diesel at around 38%, with jet fuel and marine bunkers trailing at 11% and 9%. However, in Europe, where preferential tax treatment and other policies have actively promoted diesel, it is used more intensively in personal transportation and is about two-thirds of the transport fuel market. Almost all of the transport fuel use is from oil products, as shown in table 16–3.

The breakdown in table 16–3 can be seen for all countries, groupings, and energy products in the IEA database. For more detail on the breakdown between freight and passenger transport, as well as vehicle stock and trip characteristics, individual country survey data should be consulted (for example, see ORNL [2013] or Eurostat [2013]).

Table 16–4 shows world natural gas flows. Production comes mainly from oil and gas wells and coal seams. A small amount comes from gas works in OECD countries, which is transferred into the natural gas system recorded under the From Other Sources category. Although oil products are ubiquitous, natural gas is not as widely consumed. Of the 224 countries and territories in the EIA energy database, only about one-half consume natural gas. Almost one-half of the world's natural gas goes into electricity, heat, or energy sector own use. The small bit that goes into other transformation is mostly GTLs in South Africa, Malaysia, and Canada.

In final natural gas consumption worldwide, industry claims about 21%, and the residential sector takes about two percentage points less. However, in the OECD countries with well-developed gas networks, temperate climates, and deindustrialization, the residential sector claims the larger share. In contrast, in developing Asia (outside of China), Africa, and Latin America, with milder climates, less well-developed gas networks, and the push to develop, industry typically claims more than twice the gas of the residential sector. Transport claims less than 5% worldwide. However, in places where policy has actively promoted natural gas as a transport fuel, its share has become quite significant, such as in Argentina (16%), Pakistan (21%), and Russia (32%). The share going to petrochemicals is higher in developing than in industrial countries.

Although more than a billion people worldwide do not have electricity, as with oil products, it is available in all countries. Electricity access and use is often considered an indicator of development. For example, IEA world energy balance tables show total electricity as a share of final product averages about 22% for OECD countries but only about 9% for Africa, where biomass accounts for more than one-half of end-use demand (IEA, n.d.f).

Table 16–4. World natural gas statistics, 2011 (PJ)

Production	24,704.9
From Other Sources	52.7
Imports	3,750.3
Exports (−1)	−1,612.1
Stock Changes (−1)	−378.4
Domestic Supply	**26,517.4**
Statistical Differences	−202.6
Transformation	**8,828.4**
Electricity Plants	6,953.6
CHP Plants	1,874.8
Heat Plants	0.0
Oil Refineries	0.0
Other Transformation	0.0
Energy Industry Own Use	**2,279.9**
Losses	0.0
Final Consumption	**15,206.5**
Industry	5,524.3
Transport	772.0
Residential	5,082.8
Commercial and Public Services	3,358.1
Agriculture / Forestry	0.0
Fishing	0.0
Other Non-Specified	0.0
Non-Energy Use	469.3
Petrochemical Feedstocks	414.8

Source: IEA (n.d.f).
Note: Amounts are given in petajoules (PJ) on a gross calorific value basis. One PJ is approximately 0.95 trillion Btu; (−1) indicates that although IEA labels these as positive values, the number recorded is actually minus their value.

The global sources and sectoral use of electricity measured as electricity generated with no heat rate adjustment are shown in table 16–5. Infrastructure development plays a large role in determining the evolution of electricity access and use. Although residents of high-income countries have almost universal access to electricity, the situation for the poorest countries is much more depressing. For example, the World Bank's WDI (n.d.), shows that less than one-third of the population in sub-Saharan African countries had access to electricity in 2011. Indeed, in Malawi and Democratic Republic of Congo, the access was less than

10%. With such low access, electricity use is skewed toward larger users in the industrial sector. In Africa, the 2011 industrial share of final electricity demand is 42%, and it is even higher in developing Asian countries, including China.

Table 16–5. World electricity and heat statistics, 2011

Production from:	Electricity Unit: TWh	Heat Unit: TWh
Coal and peat	9,144.3	1,565.1
Oil	1,057.9	208.0
Gas	4,852.1	1,886.0
Biofuels	330.4	119.3
Waste	91.4	98.9
Nuclear	2,583.7	8.0
Hydro	3,565.5	0.0
Geothermal	69.2	3.4
Solar PV	61.2	0.0
Solar thermal	2.2	0.1
Wind	434.2	0.0
Tide	0.6	0.0
Other sources	8.3	111.0
Total production	**22,201.0**	**3,999.7**
Imports	648.6	0.0
Exports	−649.0	−0.1
Domestic supply	**22,200.6**	**3,999.6**
Statistical differences	−19.3	−0.9
Transformation	**3.6**	**8.9**
Electricity plants	0.0	4.8
Heat plants	3.6	4.1
Energy industry own use	**1,986.9**	**496.1**
Losses	1,794.1	225.2
Final consumption	**18,396.7**	**3,268.6**
Industry	7,834.7	1,464.3
Transport	292.5	0.0
Residential	4,982.6	1,252.2
Commercial and public services	4,188.5	357.7
Agriculture/forestry/fishing	475.5	71.3
Other non-specified	622.9	123.2

Source: IEA (n.d.d).

In contrast, in the OECD, industry consumed about one-third in 2011. Since the service and government sectors tend to increase their share of the economy as countries progress along a development path, these sectors also tend to get a larger share of electricity in rich countries than in poor countries. Thus, for African and Asian developing countries including China, the combined service and government sector consumes less than 16% of electricity generated, whereas this sector consumes more than 30% in the OECD (IEA, n.d.f).

The world sources for heat generated and sectoral use in 2011 are also shown in table 16–5. As a reminder, the heat in this table is bought and sold commercially from dedicated heat plants or as a by-product of electricity generation or other industrial processes. The original IEA unit for heat was the joule (J), but the units have been converted to watt hours (Wh), with one watt hour equal to 3,600 J for easier comparison. By energy content, there is more than five times more electricity generated than heat. More than three-fourths of the heat is generated in the former Soviet Union and China, with much of the remainder generated in Europe. The share of final use going to the residential, commercial, and public sectors, presumably in the form of district heat, is about 56% in the former Soviet Union, about a percentage point lower in OECD Europe, and about 28% in China.

For each of the above uses and sectors, underlying decisions have been made on how much energy to consume and in what form. Consumers use energy for the end-use products they consume, while all other sectors use energy as an intermediate good or as a factor of production. The former can be called end-use demand, while the latter can be called factor demand. In the next sections, we will provide simple optimization models to illustrate how optimal decisions should be made for both end-use and factor demands for energy.

Household or Consumer Demand

Households should choose the bundle of goods within their constraint set that maximizes their satisfaction. In a world where consumers have n goods to choose from, we represent their satisfaction by a handy construct, a utility function, as follows:

$$U(X_1, X_2, \ldots X_n)$$

where
X_i represents consumption of the ith good for $i = 1, \ldots n$.

Now it may seem strange to represent people's happiness by a function. However, this formulation is not as strange as it first seems because each of us has preferences; we prefer some goods to others and some bundles of goods to other bundles. For example, you may prefer a Porsche to a Lamborghini. Under some fairly reasonable assumptions about preferences, we can represent such preferences by a function. The actual numerical values that the function takes on are not important, provided it correctly ranks our preferences across goods.

So you know you prefer a Porsche, but you may not rush out and buy a Porsche since you are constrained by your income. In our optimization model, we will need to include this constraint. In our simple problem, we assume consumers spend all their income. In other words, they cannot borrow or steal from the bank, save money in the bank, or lend to, borrow from, or mooch off of others. We could add in these complications but will aim for a simpler world for now.

The amount a consumer spends on the ith good is the price of the ith good (P_i) times the amount consumed of the ith good (Q_i). Income (Y) spent on all goods is the sum of the amounts spent on each good, as follows:

$$Y = P_1 Q_1 + P_2 Q_2 + \ldots, + P_n Q_n = \sum_{i=1}^{n} P_i Q_i$$

Consumers want to maximize their utility subject to their budget constraint. To keep our initial problem as simple as possible, we lump all energy goods into one product called E, with price P_E, and we lump all other goods together in one nonenergy good called N, with price P_N.

First, let's visualize this simple optimization problem by representing it in two dimensions in figure 16–5. Let the vertical axis represent consumption of energy and the horizontal axis represent nonenergy. Suppose $P_E = 4$ and $P_N = 2$ and the consumer has an income of 160. Then her budget constraint, shown in figure 16–5, is given in equation (16–1):

$$N = \frac{Y}{P_N} - \frac{P_E}{P_N} E \tag{16-1}$$

Thus, the intercept of the budget constraint equals income divided by the price of N or the amount of N that could be purchased with the given income. The slope of the budget constraint equals the price of energy divided by the price of nonenergy. Substituting in income and prices gives us the budget constraint, as shown in equation (16–2):

$$N = \frac{160}{2} - \frac{4}{2} E = 80 - 2E \tag{16-2}$$

We can graph this relationship in N-E space, as shown in figure 16–5. The N intercept is 80 and the slope of the line is –2. If you consume no E, you can have 80 N. Alternately, if you consume no N, you can have 40 E. This gives us the lower line in the figure labeled $Y = 160$.

Now notice what happens to the budget constraint if you double income to 320:

$$N = \frac{320}{2} - \frac{4}{2} = 160 - 2E$$

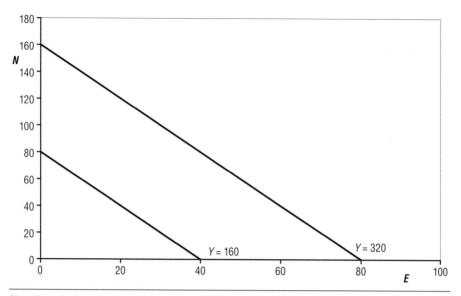

Fig. 16–5. Budget constraint: $N = Y/P_N - (P_E/P_N)E$ for $Y = 160$ and $Y = 320$

Now the budget is twice as large or twice as far from the origin. Alternatively, prices could change the budget constraint. If both prices doubled, P_E from 4 to 8 and P_N from 2 to 4 at income 320, our budget constraint becomes the following:

$$N = \frac{Y}{P_N} - \frac{P_E}{P_N}E = \frac{320}{2 \times 2} - \frac{2 \times 4}{2 \times 2}E = 80 - 2E$$

Now in real terms, we are back to equation (16–2) or the same budget as before price and income doubled. Thus, if we double income and all prices, our budget constraint does not change. If a function does not change when we double income and prices, it is said to be *homogeneous of degree zero in income and prices*. For more on functions that are homogenous see http://dahl.mines.edu/st16/st16.pdf, question 2.

See what happens in figure 16–6 if only the price of energy doubles from 4 to 8 when $Y = 320$. Then the budget constraint becomes the following:

$$N = \frac{320}{2} - \frac{8}{2}E = 160 - 4E$$

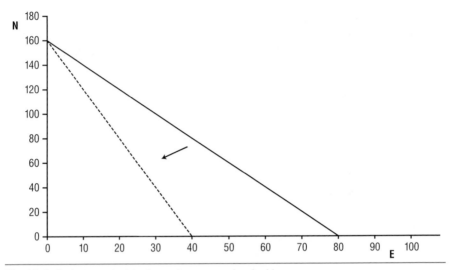

Fig. 16–6. Budget constraint when only energy price doubles

Note that not only does this price increase make energy more expensive relative to nonenergy, it also decreases the budget set everywhere except where energy consumption is zero. Thus, price changes have an effect on real income.

The consumer wants to pick the best point within her budget constraint. To continue our visual representation of the problem, let's represent our utility function with two goods as follows:

$$U = U(N,E)$$

The utility function is three dimensions: U, N, and E. To reduce it to two dimensions and graph it with the budget constraint, we hold U constant at $\overline{U}$ and define a new concept called an *indifference curve*:

$$\overline{U} = U(N,E)$$

Since the consumer has the same utility at each point along this function, they are equally happy consuming any bundles along this function or are indifferent to which specific bundle they consume. Before we actually choose a function to represent this indifference curve, we can use our old friend calculus to develop one more piece of information about the indifference curve. The total differential of the indifference curve is shown in equation (16–3) as follows:

$$d\overline{U} = \frac{\partial U}{\partial N} dN + \frac{\partial U}{\partial E} dE \qquad (16\text{–}3)$$

Since $\overline{U}$ is constant, $d\overline{U} = 0$. The interpretations of $(\partial U/\partial N)$ and $(\partial U/\partial E)$ are the change in utility from an additional unit of N and E consumed, respectively, and they are called the marginal utility of nonenergy and of energy, with dN and dE the total change in nonenergy and energy consumption. Substitute $d\bar{U} = 0$ into equation (16–3) and solve for the slope of this indifference curve (dN/dE):

$$d\bar{U} = \frac{\partial U}{\partial N} dN + \frac{\partial U}{\partial E} dE = 0 \longrightarrow \frac{dN}{dE} = -\frac{\partial U}{\partial E} \Big/ \frac{\partial U}{\partial N} = -\frac{U_E}{U_N}$$

(Note we often use the simpler, but less intuitive, notation: $\frac{\partial U}{\partial E} = U_E$ and $\frac{\partial U}{\partial N} = U_N$.)

The above equation indicates the slope of the indifference curve is minus the marginal utility of energy divided by the marginal utility of nonenergy. This ratio of marginal utilities is also called the *marginal rate of substitution* and it will have special significance when we optimize below. To get a graphical representation of this function, take a simple form of the Cobb-Douglas utility function:

$$U = E^{0.5}N^{0.5}$$

Fix utility at $U = 29$ and graph the indifference curve, as in figure 16–7, panel (I).

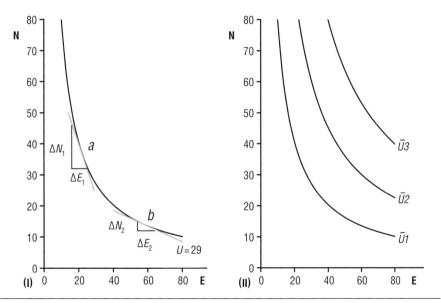

Fig. 16–7. Indifference curve representing the consumer's preferences

The slope of the indifference curve (U_E/U_N), which is minus the marginal rate of substitution, shows how much of one good the consumer is willing to trade for another. In the figure, we can see that the consumer would trade off ΔN_1 for ΔE_1 at point a. As we move down the indifference curve toward b, we can see that the tradeoff of ΔN_2 for ΔE_2 gets smaller. In other words, N is relatively more valuable and E relatively less valuable as we have less N and more E.

If we assume that people prefer more of both goods to less, bundles to the right of the isoquant are more preferred and bundles to the left are less preferred. Thus, in panel (II) of figure 16–7, the isoquant $\overline{U}_1$ represents a lower level of utility than $\overline{U}_2$, which in turn represents a lower level of utility than $\overline{U}_3$. If preferences are complete, you can imagine a whole series of indifference curves with increasing utility as you move away from the origin. Moreover, if more is preferred to less for each good, the indifference curves cannot cross.

Now we are at last ready to optimize by adding a budget constraint ($Y = 160$) to the model, as shown in figure 16–8. Find the highest indifference curve that is within the budget set. This is where the budget constraint, $Y = 160$, is tangent to $\overline{U}_1$ and consumption is E^* and N^*. Remembering that the slope of the indifference curve $= -U_E/U_N$ and the slope of the budget constraint equals $-P_E/P_N$, we have $\dfrac{U_E}{U_N} = \dfrac{P_E}{P_N}$. Thus, we should line up the ratio of our subjective values with the price ratios.

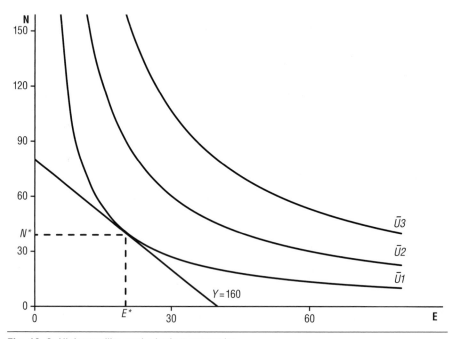

Fig. 16–8. Highest utility on the budget constraint

Alternatively, we can rewrite the optimizing relation as shown in equation (16–4):

$$\frac{U_E}{P_E} = \frac{U_N}{P_N} \tag{16–4}$$

This is a nice heuristic that tells us to consume at the point where the marginal utility of energy per dollar of energy equals the marginal utility of nonenergy per dollar of nonenergy product. Under the assumption that consumers prefer more to less, U_E and U_N are both positive. Thus, we get more utility by consuming more of each good. Further, suppose that U_{EE} and U_{NN} are negative. U_E is marginal utility, so U_{EE} is the slope of the marginal utility curve. If this slope is negative, it means marginal utility is decreasing, or economists say we have diminishing *marginal utility*. Thus, although we get more satisfaction if we consume more energy ($U_E > 0$), the last unit of energy makes us less happy than the next to the last and so on, and we make the same assumption for nonenergy. The 50th music CD we purchase gives us less satisfaction than the 49th, the 49th less satisfaction than the 48th, and so on. If this is the case, let's see what our heuristic (eq. 16–4) tells us to do. Let's suppose that at our current consumption the following applies:

$$\frac{U_E}{P_E} > \frac{U_N}{P_N}$$

Suppose that E is heat and N is listening to music CDs. At our current consumption level, we are sitting listening to CDs and it is freezing. We get more satisfaction per dollar of energy than per dollar of CD. We would rather have more energy and listen to CDs in comfort than have more CDs. However, as we allocate money away from CDs to heat up our room, the marginal utility of heat falls. As we buy fewer CDs, the marginal utility of CDs increases. We should keep reallocating until the marginal utility per dollar is equal across product. For a mathematical derivation of this heuristic, see http://dahl.mines.edu/st16/st16.pdf, question 10.

Now let's change prices and income and see what happens in our figures. First, hold income at 160 and change price. Lowering energy price from 4 to 2 to 1 moves the optimal consumption of the two goods from *a* to *b* to *c* in figure 16–9, panel (I). You can see how much E and N would be consumed at each price.

Next, graph the price against energy consumption in figure 16–9, panel (II) for our three prices, labeled to correspond to *a*, *b*, and *c*. If we fill in the values for all other prices between $2 and $20, we have derived a demand curve for energy.

We can also easily see what happens as we hold energy price constant at 4 and increase the budget from $Y = 160$ to $Y = 240$ and then to $Y = 320$, as in figure 16–10. Optimal consumption of the two goods moves out from *a* to *b* to *c*. If we connect the points along all the isoquants as we increase the budget, we trace out what is called an *income consumption curve* or *expansion path*. This curve shows us how we scale up consumption of energy and nonenergy goods as income increases. For this simple utility function, the income expansion path is a straight line.

As with our demand curve above, we can also graph income against energy consumption. By convention, income is on the horizontal axis and consumption of a good is on the vertical axis, as shown in figure 16–10, panel (II). This formulation is called an *Engel curve*. The Engel curves for energy and nonenergy are both shown in the figure. Since the utility function is symmetric in E and N, the higher Engel curve in this example is totally the result of the lower price for the composite nonenergy good.

Although we did not develop actual functions here to derive demand equations, we can imagine the general solutions for energy and nonenergy consumption as follows:

$$E = f(P_E, P_N, Y)$$

$$N = g(P_E, P_N, Y)$$

These would be the consumer demand functions, which are typically a function of own price, other prices, and income. The shapes of these functions would result from preferences and the technologies surrounding product use. We would horizontally sum all consumer demands to get market demand. In reality, we consume many goods, not just two. With a mathematical rather than a graphical solution, the problem could easily be modified to include other products and choice variables.

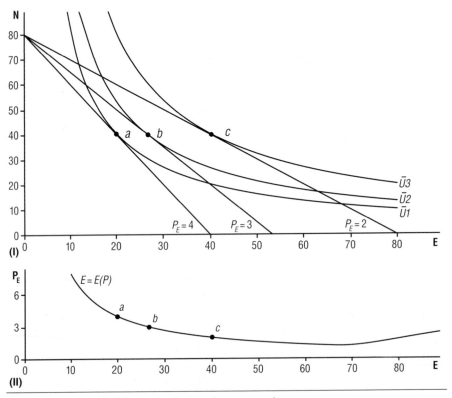

Fig. 16–9. Consumption changes with changing energy price

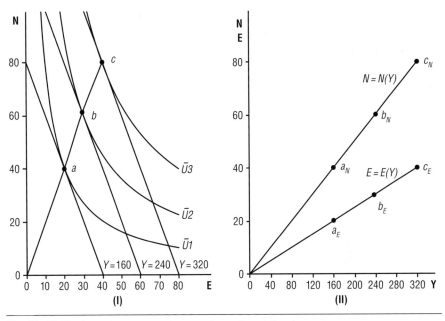

Fig. 16–10. Tracing out a consumer's expansion path and Engel curves

Consumer Demand and a Subsidy

Consider one last piece of insight on consumer demand from this graphical development of consumer optimization that has an important energy policy application. Consider the case of a per unit energy subsidy. For example, some countries subsidize kerosene to help poor consumers. Let the subsidy per unit be sb. If the initial budget constraint were:

$$P_E E + P_N N = Y$$

then the subsidized budget constraint would be the following:

$$(P_E - sb)E + P_N N = Y$$

Notice the subsidy is equivalent to a price decrease and moves the consumers' optimal bundle from point a to b in figure 16–11.

Now instead of a subsidy, give the consumer the same amount of income as the subsidy costs the government at the original prices. Represent this new budget by the dotted line that goes through point b. It has the same cost, but the consumer faces the original prices so the dotted line is parallel to the original budget before the subsidy. Notice that the consumer would choose point c under the increased income. Thus, we typically make consumers happier by giving them an equivalent amount of cash rather than a subsidy.

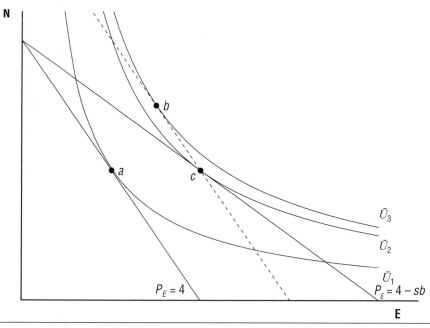

Fig. 16–11. Comparing a subsidy with an equal cost cash payment

Factor Demand for the Industrial, Commercial, and Electricity Sectors

When businesses demand energy, they want it to produce goods and services to sell. Let's suppose they sell good X. To produce X, they need energy (E) and nonenergy (N) inputs. Again, we restrict ourselves to two inputs to keep the problem as simple as possible while still allowing some choice. The technology for producing X from E and N is represented by a production function. The producer knows the price of their output (P_X), and they know the unit price of their inputs (P_E and P_N). Thus, if nonenergy includes capital, its price has been levelized to a unit price. We assume competitive input and output markets, implying the producer faces prices rather than a demand function for output and supply functions for inputs. The producer's objective function is to maximize profits, as shown in equation (16–5):

$$\pi = P_X X(N,E) - P_N N - P_E E$$

(16–5)

Note that in this case we have substituted technology into the objective function directly and do not need to add in a constraint as in the consumer case above. However, similar to the consumer case, we make some assumptions about the technology represented in the production function. X_E and X_N are greater than zero or we have positive marginal product. Adding more energy or more nonenergy input increases our output. Further, we assume that $X_{EE} < 0$, $X_{NN} < 0$, or diminishing marginal product, and $X_{EN} > 0$, or adding more nonenergy raises the marginal product of energy.

The producer chooses how much of inputs E and N to produce, and these choices determine how much output the producer sells. The first-order conditions for maximizing equation (16–5) are given in equations (16–6) and (16–7):

$$\pi_E = P_X X_E - P_E = 0 \tag{16-6}$$

$$\pi_N = P_X X_N - P_N = 0 \tag{16-7}$$

Recall from chapter 9, the economic interpretation of the first expression in equation 16–6. It is the price of output times the marginal product of energy. Since the marginal product of energy is the amount of extra output produced, the price of that extra output times the extra output is how much extra revenue an additional unit of energy brings in, called the marginal revenue product (MRP). For example, suppose that output is electricity, the price of a megawatt hour of electricity is $60, and the marginal product of an additional MMBtu of gas is 0.1 megawatt hour. Then the extra revenue bought in by an additional MMBtu of gas is 0.1 × 60 = $6. Provided the gas costs less than $6 per MMBtu, it pays to buy the gas and produce electricity.

With that explanation, let's reexamine the first-order conditions. Equation (16–6) implies that $P_X X_E = P_E$. The slope of the marginal revenue product is $\partial(P_X X_E)/\partial E = P_X X_{EE}$. With diminishing marginal product, this function slopes down, as in figure 16–12, and we can easily see the optimal purchase E_1.

Another way of looking at the first-order condition is to divide equation (16–6) by equation (16–7) and simplify to the following:

$$\frac{P_X X_E}{P_X X_N} = \frac{P_E}{P_N} \longrightarrow \frac{X_E}{X_N} = \frac{P_E}{P_N}$$

The expressions before the arrow suggests that you should hire factors up to the point where the ratio of their marginal revenue products is equal to the ratio of their prices. After canceling out the output prices, the expressions after the arrow tell us that we should hire factors up to the point where the ratio of their marginal products equals the ratio of their prices, which are also their marginal costs in a competitive industry. Rearranging even further, we get a result similar to the result for consumers:

$$\frac{X_E}{P_E} = \frac{X_N}{P_N}$$

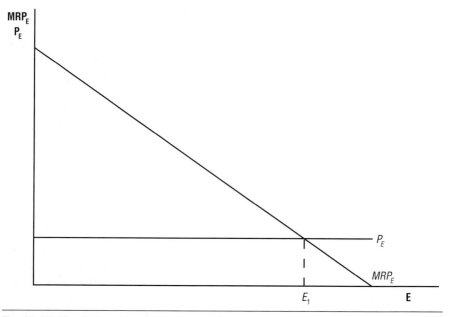

Fig. 16–12. Marginal revenue product for a producer

We should hire factors up to the point where the marginal product per dollar is equal across factors.

Although we do not have actual functions here, again we expect the solutions to equations (16–6) and (16–7) to be as follows:

$$E = f(P_e, P_N, P_X)$$

$$N = g(P_E, P_N, P_X)$$

With more than one energy input, we would also add their prices to the equations. With more than one output (as for a refinery), we would add more output prices to the equation. If we know how much output or economic activity (X) is produced, we could substitute X for its price. Factor demands that are a function of output or economic activity are called *conditional factor demands*. If we horizontally sum all individual demands for energy, we get market demand. In the next sections, we will consider how to estimate such relationships.

Econometric Estimates of Energy Demand—Picking the Functions

Many energy demand equations have been estimated on actual energy data using statistical techniques. The discipline that deals with the application of statistical techniques to estimating economic models is called *econometrics*.

In chapter 2, we considered some simple extrapolations. In this chapter, I will highlight some more statistical techniques. With univariate time series analysis, the current value of an energy variable (X_t) is a function of past values of the same variable to pick up inertia and time lags in economic adjustment in the system along with an error or multiple error terms to represent random events (ε_t):

$$X_t = \sum_{i=1}^{n} \alpha_i X_{t-i} + \sum_{i=1}^{m} \mu_i \varepsilon_{t-i}$$

Well-established statistical techniques are applied to historical data to pick the unknown coefficients (α_i, μ_i) along with lag lengths (n, m), which determine how many past values of X and random events influence the current value of the variable. This technique is more general than the exponential extrapolation, since it does not require a fixed growth rate and can accommodate cyclical behavior as well as growth over time. (For the canonical reference on univariate time series, see Box and Jenkins [1970].) Time series analysis may provide reasonably good short-term forecasts, but it does not typically forecast turning points very well. Nor can we analyze policy or understand the structure and drivers of energy consumption.

Sheik Zaki Yamani, Saudi Oil Minister from 1962 to 1986, argues that politics may govern oil prices in the short run, but economics will govern oil prices in the long run (Yergin 1991). This is likely true for other energy sources as well. Since these fundamentals change over time, they do not support models of exponential growth at a constant rate or models where current values of an energy variable are solely determined by past values.

In choosing a multivariate equation to estimate, there are a number of issues to consider. First, we need to pick our function. Theory tells us what variables we might expect to be in the equation. The simple modeling above tells us that for consumer demand, we would want to include the price of the energy product, the price of other products, and income. The most important other prices are the prices of substitutes and complements. For factor demands, we would want to include prices of factors and prices for outputs. For conditional factor demands, we would want to include prices of factors and the outputs themselves. We may also think of other variables we need to include that help shape and influence preferences and profits, such as weather, government policy, and technology.

Once we have chosen the variables, theory often tells us the direction of the relationship. Recall from chapter 3 that when own price goes up, price of a substitute goes down, or the price of a complement goes up, energy purchases are expected to go down. More consumer income raises consumption if the energy product is a normal good, or it reduces consumption if the energy product is an inferior good.

Relationships could be linear, or they could take on a whole variety of nonlinear forms, including logs, exponentials, polynomials, and inverses. Some simple examples include the following:

$$E = \alpha_1 + \alpha_2 P_E + \alpha_3 P_N + \alpha_4 Y \tag{16–8}$$

$$\ln(E) = \beta_1 + \beta_2 \ln(P_E) + \beta_3 \ln(P_N) + \alpha_4 \ln(Y) \tag{16–9}$$

$$E = X_1 + X_2 \frac{1}{P_E} + X_3 \frac{1}{P_N} + X_4 \frac{1}{Y} \tag{16–10}$$

$$E = \delta_1 + \delta_2 \ln(P_E) + \delta_3 \ln(P_N) + \delta_4 \ln(Y) \tag{16–11}$$

$$\ln(E) = \phi_1 + \phi_2 P_E + \phi_3 P_N + \phi_4 Y \tag{16–12}$$

$$E = \varphi_1 + \varphi_2 P_E + \varphi_3 P_N + \varphi_4 Y + \varphi_5 P_E P_N + \varphi_6 P_E Y + \varphi_7 P_N Y \tag{16–13}$$

$$E = \gamma_1 + \gamma_2 P_E + \gamma_3 P_N + \gamma_4 Y + \gamma_5 P_E^2 + \gamma_6 P_N^2 \gamma_7 Y^2 \tag{16–14}$$

Theory often has less to say about the exact form of the relationship than about the direction of the relationship. Choosing among the above seven functional forms, as well as others, often becomes an empirical question. Techniques exist to pick the form that has the statistical best fit. If we have one equation nested inside of another, we can fit the larger and more general model and see whether it explains a lot more than the simpler model. For example, in the above set of models, equation (16–8) is nested inside of equation (16–13), as well as in equation (16–14).

The following Box-Cox transformation nests a variety of functional forms into one variable:

$$\frac{y^\lambda - 1}{\lambda} \tag{16–15}$$

Notice when $\lambda = 1$, the function equals $y - 1$, which is linear in y. When $\lambda = -1$, the function equals $1 - 1/y$, which is a function with an inverse of y. When $\lambda = 2$, the function equals $(y^2 - 1)/2$, which contains a quadratic form of y. We have an even more interesting case when $\lambda \to 0$; the function goes to $\ln(y)$. For derivation of this last result, see http://dahl.mines.edu/st16/st16.pdf, question 4.

Another specification issue is how to capture adjustment across time. In any economic system, habit, inertia, contracts, and other constraints may prevent buyers from immediately adjusting to changes in variables. A simple way of trying to capture adjustment across time is to include lagged variables in the estimation model. A simple illustrative model is shown in equation (16–16):

$$E_t = \beta_1 + \beta_2 P_t + \beta_3 P_{t-1} + \beta_4 E_{t-1} \tag{16–16}$$

where
E_t represents quantity of energy demanded in year t,
P_t represents price in year t,
P_{t-1} represents price in year $t-1$, and
E_{t-1} represents quantity of energy demanded in year $t-1$.

Now suppose price changes by ΔP_0 and you want to know the adjustment in the current period, called the short run, and the total adjustment after all change has taken place, called the long run. The effect now in this period, period 0, of a price change is shown in equation (16–17)

$$\Delta E_0 = \beta_2 \Delta P_0 \quad \longrightarrow \quad \frac{\Delta E_0}{\Delta P_0} = \beta_2 \tag{16–17}$$

To see what happens in the next period, period 1, refer back to the two lagged terms in equation (16–16). Since both P_0 and E_0 changed, E_1 changes as follows:

$$\Delta E_1 = \beta_3 \Delta P_0 + \beta_4 \Delta E_0 = \beta_3 \Delta P_0 + \beta_4 \beta_2 \Delta P_0$$

Similarly, we can work through subsequent changes as follows:

$$\Delta E_2 = \beta_4 \Delta E_1 = \beta_4 (\beta_3 \Delta P_0 + \beta_4 \beta_2 \Delta P_0)$$

$$\Delta E_3 = \beta_4 \Delta E_2 = \beta_4 \beta_4 (\beta_3 \Delta P_0 + \beta_4 \beta_2 \Delta P_0)$$

The change for the general period t is as follows:

$$\Delta E_t = \beta_4 \Delta E_{t-1} = \beta_4^{t-1} (\beta_3 \Delta P_0 + \beta_4 \beta_2 \Delta P_0)$$

To find the total change over all periods, add all the changes to get the following:

$$\text{Total } \Delta E = \beta_2 \Delta P_0 + \sum_{i+1}^{\infty} \beta_4^{i-1} (\beta_3 \Delta P_0 + \beta_4 \beta_2 \Delta P_0)$$

Recalling that $\beta_4^0 = 1$, the right-hand side of the above equation can be simplified as shown in equation (16–18):

$$\Delta E = (\beta_3 \sum_{i+0}^{\infty} \beta_4^i + \beta_2 \sum_{i+0}^{\infty} \beta_4^i) \Delta P_0 \tag{16–18}$$

If $-1 < \beta_4 < 1$, the summations in the above equation do not explode and we can write the following:

$$\sum_{i=0}^{\infty}\beta_4^i - \sum_{i=1}^{\infty}\beta_4^i = 1 \text{ and } \sum_{i=1}^{\infty}\beta_4^i = \beta_4 \sum_{i=0}^{\infty}\beta_4^i$$

Combining these two relationships, we can solve for the summation as shown in equation (16–19):

$$\sum_{i=0}^{\infty}\beta_4^i = \frac{1}{1 - \beta_4}$$

(16–19)

Substituting equation (16–19) into equation (16–18) yields the following:

$$\frac{\Delta E_o}{\Delta P_o}\frac{P}{E} = \beta_2 \frac{P}{E} \text{ and } \frac{\Delta E}{\Delta P_0}\frac{P}{E} = \frac{(\beta_2 + \beta_3)}{1 - \beta_4}\frac{P}{E}$$

(16–20)

The short- and long-run slopes in equations (16–17) and (16–20) can be converted into short- and long-run elasticities, as follows:

$$E_t = \beta_1 + \beta_2 P_t \text{ with } \frac{\Delta E_0}{\Delta P_0} = \beta_2 \text{ and } \frac{\Delta E}{\Delta P_0} = \beta_2$$

Furthermore, equation (16–16) has a number of functions nested within it:

1. If $\beta_3 = \beta_4 = 0$, we get a simple static model:

$$E_t = \beta_1 + \beta_2 P_t + \beta_4 E_{t-1} \text{ with } \frac{\Delta E_0}{\Delta P_0} = \beta_2 \text{ and } \frac{\Delta E}{\Delta P_0} = \frac{\beta_2}{1 - \beta_4}$$

2. If $\beta_4 = 0$, we get a distributed lag model:

$$E_t = \beta_1 + \beta_2 P_t + \beta_3 P_{t-1} \text{ with } \frac{\Delta E_0}{\Delta P_0} = \beta_2 \text{ and } \frac{\Delta E}{\Delta P_0} = \beta_2 + \beta_3$$

3. If $\beta_3 = 0$, we get a partial adjustment model:

$$E_t = \beta_1 + \beta_2 P_t + \beta_4 E_{t-1} \text{ with } \frac{\Delta E_0}{\Delta P_0} = \beta_2 \text{ and } \frac{\Delta E}{\Delta P_0} = \frac{\beta_2}{1 - \beta_4}$$

4. If $\beta_2 = \beta_3 = 0$, we get an autoregressive model (AR):

$$E_t = \beta_1 + \beta_4 E_{t-1} \text{ with } \frac{\Delta E_0}{\Delta P_0} = 0 \text{ and } \frac{\Delta E}{\Delta P_0} = 0$$

5. If $\beta_2 = 0$, we get a vector autoregressive model (VAR):

$$E_t = \beta_1 + \beta_3 P_{t-1} + \beta_4 E_{t-1} \text{ with } \frac{\Delta E_0}{\Delta P_0} = 0 \text{ and } \frac{\Delta E}{\Delta P_0} = \frac{\beta_3}{1 - \beta_4}$$

6. If $\beta_2 = \beta_4 = 0$, we get a leading indicator model:

$$E_t = \beta_1 + \beta_3 P_{t-1} \text{ with } \frac{\Delta E_0}{\Delta P_0} = 0 \text{ and } \frac{\Delta E}{\Delta P_0} = \beta_3$$

The above models could be generalized by adding more variables and more lags. See Charemza and Deadman (1997) for more discussion of the above models, as well as a rearrangement into the popular error correction model.

If the above variables are measured in logs, then the βs would be elasticities. For other more complicated demand models of consumer behavior, see Deaton and Muellbauer (1980). For other more complicated models of factor demands, see Dahl (2010).

Judgment is often incorporated into other models. For example, in multivariate and econometric models, the forecaster picks the variables to include in the model. Bayesian analysis formally includes preconceived notions about the values of model parameters. This can be particularly useful if your sample is small but you have other information to include in the model. Suppose you have the following econometric model for the demand for coal from a particular mine:

$$Q_s = \alpha_o + \alpha_p P_c + \sum_{i=1}^{n} \beta_i X_i$$

where

P_c is the price of coal, and

X_i represents the n other variables that you believe influence purchases from this mine.

From long experience with the coal industry, you think that α_p is distributed as a normal variable, with mean −22 and standard deviation 0.16. Bayesian analysis will allow you to incorporate these prior beliefs into your estimate. If your priors are correct, you will be able to get better estimates than you could from the data alone. (For more discussion and references on how to use Bayesian analysis, see Greene [2012, 655–680].)

Summary

Energy is used in all facets of our economy and lives. Primary energy consumption includes coal, oil, and natural gas, as well as nonfossil-based electricity and renewables used for heat. Secondary energy includes the electricity produced from fossil fuels and products produced from fossil fuels, such as gasoline and other refined oil products and liquids from coal. Energy balances give us a broad snapshot of how an economy uses such energy by source and end use. Adjusting primary production for imports, exports, stock changes, and international bunker fuels shows how much energy is available to a country's domestic market, designated by the IEA as total primary energy supply:

$$\text{TPES} = \text{production} + \text{imports} - \text{exports} - \text{bunkers} - \text{stock changes}$$

Energy transformations show how primary sources are converted to secondary sources, as well as how much is used within the energy sector, including transportation and distribution losses. Once you adjust for all transformations (subtracting inputs and adding in outputs), you end up with final energy consumption. In more aggregate breakdowns, this final consumption is typically divided into use by the industrial, residential, commercial (which typically includes government), and transportation sectors. Nonenergy consumption includes use for petrochemical feedstocks, asphalt for roads, lubrication, and coal for coke production. Further breakdowns can be seen by major product, including coal, oil and natural gas liquid, natural gas, electricity, and renewables, as well as more detail by specific industry.

Consumers demand energy products to provide energy services based on their preferences, prices, and income. With diminishing marginal utility, simple optimization in a competitive market suggests they should consume where marginal utility per dollar of energy is equal to marginal utility per dollar for other goods, represented by good N or $U_E/P_E = U_N/P_N$.

Producers demand energy products to make profits. If producers have diminishing marginal product, and they maximize profits by buying and selling in competitive markets, their demand for a factor is their marginal revenue product curve equal to the price of output times the marginal output produced by the last unit of energy ($MRP = P_X X_E$). In a condition similar to the consumer decision, producers should hire energy up to the point where the marginal product of energy per dollar equals the marginal product per dollar of other factors represented by N or $X_E/P_E = X_N/P_N$.

The derived demand equations for both consumers and producers are likely to include the price of energy (P_E), the prices of substitutes and complements (represented by P_N), measures of output price and/or economic activity or income (represented by Y), and all other variables that affect energy demand (represented by X). We can represent such functions, which are building blocks in most of the models we have considered so far, as follows:

$$E = f(P_E.P_N.Y.X)$$

Econometric techniques exist for choosing both the functional form of the model and the lag structure. For example, the Box Cox function nests a variety of functional forms within it: $\dfrac{y^\lambda - 1}{\lambda}$. These forms include $\lambda = 1$, which is linear in

y; $\lambda = -1$, which is the inverse of y; $\lambda = 2$, which is quadratic in y; and $\lambda \rightarrow 0$, which is $\log(y)$.

Lags in adjustment can be represented in our equation by lagged variables, allowing us to distinguish between long- and short-run adjustments as follows:

$$E_t = \beta_1 + \beta_2 P_t + \beta_3 P_{t-1} + \beta_4 E_{t-1}$$

A number of functions can be developed from the above formulation depending on which βs are 0. The short- and long-run slopes with respect to price and price elasticities in the above equation are as follows:

$$\frac{\Delta E_o P}{\Delta P_o E} = \beta_2 \frac{P}{E} \quad \text{and} \quad \frac{\Delta EP}{\Delta P_o E} = \frac{(\beta_2 + \beta_3)}{1 - \beta_4} \frac{P}{E}$$

Two other econometric issues also need to be considered when estimating demand equations. The market prices and quantities are influenced by both demand and supply in the market. We need to be sure that the equation we are estimating is identified or is truly a demand function and not a supply function or a mix of the two. Additionally, when we estimate demand our random errors move us off the relationship. If these errors influence the price, we may also get a misrepresentative function.

17

Linear Programming, Refining, and Energy Transportation

Linear programming is used to allocate resources, plan production, schedule workers, plan investment portfolios and formulate marketing (and military) strategies. The versatility and economic impact of linear programming in today's industrial world is truly awesome.

—Professor Eugene Lawler, Computer Science Division,
University of California, Berkeley, 1980

Introduction

Frankel (1969) sums up many of the important properties of oil in his classic, *Essentials of Petroleum*. Oil is neither a final product nor a capital good but is needed to enjoy other goods. Since it is "a matter of life in peace and death in war," all possible means, including financial investment, political influence, and military action, are undertaken to ensure access to it. The fact that it is liquid, volatile, and composed of hydrocarbons determines how it is handled and processed.

Exploration and drilling are expensive; production is cheap. Oil transportation and refining are the most capital intensive of all. Its capital-intensive nature makes it a cyclical industry with big units working to capacity. During boom periods, it takes time to put in new equipment and take advantage of the good times; shortages make profits soar. During bust periods, companies produce as long as they cover their average variable costs; excess production causes profits to plummet. The petroleum industry's small and highly skilled labor force means that labor relations tend to be good. Petroleum is exhaustible, fugacious, and requires increasingly complex discovery techniques. Success in drilling requires technology, luck, and daring, whereas success in refining requires technical ability and organization. Refineries are located based on transport considerations and tax incidence. For example, in Europe, high tariffs on products (but not on crude oil) gave incentives to develop refineries close to markets.

437

Crude Oil Refining

Crude oil is a mix of chemical compounds. Refining separates the mix into products with various useful characteristics. Kerosene and fuel oil provide light and heat, gasoline and diesel generate power, and lubricating oil protects moving metal surfaces. Lighter products are building blocks for the petrochemical industry, and asphalt provides tough road surfaces.

Refineries take in crude oils of varying qualities and refine them using a wide range of processes. Sophisticated US refineries produce light virgin or straight-run naphtha, heavy virgin naphtha, heavy naphtha, light naphtha, light vacuum oils, heavy vacuum oils, coker gas oil, and light gas oil. These products can be further treated and blended to produce kerosene, jet fuel, gasoline, diesel fuel, distillate (gas oil), and residual fuel oil (Leffler 2000).

Figure 17–1 shows oil product consumption of the five most significant products for major world regions. The first column for each region is consumption (C); the second is production (P). In most regions, the sums of these five products are fairly balanced. North America, the Middle East, and Asia and Oceania have a small share of net imports. Europe, South America, and Central America have a small share of net exports. Eurasia, mostly Russia, with extra refining capacity, exports a considerable share of its products, while Africa is not self-sufficient in refining capacity and imports a higher share than other regions.

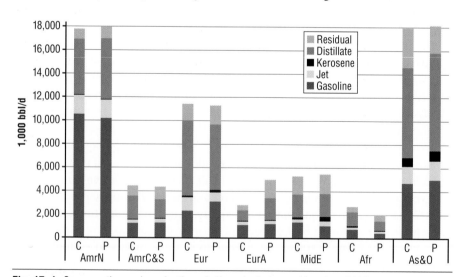

Fig. 17–1. Consumption and production of oil products by world region, 2010 (1,000 bbl/d)
Source: US EIA (n.d.d).
Note: C = consumption, P = production, AmrN = North America, AmrC&S = Central and South America, Eur = Europe, EurA = Eurasia or the FSU, Afr = Africa, and As&O = Asia and Oceania.

The figure also shows the large regional variation in product use. Gasoline has the largest share in North America (59.3%) and Eurasia (39.6%), whereas distillate, mostly in the form of diesel fuel, dominates in other regions. Distillate is particularly dominant in Europe, with more than a 55% share. Jet fuel has between 5% and 11% of the share in all regions.

Kerosene not used for jet fuel is less than 5% in all regions but is higher in developing countries in Africa and Asia, where it is used for lighting in households with no access to electricity. Residual has the highest share in the Middle East, where more than one-half of the residual used in the domestic market is for power generation.

Although many regions may be close to overall self-sufficiency in these five oil products, trade helps to balance individual products. North America has traditionally had small net imports of gasoline, jet fuel, and residual, while distillate was roughly in balance and kerosene was exported. South and Central America have been net exporters of most products but with net imports of kerosene.

Europe has exported gasoline, kerosene, and a bit of residual, and imports distillate. Eurasia has exported distillate and residual and has been close to an average balance in the three other products. The Middle East has traditionally exported all five products but gasoline. Africa exports residual and imports the other four products. Asia imports residual and has been in rough balance but trending toward exports for the other four products (EIA, n.d.d.).

However, recently some of these patterns have been changing. With increasing US production of light oil and bans on its exports with the exception of lease condensate run through a field stabilizer, product exports have been increasing. The United States became a net product exporter of all five products by 2011. Rapidly rising domestic consumption in the Middle East has increased their net imports of gasoline and decreased their net exports of other products.

Given such variations in consumption patterns, refineries around the world adapted to provide the product slates for their domestic and export markets with such processes as the following:

- *Alkylation, polymerization, and isomerization.* These processes take lighter hydrocarbons and combine them to produce heavier hydrocarbon products.
- *Thermal cracking, catalytic cracking, and hydrocracking.* These processes crack heavier hydrocarbons into lighter products.
- *Coking.* Coking converts very heavy products into coke and much lighter products.
- *Hydrotreating.* Hydrotreating cleans sulfur, nitrogen, and aromatics from products and can enhance cetane rating.
- *Reforming.* Reforming raises the octane of gasoline.

For more details on refining, see Leffler (2000) and Mushrush and Speight (1995).

The refining process begins with distillation. As the crude oil is heated, various fractions boil off, beginning with the lighter products (table 17–1). The *cut points* for the various products, or the temperatures at which oil is changed into the next product, have some slack and can be adjusted. For example, if the cut point for naphtha is changed to 225°F from 220°F, more kerosene and less naphtha would be produced.

Table 17–1. Boiling ranges for petroleum products

°F	°C	Product
<90	<32	Butane and Lighter
90–220	32–104	Straight-Run Gasoline
220–315	104–157	Naphtha
315–450	157–232	Kerosene
450–650	232–343	Light Gasoil
650–800	343–427	Heavy Gasoil
>800	>427	Straight-Run Residue

Source: Leffler (2000).

Usually heavier crudes (with more carbon) produce a higher proportion of heavy products. Heavier crudes are typically more valuable and better suited to markets that desire higher shares of residual fuel oil for power generation and industrial use. Lighter crudes are more valuable in a market where power is generated from natural gas, coal, or other substitutes, and lighter products such as gasoline and other transport fuels are more desirable. For the last decades worldwide, the share of the barrel has moved toward lighter products, making lighter crudes generally more valuable. However, distance from markets and disequilibrium sometimes alter traditional price spreads. Table 17–2 shows sample crude API gravities and their prices. Recall from chapter 2 that crudes with higher API gravities are lighter.

As mentioned previously, light crudes are generally more valuable than heavy crudes. Traditionally WTI, which is a bit lighter than Brent, has traded for $1–$2 more per barrel. However, you will notice in table 17–2 that Brent is almost $10 higher than WTI and almost $20 higher than the new North Dakota Bakken crude with an API of 41°. This anomalous result arises from the recent increases in tight, light oil production in the United States. Without infrastructure to move this new light crude into domestic and foreign markets and legal restrictions on exporting crudes, it is run in US refineries. However, US refineries have been designed to run on somewhat heavier and dirtier crudes to take advantage of their lower prices. Therefore, these refineries must be tweaked to run on these new light crudes and will only buy them at a discount. Since this light crude can't be shipped to the West Coast, heavy Alaskan crudes are competing on world markets and are more in line with foreign crudes. Further, Bakken crude is landlocked and does not yet have sufficient transportation infrastructure to move cheaply to domestic markets where it could command a higher price.

Table 17–2. Sample crude API gravities and prices, April, 2014

Crude Oil	°API	Price ($/bbl)
UK—Brent	38	106.91
Russia—Urals	32	108.06
Saudi Arab Light	34	104.87
Dubai—Fateh	32	104.68
Algeria Saharan Blend	44	108.09
Nigeria Bonny Light	37	110.19
Indonesia—Minas	34	111.12
Mexico—Isthmus	33	101.29
Alaska North Slope*	27	97.76
West Texas Sour*	34	92.00
West Texas Intermediate*	40	97.00
North Dakota Sweet*	41	87.19

Sources: OGJ Databook (2002); *OGJ* (2014a, 2014b).
Notes: * Price is for April 25, while other prices are the monthly average for April.

In addition to API gravity of crude oil, the amount of sulfur is an important determinant of quality. High-sulfur crudes (> 0.5% sulfur content) are called sour crudes, and low-sulfur crudes (< 0.5% sulfur content) are called sweet crudes. The various fractions in a barrel of crude have different economic values, all of which determine the overall value of the crude. The fractions within a barrel can be seen from a crude oil's assay report. Sample assays are shown in table 17–3.

A distillation curve (fig. 17–2) shows how much of a given crude oil boils off at various temperatures, resulting in a gain in volume as products are separated. Light products expand and take up more room when they are separated. At temperatures higher than 750°F, the heavier molecules will break, or *crack*, into lighter molecules. To prevent uncontrolled cracking of the heavier molecules, these products are sent from the distillation unit to a vacuum flasher. The flasher's lower pressure allows heavier products to separate without cracking at lower temperatures.

If we want more gasoline than comes out of the distillation process, the heavier products can be cracked to make more gasoline. With fluid catalytic cracking (FCC), the most popular of these processes, feedstocks that boil from 650°F to 1,100°F are heated along with a catalyst introduced in the form of small particles. Resulting products range from light to heavy hydrocarbons, and they are separated out in a fractionation tower. The heavy products may be sent to a thermal cracker, the light gas oil may be blended into distillate, and the cracked gasoline is blended into gasoline. By adjusting the process between gas oil and gasoline, refiners increase distillate production for the winter heating season and increase gasoline production for the summer driving season in temperate climates.

Gases coming out of the various processes are separated in gas processing units. Isobutane can be combined with propylene or butylene using a high-pressure, low-temperature process and a catalyst to form alkylate. High-octane alkylate is a valuable blending agent for gasoline. Naphtha, which lacks sufficient octane to make a good gasoline blending component, is reformed to raise its octane number by using heat and pressure in the presence of a platinum catalyst.

The residual material at the bottom of the barrel is typically processed further using high temperatures to break it into lighter products. A vacuum flasher separates out heavier products without cracking them. The most intense process is coking, which breaks down all feedstocks into lighter products and coke. *Coke,* a solid and more difficult product to handle, can be used as a refinery fuel or sold for the manufacture of electrodes, anodes, carbides, and graphite.

Hydrocracking is catalytic cracking in the presence of hydrogen. It has the advantage of producing only light products such as light distillates, gasoline, and gases. In addition to lightening the barrel, hydrocrackers are used as hydrotreaters to remove sulfur and other impurities from heavier products. These units add considerable economic flexibility to a refinery, allowing it to change yields of gasoline, distillate, and jet fuel as market conditions require and enabling it to run heavier, sour crudes (Leffler 2000). World capacities for these various processes are summarized in table 17–4.

Table 17–3. Assays for various crude oil streams

	Midland, Texas, US	Arabian Light, Saudi Arabia	Brent Blend, UK North Sea	US Alaska North Slope	La Salina, Venezuela
°API	40.80	33.4	38.00	26.4	23.5
Sulfur, wt %	0.34	1.79	0.38	1.06	1.85
Boiling Range					
°F	C1–C5	68–212	C5–75	C5–150	82–200
Vol %	4.35	8.9	6.2	2.2	3.8
°F	68–347	212–302	C5–165	150–380	200–300
Vol %	32.39	8.5	24.5	15.6	5
°F	347–563	302–455	165–235	380–650	300–350
Vol %	23.50	16.5	12.6	28.6	2.82
°F	563–650	455–650	165–350	650–840	350–400
Vol %	8.10	19.7	33.3	16.4	3.14
°F	650–1,049	650–1,049	235–350	650+	400–500
Vol %	24.30	29.8	20.7	52.4	7.53
°F	650–1,500	650+	350–550		500–550
Vol %	33.30	44.6	28.3		4.55
°F	761–1,500	1,049+	550+		550–650
Vol %	25.30	14.8	10.3		9.73
°F	878–1,500				650+
Vol %	17.95				61.91
°F	1,049–1,500				
Vol %	9.00				

Source: *OGJ Databook* (2002).
Note: C1–C5 are methane, ethane, propane, butane, and pentane. All but pentane are at 32°F (0°C).

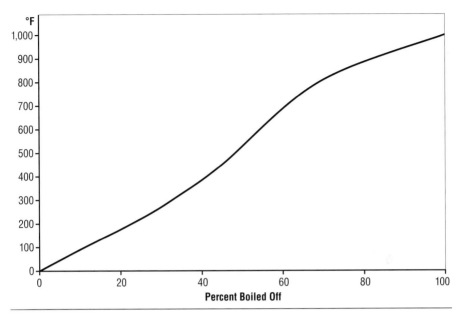

Fig. 17–2. Oklahoma sweet distillation curve
Source: Leffler (2000).
Note: °C = (5/9)(°F − 32).

Table 17–4. World refinery capacity by region, 2013

Region	No. of refineries	Crude distillation bbl/cd	Vacuum distillation bbl/cd	Catalytic cracking bbl/cd
Africa	45	3,218,085	509,504	210,380
Asia	162	25,275,612	4,662,741	3,040,668
Eastern Europe	89	10,602,308	3,946,235	868,470
Middle East	43	7,393,365	1,863,275	357,550
North America	147	21,591,067	9,499,402	6,531,656
South America	65	6,359,987	2,680,560	1,311,007
Western Europe	94	13,588,454	5,399,998	2,069,316
Total Regions	**645**	**88,028,878**	**28,561,715**	**14,389,047**
Region	**Catalytic reforming bbl/cd**	**Catalytic hydrocracking bbl/cd**	**Catalytic hydrotreating bbl/cd**	**Coke tonnes/day**
Africa	458,426	61,754	833,626	1,841
Asia	2,168,831	1,242,200	9,881,233	20,450
Eastern Europe	1,466,344	394,058	4,298,848	12,950
Middle East	630,797	566,891	2,044,063	3,300
North America	4,145,594	1,964,608	16,640,894	139,793
South America	401,638	132,400	1,689,562	20,140
Western Europe	2,006,337	1,250,364	9,581,361	12,614
Total Regions	**11,277,967**	**5,612,275**	**44,969,587**	**211,088**

Source: Breisford et al. (2013, 44).
Note: bbl/cd = barrels per calendar day.

One way to measure refinery complexity is as follows. Assign each process unit a complexity factor based on its cost relative to the cost of a distillation unit. For example, an FCC unit costs about six times more than a distillation unit per barrel, so the distillation unit is assigned a complexity factor of one and the "cat cracker" a complexity factor of six. All barrels of crude go through the distillation unit, but usually only fractions of the barrel go through other process units. Suppose there are n process units after distillation. The share of crude going to process unit i is α_i, and the complexity of process i is C_i. Then the overall complexity of a refinery C is the weighted average of all nondistillation process units' complexity factors, C_is, with the weights being the share of crude going to each process, as follows:

$$C = \sum_{i=1}^{n} \alpha_i C_i$$

Gasoline Blending

Two important qualities of gasoline are the vapor pressure and the octane number. Fuel vapor pressure must be high enough so that at low temperatures enough fuel gets into the cylinder, but not so high that oxygen cannot enter the cylinder when the engine is warm. At higher elevations, vapor pressure must not be so high that fuel vaporizes in the fuel system, causing vapor lock. Reid vapor pressure (RVP) is a measure of how quickly a fuel evaporates and varies from 5 to 15 for gasoline. RVP needs to be higher for cold than for hot weather use.

The octane number of the fuel indicates whether a fuel is likely to self-ignite. Lower octanes than required will result in self-ignition and cause "knocking." Higher octanes than required cause no such problems but do not improve engine performance and so add an unnecessary expense. Lead tetraethyl was once used extensively as a cheap additive to prevent self-ignition and to raise octane. It is now illegal in most countries worldwide because it can foul catalytic converters and can cause nerve damage in humans.

Gasolines are blends of various components that meet the required specifications for these two gasoline qualities. For example, butane and isobutane have high vapor pressures, while straight-run naphtha has a low vapor pressure. To see how blending works, take the following example from Leffler (2000). The end product goal is to have a gasoline with an RVP of 10, and the blending stocks are given in table 17–5. How much normal butane must be blended?

Table 17–5. Reid vapor pressure blending problem

Component	Barrels	RVP	Motor Octane
Straight-Run Gasoline	4,000	1.0	61.6
Reformate	6,000	2.8	84.4
Light Hydrocrackate	1,000	4.6	73.7
Cat-Cracked Gasoline	8,000	4.4	76.8
Normal Butane	X	52.0	92.0

Source: Leffler (2000).

Gasoline Reid vapor pressure (RVP) is a simple weighted average of the RVPs for all the blending stocks:

$$RVP = \left.\sum_{i=1}^{n}(RPV_i)(B_i)\middle/\sum_{i=1}^{n}(B_i)\right.$$

where
RVP_i is the Reid vapor pressure of product i,
B_i is the number of barrels of product i blended in, and
n is the number of products blended.

For the gasoline products blended in table 17–5, the gasoline RVP is as follows:

$$RVP = \frac{4{,}000 \times 1 + 6{,}000 \times 2.8 + 1{,}000 \times 4.6 + 8{,}000 \times 4.4 + x \times 52}{(4{,}000 + 6{,}000 + 1{,}000 + 8{,}000 + x)} = 10$$

or:
$$60{,}600 + 52x = 10(19{,}000 + x)$$

Solving the above equation, we find that the required quantity of normal butane is $x = 3{,}081$ barrels. Total gasoline production would be $19{,}000 + x = 19{,}000 + 3{,}081 = 22{,}081$ barrels. Blending for octane requirements can be done in the same way. Blending for both RVP and octane at the same time would require the simultaneous solution of two equations.

Isobutane, isopentane, and isohexane have higher octane numbers than butane, pentane, and hexane, but the isos are more expensive. However, if refiners have trouble meeting their octane requirements, they may convert butane, pentane, and hexane to their iso equivalents through a process called *isomerization*. Additionally, polymerization converts gases to liquid fuels.

In some large US cities, the federal government requires that oxygen be added to gasoline so that CO_2 rather than CO is exhausted. Historically, the two most common blending agents to raise oxygen levels have been ethanol and methyl tertiary butyl ether (MTBE). However, public opposition and legislative threat to MTBE caused US refiners to stop using it in 2006. Worldwide, ether oxygenates account for 94% of gasoline blending agents (*OGJ* 2013a).

Unlike gasoline, diesel fuel is required to be self-igniting. The *cetane rating* is a measure of the self-ignition properties of a fuel (the ignition delay). The shorter the delay, the better. As with octane numbers, cetane ratings can be increased through the blending of products.

Furnace oil, also called *gas oil* and *number two distillate*, is very similar to diesel but has more latitude in specification, since it does not have to meet cetane requirements. Residual fuel oil (*resid*) is the heavy bottom end of the barrel. It must be heated before it can be transported and typically contains contaminants like sulfur and heavy metals. Hydrotreating can be used to remove sulfur, nitrogen, and metals to make resid more environmentally acceptable. Typically, the hydrogen used in this process is a by-product of a reformer.

Linear Programming to Optimize Refinery Profits

A typical refinery uses many of these processes and varying crude types to produce a wide range of products. The goal of the refinery is to pick a crude oil and product slate that maximizes profits subject to a variety of capacity and quality constraints on inputs and outputs. Often the objective and constraints can be represented as linear functions, and we can apply linear programming to their solution. Symonds (1955) first modeled refineries in a linear programming framework. To see how linear programming works, let's begin with a simple blending problem based on Kaplan (1983).

In this example, two processes produce two grades of gasoline. Process one produces straight-run gasoline (u_1) with a maximum capacity of 100,000, and process two produces cracked gasoline (u_2) with a maximum capacity of 140,000. The two products (u_1 and u_2) can be blended into two grades of gasoline (X_1 and X_2). The blending process can be represented by the following Leontief fixed coefficient production functions:

$$\text{Grade one: } X_1 = 2.5 \min(u_1, u_2/2)$$

$$\text{Grade two: } X_2 = 2(u_1, u_2)$$

Thus, if $u_1 = 1$ and $u_2 = 2$, we get $2.5 \times \min(1, 2/2) = 2.5$ gallons of grade X_1. However, if we add two more units of u_2 without any more u_1, we will not get any more of grade one gasoline, or we must blend in a ratio of one to two. We can represent these production functions by *isoquants*, or lines that represent equal quantities. For example, two isoquants for grade one gasoline are shown in figure 17–3.

I_1 represents 2.5 gallons of grade one gasoline everywhere along it. I_2 represents the sets of inputs that will produce 5 gallons of grade two gasoline everywhere along it.

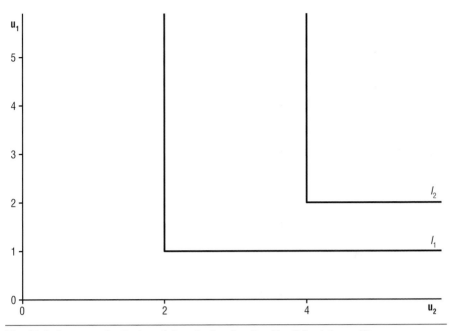

Fig. 17–3. Isoquants for the Leontief production function $X_1 = 2.5 \min(u_1, u_2/2)$

In figure 17–3, start at $u_1 = 1$ and $u_2 = 2$. Note that adding more u_1 without any u_2 leaves us on the same isoquant; similarly, adding more u_2 without any more u_1 also leaves us on the same isoquant.

Given that profits for grade one gasoline are $\pi_1 = \$0.08$/gallon and for grade two are $\pi_2 = \$0.09$/gallon, the objective function is to maximize total profits, as follows:

$$\pi = \$0.08X_1 + \$0.09X_2 \qquad \text{(17–1)}$$

This objective function is maximized subject to the capacity constraints for the two blending agents, u_1 and u_2. To develop these constraints, note that 2.5 gallons of grade X_1 requires 1 gallon of u_1. The gallons of u_1 per gallon of $X_1 = 1/2.5 = 0.4$. Similarly, for X_2, we know that 2 gallons of grade X_2 requires 1 gallon of u_1. The gallons of u_1 per gallon of $X_2 = 1/2 = 0.5$. Total requirements of u_1 for X_1 and X_2 have to be less than 100,000, giving the u_1 constraint as $0.4X_1 + 0.5X_2 \le 100{,}000$. For the u_2 constraint, notice that 2.5 gallons of grade X_2 requires 2 gallons of u_1. The gallons of u_1 per gallon of $X_2 = 2/2.5 = 0.8$. Similarly, for X_2, we know that 1 gallon of grade X_2 requires 1 gallon of u_2. The gallons of u_2 per gallon of $X_2 = 1/2 = 0.5$. Total requirements of u_2 for X_1 and X_2 have to be less than 140,000, giving the u_2 constraint as $0.8X_1 + 0.5X_2 \le 140{,}000$. For the general case, if $X_i = A\min(au_1, bu_2)$, then A gallons of grade X_i requires $\dfrac{1}{a}$ gallons of u_1. The gallons of u_1 per gallon of $X_i = \dfrac{1}{a}\Big/A$.

Then the whole optimization problem is the following:

Maximize $\qquad\qquad\qquad \pi = \$0.08X_1 + \$0.09X_2$

Subject to $\qquad\qquad 0.4X_1 + 0.5X_2 \leq 100{,}000 \qquad\qquad$ (straight run)

$\qquad\qquad\qquad\quad\; 0.8X_1 + 0.5X_2 \leq 140{,}000 \qquad\qquad$ (cracked)

The easiest way to see the solution to this problem is to graph it in X_1X_2 space. With linear constraints, we can find two points for each function and connect them. For constraint one:

$$X_1 = 0 \rightarrow X_2 = 175{,}000 \text{ and } X_2 = 0 \rightarrow X_1 = 250{,}000$$

For constraint two:

$$X_1 = 0, X_2 = 280{,}000 \text{ and } X_2 = 0, X_1 = 175{,}000$$

The graphs for these two functions are shown in figure 17–4. The constraint set for this problem is $0ABC$. We need to find where in this set profits are highest. Rewrite eq. (17–1) as follows:

$$X_2 = \pi/0.09 - (0.08/0.09)X_1 = \pi/0.09 - 0.889X_1$$

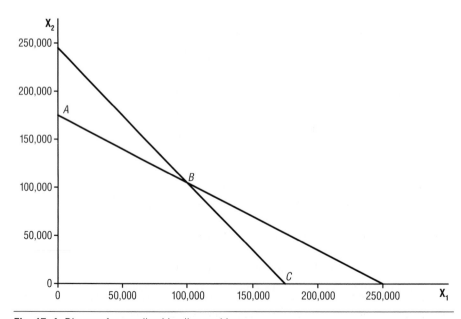

Fig. 17–4. Diagram for gasoline blending problem

We need to find the line with the above slope that is further from the origin but still in the constraint set. In other words, find the highest line with slope $dX_2/dX_1 = -0.889$ that touches one of the points A, B, or C.

One way of solving this problem is to check the profits at each point. Profits at points A and C are the following:

$$\pi_A = 0.08X_1 + 0.09X_2 = 0.08(0) + 0.09(200,000) = 18,000$$

$$\pi_C = 0.08X_1 + 0.09X_2 = 0.09(175,000) + 0.09(0) = 14,000$$

To find profits at B, we need to solve for the values of X_1 and X_2 or solve the two constraints simultaneously:

$$0.4X_1 + 0.5X_2 = 100,000$$

$$0.8X_1 + 0.5X_2 = 140,000$$

Subtracting the second equation from the first and solving for X_1 yields the following:

$$-0.4X_1 = -40,000 \rightarrow X_1 = 100,000$$

From the second constraint, we can solve for X_2 to be the following:

$$X_2 = 140,000/0.5 - 0.8X_1/0.5 = 280,000 - 1.6X_1$$

Substituting in the value for X_1 gives us the following value for X_2:

$$X_2 = 280,000 - 1.6(100,000) = 120,000$$

Alternatively, by linear algebra, this system can be written as follows:

$$\begin{bmatrix} 0.4 & 0.5 \\ 0.8 & 0.5 \end{bmatrix} \begin{bmatrix} X_1 \\ X_2 \end{bmatrix} = \begin{bmatrix} 100,000 \\ 140,000 \end{bmatrix}$$

The solution is as follows:

$$\begin{bmatrix} X_1 \\ X_2 \end{bmatrix} = \begin{bmatrix} 0.4 & 0.5 \\ 0.8 & 0.5 \end{bmatrix}^{-1} \begin{bmatrix} 100,000 \\ 140,000 \end{bmatrix} = \begin{bmatrix} -2.5 & 2.5 \\ 4.0 & -2.0 \end{bmatrix} \begin{bmatrix} 100,000 \\ 140,000 \end{bmatrix} = \begin{bmatrix} 100,000 \\ 120,000 \end{bmatrix}$$

So profits at solution point B are the following:

$$\pi_B = 0.08(100,000) + 0.09(120,000) = 18,800$$

Since these profits at B are higher than at points A and C, we would want to produce 100,000 gallons of grade one and 120,000 gallons of grade two.

To see how much u_1 is used in each grade, go back to constraint one for straight-run gasoline and substitute in the quantities for each grade produced:

$$0.4(100,000) + 0.5(120,000) = 100,000$$

Since the first term in the above constraint is the amount of straight-run gasoline used in grade one and the second term is the amount of straight-run gasoline used in grade two, then the following are the values for u_1:

$$u_1 = 40,000 \text{ for grade one and 60,000 for grade two}$$

Substitute the grade amounts into the constraint for cracked gasoline:

$$0.8(100,000) + 0.5(120,000) = 140,000$$

Again, the first term in the above expression is the amount of cracked gasoline used in grade one gasoline and the second term is the amount of cracked gasoline used in grade two:

$$u_2 = 80,000 \text{ for grade one and 60,000 for grade two}$$

These techniques can be extended to more complicated problems that involve running processes and blending, as in the following classic problem from Symonds (1955).

Suppose that a refinery has four crude oils, A, B, C, and D, available in quantities of 100, 100, 200, and 100. Crude C can be run in process C_1 to produce gasoline (G), heating oil (H), and fuel oils (F), and in process C_2 to produce G, H, F, and lubes (L). The four products G, H, F, and L are required in the following quantities: 170, 85, 85, and 20. Profits on crudes A, B, C_1, C_2, and D are 10, 20, 15, 25, and 7.

Thus, a barrel of crude A yields 0.6 barrels of gasoline, 0.2 barrels of heating oil, etc. The product slates for a barrel of each crude oil and process are given in table 17–6.

Table 17–6. Summary of refinery problem

Profits on Crudes	Type of Crude or Process					Product Demand
	A 10	B 20	C_1 15	C_2 25	D 7	
Products	Product Slate for Crude or Process					
G	0.6	0.5	0.4	0.4	0.3	170
H	0.2	0.2	0.3	0.1	0.3	85
F	0.1	0.2	0.2	0.2	0.3	85
L	0.0	0.0	0.0	0.2	0.0	20
Total Crude	100	100	$C_1 + C_2 = 200$		100	

Source: Symonds (1955).

The objective function we maximize for this problem comes from multiplying each product by its profits and summing up:

$$\pi = 10A + 20B + 15C_1 + 25C_2 + 7D$$

Total amount of gasoline from crude A is 0.6A, from B is 0.5B, etc. To make sure we satisfy the gasoline market, our gasoline constraint must be the following:

$$170 < 0.6A + 0.5B + 0.4C_1 + 0.4C_2 + 0.3D$$

Similarly, our constraints for heating oil, fuel oil, and lubricants are the following:

$$85 \leq 0.2A + 0.2B + 0.3C_1 + 0.1C_2 + 0.3D$$

$$85 \leq 0.1A + 0.2B + 0.2C_1 + 0.2C_2 + 0.3D$$

$$20 < 0.2C_2$$

In addition, because we cannot use more crude oil than we have at our disposal, the crude oil constraints are the following:

$$A < 100$$

$$B < 100$$

$$C_1 + C_2 < 200$$

$$D < 100$$

All variables must also be constrained to be nonnegative. To see how to solve such a problem using Excel Solver, go to http://dahl.mines.edu/b1401.pdf.

Energy Transportation

Energy is transported by a variety of means over the earth's surface. Most electric power is transported in overhead insulated copper or aluminum power lines or in buried cables. After the battle of the currents between Westinghouse, who supported alternating current or AC, and Edison, who supported direct current or DC, in 1880s and 1890s, AC generally became the accepted mass-produced and transported commercial type of electricity worldwide (King 2011). Since line losses are less at higher voltages and increasing and decreasing voltage is relatively inexpensive, voltage is increased for long-distance transport. In 1903, a high-voltage AC (HVAC) transmission line of 50 kV was built in Canada. Other milestones in

HVAC electricity transmission technology include a 287 kV line from Boulder Dam in the United States in 1936, a Swedish 400 kV line in 1952, a Quebec Canada line of 735 kV in 1964, and a 765 kV line from the US Pacific Northwest to Los Angeles in 1968. The standard capacity for a 400 kV line is 600 megawatts (MW) and the standard capacity for a 765 kV line is 2,000 MW (Historical Archive 2007; Automation and Power Technologies 2013a; Automation and Power Technologies 2013b). China completed its first 1,000 kV (5,000 MW) line in 2009. AC lines with capacities larger than 800 kV are considered ultra-high voltage (UHVAC) (Global Transmission Report 2009).

AC power lines in the United States above 100 kV are considered high-voltage transmission lines. However, of the 168,768 circuit miles (1 kilometer = 0.625 miles) of AC transmission in the United States in 2010, none were below 200 kV. About 49% were in the 230 kV range, about 35% were in the 345 kV range, about 15% were in the 500 kV range, and less than 2% were in the 765 kV range (US EIA 2011c). Since the United States and Canada are well connected, they are one large market, with more than 210,000 miles of high-voltage transmission lines (Energy Library 2009).

In Europe, 35 countries are represented by the European Network of Transmission System Operators for Electricity. Of the more than 175, 000 total circuit miles of transmission lines 200 kV or higher in these countries, about 48% are 220 kV and about 50% are 400 kV (ENTSOE, n.d.). China, with its rapidly growing grid, has an even larger network of about 315,000 miles (507,000 km) of lines 220 kV or higher. However, it is not yet nearly as interconnected as the North American and European grids (TBEA 2013).

Although AC power transmission came to dominate the grid for many decades, DC has advantages in certain cases. DC power lines are cheaper to build and have lower line loss. However, to transport power by DC, it must be converted at the power plant from AC to DC with a rectifier and then converted back with an inverter before it can be used. The conversions eat up the transport gains of DC for distances of less than around 300 miles overland or 30 miles in submarine cables for a 400 kV line. Thus, almost all submarine power cables are high-voltage direct current (HVDC).

The HVDC technology developed in Sweden and Germany with the first 100 kV line built from the Swedish mainland to Gotland Island in 1954. Other HVAC milestones include the 250 kV cable from New Zealand's South Island to North Island in 1965, the 500 kV Pacific Northwest to Los Angeles line in 1963, and the 600 kV line from Itaipu to São Paulo, Brazil in 1985. DC lines greater than 600 kV are considered ultrahigh voltage DC lines (UHVDC). An 800 kV (7,300 MW) line was completed in China in 2013 between Nuozhadu and Guangdong (Lantau Group 2014). Currently, the two largest producers of HVDC, with 80% of the market, are ABB and Siemens. Information on new projects and technologies is available at their websites.

With AC power, the electrons move back and forth in cycles, with the hertz being the number of cycles per second. The two most common cycles are 60 hertz, used in North America, and the more common 50 hertz, used in the European Union, Russia, China, India, and elsewhere. If two grids are to be connected with an AC power line, the cycles need to be synchronized. DC transmission lines can be used to connect two grids with the same hertz that are out of sync or even two grids that have different hertz, as on the main island of Japan. The grid on the 50 hertz side of the island was built by a German company, and the grid on the 60 hertz side of the island was built by an American company, hence the incompatibility.

Total world high-voltage capacity is still relatively small. It is less than 3% of circuit miles in the North American grid and an even smaller proportion in Europe. However, increases in the bulk power market, increasing market integration, improved technology, and DC generation from solar photovoltaics are making HVDC more attractive. Large capacity additions in the UHVDC as well as UHVAC markets are being planned in China and India. More than one-half of new large-scale transmission growth in the coming decade is expected to be HVDC and UHVDC (Martin and Murach 2012).

Increasing desire to coordinate and increase efficiency on ever-larger and more diverse systems has increased interest in smart grids. Such grids add real time digital communication across the grid among generation companies (gencos), transmission companies (transcos), distribution companies (discos), consumers, and related equipment. They can signal congestion and surplus power, and they can turn appliances on and off to keep the grid running smoothly. Meters can run backward or forward as distributed power is brought online or power is taken from the grid (US DOE c. 2008).

Oil has a long history of transport by water. In 1861, oil was first shipped in barrels on a sailing ship from Philadelphia to London. Oil was also loaded in barrels and shipped by horse and wagon on land. The measurement of oil in barrels has lingered to the present day, although oil is no longer shipped in barrels. Originally, barrels held 50 gallons, but contracts were designated as 42 gallons/barrel to allow for spillage. This anachronism remains as well.

The first bulk crude ocean carrier began commercial life in 1863. Such carriers require bulkheads (tanks) to separate oil to keep it from sloshing and destabilizing the vessel. The most often used commercial measure for a tanker is a deadweight ton, and it indicates total tonnage, including cargo, fuel, provisions, and crew. It is usually measured in metric tons. After World War II, tankers up to 30,000 deadweight tons (dwt) were built.

Since surface area increases as a square, and volume goes up as a cube, there are significant economies of scale in tankers, and so tanker size has continued to increase. (For pipelines, the volume increases as the square of the radius.)

After 1956 and a shift to longer voyages, the size of tankers rapidly escalated. By 1966, the largest tanker was 210,000 dwt, with plans for tankers up to 500,000 dwt. Tankers below 200,000 are often categorized as follows:

- Panamax (50,000–70,000 dwt), the largest size that can go through the Panama Canal
- Aframax (80,000–120,000 dwt), a size based on the Average Freight Rate Assessment System that can fit in most ports around the world.
- Suezmax (120,000–200,000 dwt), the largest size that can go through the Suez Canal

Currently, tankers between 200,000 and 320,000 dwt are designated as very large crude carriers (VLCCs), and those over 320,000 dwt are designated ultra large crude carriers (ULCCs). The largest currently operating tankers are two T1-class tankers built in 2003 and 2004. They are 441,585 dwt and carry 3,166,353 barrels each (Maritime Connector, n.d.). The largest tanker ever built and now scrapped was 565,000 dwt and carried more than 4 million barrels of oil. Table 17–7 shows the size distribution of the world tanker fleet by the average freight weight assessment (AFRA) size classification.

Table 17–7. World tanker fleet by size, capacity, freight rate, 2012

Vessel Class	Vessel Size (1,000 dwt)	Sample Capacity in Barrels	# of Vessels (2012)	Total dwt	Spot Rate Worldscale
General Purpose (GP)	16.5–24.9	20,000 dwt = 150,000 bbls	1,456	11,985.9	151.8
Medium Range (MR)	25.0–44.9	35,000 dwt = 280,000 bbls	2,476	52,576.7	159.4
Long Range 1 (LR1)	45.0–77.9	70,000 dwt = 500,000 bbls	2,445	68,040.4	103.9
Long Range 2 (LR2)	80.0–159.9	120,000 dwt = 750,000 bbls	2,594	152,745.1	64.9
Very Large Crude Carriers (VLCCs)	160.0–319.99	200,000 dwt = 1,000,0000 bbls	1,406	200,121.6	49.3
Ultra Large Crude Carriers (ULCCs)	>320	320,000 dwt = 2,800,000 bbls	–	–	–

Sources: OPEC (2013); Geography of Transport Systems (2013).
Note: World tanker fleet includes commercial vessels of 10,000 dwt owned by independent oil companies as well as governments.

Since supertankers require super ports, ports in many parts of the world have been deepened or new ones have been built to accommodate these large ships. Because deepwater ports for VLCCs and ULCCs require expensive special conditions and equipment, numerous single-point mooring systems have also been set up. The world leader in their construction is SBM Offshore (2013). These buoys, held in place by anchor chains, hold ships further out to sea while hoses deliver oil to shore. This system can accommodate larger tankers than many ports can handle. It is economical because ships remain in open waters, not requiring jetties and breakwaters. Tankers can moor and still move 360° around these buoys. Thus, tankers can maneuver more easily than in a fixed dock and are less likely to spill oil.

The only US port that can take the VLCCs and ULCCs is the Louisiana Offshore Oil Port (LOOP), 18 miles offshore from Grand Isle. In 2011, about 11% of US imports were offloaded at LOOP. It is connected to a pipeline system that can take crude oil to refineries representing almost one-half of US refining capacity (LOOP 2013; Reuters 2012a).

When limited by port size, another alternative is to transship, or load oil onto large ships for long distances and then offload onto smaller ships before reaching the final destination. Deepwater ports in the Caribbean, such as at Trinidad and the Bahamas, have been popular transshipment points for crude oil going into shallower US ports.

In 2010, almost 60% of crude oil production was exported, and figure 17–5 shows the global trade patterns for these oil flows, along with the other fossil fuels. More than one-half of these exports were transported by tanker (US EIA, n.d.d).

Bottlenecks, or areas where traffic jams affect movement of energy products, remain a problem in shipping lanes. The estimated annual volumes of oil and products passing through the seven major world bottlenecks in 2011 (US EIA 2012c) are as follows:

- 17 million bbl/d through the Straits of Hormuz between Oman and Iran, which connects the Gulf of Aden and the Arabian Sea
- 15.2 million bbl/d through the Straits of Malacca below Singapore, which connects the Indian Ocean and the South China Sea
- 3.4 million bbl/d through Bab el-Mandeb between Africa and Yemen, which connects the Red Sea with the Gulf of Aden and the Arabian Sea
- 3.8 million bbl/d through the Suez Canal and Sumed Pipeline, which connect the Red Sea with the Gulf of Suez
- 2.9 million bbl/d (in 2010) through the Bosporus/Turkish Straits, which connects the Black Sea and the Mediterranean
- 0.8 million bbl/d through the Panama Canal, which connects the Caribbean Sea with the Pacific Ocean
- 3.0 million bbl/d (in 2010) through the Danish Straits, which are three channels that connect the Baltic and the North Sea between Sweden and Denmark and across Denmark

Oil companies own tankers and also charter tankers to move their oil. Around 80% of the world's tanker capacity is owned by independent tanker owners. This way, if an oil company does not need a tanker, it does not sit idle. The independents absorb the market fluctuation and spread the tankers over other users. Independents may also have other shipping interests to offset a lull in tanker use. For more information on independent tanker owners, go to International Association of Independent Tanker Owners (INTERTANKO, n.d.).

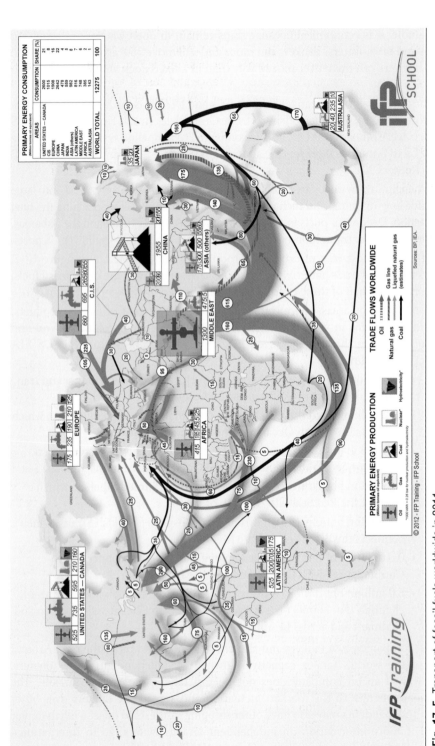

Fig. 17–5. Transport of fossil fuels worldwide in 2011

Source: Used with permission from IFP School.

Charter rates are typically quoted in an index called Worldscale. These rates, published annually, list the cost per ton to carry oil between designated ports in a 19,500 dwt tanker traveling at 14 knots. The base rate for any voyage is designated as Worldscale 100, which depends on distance and cost of inputs for the voyage. See some sample distances for major tanker routes in table 17–8.

Table 17–8. Representative tanker distances in nautical miles

From	To	Nautical Miles
Arabian Gulf	Rotterdam, Netherlands via Cape	11,170
Arabian Gulf	Rotterdam, Netherlands via Suez	6,396
Arabian Gulf	Fos Sur Mer, France (Mediterranean) via Cape	10,784
Arabian Gulf	Fos Sur Mer, France (Mediterranean) via Suez	4,499
Arabian Gulf	New York, USA via Cape	11,918
Arabian Gulf	New York, USA via Suez	8,318
Arabian Gulf	San Francisco, USA	11,181
Arabian Gulf	Yokohama, Japan	6,717
Arabian Gulf	Western Australia	6,921
Arabian Gulf	La Plata, Argentina	8,782
Tripoli, Libya	Isle of Grain, UK	4,777
Tripoli, Libya	New York, USA	5,044
New Orleans, USA	New York, USA	1,707
New Orleans, USA	San Francisco, USA via Panama	4,649

Source: Jenkins (1989).
Note: 1 nautical mile = 1.15 English mile = 1.85 km.

A larger ship can transport crude more cheaply than a smaller one, so its rate will be a percent of Worldscale. For example, "Worldscale 45" indicates that the price per barrel is 45% of the published Worldscale rate. Rates vary with the size of the ship and market conditions. Rates will be lower when the tanker market is weak than when it is tight. Some sample average spot rates by size category are shown in table 17–7.

Dirty tankers are those carrying black cargoes, such as crude and resid. They do not require as much cleaning between voyages as *clean tankers*, which carry so-called white product, encompassing the lighter, more valuable products. White cargoes are more expensive to transport because contamination must be avoided and because much smaller ships typically carry these products. LPG (butane and propane) and ethylene must be transported in high-pressure containers and are not included in the designation of clean or dirty tankers.

Tankers sometimes sail under flags of convenience, meaning they are registered in a country where tax rates, operating standards, and environmental requirements are more lax than the home country of the chartering company. The two most popular flags of convenience are Panama and Liberia. Together they accounted for more than 20% of the world tankers, and more than 80% of their merchant marines are foreign owned (US CIA 2010). Because these tankers are generally larger ships, their tonnage represents about one-third of world tanker tonnage. Panama's share is slightly higher than Liberia's (API 2012).

Inland transportation of oil and products includes pipelines, rail, and truck. Water transport consists of self-propelled vessels, such as tankers on the coast, and barges, which are moved by tug or push boats. Oil barges were the first to use the US Gulf Intracoastal Waterway, which runs from Texas to Florida. Transportation of crude oil and products by share within the United States is shown in table 17–9, with a comparison for crude shipments in China. Although tankers, of course, heavily dominate crude imports into the two countries, pipelines heavily dominate transport within both countries. Rail, which comes second in China, is responsible for transporting a very small share in the United States. This share has been growing with the new tight light oil production and pipeline bottlenecks. Although products also are largely transported by pipeline in the United States, local distribution relies more heavily on motor and rail transport.

Table 17–9. Domestic transport of crude oil, oil products, and coal by modal share

Product/Country	Crude in United States	Crude in China	Products in United States	US Coal
Year	2009	2011	2009	2012
Pipelines	79.8	73.8	63.3	—
Water	19.4	11.6	25.7	11.6
Motor	0.5	—	6.8	11.4
Rail	0.3	14.6	4.2	69.6
Tramway/Conveyer	—	—	—	7.4

Sources: US BTS (n.d.); US NMA (n.d.); personal conversation with Professor Lei Zhang, Chinese University of Mining, Xuzhou, China.

Receiving terminals for crude oil and petroleum products have large tanks for temporary storage, called *tank farms*, where crude and products are segregated by owners. Various grades of crude and products must also be segregated into batches in the pipeline. Scheduling and dispatching of batches is done by computer, with a single batch having a minimum volume. Unless products are physically separated in a pipeline (e.g., by large rubber balls), there is some mixing at the interfaces. If product specification is not too tight, the products at the interfaces can be used; otherwise, they need to be reprocessed (Van Dyke 1997).

Pipelines are manufactured in various standard sizes up to 56 inches (142 centimeters or cm) in diameter. Pumping stations keep the crude moving. For example, along the 48-inch, 800-mile (1,290 km) Alaska pipeline from the North Slope to Valdez, there are currently nine operating pumping stations. The oil travels at 5.4 mph and takes almost six days to reach the terminal at Valdez. The highest throughput was more than 2 million barrels per day in 1989. With falling production, throughput has fallen, as well, and 2013 throughput averaged 534,480 barrels per day (Alyeska Pipeline Service Company n.d.).

To increase throughput, either the diameter of the pipe or the power of the pumping station can be increased (Miesner and Leffler 2006). Table 17–10 shows the sizes of recently laid pipe in the United States, along with throughput for 10,000 and 20,000 horsepower of pumping capacity. Pumping stations and equipment are an additional 25%.

Table 17–10. Sample oil pipeline diameters, construction costs, and capacities, 2012

Inner Diameter Pipeline	$/Mile	Throughput bbl/d w/10,000 HP	Throughput bbl/d w/20,000 HP
12	1,548,033	47,000	60,000
16	1,197,394	75,000	100,000
20	2,141,235	120,000	150,000
24	1,922,659	160,000	210,000
30	2,873,543	185,000	310,000
36	5,554,192	NA	NA
42	7,286,381	NA	NA

Source: OGJ Editors (2013b, table 4); Cookenboo (1980).
Note: NA = not available; bbl/d = barrels per day; and HP = horsepower.

Both oil and its products play a large role in international trade. Around 58% of the world's oil production was exported in 2010, and product exports are approximately 60% as large as crude oil exports (BP 2014). The world's largest exporters and importers of oil products are shown in table 17–11.

Table 17–11. Largest net exporters and importers of refined petroleum products in 2010

Rank/Net Exporters	Net Exports of Refined Products, 1,000 bbl/d	Net Exports as % of Domestic Consumption	Net Importers	Net Imports of Refined Products, 1,000 bbl/d	Net Imports as % of Domestic Consumption
1/Russia	2,172.0	72.6%	Japan	960.9	21.7%
2/S. Arabia	1,274.1	53.7%	Mexico	418.2	20.1%
3/India	867.0	26.6%	Hong Kong	382.1	99.9%
4/Canada	823.7	36.5%	Germany	381.5	15.4%
5/Kuwait	656.1	171.1%	Indonesia	372.8	26.9%
6/Venezuela	463.3	64.5%	France	370.6	20.2%
7/Algeria	454.6	140.6%	China	348.1	3.7%
8/US Virgin Is.	310.4	273.5%	Spain	316.7	22.0%
9/Norway	249.6	113.1%	Brazil	299.0	11.4%
10/Italy	234.7	15.2%	US	269.0	1.4%

Source: US EIA (n.d.d).
Note: US Virgin Is. = US Virgin Islands. For data on additional countries, see http://dahl.mines.edu/t1711.pdf.

A number of these countries have set up export refinery industries, with exports greater than the amount they supply to their domestic markets. They include Kuwait, Venezuela, Algeria, Aruba, and the US Virgin Islands. At net exports of more than a million barrels a day in 2012, the United States will likely be in the top eight list, once comparable international data is available.

Natural gas is transported primarily by pipeline with a small (10%) but growing amount (EIA n.d.d.) transported as LNG. Gas transportation costs are considerably higher than for oil, and vary considerably by distance and conditions (see fig. 17–6).

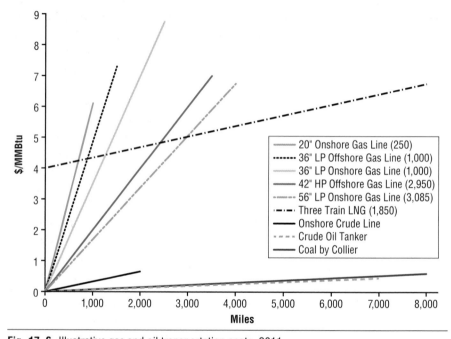

Fig. 17–6. Illustrative gas and oil transportation costs, 2011

Source: Used with permission of James Jensen, Jensen and Associates, Inc.

Note: LP = low pressure; HP = high pressure (in this figure); " = inch. Natural gas delivery capability in million cubic feet per day is given in parentheses. Divide by 35.3 to obtain million cubic meters per day. Divide miles by 0.621 to convert to kilometers and divide by 1.055 to obtain price in dollars per million kilojoules. Multiply by 2.54 to convert inches to centimeters. Numbers in parentheses indicate natural gas deliverability in millions of cubic feet per day.

In figure 17–6, the left-hand axis represents transport costs in dollars per MMBtu or energy equivalent transported. Oil tankers and coal bulk carriers are the cheapest method of transport, followed by oil pipelines. To convert oil transport costs to per-barrel costs, multiply by 5.8. To convert coal to transport costs per US metric tonnes of coal, multiply by 22.2. Gas onshore is cheaper than gas offshore, and gas transport costs are, of course, cheaper as pipeline diameters increase. Shipping LNG becomes relatively cheaper than shipping natural gas by small pipeline in as little as 750 miles. However, if the market is large enough to warrant 56 inches (142 cm) onshore pipelines, LNG transport will not become competitive until distances are more than 3,000 miles (4,800 km).

Because of these high transport costs, gas that is far from a market, particularly small-scale finds, may not be commercial. For this reason, around 3% of gross natural gas production worldwide was vented or flared in 2012, and another 11% was reinjected to maintain pressure in oil wells. Less than 1% was vented or flared in North America and Europe. Around 4.6% of the gas in the Middle East was vented or flared, and another 12% was injected. In 1999, Nigeria vented or flared more than one-half of its gas and injected another 13%. With the large LNG project that came onstream in 1999, by 2009, Nigeria vented and flared only about one-fourth of gross natural gas production. Ghana, a relatively new gas producer in 2009, flared or vented almost 90% of its gross natural gas production but only about a quarter by 2012 (US EIA, n.d.d).

There are three types of gas transportation in the United States. At the field, small gathering pipelines transport gas to long-distance transportation pipelines. Distribution pipelines then transport gas from large, long-distance pipelines to the end user. In 2012, the United States had 1,246,463 miles of distribution pipelines and 303,303 miles of transmission and 16,729 gathering pipeline operated by more than 1,000 companies (US BTS, n.d.). The company operating the largest transmission system in the world is Gazprom. It operates the 102,300 miles of the Unified Gas Supply System of Russia (Gazprom 2014).

Gas pipelines require compressor stations. As with all pipelines, more gas can be transported by increasing pipe diameters or by increasing the horsepower of the compressors. In the Netherlands, for example, gas may travel at 5 kilometers per hour (km/hr) during slack summer months and more than 50 km/hr in peak winter periods. Gas can also be stored in pipelines by increasing line pack. Compressor stations built in the United States in 2011 ranged in size from 350 to 57,000 HP, with costs generally running between $2,000 and $5,000 per horsepower (*OGJ* Editors 2013b, table 5). Even larger stations are being built in Russia, with one at Portovaya to accompany the Gryazovets–Vyborg Pipeline rated at more than 100,000 HP (Pipeline International 2011). Compressor stations are more typically rated in megawatts in Russia, with 1 MW = 1,341 HP.

More than one-half of the world's coal is used to produce electricity and more than 60% of this utility use is located within 38.5 miles (60 km) of the mine mouth. Only about 15% of the world's coal is traded internationally. Nevertheless, coal transport is still big business. Transport of coal by sea constitutes the world's largest dry cargo. Total coal trade was more than 1 billion tons a year in 2011, with steam coal approximately 75% of the total in 2011 (World Coal Institute, n.d.). In 2010, crude oil and petroleum products were about one-third of total world shipping tonnage. Steam coal is about one-half the tonnage of oil (UN Conference on Trade and Development Secretariat 2011). Major shipping routes for coal are the black lines seen in figure 17–5. As in the independent tanker trade, independent bulk shippers have no fixed schedule but go where needed, with deals typically set up by brokers.

Coal is shipped not only by bulk cargo ship, but also by rail, barge, truck, and conveyor. Although rail costs vary considerably by distance, length of trains, and other factors, they are typically two to three times the cost per tonne of seaborne transport. The majority of domestically used US coal is transported by rail, accounting for more than 40% of freight tonnes transported for American railroads and about 22% of their revenues in 2012 (Association of American Railroads 2013).

The cheapest way to transport coal by rail is by dedicated unit trains. These trains often have more than 100 cars, with capacities of 100 short tons each. They move back and forth between a mine and a power plant without uncoupling. Instead, they are loaded from the top by moving under a chute and unload by dropping their coal out the bottom at their destination.

Table 7–9 shows between 11% and 12% of US coal moves over lakes, along inland waterways, or along the coasts. The Jones Act, passed in 1920 to protect US shipping, requires that this coal, as well as all other goods shipped between US ports, must be transported on US-built, -staffed, and -owned vessels. Since US vessels tend to be more expensive, this act raises the cost of waterborne shipping within the United States and gives a slight transport cost advantage to foreign coal.

Conveyor belts and tramways, most often between the mine and a mine-mouth power plant, transport another 7.4% of US coal. A somewhat higher percentage is transported by trucks, usually from mine mouth to preparation plants, to mine-mouth power plants, or to loading points for transfer by rail or water. By contrast, the domestic transport situation has evolved somewhat differently in China, the world's largest coal consumer. In 1980, rail transported around 70% of Chinese domestic coal. However, by 2010, with major bottlenecks, rail transport accounted for less than one-half of these shipments and water accounted for around one-third (Tu and Johnson-Reiser 2012).

When a company decides among the various transport modes discussed above, they must consider costs along with transportation requirements and constraints. Symonds (1955) demonstrates another use for linear programming, which is to minimize transport costs for a crude oil problem. Suppose that you have m oil supply points, with X_i being the quantity of oil available at the ith point. You have n refineries, with Y_j being the oil requirements at the jth refinery. Let A_{ij} be the amount of oil transferred from supply point i to refinery j and C_{ij} be the unit cost to transfer a barrel of oil from i to j. This unit cost can be computed using the levelized costs from chapter 15 plus operating costs.

The objective function is total transport costs, which can be easily computed. Total cost for transferring oil from supply point one to refinery one is $C_{11} \times A_{11}$, or cost per barrel times the number of barrels transported. Similarly, total cost for transferring oil from supply point i to refinery j is $C_{ij} \times A_{ij}$. Total transport cost (TTC) to be minimized is the sum of all costs from each supply point to each refinery:

$$TTC = \sum_{j=1}^{n} \sum_{i=1}^{m} C_{ij} A_{ij}$$

The above function is minimized subject to nonnegativity as well as other constraints. No suppliers can ship more than they have available:

$$\sum_{j=1}^{n} A_{ij} \leq Y_i$$

Each refinery must satisfy its crude oil requirements:

$$\sum_{j=1}^{n} A_{ij} = X_j$$

Computer software can solve such models and can handle models with millions of variables and hundreds of thousands of constraints. For links to more online information about linear programs, see http://dahl.mines.edu/st17/st17.pdf, question 25.

Another common application for linear programming in an energy context is to pick energy sources that will minimize the cost of supplying energy services. For example, suppose a power generator supplies electricity to a particular service area. Linear programming (or more complicated nonlinear or integer programming, in which all the functions in the model are not linear or continuous) can help determine whether coal, natural gas, oil, or wind is the cheapest way to satisfy power demands.

Summary

Crude oil is seldom used directly but rather is refined into a whole slate of products, ranging from coke and resid at the bottom of the barrel up to refinery gases at the top end. Distillation divides the barrel into a number of products, such as gases, gasoline, kerosene, light gas oil, heavy gas oil, and resid, depending on the original composition of the crude oil. These products can be further processed to make more valuable products. Alkylation, polymerization, and isomerization make lighter products into heavier products. Thermal cracking, catalytic cracking, visbreaking, and coking make lighter products from heavier products, but leave an even heavier residue. Hydrocracking is cracking in the presence of hydrogen, which leaves no heavier residue, while hydrotreating is used to remove sulfur and other impurities.

RVP and octane are the most important characteristics of gasoline, while cetane is an important characteristic of diesel. Refiners blend various products to meet the appropriate values for these characteristics to ensure good engine performance while meeting environmental regulations. In certain areas at certain times, US government regulations require refiners to raise the oxygen content of fuels to reduce CO production, which is now satisfied with ethanol. The most common oxygenates worldwide are ethanol and ethers such as MTBE.

Oil companies run refineries to maximize profits. To do so, they should choose optimal refinery locations depending on tax and transport costs, while choosing a crude and product slate depending on market requirements and constraints. Linear programming helps make these decisions by optimizing a linear objective function subject to numerous constraints.

A second popular use of linear programming is to solve transportation problems. Energy is often transported over long distances with a choice among transit routes or modes. Electricity always travels by wire, but voltage and current type need to be chosen. Coal travels by rail, barge, tanker, conveyor belt, and truck. Oil and oil products travel by pipeline, tanker, barge, rail, and truck. Gas travels by pipeline or can be liquefied and sent by tanker or truck. In addition, there is the choice of which producer is to satisfy which consumer. Transport costs should be minimized for all these choices, subject to all capacity and demand requirements.

Often a given energy service can be supplied by a different energy product. Truck freight transport can be fueled by gasoline or diesel, cooking can be fueled by electricity or gas, and electricity can be made from fossil fuels, renewables, or nuclear sources. Linear and nonlinear programming can be useful in deciding which energy product will be used to supply which energy service to minimize costs, as well in the energy problems noted here.

18

Energy Futures Markets for Managing Risk

The only certainty is uncertainty.

—Pliny the Elder

Introduction

Energy is often thought to be a risky business: a chance of gain balanced with a chance of loss. Energy-related personal risks include the possibility of accidents that have safety and health implications, such as a nuclear power plant meltdown, an oil spill, an LNG explosion, or air and water pollution. Energy economic risks include possible losses in real assets from falling crude oil prices, rising drilling rig costs, expropriation of a coal mine by a foreign government, and increasing environmental regulations on gasoline. Possible losses could also be financial, rather than from real assets.

To help pay for real assets, energy firms issue claims on the profits from them. For example, claims for Exxon-Mobil include stocks, bonds, and other financial agreements, which get their value from the value of Exxon-Mobil's underlying assets: oil leases, refineries, tankers, pipelines, and so forth. Energy financial risk relates to possible losses from changes in the value of these financial assets issued by energy companies (Allen 2012).

Jorion (1996) cites five types of financial risks that energy companies face:

- *Market risk* relates to price changes of financial assets and liabilities.
- *Credit risk* relates to defaults on contractual obligations.
- *Liquidity risk* relates to the lack of market activity or to a failure to meet a cash flow obligation.
- *Operational risk* relates to technical problems with financial trading systems and fraud.
- *Legal risk* relates to losses from failures to comply with the law or adverse regulatory changes.

Risk in an energy investment context relates not just to losses but also to the volatility of a return or price, usually defined as the *variance* (σ^2) or its square root (σ), designated as its standard deviation. Suppose you want to measure price variance of an energy asset. You have observations of its price for n periods $\{x_1,...,x_n\}$. A measure of variance for this price is the following:

$$s^2 = \frac{\sum_{i=1}^{n}(x_i - \bar{x})^2}{n-1}$$

where the mean return of the portfolio is $\bar{x} = \dfrac{\sum_{i=1}^{n} x_i}{n}$.

The popular notion of risk as losses under this definition would be *downside risk*. Because people tend to be risk averse, they often require a higher rate of return to compensate them for taking more risk, and they may be willing to pay to eliminate some risk. For example, an individual may prefer a sure $100 to an asset with a 10% chance of paying $1,000 (and a 90% chance of paying nothing). Although the expected values of the assets are equal ($100 and 0.10 × $1,000 + 0.9 × $0 = $100), a risk-averse person prefers the sure thing to a riskier venture (higher variance asset) with the same expected value. A risk-neutral person would be indifferent between the sure thing and the risky asset, while a risk-taker would value the risky asset more. Sometimes an individual may be risk averse to a loss but a risk-taker for a gain.

High volatility in energy prices since 1973 has caused some energy firms to seek ways to reduce risk. Figure 18–1 demonstrates this volatility for daily spot prices for West Texas Intermediate (WTI) at Cushing, Oklahoma, measured in dollars per barrel ($/bbl), and for natural gas at Henry Hub in Louisiana in dollars per million British thermal units ($/MMBtu) from 1997 to mid-2014. Over this period, crude oil price averaged $56.41/bbl, and natural gas price averaged $4.70/MMBtu. The respective standard deviations for these prices were $31.90/bbl and $2.32/MMBtu. Thus, crude oil's average variation from its mean was about 57% of its average price, and natural gas's average variation was about one-half of its average price.

Risk-averse players in the market may prefer not to face such variations in price. Thus, high volatility, coupled with developments in information technology, financial theory, and a political climate favoring market over government solutions, have led to financial markets that allow the transfer of risk through the buying and selling of financial derivatives. *Financial derivatives* are financial assets that derive their value from an underlying asset. They represent a way of transferring risk from parties who want less risk to those who are willing to take on the risk for a price (Jorion 2009).

Fig. 18–1. Daily WTI crude oil and Henry Hub natural gas prices
Source: US EIA (n.d.d).

Thus, an energy financial derivative is a financial instrument with a value based on some underlying energy asset. Most energy derivatives are built from four basic instruments:

• futures
• forwards
• options
• swaps

An *energy futures contract*, the topic of this chapter, is an agreement to buy or sell a specific energy asset at some future point in time. The contract is purchased through an organized exchange, with a standardized contract that can be resold on the exchange. The contract specifies the exact day that trading is closed and delivery is required. If you buy a June crude oil futures contract on the Intercontinental Exchange (ICE) for $90 per barrel, you are agreeing to buy 1,000 barrels of crude oil at $90 per barrel in June. You pay a small transaction fee and put up a margin of typically less than 5% of the value of the contract.

A *forward contract* resembles a futures contract in that it is an agreement to purchase some asset in the future. However, the contract is not purchased on an exchange, nor can the contract typically be resold without the agreement of both parties. Because it is a bilateral agreement that is not standardized, it can be tailored to the individual customer's needs. A 30-day forward contract to buy gasoil in Rotterdam would be a contract to buy gasoil in 30 days. Such bilateral contracts not traded on organized exchanges are referred to as *over-the-counter* (OTC) transactions.

An *option*, the topic of the next chapter, is the right to buy (a call option) or sell (a put option) an asset at an underlying price called the *strike price*. Energy options are usually written on futures contracts. If you own a call option to buy a natural gas futures at $4 per Mcf on NYMEX (now part of CME Group), you have the right (but not the obligation) to buy natural gas at $4 per Mcf.

A *swap* is an OTC derivative that exchanges one cash flow based on an underlying asset for another. For example, an airline may do a swap with Morgan Stanley, trading a variable for a fixed price on jet fuel. In essence, the airline pays the contracted fixed price and Morgan Stanley makes up the difference, paying out if the market price is higher and receiving payment if the market price is lower than the contracted fixed price.

Energy Futures Contracts

For energy futures contracts, the buyer of the futures has a long position and the seller has a short position. The contract date is designated by the delivery month; the exact delivery date is specified by the rules of a standard contract and varies by product.

For example, CME Group oil futures stop trading three business days prior to the 25th of the month. On the 25th of the month, pipeline nominations are made for the next month. *Nominations* are the schedules for shipment for the next month. Thus, spot transactions for crude oil are really for delivery next month because of the time it takes to get in the pipeline queue. This means that spot and one-month futures price are almost synonymous, and the one-month future is often used as a quote for the spot price or is designated as the spot price in contracts.

Other contracts have different last trading days, which vary from contract to contract. Samples of other contract information for energy futures and options can be found at the two largest energy derivative markets: NYMEX (CME Group, n.d.a) and ICE (ICE 2013a).

The New York Mercantile Exchange (NYMEX) evolved out of a dairy exchange way back in 1882 and only diversified into its first energy contract in 1978, with a heating oil futures contract at New York Harbor (NYH). It introduced its most heavily traded contract for WTI with delivery at Cushing, Oklahoma in 1983. Trading was done by open outcry in a trading pit, and it soon became the largest energy futures and options market in the world. After hours trading was introduced in 1993 through NYMEX ACCESS (CME Group 2012b).

In 2000, the much younger Atlanta-based ICE was created by energy traders. Its stated goal was to introduce electronic trading to facilitate OTC trades in energy derivatives. It acquired energy futures contracts when it purchased the London-based International Petroleum Exchange (IPE) in 2001, the second largest energy futures and options market at the time. IPE was introduced by energy traders in 1980 and launched its first gasoil futures contract in 1981 and its most famous Brent futures contract in 1988. By 2005, ICE had gone public, and IPE had abandoned open outcry trading in London, leaving only electronic trading. ICE began partnering with Canada-based Natural Gas Exchange (NGX) in 2008 to allow electronic OTC spot and financial trading of North America natural gas and electricity (ICE 2012b). It purchased the London-based Climate Exchange in 2010, which traded in carbon derivatives. ICE abandoned open pit trading in New York entirely in 2012. The ICE trading platform can be accessed through hubs, the Internet, and mobile devices.

ICE rapidly acquired market share on its electronic platform from NYMEX. NYMEX responded by introducing the ClearPort platform and related services that allow OTC trading. NYMEX also acquired an electronic trading platform in 2006 (CME Globex), which the Chicago Mercantile Exchange (CME) had introduced in the 1990s. NYMEX was subsequently purchased and became part of CME Group in 2008. Although CME Group still operates an open outcry market, its share had fallen to less than 1% of trades by 2012, and it is likely to be phased out. A variety of energy futures contracts are traded on organized exchanges, with a sample shown in table 18–1.

Asia has lagged behind Europe and North America in developing energy derivative markets. Fusaro (2005) maintains that the Asian markets were more stable due to greater regulation and state intervention that has limited price volatility and the need for organized derivative markets. Trading activity in China took off in 1993 and 1994 only to be banned by a suspicious government. However, rumors have been floating around that China is considering introducing an oil futures contract on the Shanghai Futures Market (Wong 2014)

Not all contracts introduced make it. The Singapore International Monetary Exchange (SIMEX), which evolved into the Singapore Exchange (SGX), began trading Brent Crude electronically with IPE in 1995, and other energy derivatives with NYMEX in 1996. It has since delisted all of its energy contracts except for a new fuel oil contract introduced in 2010 that has not yet proved to be very actively traded. SGX is also contemplating the launch of a new electricity futures contract in 2014. India's largest commodity futures market, Multi Commodity Exchange (MCX), has introduced a number of energy futures contracts within the last decade that seem to be very actively traded.

An important feature of energy futures contracts is that they are *marked-to-market* every day. Thus, contract gains and losses are settled at the end of each trading day, and the contract is rewritten at the closing futures price. For example, suppose you buy a crude oil futures contract for $70/bbl for June delivery and you have posted a margin of $3,375. Under this contract you have agreed to buy crude oil at $70/bbl in June. Someone else sells that contract for $70/bbl for June delivery and has also posted a margin of $3,375. That person has also contracted to sell crude oil for $70/bbl.

Table 18–1. Sample energy futures contracts

Product	Started Year/Month	Delivery	Exchange	Web Page, http://
Heating Oil (#2)	1978/10	NYH	CME, US	www.cmegroup.com
Crude Oil, Light Sweet	1983/3	Cushing, Oklahoma	CME, US	www.cmegroup.com
Gasoline, Unleaded	1984/12	NYH	CME, US	www.cmegroup.com
Propane	1987/8	TEPPCO	CME, US	www.cmegroup.com
Natural Gas	1990/4	Henry Hub, Louisiana	CME, US	www.cmegroup.com
Coal	2001/7	Ohio/BSSR	CME, US	www.cmegroup.com
CBOE Ethanol	2005/3	Chicago	CME, US	www.cmegroup.com
RBOB Gasoline (replaced unleaded)	2005/10	NYH	CME, US	www.cmegroup.com
ULS Diesel (replaced heating oil)	2013/4	NYH	CME, US	www.cmegroup.com
Gas Oil	1981/4	ARA	ICE Futures Europe	www.theice.com
Brent Crude Oil	1988/6	Sullom Voe	ICE Futures Europe	www.theice.com
Natural Gas	1997/1	NBP	ICE Futures Europe	www.theice.com
EU Emission Allowance	2005/4	National registries	ICE Futures Europe	www.theice.com
Nordpool Electricity	1995	No Delivery	NASDAQ OMX	www.nasdaqomx.com
DME Oman Crude Oil	2007/6	FOB Loading Port	DME	www.dubaimerc.com
Gasoline	2004/7	Tokyo	TOCOM	www.tocom.or.jp
Electricity, NSW-peak	2002/9	No Delivery	SFE, Sydney	www.asx.com.au
Electricity, NSW-base	2002/9	No Delivery	SFE, Sydney	www.asx.com.au
Natural Gas	2006/7	Hazira Hub, India	MCX	www.mcxindia.com
Brent Crude Oil	2005/6	Mumbai, India	MCX	www.mcxindia.com
Marine Grade Fuel Oil	2010/2	Singapore	SGX	www.sgx.com

Sources: Exchange homepages and other Internet resources.
Note: ARA = Amsterdam, Rotterdam, Antwerp; BSSR = Big Sandy Shoal River; CBOE = Chicago Board Options Exchange; CME = Chicago Mercantile Exchange; DME = Dubai Mercantile Exchange; FOB = free on board and indicates the price after it has been loaded; MCX = Multi Commodity Exchange of India; NBP = Transco's National Balancing Point on the UK Grid; Nordpool = the Nordic Power Pool between Denmark, Finland, Norway, and Sweden; NSW = New South Wales, Australia; NYH = New York Harbor; RBOB = reformulated gasoline blendstock for oxygen blending; SFE = Sydney Futures Exchange; SGX = Singapore Exchange; TEPPCO = Texas Eastern Product Pipeline Company; TOCOM = Tokyo Commodity Exchange; and ULS = ultra low sulfur diesel.

Suppose that tomorrow, the price of crude oil for next month's delivery closes at $70.50. Since you have contracted to buy at $70, your contract gains value equal to $0.50 per barrel for 1,000 barrels (a total of $500), whereas the contract that sells the crude for $70/bbl loses $500 in value. The $500 gain would be credited to your margin account and the $500 loss would be debited from the seller's margin account. If your margin account goes below the daily minimum, you will have to bring it up to the required level, or your position will be liquidated. Such marking to market transactions decrease the risk of default for derivative contracts and make it possible to hold contracts on small margins (Hull 2000).

In this example, if you were buying crude and closed out your futures contract on that day, you would have had the extra $500 to pay for the extra $0.50 a barrel exactly as though you had bought oil at $70 minus a small transaction cost. Thus, any day you want to get out of the market, you can close out your position by taking the opposite position at the prevailing futures price and realize any gain or loss from the original contract. A person selling the crude and closing out their account would sell at $70.50 but would lose $500 in their account, so in reality they would be selling at $70 as contracted.

Energy futures markets are not new. You will recall from figure 7–7, the history of US oil prices. There was dramatic price volatility at the beginning of the market in the 1860s, which led to the first oil futures contracts in the form of pipeline certificates in the 1870s. During the next 20 years, more than 10 exchanges in the United States, Canada, and Europe traded crude oil futures. However, with the Rockefeller monopoly and eventual vertical integration and multinational control of the market, prices stabilized and oil futures markets dried up (Weiner 1991).

It took a new round of volatility in energy prices that began with oil shocks in the 1970s, and continued on into deregulation in gas and electricity markets, to create markets for energy risk management. Numerous energy futures contracts have been devised, as noted above.

Futures exchanges are clearinghouses that act as intermediaries for futures contracts. They ensure performance by buying the contracts from the sellers and selling the contracts to the buyers, keeping buyers and sellers anonymous to one another. If one party in a contract fails to perform, the clearinghouse provides insurance by honoring the contract. The clearinghouse matches buyers to sellers with open contracts at the close of trading so deliveries can be made. Forward contracts generally take delivery; however, futures contracts rarely take delivery (fewer than 5% of contracts). Rather, these financial contracts are used to manage price risk. Instead of taking delivery, a futures contract holder will take the opposite position in the market so the contracts net out (e.g., a buyer of a crude futures contract [long position] sells a crude futures contract [short position]). Because so few deliveries are made, these futures contracts are sometimes referred to as *paper barrels* and the CME petroleum futures market as a *Wall Street paper refinery*.

Most exchanges provide information on current and historical futures prices. Some of the more heavily traded energy futures are quoted in the popular press. For example, the *Wall Street Journal* lists price quotes for CME futures, including crude oil, RBOB gasoline, heating oil, and natural gas. Table 18–2 contains a sample futures quote for RBOB gasoline delivered to New York Harbor and traded on CME. The contract size is 42,000 gallons, and the price is in dollars per gallon. (RBOB stands for reformulated gasoline blendstock for oxygen blending.)

Table 18–2. Sample futures quotes for heating oil on CME

Gasoline: NY RBOB (NYM); 42,000 gal.; $/gal.

Month	Open	High	Low	Settle	Change	Lifetime High	Lifetime Low	Open Interest
13-Feb	2.8719	2.9442	2.8590	2.9348	0.0594	2.9442	2.6915	27,329
13-Mar	2.8898	2.9531	2.8725	2.9409	0.0511	2.9550	2.7117	128,433
13-May	3.0244	3.0711	3.0172	3.0597	0.0277	3.0711	2.8757	34,722
13-Jun	2.9905	3.0261	2.9786	3.0137	0.0210	3.0261	2.8438	27,073
13-Aug	2.9123	2.9278	2.8932	2.9181	0.0131	2.9278	2.7650	9,196
13-Oct	2.6949	2.7011	2.6824	2.6993	0.0074	2.7011	2.5854	8,673
13-Nov	2.6401	2.6574	2.6316	2.6543	0.0078	2.6574	2.5483	12,428
13-Dec	2.6065	2.6289	2.6000	2.6232	0.0068	2.6289	2.5244	7,481
14-Jan	2.6102	2.6112	2.6042	2.6110	0.0068	2.6112	2.5164	2,056
14-Feb	2.6052	2.6052	2.6052	2.6127	0.0066	2.6150	2.5382	712
14-Jun	2.7428	2.7428	2.7428	2.7365	0.0082	2.7428	2.6620	170

Sources: WSJ Market Data Group (n.d.).

US environmental regulations require oxygenates to be blended into gasoline to make it burn cleaner. Given environmental constraints, ethanol has become the blend stock of choice. However, ethanol has a great affinity for water, making it corrosive. Further, different regions have different blending specifications. RBOB is the basic blending stock for reformulated gasoline, with the ethanol splash blended in at the terminal rack and then dispersed by truck to local gasoline stations.

The first column in table 18–2 is the month of the delivery date. "Open" indicates the first price for the day, "High" is the high price for the day, "Low" is the low price for the day, and "Settle" is the last price for the day. "Change" is the difference in the settle price from yesterday to today. Since a May contract was up $0.0277 from yesterday, it must have closed yesterday at $3.0320. Lifetime "High" and "Low" represent the highest and lowest prices ever attained for the particular contract. "Open Interest" is the number of this contract outstanding at the close of the day.

Another somewhat similar energy derivative is the forward contract, operating on the OTC markets. The name OTC arose because they traditionally were not traded or guaranteed on organized exchanges. Such trades are *bilateral*, or between two individual entities. They are typically not standardized but represent an agreement to buy or sell an energy product in the future at an agreed-upon price. Conditions of the trade are negotiated by the parties, and typically the commodities are taken for delivery.

A Brent crude oil forward market developed in the 1980s because there was no European crude oil futures market. (Brent forward contracts are unusual because the contracts *are* standardized.) Bilateral trades are made via telephone, telex, and fax. With no clearinghouse, all transactions and transacting parties must be tracked individually.

The series of bilateral agreements in this market (called *daisy chains*) are riskier than futures contracts because a default by one party can cause a whole series of defaults with no clearinghouse to provide insurance. Such a default happened in 1986, when crude oil prices fell from around $30 to under $10. Some parties refused to take delivery and pay the higher contracted price. This refusal caused defaults down the rest of the chain. However, the Brent forward market survived.

There is less price transparency in the crude forward market than in the crude futures market, although *Platt's*, *Petroleum Argus*, and other news services survey participants and report prices daily. Brent delivery is at the Sulom Voe terminal in the North Sea. Once a cargo has been assigned a loading date it is called *dated Brent* or a *wet barrel* as opposed to a *paper barrel*. Dated Brent can still be traded, but on the spot market. Since 21 days' notice is generally required to schedule tankers to pick up the crude, contracts more than 21 days out are considered in the Brent forward market, whereas contracts less than 21 days out are in the *Brent 21-day market*. It was formerly called the 15-day market when shorter notice was required (Slade et al. 1993; Weiner 1991). Spot prices quoted for the dated Brent market, which have become market prices for many crude contacts, are based on a window of dated Brent prices that has changed over time. In 2012, Platts changed the definition to those crudes loading from 10–25 days in the future to 10–21 days to increase the number of cargoes used to indicate the dated Brent price (Lawler 2011).

Because of declining production, two similar substitute North Sea crudes were added as substitutes for Brent delivery in 2002—Forties and Oseberg, and a third, Ekofisk, was added in 2007. Thus, Brent is sometimes referred to as BFOE (Dunn and Holloway 2012). There are also Brent futures contracts on both NYMEX and ICE that can be used in crude contracts, but the dated Brent of the forward contracts seems to be the more commonly employed marker price.

Hedging with Energy Futures

Energy futures contracts can be used to hedge (reduce the effects of) risk, to *speculate* (take on risk for an expected profit), and to plan cash flows for storage, transport, and processing. A *hedger* is a buyer or seller of a real commodity (for example, an oil producer or refiner) who takes a position opposite of the forward or futures market. A producer holding crude is said to be long on crude, so he sells, or shorts, a forward; a refiner who is short on crude buys a future; a speculator who is neither a buyer nor a seller of the product takes on risk in hopes of a profit.

To understand how this works, look at the following example. (We will ignore transaction and storage costs.) A trader has a barrel of crude in transit. Her cost, including a normal profit for herself, is S_t = $80/bbl. Upon delivery at time T, she receives the spot price (S_T). If the price has risen to $81, she gains $1; if it is $79, she loses $1. Thus, she can lose or gain money or make a normal return depending on S_T, as shown by three sample delivery prices in table 18–3.

Table 18–3. Gains and losses in the spot market at various prices

S_t	S_T	Gain or Loss $S_T - 80$
$80	$79	–$1
$80	$80	$0
$80	$81	$1

Now, if the trader wants to hedge, she sells a forward contract at t for delivery at T at a price of F_t^T. (Forward examples are used instead of futures because marking to market introduces some additional complexities that do little to change the value of F_t^T.) Suppose the contracted forward price is $F_t^T = \$80$ to deliver oil at time T. This means the trader has contracted to sell oil at $80 per barrel at time T. At T, the contract is worth $(F_t^T - S_T = \$80 - S_T)$ per barrel.

Thus, if the trader has contracted to sell oil at $80 but others are selling oil for only $79, the contract is worth $1; if the price rises to $81 and she has contracted to sell at $80, she will lose $1. Now compare the value of the contract barrel with the value of a barrel of cargo at the three delivery prices above ($S_T = \$79, \$80, \$81$) in table 18–4.

Table 18–4. Gains and losses in the spot and forward markets at various delivery prices

S_T	$S_T - S_t$ Spot Market	$F_t^T - S_T$ Forward Market	Combined Market
$79	–$1	$1	$0
$80	$0	$0	$0
$81	$1	–$1	$0

Note that what she gains or loses in the real market is offset in the futures market. Thus, she has been able to reduce the affective price volatility to herself. The cost has been her transaction cost for entering into the contract.

The purchaser of the futures contract could have been another hedger, such as a refinery, or it could have been a speculator. If the buyer is a speculator, he experiences the full risk from price changes in the market; speculators take on risk, and hedgers shed risk. The US Commodity Futures Trading Commission (CFTC) reports commitments of traders (COT) by exchange and product. The *commercials*, or those who also deal in the commodity itself, are considered hedgers. The *noncommercials*, such as swap dealers, hedge funds, and investment banks that trade futures but do not handle the physical product, are considered speculators (US CFTC, n.d.).

For example, the CFTC indicated that in November 2012, 49.0% of open interest for CME WTI was held by commercials, who held net short positions, and 47.1%

was held by noncommercials, who held net long positions. The remainder was held by those small enough they are not required to report (*nonreportables*). Thus, the commercials are tending to hedge falling prices, while the noncommercials are tending to speculate that prices will rise. The same numbers for CME Henry Hub natural gas are somewhat different: 31.0% of open interest was held by commercials, who held net long positions, and 64.1% was held by noncommercials, who held net short positions (US CFTC 2012).

In the simple numerical example above, the product for the spot and the futures market were the same. However, the two products do not have to be the same to use one to hedge the other; only the prices need be highly correlated. For an example where a different product is used to hedge, see http://dahl.mines.edu/st18/st18.pdf, question 35.

Arbitrage

An important phenomenon influencing futures prices is *arbitrage*, which is the simultaneous buying and selling of the same product to make a profit without any cash requirement or risk. For example, if Chinese yuan are trading for $0.16 in one market and $0.165 in another market, you can buy them for $0.16 and resell them simultaneously for $0.165 in the other and make a profit (unless transaction costs are more than $0.005). Arbitrage assures that in competitive transparent markets, where everyone knows the prices, the price of the same product should differ across markets by no more than transportation and transaction costs. For an example to illustrate how arbitrage promotes market efficiency, see http://dahl.mines.edu/b1801.pdf.

Now let's see how arbitrage and the underlying spot price influence a future price. We start with the simplest case, the future price of an asset sold at discount with no income. For example, you buy a US Treasury bill (T-bill) at $9,700. It redeems for $10,000 at maturity in 12 months. You receive no income from the bill, and you get the difference between what you paid and the redemption value. The price of this security will vary over time as the interest rate changes until maturity but will approach the redemption price as the asset approaches maturity.

Thus, if you hold the asset to maturity, you have no price risk and know exactly what price you will receive. However, if you need cash and plan to sell the asset before maturity, you have price risk. One way of mitigating this price risk is to buy a futures contract. However, before you do so, you want to know if the futures contract is properly priced.

Let's see what the future price for such an asset should be. Suppose you will sell the bill at time T, where T is some date before maturity of the bill. Take a futures contract for time T. S_T is the spot price of the bill at time T, which is unknown. S_t is the current spot price. F_t^T is the future price at time t to be paid at T, and r is the risk-free rate of return, considered here to be your cost of carry. Then the following applies in a simple world:

$$F_t^T = S_t e^{r(T-t)}$$

The price for a futures contract should be equal to the principal and accrued interest over the time period $(T - t)$. To see why, suppose that $F_t^T > S_t e^{r(T-t)}$. You will pay more for the asset at T by holding the futures contract than if you purchased the asset at time t and held it until T. Arbitrageurs would want to sell the futures contracts F_t^T and buy spot at S_t, thus lowering F_t^T and raising S_t. You can use the risk-free rate in this context to lock in a certain profit. You lock in the current price when you buy at S_t, and you lock in the future price when you sell F_t^T, as well.

Consider the numerical example from Hull (2000, 38). You own a newly issued one-year discount bond that you will want to sell in six months. The risk-free rate is 6% per annum. The current bond price is $930. The formula says the futures price negotiated today should be the following:

$$F_t^T = 930e^{0.06-0.5} = \$958.3$$

Now suppose that $F_t^T < 958.3$. Let $F_t^T = \$950$. Since the futures contact is cheaper than the security at T, the arbitrageur should buy the futures for $950 and sell the bond for $930. Putting the $930 into the market at 0.06%, he would make [$930e^{0.5\times0.06} = \$958.3$] at time T. The futures contract says he has agreed to buy the bond at $950. He will have enough to buy the bond with a profit of $8.30. If he does not own the bond, he can sell short (borrowing the bond and selling it, and then purchasing it back at a later time at the cheaper forward price of $950) and still make a profit of $8.30. However, as arbitrageurs buy the futures, they will bid its price up; as they sell the bond, they will lower its price until no money can be made from arbitrage and *voila*:

$$F_t^T = S_t e^{0.5\times0.06}$$

What Determines Energy Future Prices on Commodities?

Modeling future prices for energy commodities is slightly different than for energy stocks and bonds because commodities incur storage costs in addition to interest costs. Suppose we have a barrel of oil today at S_t. The cost of delivery at T would be $(S_t + U_t)\exp^{r(T-t)}$, where U_t is the unit storage cost at time t, which we assume is paid up front for the whole time period. Alternatively, storage could be paid out over the storage period, perhaps once a month. In such a case, U_t would be the present value of storage costs at time t. Initially, we expect the forward price at time t for delivery at T would be the following:

$$F_t^T = (S_t + U_t)e^{r(T-t)}$$

Often the convenient assumption is made that storage costs are some constant percent of the spot price (μ). Then we would expect the futures price at time t for delivery at T to be:

$$F_t^T = S_t e^{(r+\mu)(T-t)}$$

For commodities bought and sold for actual consumption instead of just investment, there may be some additional benefits from holding inventories. This additional benefit is called a *convenience yield* (δ). Inventories can stabilize the production process by filling in during shortages or when there is higher-than-anticipated demand. In such periods, prices are likely to be higher; if you are holding stocks during these periods, you can benefit from the higher prices. Alternatively, in periods of weak demand or supply surpluses, prices may fall and inventory holders may have less benefit from having stocks. Note in an equilibrium market, convenience yield should not become negative, as stocks that lose you money should be sold (i.e., $\delta \geq 0$). The net benefits that result from holding inventories do not accrue to those who only hold futures contracts. If there are additional net benefits to the holder of stocks (δ), we can subtract this benefit from the costs of holding the inventory. The formula for the futures price then becomes the following:

$$F_t^T = S_t e^{(r+\mu-\delta)(T-t)} \qquad\qquad (18\text{--}1)$$

If $r + \mu - \delta > 0$, then $r + \mu > \delta$. This shows that the convenience yield is smaller than carrying costs, which are interest and storage. Such a market is called a *normal* or *contango market*, and F_t^T (for further out contracts) increases (fig. 18–2).

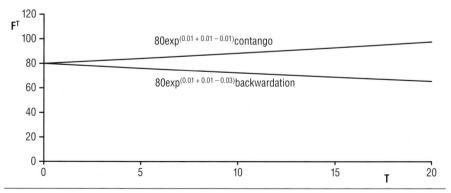

Fig. 18–2. Future prices today ($t = 0$) by maturity date

To simplify the notation in this example, let $t = 0$ and drop it from the formulas. In figure 18–2 in the top line, $r = 1\%$, $\mu = 1\%$, $\delta = 1\%$, and current $S = 80$, then by substitution into equation (18–1):

$$F^T = Se^{(r+\mu-\delta)T} = 80e^{(0.01+0.01-0.01)\times T}$$

$$T = 5: F^5 = 80e^{(0.01+0.01-0.01)5} = \$84.102$$

$$T = 10: F^{10} = 80e^{(0.01+0.01-0.01)10} = \$88.414$$

If $r + \mu - \delta < 0$, then $r + \mu < \delta$, and the convenience yield is larger than carrying costs (interest and storage). Such a market is called a *backward* or *inverted market* and further-out futures prices are lower. Changing δ in the above example to 0.03 gives backwardation and results in the lower forward curve in figure 18–2. If there is a supply or demand shock with low inventories, we expect δ to be high, and the market is more likely to be in backwardation. With high inventories, δ is more likely to be low and the market would be normal or in contango.

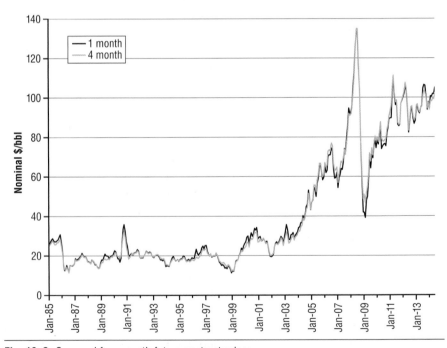

Fig. 18–3. One- and four-month future contract prices
Source: US EIA (n.d.b).

Figure 18–3 shows the one-month (black line) and four-month (gray line) CME WTI future contracts on monthly data from 1995 to mid-2014. If the white line is above the black, the market is in contango (at least for the four-month futures), indicating convenience yield is relatively low. If the white line is below the black, the market is in backwardation, indicating the convenience yield is relatively high.

Over the whole period in figure 18–3, the market has been in backwardation a bit more often (55%) than in contango. In the weak markets from 1985 through 1989, the market was in backwardation much of the time (85%). During the 1990s, the market months for contango and backwardation were more balanced, with backwardation slightly dominating (56%). During the price run up and commencement of the first Gulf War, August 1990 to March 1991, the market was in backwardation. Since the turn of the century, oil-price volatility has taken an exciting turn and contango months have been somewhat more prominent (56%). The market was in backwardation during the price increases of 2003 through November 2004. It was in contango from early 2005 through mid-2007 as spot prices first increased and then decreased. It remained in backwardation through much of the next dramatic spot price run up until summer of 2008. Contango then strongly dominated through April 2013, when the market went into backwardation for most of the months thereafter through mid-2014.

Convenience yield influences the forward curve and gives us some information on market expectations and the market's valuation of having inventories. The most common way to measure convenience yield is to apply equation (18–1) using spot prices, future prices, and storage costs. The computation is quite straightforward. First, take the natural log and then solve for δ:

$$\ln F_t^T = \ln S_t + (r + \mu - \delta)(T - t) \rightarrow \delta = r + \mu + (\ln s_t - \ln F_t^F)/(T - t) \quad \textbf{(18–2)}$$

Convenience yield for energy commodities can be estimated using equation (18–2). All the variables in the equation are easily accessible except for the parameter to compute storage costs (μ). If the data is monthly, μ is the monthly storage cost as a percent of the spot price. Such costs vary considerably across storage locations and market situations. From a search of the literature and discussion with industry professionals, I found some representative anecdotal cost information that suggests μ averages 0.012. The black line in figure 18–4 is the computed convenience yield using the one-month future as the spot price and the four-month future. Daily data is used, but costs and interest rates are measured by month. The white line is US commercial stocks of crude oil, so it does not include government stocks held in the Strategic Petroleum Reserves (SPR).

Convenience yields averaged 0.02 or 2% of the monthly price over the whole sample, but there has been considerable variation across time. Saudi Arabia instituted netback pricing and production increases near the end of 1985, causing nominal prices to spiral down almost 60% from January to August 1986. Convenience yields fell and even became negative briefly in March 1986.

A negative convenience yield makes no economic sense. If inventories confer negative benefits, there should be no inventories. However, recall that storage costs for the computations are based on averages. During periods of high inventories (see the white line in the figure), they are likely to underestimate storage costs and bias convenience yield down. During periods of low inventories, they are likely to overestimate storage cost and bias convenience yield up. You will notice in the figure that spikes into the negative direction tend to occur during periods of high inventories, while spikes in the positive direction occur during periods of low inventories. Further, the averages themselves are based on limited evidence, while the market may also be caught unawares and be in disequilibrium.

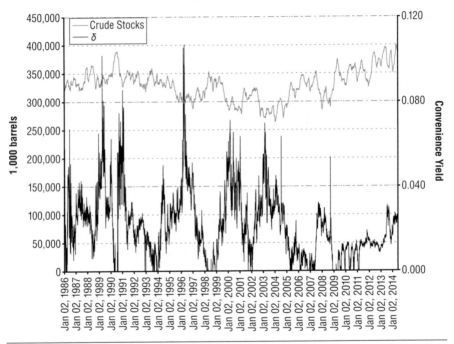

Fig. 18–4. Three-month convenience yield for US light sweet crude oil, January 1986 to June 24, 2014, and US stocks of crude oil

Sources: US EIA (n.d.b.); Federal Reserve Bank of St. Louis (n.d.); Data for storage cost from US CBO (1982); Adelman (1992); personal conversation with Thorsten Viertel at Shell Oil; and Metalaugmentor (2010).

Convenience yields rebounded and spiked in June 1986. With OPEC's failure to reach a production agreement, they generally trended down through the rest of 1986. Prices were a bit higher in 1987, averaging around $19, and convenience yields were a bit stronger as well. Lower prices in 1988 were also reflected in lower convenience yields. In late 1988, OPEC reached a production agreement, and markets tightened. Oil prices and convenience yields increased.

Prices generally trended up through 1989, but markets started to weaken in March 1990, and inventories piled up. Convenience yields fell and remained very low from May until August 2, 1990, when Iraq invaded Kuwait. They rebounded, reaching a peak in February 1991, when it was clear that Saddam Hussein's forces would be routed. Convenience yields then trended down with a weak economy and increasing production from OPEC and the North Sea, becoming negative briefly in December 1993. The markets were in backwardation throughout the Gulf crisis until May 1991.

Oil prices fluctuated around $20 per barrel throughout 1991 to 1993. Iraq was out of the market; Kuwait tried to regain prewar production. Turmoil in the former Soviet Union cut its oil exports. Convenience yields were high but fell as production began to recover. By 1993, OPEC and North Sea production increases were again putting downward pressure on oil prices. The market was in contango throughout 1993 as oil prices and convenience yields drifted down. Through 1994 and 1995, OPEC did not increase its quota, with prices and yields edging upward.

Through 1996, a strengthening world economy improved prices and convenience yields. The market was in backwardation, as is common when markets are tighter. Prices and convenience yields continued to fall through 1997 and most of 1998 as the Asian economy sank into deep recession. Again, we see some spiking negative convenience yields when inventories are high. Iraq sold oil for humanitarian revenue, and OPEC produced over quota. The market again went into contango. In March 1999, OPEC agreed to cut production; Norway, Mexico, and Russia agreed to cuts as well. With these cuts and a recovering Asian economy, prices and convenience yields began to rise, and the market went into backwardation.

With the terrorist attacks on the US World Trade Center on September 11, 2001, and the ensuing weakened US economy, the market again went into contango, and convenience yields fell. By March 2002, oil production cuts, a strengthening US economy, and the US-Iraqi sparring over weapons inspections raised oil prices, increased convenience yields, and again put the market into backwardation. The strong market in 2004, especially from oil growth in China, came as a surprise to most. The market remained in backwardation through most of 2004. With continued strong growth of the world economy, shortages of personnel and materials for the oil industry, and little spare capacity in OPEC, oil prices continued to ratchet up through 2005 and part of 2007. There were some price spikes from hurricanes, but convenience yields were relatively low, and the market was in contango much of the time. After a big dip at the end of 2006, oil prices begin their spectacular climb, peaking at more than $140 per barrel in mid-summer of 2008. Convenience yields were generally higher and inventories generally lower. However, for the remainder of the year, oil prices generally fell, especially after September as the United States plunged into a financial crisis. The oil market was in backwardation over most of the price run-up. Convenience yields rose with the price run-up but slipped to very low levels as price plunged below $40 per barrel by the end of 2008 and the beginning of 2009. They generally recovered through mid-2014. The market had been in contango, and convenience yields had remained generally below the historical average through mid-2013. From mid-2013 to mid-2014, the market was generally in backwardation and convenience yields were higher than for most of the previous three years.

Efficient Market Hypothesis

One of the oft-cited advantages of energy futures markets is that they provide price transparency and reveal information on energy price expectations. The efficient market hypothesis suggests that the future price (F_t^T) plus any necessary risk premium (RP) is a good predictor of the spot price. Therefore, the expected spot price for electricity, $E(S_T)$, at PJM (the Pennsylvania–New Jersey–Maryland ISO) in an efficient market would be the following:

$$E(S_T) = F_t^T + RP_t$$

Absence of a risk premium would suggest that hedgers dominate the market on both sides in roughly equal numbers, with no need for speculators. Significant speculators in the market would require an expected return, and the risk premium would then not be zero. A statistical test of the efficient market hypothesis under the assumption of no risk premium could be done with the following function:

$$S_T = \alpha + \beta F_t^T$$

The null hypothesis of no risk premium and market efficiency against the alternative of an inefficient market or a risk premium, or both, would be as follows:

Null hypothesis H_0: $\alpha = 0$ and $\beta = 1$

Alternative hypothesis H_1: $\alpha \neq 0$ and/or $\beta \neq 1$

Herbert (1993) performed this test for the US natural gas futures market and was able to reject the null hypothesis (that $\alpha = 0$ and $\beta = 1$) for November 1992 through February 1993. He concluded that the gas market was inefficient and also noted that other studies found such inefficiency in the early years of the US crude oil futures markets. What is more likely is that the risk premium is not equal to zero, since, if markets were inefficient, there would be ways to systematically make profits. For example, suppose the following:

$$E(S_T) > F_t^T + RP_t$$

If so, then you should buy the future, since it would go up in value as you approached the end of trading. As all players tried to make profits, the profits would be dissipated. Indeed, Deaves and Krinsky (1992) and Jalali-Naini and Manesh (2006) find a risk premium in the crude oil market that varies over time.

Crack and Spark Spreads

Various derivatives can be combined into trading strategies. One such popular strategy is a *crack spread*. A refiner buying crude oil and selling products makes money on price differentials and is more interested in the difference between the crude and product prices than in the absolute price level of each. To lock in a spread, a hedger can simultaneously buy crude futures and sell product futures. To illustrate a crack spread, consider the sample refinery output for the four most important products, along with sample prices, shown in table 18–5.

Table 18–5. Sample US refinery production and prices

	Production (1,000 bbl/d)	Price ($/bbl)
Gasoline	8,754	$141.12
Jet fuel	1,420	$134.86
Distillate	4,484	$141.29
Residual	395	$104.83
Crude*		$100.02

Note: * = refinery acquisition price per barrel; bbl/d = barrels per day.

The weighted average value for products in this example would be the following:

$$\frac{141.12 \times 8,754 + 134.86 \times 1,420 + 141.29 \times 4,484 + 104.83 \times 395}{8,754 + 1,420 + 4,484 + 395} = \$139.63$$

The crack spread is the weighted value of the products minus the price of the crude oil and equals the following:

$$\$139.627 - \$100.02 = \$39.63$$

In reality, the crack spread is slightly different from the above value because all the products from a barrel have not been accounted for. In the futures market, a crack spread will ignore an even larger share of the barrel.

A typical crack spread will be a 3-2-1, with three barrels of crude futures for every two barrels of gasoline and one barrel of heating oil (distillate) futures sold. Because price variations on spreads are typically less than on individual contracts, margins per contract will be smaller with a spread. A refinery seeking to hedge would sell a crack spread to buy crude oil and sell product futures. If you hedge for a refinery, you sell the spread closest to your product slate. Buying a crack spread indicates the purchaser is buying the products and selling the crude (see also CME Group [2012a]).

A newer trading strategy is a *spark spread*, which simulates the profits from a power plant. It was possible from 1996 to 2002 to hedge or sell such a spread on NYMEX by shorting electric power and buying the fuel used for generation (fuel oil, natural gas, or coal). However, in May 2002, the electricity contracts including spark spreads on NYMEX moved to the OTC market. Such spreads have also started trading over the counter in Europe.

As discussed previously, hedging in both futures and forward markets stabilizes prices for the hedger. Although they minimize downside risk, they also minimize upside profits for the hedger. However, the profits and losses can be spectacular for speculators without the offsetting assets in the cash market. For example, in 1992, Showa Shell, a Japanese refiner and distributor 50% owned by Shell Oil, lost more than $1 billion in foreign exchange markets. Traders bought billions of unauthorized dollars forward in 1989 at an average exchange rate of ¥145. The traders rolled losses over from year to year without reporting them and were not found out until 1992. By the end, the dollar was trading at ¥125 (Giddy, n.d.). Kashima Oil, another Japanese energy company, also lost more than $1 billion speculating in dollar forward markets.

In a more complicated trading strategy, MG Refining and Marketing (MGRM), a subsidiary of Metallgesellschaft AG, Germany's 14th largest industrial firm in 1993 and 1994, contracted to supply gasoline and heating oil at fixed prices for 10 years. It then hedged these forward contracts with futures and other OTC assets (so far, so good). But MGRM had a problem. The hedging was short-term, but the forward contracts were long-term. Thus, the short-term assets had to be continuously rolled over. Such a strategy can work when markets are in backwardation or prices are falling, but not when they are in contango or prices are rising, as they were through 1993 and the beginning of 1994. When the dust settled, MGRM had lost about $1.5 billion (Barth and McCarthy 2012).

Speculation and High Prices

The large price run-up from 2007 to mid-2008, as shown in figure 18–1, could have resulted from underlying fundamentals or shifts in supply and demand in the spot market. However, many believed it resulted from greedy speculators bidding up futures prices, which in turn bid up the spot price. So let's consider this possibility. Represent the spot market by supply and demand in figure 18–5 panel (a). Ignore any complications of market power on supply and assume two periods: one now and a later period to keep the diagram and discussion as simple as possible. The spot price is now determined by the intersection of supply and demand at P_e, Q_e in the diagram. The only way the futures price can affect the spot price now is if it shifts D or S in the spot market. Let's examine how this might happen.

Suppose speculators are bidding up the futures price starting in early 2007. If refineries expect markets to be tighter and crude prices to rise in the future, they may buy crude and put it into inventories, increasing demand and spot prices as in panel (b). If suppliers expect markets to be tighter and crude prices to rise in the future, they may withhold supplies now, which will increase price as in panel (c). Later when spot prices do increase, refineries may reduce inventories, thus reducing demand; producers may produce more, increasing supply. Prices will then be lower in the later period than they would have been. If the market does become tighter with higher prices, as signaled by the futures prices, the effect of the higher futures prices would have increased current prices and reduced futures prices, which would increase price stability in the market.

Thus, speculators have done us a favor and improved market efficiency by moving some crude oil consumption from the current period to the future period, where it has more value. In the process, the speculators would have made some money by buying before the price run-up and benefiting from the price run-up. Alternatively, if the speculators were wrong and bid up the futures prices, but the spot price next period actually fell, they would make the price less stable by raising price now and making the price next period even lower. They would also lose money. Thus, there is a market correcting tendency for speculators that are wrong too often to be driven from the market.

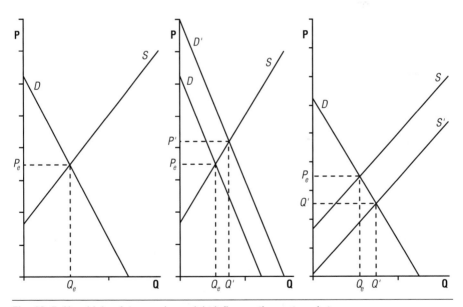

Fig. 18–5. How higher futures prices might influence the spot market

So are speculators responsible for the high prices in the spot market from 2004 to 2008? The short survey of futures markets research by Büyüksahin and Robe (2012) indicates that futures market activity has increased dramatically in energy and other commodity markets since 2000. Given the high volatility, this increased activity comes as no surprise. Further, the amount of trading by financial institutions also became a larger share of the trading. Since such institutions are considered speculators, studies have considered their behavior. If they caused the dramatic price run-up through 2008 and the subsequent collapse, they should have been buying futures when prices were rising and selling futures when prices were falling. Büyüksahin and Robe (2012) noted that speculators such as hedge funds were more often selling when prices were rising and buying when prices were falling during the years 2004 to 2009. This suggests that speculators were more likely to have had a stabilizing effect on the market.

So what else was happening? Observed prices are the result of shifts in supply and demand that may be difficult to untangle. Further, we don't have information on all inventories and motives. With these caveats, consider what the price, inventory, and production data suggest. From figure 18–5, we can observe the following general price trends by quarters. Prices tended to: rise 2004:I–2006:II; fall 2006:III–2006:IV; rise sharply 2007:I–2008:II; fall sharply 2008:III–2009; and trend up through 2011.

Now consider inventories during this turbulent time. Figure 18–6 shows monthly inventory stocks for total commercial petroleum inventories (including crude and products) for the United States and the whole OECD from January 1993 to September 2012, and crude oil stocks for OPEC beginning in 2002.

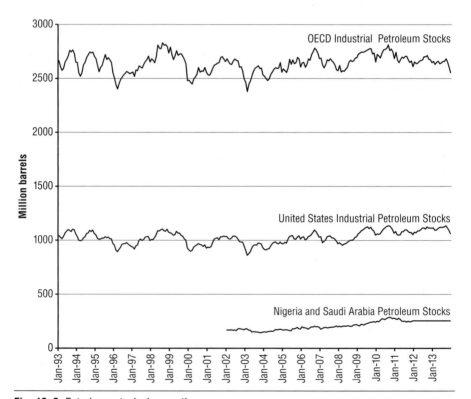

Fig. 18–6. Petroleum stocks by month

Sources: OECD and US stocks, US EIA (n.d.d); Nigeria and Saudi Arabia, JODI (n.d.).

The first thing to notice in the commercial petroleum inventories is the seasonal effects. Simple regression suggests a pattern for commercial stocks. They tend to be lowest in February and March, build to a peak in July and August and then slowly trend down with a sharper drop from November to February. Petroleum inventories are available for only two OPEC countries that span the 2004–2008 period. Although inventories fluctuate and generally trended up as they increased local consumption, no regular statistical seasonality could be discerned for these two tropical countries.

Since OECD and US inventories generally trended down during the biggest price run-ups, the evidence is not so strong that speculators were bidding up prices from 2007 to 2008. However, there were a number of fundamentals that came into play. There is general agreement that strong world demand was putting upward pressure on prices until the bottom fell out in the second half of 2008. Over this period, commercial inventories tended to be generally rising through 2004 to 2006, increasing upward pressure on price. However, with the big run-up through mid-2008, commercial inventories were more generally falling, suggesting they were slowing the price run-up a bit, as were the speculators. After the price collapse, rising inventories tempered the price fall, as did the speculators.

We can look for signs that OPEC orchestrated the run-up by production cut-backs in figure 18–7. OPEC production generally increased over periods of rising prices, although they were somewhat limited in spare capacity. (EIA defines *spare capacity* as the amount of production that can come online within a month and stay online for at least three months.) Spare capacity for OPEC averaged 2.2 million bbl/d from 2001 to 2012. However, with the unexpected strong demand in 2004, OPEC increased production, and spare capacity fell nearer to 1 million bbl/d, where it hovered until late 2006. Then production was reduced and spare capacity increased, presumably to shore up falling prices. However, as prices rebounded, OPEC increased output, and spare capacity began a decline back to 1 million bbl/d in time for the price to plummet in the last half of 2008. Production then fell again to help slow the decline in prices but has then trended up, as has price.

Adding production and spare capacity yields an estimate for total capacity. OPEC has continued to increase capacity over the whole period. Since markets for equipment and engineers to develop capacity were quite tight, development costs shot up and development slowed. IHS estimates that the cost of upstream oil and gas development increased 230% from 2000 to 2008 and had almost recovered to this previous high in the second half of 2012 (Tippee 2012).

Such cost increases were another fundamental that was pushing up the price. With the delays, the graph shows the biggest increase in capacity to come online after the price collapse in later 2009. The capacity and production dip in 2011 reflects the revolution in Libya. However, the rest of OPEC was able to rapidly make up the production shortfall, while Libya had not quite returned to prerevolution production and capacity by late 2012.

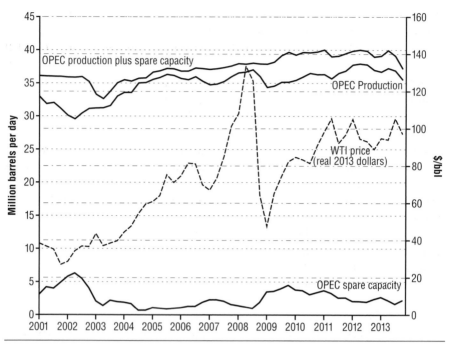

Fig. 18–7. Real WTI price and OPEC crude capacity, production, and spare capacity
Sources: OPEC statistics, US EIA (n.d.c); price, US EIA (n.d.b).

Thus, the evidence suggests that strong world demand for oil products, engineering equipment and personnel shortages, increasing costs of development, and low OPEC excess capacity were strong contributors. Speculators were more likely to have had a stabilizing influence, whereas inventory policy in the price run-up from 2004 to 2005 increased the run-up, but thereafter was more likely to have been a stabilizing influence as well.

Summary

Market risk, or volatility in energy prices and returns, is a fact of life for the corporate manager. Success requires that risk be properly managed. Sometimes companies may want to shed risk; at other times, they may want to limit risk; at still other times, they may take on more risk. In such cases, energy financial derivatives can help. Futures, options, forwards, and swaps all can be used to transfer and manage energy market risk. They increase market efficiency by breaking up risk into pieces and transferring the pieces to those who will accept them at least cost. A corporation or other entity can accept the risks it wants and transfer the remainder to others. Thus, those who have a comparative advantage in dealing with a particular form of risk will take it on, and as in any trading situation, such a trade will tend to make both parties better off.

A hedger deals in the real commodity and uses the futures markets to lock in a price and transfer risk. A speculator does not deal in the commodity itself but trades derivatives to try to make a profit, and in the process provides liquidity to the market. But care must be exercised. Unwise speculators can put themselves at great risk; the landscape is littered with billion-dollar mistakes in the derivative markets.

Energy price derivative schemes have come and gone during the past 150 years as price volatility has waxed and waned. With higher energy price volatility since 1973, a number of futures and options contracts on crude oil, oil products, natural gas, electricity, and coal have been developed and are traded on organized exchanges, including CME and ICE. A modern futures contract locks in a future price on an energy commodity. Going long in a futures contract locks in a future price to buy the underlying commodity, while going short locks in a future price to sell it. Most futures contacts are not held until delivery but are resold and used to lock in the price of a future transaction. A *clearinghouse*, an intermediary for derivatives sold on organized exchanges, guarantees contract performance by buying contracts from sellers and selling contracts to buyers, with buyers and sellers anonymous to each other. The clearinghouse concept, and the fact that futures contracts are marked to market every day, keeps transaction costs and default risk low.

In order to use futures to increase risk-management efficiency, it is important to know what a futures price should be. The futures price of a commodity depends on the current commodity price, the discount rate, storage costs, and any convenience yield that results from holding the commodity according to the following:

$$F_t^T = S_t e^{(r+\mu)(T-t)}$$

A convenience yield that is smaller than carrying costs means the market is normal, also called in contango. Here, futures prices are higher than the current spot price. The opposite is called a backward or inverted market, in which futures prices are lower than the current spot price. High convenience yields indicate the market is tight and inventories are valuable now. Low convenience yields suggest the market is weak and inventories convey a lower benefit.

Advantages of futures markets are that they provide price transparency and reveal information about price expectations. If futures markets are efficient, the expected spot price is equal to the futures price plus a risk premium. If the market is predominantly composed of hedgers on different sides of the market, the risk premium may be very small.

Spreads are trading strategies that include more than one financial asset. Selling a crack spread (buying a crude future and selling product futures) allows a refinery to hedge its refinery margin. Selling a spark spread (selling electricity futures and buying fuel futures) allows an electric utility to lock in a profit margin.

The large crude oil price run-ups followed by the crash in 2007 to 2008 have puzzled many. This is not so surprising, since the oil market is influenced by economic, political, and seasonal effects, making it hard to disentangle and isolate root cause. Many blamed OPEC. OPEC blamed speculators. However, examination of prices, inventories, speculative positions, and OPEC productions do not provide strong support for either of these positions. Rather, fundamentals of demand, cost, and inventory behavior seem to have stronger support.

19

Energy Options for Managing Risk

Betting on a horse, that's gambling, betting you can make three spades, that's entertainment, betting that oil price will go up three points, that's business. See the difference?

—Modified from Robert Pardo, 1985

Introduction

In the last chapter, we considered how two financial instruments, futures and forwards, can be used to transfer price risk. For both of these derivatives, the owners are obligated to buy or sell an energy product at a set price. They effectively set the price for the given product at some point in the future. However, sometimes market participants only want to protect themselves from or take advantage of price increases or price decreases. For example, an electric generator that uses gas as a fuel may want to put an upper bound on the price it has to pay for gas and put a lower bound on the price it receives for electricity. In such cases, the power generator would want to turn to options markets.

An option allows but does not obligate the contract holder to buy or sell an asset at a set price (known as the *strike* or *exercise price*) by a certain date. For energy futures options, the underlying asset is a futures contract rather than the physical commodity itself. A *call option* gives the holder the right to buy a futures contract, and a *put option* gives the holder the right to sell the futures contract. In this chapter, we will see how to price options, what variables affect their price, and how to put options together into trading strategies.

For example, suppose that our electricity generator is holding a CME June call option on a futures contract for natural gas at Henry Hub with a strike price of $4.50 per MMBtu (per 1.055 gigajoules [GJ]). This option gives the generator the right to buy 10,000 MMBtu for $4.50/MMBtu any time between now and four business days prior to the delivery month. Thus, the generator will not have to pay more than $4.50/MMBtu for natural gas. Buying the call option has effectively locked in a maximum price it will have to pay for natural gas. If the same power generator bought a put option on electricity futures for $50/MWh, it has the right to sell its power for $50/MWh. By buying a put option, the generator has locked in a minimum price it will receive for electric power.

Options are not new. Options on tulips were traded as long ago as the 1600s in Holland; options on stocks were traded on the London Stock Exchange in the 1820s and over the counter (OTC) in the United States in the 1860s. Exchange trading of US stock options started in 1973 on the Chicago Board Options Exchange. The first energy futures option was for light sweet crude oil on the NYMEX in 1986. Examples of other option contracts and when they began trading can be seen in table 19–1.

Table 19–1. Sample options contracts

Product	Traded Since Year/ Month	Delivery	Exchange	http://
Electricity, Baseload (European)	1989/9	No Delivery	NASDAQ OMX	www.nasdaqomx.com
Crude Oil Light	1986/11	Cushing	CME, US	www.cmegroup.com
Heating Oil	1987/6	NYH	CME, US	www.cmegroup.com
Gasoline Unleaded	1989/3	NYH	CME, US	www.cmegroup.com
Natural Gas	1992/10	Henry Hub	CME, US	www.cmegroup.com
Heating Degree Days	1999/9	No Delivery	CME, US	www.cmegroup.com
RBOB Gasoline*	2005/10	NYH	CME, US	www.cmegroup.com
ULS Diesel#	2013/4	NYH	CME, US	www.cmegroup.com
Brent Crude	1989/5	Sullom Voe	ICE	www.theice.com
Gasoil	1987/7	ARA	ICE	www.theice.com
EU Carbon Allowance (EUA)	2005/1	No Delivery	ICE	www.theice.com

Sources: Exchange home pages and other Internet sources.
Note: NYH = New York Harbor; ARA = Antwerp, Rotterdam, Amsterdam; ICE = Intercontinental Exchange, which took over the International Petroleum Exchange; NASDAQ OMX took over Nordpool. CME Group took over the New York Mercantile Exchange (NYMEX) and the Chicago Board of Trade (CBOT); * replaced unleaded; # replaced heating oil.

Not every option contract that is launched succeeds. If the derivative trades lightly or does not have enough liquidity to prove profitable to the exchange, products may be delisted. It can be abandoned entirely or moved to the OTC market. For example, in February 2002, a number of electricity options were delisted on the NYMEX and reintroduced as OTC instruments later in the year. Also following new regulations, ICE and CME transitioned all cleared energy swaps to futures and options in October 2012 in both the United States and Europe (ICE 2012a). Thus, these markets would be cleared on exchanges and become more transparent, as called for by the G20 in 2009 and as required by the US Dodd-Frank Act by the end of 2012 (Felsenthal et al. 2011).

Options are not "marked to market," but as with futures, they are traded three ways. They trade on organized exchanges with open outcry but in ever-decreasing amounts. They also trade electronically on exchanges and over the counter using telephones and the Internet. Options come in two varieties: puts and calls. An American put or call can be exercised any time up to expiration, but a European put or call option can only be exercised on the expiration date. Whether the option is American or European depends on the rules of the exchange, and not on whether the exchange is in the Americas or Europe. American options tend to dominate exchanges worldwide, especially for stocks and equities, but since European options are much simpler to value, they are the usual starting point for a discussion of options (Hull 2011).

Most OTC options are European. Basic put and call options discussed in this chapter are called *plain vanilla* options. There are other more complicated options called *exotics*. For example, an Asian option depends on the average price of the underlying asset over some period of time. These are called *average price* options on CME and ICE.

Both American and European energy options are offered on CME and ICE. Of the more than 400 energy futures contracts with open interest listed on CME in July of 2014, 68 of them are listed as having options trading. Many of the more popular contracts trade as both American and European options.

Concerning the two most popular energy contracts on CME, three observations may be made (CME, n.d.b):

- The European Henry Hub contract was vastly more popular than the American, with more than 40 times the open interest by contract numbers.
- The reverse was true for the WTI (light sweet crude) contracts, with the American-style contract having more than 19 times the open interest of the European contracts.
- The American WTI (light sweet crude) contracts were more than 50 times more popular than the Brent contracts, which were only traded as European options (CME n.d.b).

Pricing Options

To help us understand options, Kolb (1995) begins with a European call option, which can only be exercised at the expiration date. Since the call option gives the owner the right to buy the underlying asset at price K, the value of a European call at expiration depends on the spot price at expiration, S_T, of the underlying asset and the strike price K. If the price of the underlying asset exceeds K, the call owner can buy the asset for K and resell it for S_T, making a profit of $S_T - K$, as shown in figure 19–1. Below K, the option has no value and will not be exercised.

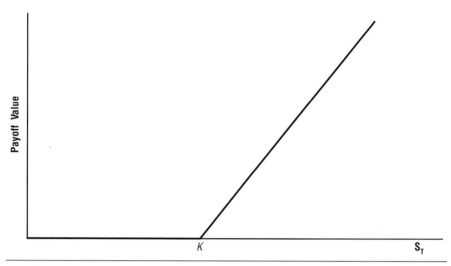

Fig. 19–1. Payoff of European long call at expiration

Options, whether they can only be exercised at time T or not, may have value at other times as well. A call is said to be "in the money" at time t if $S_t > K$; "at the money" if $S_t = K$; and "out of the money" if $S_t < K$. If S_t is much higher than K, the option is said to be "deep in the money," and if S_t is much less than K, the option is said to be "deep out of the money."

The purchase of a put option gives someone the right to sell the underlying commodity at price K. The value of a put at expiration also depends on S_T and K. If the price of the underlying asset falls below K, the put owner can buy the asset for S_T and resell it for K, making a profit of $K - S_T$, as graphed in figure 19–2.

At prices for the underlying asset above K, the put has no value and will be allowed to expire. A put is said to be in the money at time t if $S_t < K$, at the money if $S_t = K$, and out of the money if $S_t > K$. The four possible European options and their payoffs at expiration spot price S_T are as follows:

- long call with payoff = max $[S_T - K, 0]$
- long put with payoff = max $[K - S_T, 0]$
- short call with payoff = $-$max $[S_T - K, 0]$
- short put with payoff = $-$max $[K - S_T, 0]$

The long payoff is for the purchaser and the short is for the seller, called the *writer of the option*. Notice that the short position is the negative of the corresponding long position. Also, since S_T is not bounded, the possible payoffs are infinite with a long call, and the corresponding possible losses are infinite for a short call. Thus, writers of calls may be putting themselves at considerable risk. The writer of an option must maintain a margin to make sure she does not default if the buyer exercises the option.

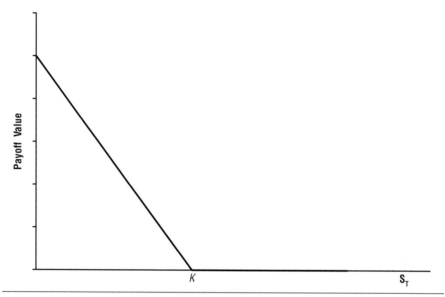

Fig. 19–2. Payoff of European long put at expiration

Typically, only large players with deep pockets write options. They include trading companies and branches of large energy companies such as BP, E.ON, ExxonMobil, Koch, Cargill, RWE, Total, and Shell, along with large investment banks such as Barclays, Goldman Sachs, JPMorgan Chase, Merrill Lynch, Morgan Stanley, and Credit Suisse.

The above formulas show the payoff of the option at various expiration spot prices but do not include what is paid for the option. Suppose we pay C_t for the call and P_t for the put at time t. The option expires at T, so it is held for $T - t$ periods. The amount invested at T will be the initial price paid plus interest or $C_T = (1 + r)$ $(T-t) C_t$ and $P_T = (1 + r)(T-t) P_t$, where r is the interest rate for the time period ($T - t$). Then the net payoffs of the above put and call at various spot prices at expiration are the following:

- long call with payoff = max $[S_T - K - C_T, - C_T]$
- long put with payoff = max $[K - S_T - P_T, - P_T]$
- short call with payoff = −max $[S_T - K - C_T, - C_T]$
- short put with payoff = −max $[K - S_T - P_T, - P_T]$

Options Quotes

The price of an option is what it costs to buy it, or the C_t and P_t from the last section. Prices of some of the more popular options are quoted regularly in the financial press. For example, the option quote in table 19–2 is for natural gas at

Henry Hub traded on the CME during 2013. The contract size is 10,000 MMBtu, measured in dollars per million British thermal units. The first column is the strike price measured in cents per million British thermal units. The next six columns are the price of puts and calls for various months measured in dollars per million British thermal units. To find the total cost of the option, multiply the option price times the contract size. For example, to have the right to buy 10,000 MMBtu in April for $3.45/MMBtu would cost you $0.103 × 10,000 = $1,030. The buyer is obligated to pay for the full cost of the option and cannot buy an option on margin. The last trading day is three business days before the first day of the next month. The seller of the options always gets to keep the cost of the option, whether the buyer exercises the option or lets it expire.

Table 19–2. Energy futures options quotes

Natural Gas Henry Hub
10,000 MMBtu ($/MMBtu)

Strike Price	Calls			Puts		
	Mar	Apr	May	Mar	Apr	May
3,350	0.059	0.145	0.223	0.137	0.158	0.165
3,400	0.043	0.123	0.198	0.171	0.186	0.190
3,450	0.031	0.103	0.174	0.209	0.216	0.216
3,500	0.022	0.086	0.153	0.250	0.249	0.245
3,550	0.015	0.072	0.134	0.293	0.284	0.276
3,600	0.011	0.059	0.117	0.339	0.322	0.309
3,650	0.007	0.048	0.101	0.385	0.361	0.343
Volume	2,282	168	51	0	61	35
Open Interest	11,422	2,846	1,747	0	1,020	1,398

Source: CME Group (2013).
Note: The strike price is in tenths of a cents, but the price of puts and calls is in dollars. Each of these contracts also traded at other strike prices not listed here. Open interest from prior day and volume is for these six sample strike prices. These trades represent less than one-half of the total traded option volume for Henry Hub gas for these three months.

Notice the much larger number of calls than puts. This suggests that the market may be expecting prices to rise above these strike prices, with more buyers trying to put a cap on the price they pay than sellers trying to establish a floor.

The above market quotes tell us what C_t and P_t are on a particular day. In the next section, we will use the binomial pricing model to learn what C_t and P_t should be.

Valuing Options with Replicating Formulas

To use financial derivatives efficiently for hedging and speculation, it is important to know their values at time t, or C_t and P_t. To learn how to value options, we will start with a one-period binomial pricing model from Kolb and Overdahl (2007). Suppose there is an underlying asset worth $100 now at period t. We assume the price of the asset pays no interest and can go up to $110 or fall to $90 next period, T, when the option expires. If we have a call option on this asset with a strike price of $100, the option will be worth either $10 or $0 next period, as shown in figure 19–3.

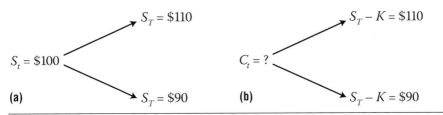

Fig. 19–3. Valuing a call from an underlying asset

We want to know the value of the call option today, or C_t. One trick is to find a portfolio that is equivalent to the call option. Then arbitrage assures that portfolios with the same payoff will have the same value. To begin, we will look at such a portfolio and then see how to derive it. Buy one-half of the above asset. It will be equal to $55 or $45 at time T depending on whether the price goes up or down, as shown in figure 19–4, panel (a).

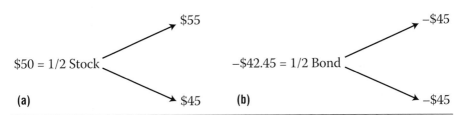

Fig. 19–4. Value of one-half of an asset (a) and a bond (b) in one period

Next, sell a bond (equivalent to borrowing money) worth $42.45 at a risk-free rate of 6%. It is worth –$42.45 now. In the next period, T, you will have to pay back the following: $-(1 + 0.06) \times 42.45 = -\$45.00$, as shown in figure 19–4, panel (b).

After a year, if the stock goes up, you will have $55 – $45 = $10. If the stock goes down, you will have $45 – $45 = $0. These payoffs are equivalent to the payoffs for the call option. So this portfolio must have the same value as the call, or else there would be an arbitrage opportunity. The nice thing about this portfolio is that we know its value right now, which is:

$$0.5 S_t - 42.45 = \$50 - \$42.45 = \$7.55$$

So now we know that the value of the above call option is $7.55. Since we can always find a replicating formula composed of the underlying asset and a risk-free bond, we can always value the option. It is quite easy to find a replicating portfolio (*Port*) at time *t*. Let *N* equal the number of shares of stock, B_t equal the value of the bonds purchased, and *r* equal the risk-free interest rate. The portfolio is now equal to the following:

$$Port = N \times S_t + B_t$$

At time *T*, the debt equals the amount borrowed plus the interest or $B_T = (1 + r)B_t$. The stock price can go up or down. Assume the stock price goes up or down by the fraction *n*. Let $U = 1 + n$, $D = 1 - n$, and $R = 1 + r$. The portfolio value at expiration if the stock price goes up is the following:

$$Port_{u,T} = N \times U \times S_t + R \times B_t$$

The portfolio value at expiration if the stock price drops is as follows:

$$Port_{d,T} = N \times D \times S_t + R \times B_t$$

Designate the two values of the call option (10 and 0 in the above example) to be c_u and c_d. Thus, c_u is the value of the call when the stock goes up, and c_d is the value of the call when the stock price goes down. Next, set them equal to the portfolio, as follows:

$$Port_{uT} = N \times U \times S_t + R \times B_t = S_T - K = 10$$

$$Port_{dT} = N \times D \times S_t + R \times B_t = c_d = 0$$

Solving for *N* and B_t tells you how many shares of the stock or underlying asset to buy and the value of the bonds you must buy:

$$N = \frac{c_u - c_d}{(U - D)S_t}$$

$$B_t = \frac{c_u \times D - c_d \times U}{(U - D) \times (-R)}$$

Once we know how many stocks and bonds to buy or sell, we can value the portfolio and, hence, the call option. Thus:

$$N = \frac{10 - 0}{(1.1 - 0.9) \times 100} = 0.5$$

$$B_t = \frac{10 \times 0.9 - 0 \times 1.1}{(1.1 - 0.90) \times (-1.06)} = -42.5$$

The solution tells us to buy (+) half a stock and sell (−) $42.45 worth of bonds. It should come as no surprise that the value of the portfolio is, as before:

$$Port = N \times S_t + B_t = 0.5 \times 100 + (-42.45) = \$7.55$$

Creating Probabilities for a Binomial Lattice Model

What is interesting about these problems is that we do not need to know the probability that the stock price goes up or down to value the option. Nor do we need to know the investor's risk preferences. Whether the investor is risk averse, risk neutral, or a risk lover never enters the computation. However, if we assume risk neutrality, we can convert the above value into probabilities. We will find such a conversion useful when we move to the multiperiod binomial model. Further, since the valuation is not affected by the risk preferences of the investor, it does no harm to assume risk neutrality, and it gives us some valuable results.

So let's see how to compute such probabilities. Let p equal the probability that the stock goes up to 110, and $(1 - p)$ equal the probability that the stock price goes down to $90. The expected value of the stock in one year is $p \times 110 + (1 - p) \times 90$. A bond at the risk-free rate would be equal to 106 with certainty. If an investor is risk neutral, the investor would find the stock and the bond portfolio equal in value, otherwise prices would change to make them equal in value. Hence, we can estimate a probability p that would make these two portfolios equal in value as follows:

$$p \times 110 + (1 - p) \times (90) = (1 + r) \times 100 = (1.06) \times 100$$

Solving yields p as follows:

$$p \times 110 - p \times 90 = 106 - 90 \rightarrow p = 0.8$$

So the probability that the stock price goes up can be represented by 80%. Then the probability that the stock price goes down would be 20%. Using these values, we can also value the option as follows:

$$\frac{0.8 \times (10) + 0.2 \times (0)}{1.06} = \$7.55$$

The numerator is the expected value of the stock, and the denominator discounts that value of the stock back to the present. Note that this trick, which will prove useful in the multiperiod valuation, magically yields the same value as earlier.

Similarly, the general formulas for probability under risk neutrality can be derived as follows:

$$p \times U \times S_t + (1 - p) \times D \times S_t = (1 + r) \times S_t = R \times S_t$$

Solving, we get equation (19–1):

$$p = \frac{R - D}{U - D} \tag{19-1}$$

What if there are two periods to maturity? To see what happens in this more complicated case, use a European put option, which expires after two periods. Let the spot price of the stock be 100 and the strike price be 101. The price increases and decreases are 10% and the risk-free rate is 6%. Now the problem seems more complicated, but we can use the tools developed above to value the put option. First, set up the lattice to see the various possible spot prices for the underlying asset shown below in figure 19–5. The lattice has two nodes in period $t + 1$. B is the node when the underlying asset goes up ($S_t + 1 = (1 + 0.10) \times 100 = 110$) and C is the price when the underlying asset goes down ($S_t + 1 = (1 - 0.10) \times 100 = 90$).

$A\ (S_A = 100)$

$B\ (S_{t+1} = 1.1 \times 100 = 110)$

$C\ (S_{t+1} = 0.9 \times 100 = 90)$

$D\ (S_{t+2} = 1.1^2 \times 100 = 121)$

$E\ (S_{t+2} = 1.1 \times 0.9 \times 100 = 99)$

$F\ (S_{t+2} = 0.9^2 \times 100 = 81)$

Fig. 19–5. Value of an underlying asset in a binomial lattice

There are three nodes in $t + 2$. D is the node when the price goes up in $t + 1$ and $t + 2$ ($S_t + 2 = 1.1 \times 1.1 \times 100 = 121$). E is the node when the price has gone up in $t + 1$ period and down in $t + 2$ or down in $t + 1$ and up in $t + 2$ ($S_t + 2 = 1.1 \times 0.9 \times 100 = 99$). F is the node where price has fallen in $t + 1$ and $t + 2$ ($S_t + 2 = 0.9 \times 0.9 \times 100 = 81$). The above nodes also have implied values (P) for our put option, which can be seen in figure 19–6.

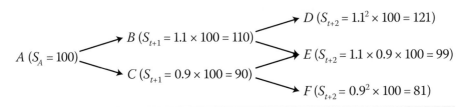

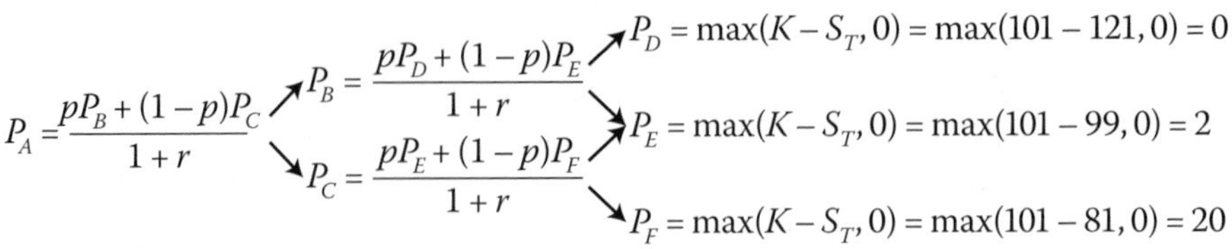

$$P_A = \frac{pP_B + (1-p)P_C}{1+r}$$

$$P_B = \frac{pP_D + (1-p)P_E}{1+r}$$

$$P_C = \frac{pP_E + (1-p)P_F}{1+r}$$

$$P_D = \max(K - S_T, 0) = \max(101 - 121, 0) = 0$$

$$P_E = \max(K - S_T, 0) = \max(101 - 99, 0) = 2$$

$$P_F = \max(K - S_T, 0) = \max(101 - 81, 0) = 20$$

Fig. 19–6. Value of a put option in a binomial lattice

For example, at node D, the stock value is 121. We have the right to sell the stock value at 101. Our put option is worth nothing: $(P_D) = 0$. At node E, we have the right to sell our option at 101, and the price is 99. Our put option is worth $P_E = 101 - 99 = 2$, and at F our put is worth $P_F = 101 - 81 = 20$. If we could find a replicating portfolio that gives us 0 when the stock price is 121, 2 when the stock price is 99, and 20 when the stock price is 81, then we could value the put option. However, that is more difficult now than in the one period case. Let's continue to value the option at other nodes before we see how to value the option in the current period using probabilities. At node C, the situation is a bit more complicated. At first we might guess that the stock option is worth $101 - 90$. However, this is a European option and we cannot exercise it at C; we can only hold it until the stock goes up to 99 or down to 81, or we can sell it to someone else. But they can't exercise it until $t + 2$ either. However, if we know the probabilities of the stock price going up to 99 (p) or down to 81 ($1 - p$), it is quite easy to value the option at node C:

$$P_C = \frac{p \times 2 + (1-p) \times 20}{1+r}$$

Similarly, we can easily price the option at node B:

$$P_B = \frac{p \times 0 + (1-p) \times 2}{1+r}$$

To get the present value of the put now at time t, represented by node A, we follow the same procedure:

$$P_A = \frac{p \times P_B + (1-p) \times P_C}{1+r}$$

We now know how to compute the value of the European put at each node in the lattice. What remains to be done is to compute the p values, which were computed above using equation (19–1) as $p = 0.8$ and $1 - p = 0.2$.

With these probabilities, we can compute the probability of the underlying asset being at each end node, the probability of the put at each end node, and the value of the put at each node. For example, if the price goes up from now to the first period (from node A to node B) $S_B = 1.1 \times 100 = 110$. The probability of a price increase is 0.80. To go from node B to node D, the price would have to increase twice, so the stock value would be $1.1 \times 1.1 \times 100 = 121$. The probability would be equal to $0.80 \times 0.80 = 0.64$. To go from node C to node F, the stock is worth $0.9 \times 0.9 \times 100 = 81$. The probability of going from C to F would be $0.20 \times 0.20 = 0.04$. Since there are two ways to get to E from A, you would add together the probabilities or $0.80 \times 0.20 + 0.20 \times 0.80 = 0.32$. Applying the formulas developed in figure 19–6, the value of the stock at nodes of a two-period lattice and the likelihood of being at each end node are shown in figure 19–7.

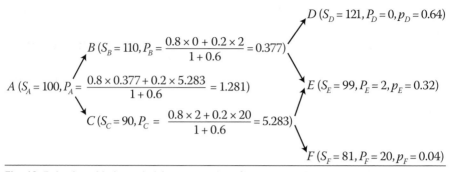

$D (S_D = 121, P_D = 0, p_D = 0.64)$

$B (S_B = 110, P_B = \dfrac{0.8 \times 0 + 0.2 \times 2}{1 + 0.6} = 0.377)$

$A (S_A = 100, P_A = \dfrac{0.8 \times 0.377 + 0.2 \times 5.283}{1 + 0.6} = 1.281)$

$E (S_E = 99, P_E = 2, p_E = 0.32)$

$C (S_C = 90, P_C = \dfrac{0.8 \times 2 + 0.2 \times 20}{1 + 0.6} = 5.283)$

$F (S_F = 81, P_F = 20, p_F = 0.04)$

Fig. 19–7. Lattice with the underlying asset value (S_i), put value (P_i), and probability at each node (p_i)

Since you can only exercise at maturity and you know the put value at various stock prices at expirations, you compute by working backward. Thus, if you are at B, the only two possibilities would be for the put to go to D with probability 0.80 or to value E with probability 0.20. The value of the put at B would then be (0.80 × 0 + 0.20 × 2)/(1.06) = \$0.377. The value at node C would be (0.80 × 2 + 0.20 × 20)/(1.06) = \$5.283. From A, you could reach all the possible end nodes, and the value would be (0.64 × 0 + 0.32 × 2 + 0.04 × 20)/(1.06)^2 =\$1.281. Alternatively, you could get the value by discounting back from the values at B and C as follows: (0.80 × 0.377 + 0.20 × 5.283)/1.06 = \$1.281. Thus, the current value of a European put for the above stock is \$1.281.

For an American option, the expected value of the put at nodes D, E, and F would be the same as for the European option, or the max $(0, K - S_T)$. At nodes A, B, and C, they might vary because you can exercise the put at B and C, if it is more profitable than holding it to maturity. See http://dahl.mines.edu/st19/st19.pdf, question 32 for an example of pricing with an American option.

We have assumed no dividends are paid in the above model, which is the case for an energy future. However, dividends are quite easy to incorporate. When dividends are paid, the stock price falls by the amount of the dividend. Thus, if there were dividends, you would merely have to reduce the stock price for dividends paid in the preceding period at each node.

Up to now, we have assumed that we know how much the price goes up or down in a future period, which we do not know for an asset. However, under some reasonable assumptions about how price changes are distributed, Kolb and Overdahl (2007) show how to estimate U and D using historical price data. Let's make such an estimate using daily Brent crude oil spot prices from 1991 to mid-2014. First translate price changes into rates of return. Remember, the rate of return on oil for a day is any capital gains or losses from one day to the next divided by the original value of the asset = $(S_{t+1} - S_t)/S_t = \Delta S_t / S_t$. Thus, if oil on Tuesday is \$105/bbl, and oil on Wednesday is \$107.10/bbl, the rate of return from Tuesday to Wednesday is (107.10 − 105)/105 = 0.02, or 2%.

Next, we approximate price changes by exponential functions for easier computation. Let $S_t = e^{\mu t} S_0$, where μ shows us how price changes from period 0 to t. If μ is positive, price increases; if μ is negative, price decreases. It is easy to see that μ is the continuous rate of return from period 0 to t by the following. For very short time periods, we can approximate ΔS_t by its derivative, $dS_t/dt = \mu e^{\mu t} S_0$. The continuous rate of return is then the following:

$$\frac{dS_t\big/dt}{S_t} = \frac{\mu e^{\mu t} S_0}{e^{\mu t} S_0} = \mu$$

Logs are often used to estimate this return as follows. Taking logs of the formula $S_t = e^{\mu t} S_0$ yields $\ln(S_t) = \mu t + \ln(S_0)$. Then $\mu = (\ln(S_t) - \ln(S_0))/t$. The continuous rate of return from day $i - 1$ to i is $\mu_i = \ln S_i - \ln S_{i-1}$. The estimated variance for the daily rate of return is the following:

$$\overset{\wedge}{\sigma}{}^2 = \frac{\sum\limits_{i=1}^{n} (\mu_i - \bar{\mu})^2}{n-1}$$

where $\bar{\mu}$ is the mean of the returns, which equals $\sum_{i=1}^{n} \frac{\mu_i}{n}$.

The estimated variance for the daily Brent crude oil (EIA, n.d.b) rate of return over the sample is $\hat{\sigma}^2 = 0.00051$, and the annual average risk-free rate from three-month T-bills (Federal Reserve Bank of St. Louis, n.d.) over this period is 0.01886. If your lattice has a daily period, then you can compute your D and U as follows:

$$U = e^{\sqrt{\hat{\sigma}^2}} = e^{\sqrt{0.00051}} = 1.02284 \qquad \text{(19–2)}$$

$$D = \frac{1}{U} = 0.97767 \qquad \text{(19–3)}$$

$$R = 1 + \frac{r}{365} = 1.00005$$

The risk-free probability can be computed from equation (19–1):

$$p = \frac{R-D}{U-D} = \frac{1.00005 - 0.97767}{1.02284 - 0.97767} = 0.50232$$

Once you have U, D, R, and p, you can then use the binomial lattice method above to compute the value of European and American puts and calls that expire any number of trading days out using the methodology discussed above. However, since there are approximately 250 trading days in a year, using a daily lattice for an option one year out would require about a 250 period lattice. This can be quite cumbersome to compute, since the number of nodes at the end of the year would be 2^{250}. So lattices often are computed using longer time periods, such as a week or a month. If your lattice period is a week, you convert your variance to weekly data by multiplying by 5, the number of trading days in a week, and use the weekly interest rate = $r/52$. If your lattice period is a month, you convert your variance to monthly data by multiplying by 20, the average number of trading days per month, and use the monthly interest rate = $r/12$.

Now complete the example using weekly data. First, convert your variance to weekly data:

$$\sqrt{\hat{\sigma}_w^2} = \sqrt{5\hat{\sigma}^2} = \sqrt{5 \times 0.00051} = 0.05050$$

Again compute U and D:

$$U = e^{\sqrt{\hat{\sigma}_w^2}} = e^{0.05050} = 1.05180$$

$$D = \frac{1}{U} = \frac{1}{1.05180} = 0.95075$$

The weekly interest rate is $r_w = 0.01886/52 = 0.00036$. The risk-free probability is the following:

$$p = \frac{R-D}{U-D} = \frac{1.00036 - 0.95075}{1.05180 - 0.95075} = 0.49094$$

Now p can be used as in the lattice example for above.

To sum up this procedure for valuing options:

1. Use historical data to calculate the variance of returns for the underlying asset. For energy options, it would be a futures price.
2. Choose the unit period for your lattice: daily, weekly, monthly, etc.
3. Convert your variance to correspond to the unit period of your lattice if necessary.
4. Compute U and D using equations (19–2) and (19–3).
5. Find the risk-free interest rate for the periodicity of your lattice.
6. Compute p using equation (19–1).
7. With U, D, and p, set up the lattice for the underlying asset.
8. Using the underlying asset, compute the value of the option at each expiration node.
9. For a European option, compute nodes prior to expiration by taking the discounted present value of the option at the next nodes.
10. For an American option, compute nodes prior to expiration by choosing the higher of the discounted present value of holding the option with the value of exercising.

Variables that Affect Option Prices

Kolb and Overdahl (2007) discuss a number of variables that should affect the value of an energy option. These effects are summarized in table 19–3 for American-style options. A call increases in value if the underlying asset price increases, whereas the reverse is true for a put. At a higher exercise price, the "in the money" region is smaller for a call and larger for a put, lowering the value for the former and raising the value for the latter. If there is more risk or volatility for the underlying asset, there is more chance that the underlying asset will take on extreme values or more chance that the option will be deep in or out of the money. However, the downside is limited to zero, so both puts and calls increase in value. If expiration is more distant, there is more time and more possibility to be in the money, so both options increase in value.

If the interest rate increases, their argument is a bit more complicated. You can think of a call as a potential liability, which you will have to pay when exercising the option or buying the underlying asset. The higher the interest rate is, the lower the value of this future liability and the higher the call price. The opposite argument is made for a put. The put represents an asset or a payment received when exercised at some future point. When discounted back at a higher interest rate, the value of this asset drops, reducing the value of the put. For more detail and arbitrage examples to demonstrate these principles in the no-dividend case, see Kolb and Overdahl (2007).

Table 19–3. Variables that affect American option values before expiration

Variable	Value of Call	Value of Put
1. Underlying Asset Price ↑	Up	Down
2. Exercise Price ↑	Down	Up
3. Asset Risk ↑	Up	Up
4. Time until Expiration ↑	Up	Up
5. Interest Rate ↑	Up	Down

Source: Kolb and Overdahl (2007).

Option Trading Strategies

As with future contracts, you can put together option contracts into trading strategies. The contracts can involve a real asset and an option such as a covered call (in which you are long in crude and you short a call). Alternatively, strategies may involve more than one option, called a *spread*. To see the value of combining options, we will focus on the value of European options at expiration.

Suppose a put costs $3.00 per barrel and a call costs $5.00 per barrel on oil futures, with both costs including interest from purchase time (t) to expiration (T).

Both have strike price of 75. You want to value this trading strategy at expiration for underlying asset prices S_T. Such a trading strategy, called a *long straddle*, is depicted in figure 19–8.

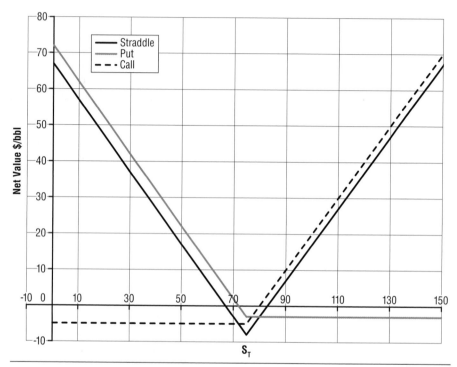

Fig. 19–8. Net value of a European long straddle at expiration

To value trading strategies, first value them separately. The call value is the dotted black line; the put value is the solid gray line. These are added together for the value of the straddle, represented by the solid black line. Notice this spread is in the money at low prices and high prices but is out of the money at prices near the strike price. It might be a winning strategy if you expect a lot of volatility in the underlying asset, making it likely that the price will not be near the strike price at expiration. To price such a spread, you could apply the lattice technique as well. You would still get the U, D, and p from the underlying asset. However, instead of using the value of the put as in the above example, you would now substitute the value of the spread.

Kolb and Overdahl (2007) have a number of other spreads with two or more derivatives. Try to imagine what the values of these spreads might be at different strike prices. With a *short strangle* you sell a call above and a put below market price. With a *put ratio vertical*, you buy a put at higher strike price and short two puts with a lower strike price, all with the same expiration date. You can do spark and crack spreads with options as well. You can use the file http://dahl.mines.edu/ch19m.xlsx to graph the value of spreads using one, two, three, and four option contracts.

Energy Swaps

The last financial derivatives briefly mentioned in this chapter are swaps. They are the most recent derivatives in financial markets, dating back to the early 1980s. A *swap* is an agreement to exchange cash flows in the future according to some agreed-upon formula. As with forward markets, these have traditionally been bilateral OTC agreements, with no regulatory oversight or reporting requirements. Often the swap is an exchange of cash flows, with one at a fixed rate and the other at a floating rate, called a *plain vanilla swap*. The underlying asset for a swap, and its estimated *notional* global amount outstanding in trillions of dollars as of June 2012, were the following: foreign exchange ($53.150), interest rates ($451.831), equities ($6.260), credit default swaps ($30.261), and commodities ($2.852) (Bank for International Settlements 2013).

To illustrate a swap, take the following example. A US independent producer swaps a floating rate for a fixed natural gas price of $3.54 with JPMorgan. Thus, JPMorgan agrees to accept the differential from a fixed price of $3.54 per Mcf for five years. Under this agreement, if the price fell below $3.54, JPMorgan would pay the independent the difference between $3.54 and the market price, and if the market price went above $3.54, the independent would have to pay JPMorgan the difference. In this way, the independent has a fixed price of $3.54 for gas, while JPMorgan bears the risk of the price differentials. The notional value of this contract would be the total principal amount that is the basis of the cash flows, or $3.54. Notice this is typically much smaller than the amount of price risk that the speculator is taking on.

Traditionally a swap agreement, which did not require upfront payments, stipulated the amount of the commodity covered by the swap, the length of the agreement, the price index that served as the basis for the financial swap, and the frequency of payment. For swaps, short-term is considered less than three months and long-term is six months to thirty years. Thus, swaps could provide much longer term protection than futures contracts. Swap agreements, which are bilateral agreements, had not been publicly disclosed. However, eventually some of the news services such as Reuters, Dow Jones Markets, and Bloomberg began to provide energy swap quotes. At the beginning, they were not typically traded on exchanges. However, ICE started as an OTC market and developed a thriving business, and NYMEX introduced swaps trading for 25 common products in 2002.

Although the swaps were bilateral, the exchanges began providing some standardized products, along with clearing services and default insurance, by guaranteeing both sides of the transactions. This was accomplished through margin requirements and marking to market. For more information and a list of member companies, see the International Swaps and Derivatives Association (2013). You will note that the majority of the members are large banks or other financial intermediaries. However, trading companies affiliated with notable energy companies involved in the swap market include EDF Trading Limited; Hess Energy Trading Company, LLC; RWE Supply & Trading GmbH; Shell Energy North America, LP; and TOTSA Total Oil Trading SA.

After the financial crisis in the United States in 2008, pressure increased for more regulation in the swap market. In 2009, the G20 leaders called for more transparency with standardized contracts reported, traded, and cleared on exchanges or electronic platforms (Chance 2010). The US Dodd-Frank Act was enacted in 2010, requiring swap dealers conducting more than $8 billion per year to be registered and to release real-time data to swap data repositories (SDRs). Reporting for such data began in February 2013 (US CFTC 2013).

To avoid the higher costs associated with becoming a swap dealer, more than one-half of the notional $18 million daily energy swap trading value had migrated to the futures market by the end of January 2013. One way of doing so is to conduct a privately negotiated block trade that is cleared in the CME or ICE futures market (Leising 2013).

Summary

Options on futures are another way to manage financial risk in energy markets. A futures option gives the holder the right to either buy or sell a futures contract on an energy commodity at a specified price. A call option allows the purchase of a futures contract and a put option allows the sale. An American option can be exercised at any time up to its expiration date, while a European option can only be exercised at expiration. An option price depends on the underlying asset price, the exercise price, the asset risk, the time until the expiration, and the interest rate.

An option can be priced by finding a portfolio consisting of the underlying asset and risk-free bonds that have the same payout. Arbitrage arguments suggest that the option should then have the same price as the portfolio. This price does not depend on the probability that the underlying asset price goes up or down or on the investor's risk preferences. Therefore, we can assume that the investor is risk free and create probabilities that the underlying asset prices will go up or down. Using the variance of the underlying asset, a binomial lattice model, and the computed probabilities, we can compute the value of an option for discrete time periods. These computations become more complicated as the discrete time periods grow shorter and the time until expiration lengthens. Companies who write options and make a living managing risk have complex computer programs to help them compute option values.

Options can be used to hedge and speculate as well as be combined into trading strategies that include more than one derivative at a time for the same purchases.

Depending on how you think prices will evolve, various spreads including two or more options contracts may help you to manage risk.

20

Climbing the Energy/Development Ladder to Sustainability

Sustainability means not turning resources into junk faster than nature can turn junk into resources.

—Steve Goldfinger, Science Comedian

Introduction

Humans may have used controlled fire as long as a million years ago (Choi 2012). Thus, for millennia humans have created heat and light by burning wood and other biomatter. Unfortunately, many still use such traditional fuels. However, as people climb the development ladder, they tend to climb the energy ladder. They transition from inefficient and often dirty fuel sources, such as dung, crop residue, and wood, to more modern, concentrated, convenient, and efficient fossil fuels and electricity, as seen in figure 20–1. The climb may not be smooth. In a process called *fuel stacking*, the very poor may use a combination of fuels ranging from traditional to modern as their budgets and sporadic market interruptions dictate. However, regardless of the method, all aspire to clean, cheap, efficient, and safe sources of energy.

We can also see this inexorable climb in figure 20–2. It gives estimates of the dramatic global move into fossil and other modern energy sources in the last 150 years. Although we can question the accuracy of planetary numbers in figure 20–2 with birthdays more than a century ago, they nevertheless illustrate a story of changing fuel use and technologies. In 1850, wood and other traditional sources were the dominant fuels (estimated at more than an 85% share), with the remainder being coal. Electricity was decades away; oil and gas of any significance were also yet to come. Horsepower was still of the equine variety.

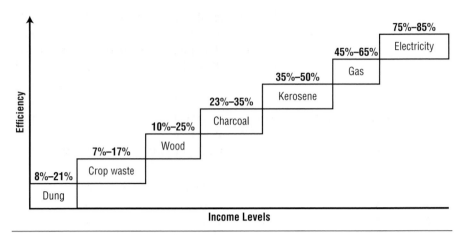

Fig. 20–1. Energy ladder for household energy use

Sources: Figure adapted from Chambwera (2004); efficiencies of wood, charcoal, kerosene, gas, and electricity from van der Plas (1995); efficiencies of dung and crop waste prorated from wood using efficiencies in Kennes et al. (1984).

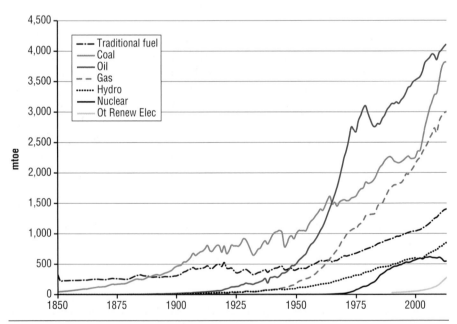

Fig. 20–2. World consumption of energy by source, 1850–2013

Sources: 1850–1994 values were developed by Nakićenović (1984) and updated for use in Grübler (1998) for most series and were downloaded from Grübler (n.d.). I updated the series to 2013 by extrapolating with data from BP (2014). Traditional fuels for 1850 to 1970 were derived by adjustments to wood values from Grübler (1998) with an adjustment for nonwood traditional fuels from Nakićenović et al. (1998, figure 3.1, 13). They were updated by extrapolating from data in World Bank's WDI (n.d.) on combustible renewable and waste data. Generation of electricity from other renewables was taken directly from BP (2014).

Although the figure shows that traditional fuel consumption has gradually increased since 1850, all the fossil-fuel sources have since passed the traditional sources. This occurred with coal around 1890, oil around 1950, and natural gas around 1965. Electricity is measured as delivered energy and so is not adjusted for heat rate. While hydro and nuclear have probably passed wood, they are not thought to have surpassed all traditional sources yet unless we adjust for heat content. Traditional sources are estimated to have fallen to below 10% of primary energy. The portion supplied by wood accounts for an estimated one-third of traditional fuels and about 5% of global energy consumption (World Bank, n.d.; UNFAO, n.d.).

The transition into fossil fuels in the West began in England in the 1500s with wood shortages. Coal in England, which was less expensive, appears to have surpassed wood as a fuel source sometime in the 1600s. Two centuries later in 1850, wood had been largely replaced by coal (Tverberg 2012). Income at that time is estimated to have been around $4,000 per capita in 2013 international dollars ([Maddison Project Database 2010] converted from 1990 to 2013 using the US CPI [US BLS, n.d]). Wood lasted longer in places where it was not in short supply. Coal did not pass wood in the United States until the 1880s, when income was about $6,000 per capita. At that time, almost three-fourths of the US population lived in rural areas (US DOC 1975). However, with plentiful fossil fuels and the ability to borrow technology from the mother country, the transition away from wood in the United States was faster. Although US GDP per capita remained around 75% that of the United Kingdom through much of the 1800s, by the 1900s, the United States had caught up and has generally remained above since. By 1950, the energy share contributed by biomass had fallen to less than 5% in the United States. Rural population had fallen to 36%, and per capita income had risen to $16,800.

The World Development Indicators, World Bank (n.d.), contains the energy share of combustible renewables and waste for 11 country groupings by region or income level in 1960. (Since combustible renewables include biomass, biogas, and bioliquids, I will hereafter refer to combustible renewables as bioenergy and refer to their category as bioenergy and waste [BioE&W].) WDI low-income (≤$1,040 per capita in 2013$) and middle-income countries ($1,041 to $12,658 per capita in 2013$) consumed BioE&W for more than one-half of their primary energy supply. In contrast, high-income countries (>$12,658 per capita in 2013$) consumed less than 5%.

Thus, BioE&W use is largely a fuel for the poor. This notion is reinforced in table 20–1, which shows total primary energy, BioE&W share of total domestic energy supply, and its use share by sector and region in 2011 from the IEA. IEA's term for BioE&W is biofuels and waste. I prefer to use the term bioenergy since in some sources, biofuels are exclusively liquid fuels.

Table 20–1. Population, primary domestic energy supply, bioenergy and waste share, and breakdown by sectoral use, 2011

Country/ Region	Primary Energy Supply (mtoe)	Pop. 10^6	BioE&W % Energy	BioE&W Percent of Use by Sector							
				% Elec & Heat	% Ot Trsfm & Ener.	% Ind.	% Trans.	% Resid.	% Com & Public	% Ag /For/ Fish	% Ot Non Spec
World	13,201,756	6,958.0	10.0	10.0	5.0	15.1	4.5	63.0	1.4	0.5	0.3
OECD	5,304,780	1,240.5	5.0	31.6	0.2	27.1	15.7	22.3	1.8	0.7	0.4
EU (27)	1,654,009	503.4	7.8	38.2	0.2	19.0	10.8	28.0	1.5	1.3	1.0
US	2,191,193	312.0	4.2	22.6	0.0	34.9	28.2	11.8	2.2	0.4	0.0
FSU–OECD	1,175,938	339.6	1.7	30.4	1.0	7.9	1.6	50.1	7.0	1.4	0.6
Mid East	646,794	208.6	0.1	0.6	12.5	0.0	0.0	17.9	0.0	0.0	69.0
Amr–OECD	589,209	460.2	19.0	8.7	14.4	37.4	12.4	23.2	0.6	3.1	0.2
Brazil	270,028	196.7	28.9	7.5	19.4	44.2	16.6	9.0	0.2	3.1	0.0
Asia–China	1,593,017	2,312.7	22.6	5.0	3.1	14.8	74.3	2.0	0.0	0.3	0.0
China	2,727,728	1,344.1	7.9	7.3	0.0	0.0	0.5	92.2	0.0	0.0	0.0
India	749,447	1,241.5	24.7	6.8	0.0	16.1	0.1	73.4	3.5	0.0	0.0
Africa	700,333	1,045.0	48.0	0.1	11.2	8.6	0.0	78.3	1.1	0.4	0.2

Sources: IEA (n.d.f).
Note: Eu (27) = the 27 countries in the EU as of 2007; FSU–OECD = countries of the former Soviet Union except the Baltic countries, which are now part of the OECD; Asia–China = Asian countries excluding China, and Amr–OECD = Non-OECD countries in the Americas, which include the Carribean and Central and South America except for Chile; Pop. = population in millions; BioE&W = bioenergy and waste; Elec & Heat = use electricity and district heat generation; Ot Trsfm & Ener. = use for other tranformation and energy industry's own use; Ind. = industrial use; Transp. = transportation uses; Resid. = residential use; Com & Public = use in the commercial and government sectors; Ag/For/Fish = use in agriculture, forestry, and fishing; Ot Non Spec = other nonspecified use.

Column 4 in the table shows an estimate of BioE&W share of total primary energy. Notice that its share is 5% in the relatively rich OECD, but even lower in the fossil-rich Middle East and non-OECD former Soviet Union countries (FSU–OECD). The European Union has heavily promoted bioenergy, especially recently. Its BioE&W share has risen to 7.8% from below 2% during most of the 1970s. In 2009, the European Union adopted an even more ambitious goal of 20% for all renewables, including bioenergy and waste, by 2020 (EU 2009c).

China's 7.9% share is of relatively recent origin. With rapid income growth, urbanization, and industrialization, bioenergy share has trended down for the last four decades. In the early 1990s, it was very near the other Asia–China countries (Asia less China) of 22.6%. India represents about one-half of the economic activity of this group, and its biomass share is near the average for the whole group (World Bank n.d.).

Americas–OECD, which contains the countries in the Americas except the OECD countries, includes the Caribbean along with South and Central American countries less Chile. It has a BioE&W share a few percentage points below the Asia–China group. Brazil is the economic powerhouse in Latin America, with around 45% of the economic activity (World Bank, n.d.) and around the same share of BioE&W. Brazil's unique ethanol program, with flex fuel vehicles and use of *bagasse* (cane waste) as a fuel source, puts its industrial and fuel use share at the highest in the table 20–1 groupings outside of Africa.

Africa, with the lowest average per capita income of any continent, still relies on BioE&W for almost one-half of its energy (48%), largely in the form of traditional biomass. If we omit the desert countries to the North, with little forest cover, and consider only Sub-Saharan Africa, the BioE&W use is 10% higher. Taking away South Africa raises the percentage even more. Even in fossil-rich Angola and Nigeria, BioE&W exceeds 50% of the share of total fuel (World Bank n.d.).

Before focusing attention on the countries with high BioE&W use, let's consider a somewhat broader view of BioE&W. The IEA includes the following in their BioE&W category: municipal waste (2.3% of global BioE&W consumption); industrial waste (1.7%); solid fuels (93.7%), biogas (2.1%), and liquid biofuels (0.2%). The values in parenthesis are the computed global share of BioE&W in 2011. Table 20–2 shows that the solid fuels category, which includes wood and agricultural waste, is by far the largest. It also shows that the other four categories are dominated by richer countries. Municipal waste is mostly used to generate electricity in the OECD countries, with a small amount also used in Singapore and Taiwan. Presumably, some of this was originally motivated more for disposal of waste than for the generation of electricity.

Table 20–2. Domestic supply of total bioenergy and waste in terajoules and by source share, 2011

	Bioenergy and Waste TJ	Municipal Waste %	Industrial Waste %	Primary Solid Bioenergy %	Biogas %	Liquid Biofuels %
World	52,503,375	2.3	1.7	93.7	2.1	0.2
OECD Total	9,411,211	12.1	5.7	74.1	7.5	0.6
United States	2,781,685	10.6	5.0	74.6	8.3	1.4
European Union—27	4,756,087	14.4	5.1	71.0	9.0	0.4
Non-OECD Europe and Eurasia	887,121	0.4	20.3	78.9	0.3	0.0
Middle East	23,686	0.0	0.0	99.3	0.7	0.0
Latin America	4,118,691	0.0	0.0	99.3	0.1	0.6
Brazil	2,735,945	0.0	0.0	99.1	0.2	0.7
Asia excluding China	14,989,439	0.5	0.0	99.2	0.3	0.0
China, People's Republic of	9,015,148	0.0	2.1	94.1	3.8	0.0
India	7,731,493	0.0	0.0	100.0	0.0	0.0
Africa	14,099,994	0.0	0.0	100.0	0.0	0.0
OECD Asia Oceania	791,857	11.6	17.1	67.4	3.8	0.1

Sources: IEA (n.d.f); IEA (n.d.p); for conversions, see IEA (2005).
Note: All sources are given in terajoules (TJ) except liquid biofuels, which are given in tonnes. Liquid biofuels are computed here as the residual between total biofuels and waste converted from kilotonnes oil equivalent (ktoe) to terajoules and the other four categories. One TJ = 23.884 toe or 947.817 million Btu.

Almost 60% of industrial waste is used in the OECD, with much of the rest used in the former Soviet Union and China. About 55% of this waste is used to generate electricity and heat, with most of the remainder used in the industrial sector. Almost two-thirds of the biogas is produced in the OECD, where the bulk is used for electricity and heat plants, with much of the remainder coming from China, where it is used in rural households.

The Chinese government has promoted the use of biogas for many decades. Mao encouraged its use during the Great Leap Forward (1958–1960), though its increase in use then was more of a limp than a leap. But with lessons learned, better technology, better geographical choices, and popularization programs and subsidies to build the digesters, recent programs have been more successful. A few of the projects have even been certified under the Kyoto Clean Development Mechanism (CDM). It is estimated that 30 million Chinese rural homes now use biogas, with the government target of 80 million households by 2020 (Energypedia 2012).

Biogas has a number of advantages. Biogas digesters that use animal and human manure along with other biomatter are relatively easy to build, with straightforward technology (for example, see Gregory 2010). Bacterial action takes the carbon out of the refuse and combines it with water in the following reaction:

$$2C + 2H_2O = CH_4 + CO_2$$

Biogas has a number of advantages, and it can be used to achieve the following:
- It reduces emissions of methane, which is a potent greenhouse gas.
- It reduces fecal waste runoff into water supplies.
- It can increase availability of straw for use as animal feed and bedding.
- It provides a clean-burning and more efficient household fuel. (Traditional stoves may have an efficiency of 10%, while a biogas stove may have an efficiency of 60%.)
- It reduces cooking time and effort.
- It reduces the time needed to collect wood.
- It provides a valuable fertilizer in the refuse, with higher retention of nitrogen, ammonia, and phosphorus than ordinary composting.
- It reduces pathogens in the biowaste to improve hygiene.

Liquid biofuels (often referred to as biofuels) include ethanol and biodiesel, which are mostly used in the transport sector. Ethanol is produced from sugar and starch-based crops such as sugar cane (Brazil) and corn (United States) (ORNL 2011). Biodiesel fuels are based on a variety of vegetable oil crops (e.g., palm oil in Indonesia, rapeseed in Germany, and soybeans in Argentina) or animal and vegetable fat wastes. These fuels provide energy mostly for the wealthier nations. About two-thirds of consumption is in the United States and Brazil with much of the remainder in Europe. This use is largely the result of policies that include tax concessions, direct subsidies, blending requirements, and special treatment for biofuel vehicles. For example, at least 52 countries had biofuel mandates as of 2012 (Lane 2013).

China has provided direct subsidies to ethanol producers, tax exemptions, low interest loans, and subsidies for growing feedstock on marginal lands (Global Subsidies Initiative of the International Institute for Sustainable Development 2008). As a result, total biofuels use posted double-digit growth rates in the first decade of the 21st century in the Americas, Europe, and China. In 2003, the European Union set a goal of having 2% of transport fuel use from renewables in 2005, increasing to 5.75% in 2011. However, these goals were not mandatory and were not met in aggregate (Flach et al. 2011). Biofuel use was 4.4 % of total transport energy fuel in the European Union as measured by energy content in 2011. However, recent concerns over deforestation from increasing palm oil plantations and biofuel effects on food supply have caused the European Union to reduce the 10% biofuel mandate for 2020 to 6%. (Lane 2013). Although biofuels are around 17% of transport fuel use in Brazil and consequently around 10% in South and Central America combined, they do not attain the 1% level in other developing regions or the 5% level in OECD countries (IEA, n.d.f)

The patterns vary somewhat across the two main fuels. Ethanol got an earlier start than diesel, with government promotion in Brazil and later in the United States. It is still the largest share (70.7%) of liquid biofuels measured by energy content in the United States and Brazil, and continues to dominate global use. In 2011, the United States and Brazil accounted for about 83% of global fuel ethanol consumption and 87% of global production. Europe consumes an additional 7.5% and Canada adds another 3%. These latter two import about 30% of their consumption, while China produces and consumes an additional 3% (EIA, n.d.d).

Biodiesel had a later start. The earliest statistics for use of biodiesel that I have found are for France and Czech Republic in 1992 (UN, n.d.a). The European Union is still the leader with about 58% of biodiesel use. About one-fourth of EU use comes from imports. Germany is the largest consumer, producer, and exporter of biodiesel in Europe. Spain and Italy are the largest importers (US EIA, n.d.d).

The United States and Brazil are still among the largest individual consumers and producers, but for biodiesel, their combined consumption only accounts for another 25% of global consumption. Africa, the Middle East, and the former Soviet Union each account for less than 1% of global consumption. The two largest exporters, Argentina and Indonesia, each export more than they consume, and as a result, Asia and South America are the largest exporting regions. Asia and Oceania account for another 9% of global consumption. The largest consumer in Asia is Thailand, with about 2.5% of global consumption (EIA, n.d.d).

Thus, these first generation biofuels, for the most part, still make only a rather small contribution to transportation compared to gasoline, diesel, and jet fuel.

To analyze the current potential of biofuels, we consider the United States, which makes ethanol largely from corn and the majority of its biodiesel from soybeans. The

US ethanol yield from corn is about 2.77 gallons per bushel (412.8 kl/tonne), and the biodiesel yield from soybeans is about 1.46 gallons per bushel (1,080.2 kl/tonne) (Food and Agricultural Policy Research Institute 2006; Beuerlein, n.d.). US EIA (n.d.d) gives US gasoline and distillate consumptions in 2010 as 9.0586 and 4.223 million barrels per day, respectively. IEA (n.d.m) indicates about two-thirds of distillate is for transportation, which should be the diesel fuel component. The United States produced about 10 billion bushels of corn in 2010 and about 3.33 billion bushels of soybeans. Further, a gallon of ethanol has about two-thirds the energy content of gasoline, and biodiesel has about 90% the energy content of diesel.

From these statistics, we find that if the whole US corn crop had been used to produce ethanol in 2010, it would have been equivalent to only about 13% of US gasoline consumption. Similarly, if the whole US soybean crop had been used to produce biodiesel, it also would have been equivalent to about 13% of the US diesel consumption. To put these numbers in a global context, in 2010, the United States produced about 37% of the world's corn crop, accounted for about 47% of global corn exports, and consumed about 39% of the world's gasoline. It produced about 33% of the world's soybean crop, accounted for about 45% of soybean exports, and consumed about 16% of the world's diesel fuel (UNFAO, n.d.; IEA, n.d.m).

Most would agree that many of these first generation biofuels from food crops may make marginal contributions to transport fuels but are not the long-term solution. The economics of ethanol from sugar cane where irrigation is not needed, such as in Brazil, are much better. The energy output to input for sugar cane is 8:1, whereas for US corn or European sugar beets, it is less than 2:1 (Stillman 2006). But even sugar cane and palm oil with better economics are unlikely to be global solutions with current technology at current consumption rates. Replacing one-fourth of current transportation fuel using sugar cane as a feedstock is estimated to require about 17% of the globally available land, and palm oil would require about 22% (Cramer 2007). Therefore, much research is being done on nonfood crops that can be grown in marginal areas, such as switchgrass and other grasses for ethanol and algae for biodiesel.

Combustible Biomass and the World's Poor

Solid bioenergy (biomass) such as wood (some of which is consumed as charcoal), straw, and other crop residue have now been largely surpassed by fossil fuels in many parts of the world. However, they are still a mainstay for the lowest income countries. In this section, I consider the situation in some of these countries.

Traditional biomass use may be sustainable in the sense that its production can continue to provide fuel sources. Indeed, for millennia, many parts of Africa have received much of their energy from traditional biomass. Take the examples of South Africa and Ethiopia, two countries at opposite ends of the African income spectrum. Both consumed more bioenergy and waste in 2011 than in 1970 (World

Bank, n.d.) Thus it seems that at least for these four decades, production has been more than sustained.

There were concerns in the 1970s of an imminent global fuel wood crisis leading to deforestation, erosion, and desertification. Subsequent events and analysis suggest that locally gathered fuel wood and other biomass that is burned do not necessarily result in deforestation. The wood gathered and burned is often gathered by women and children and typically does not involve tree removal. More often, deforestation is the result of converting forests to cropland or is localized near cities, where trees provide commercial fuel or charcoal for nearby city dwellers (Arnold et al. 2006). However, even if traditional fuel use does not cause deforestation, it does have some other disadvantages that encourage the move to higher quality fuels. Burning such fuels emits health-damaging pollutants that include carbon monoxide, particulates, nitrogen dioxide, and aromatics. These pollutants are irritating and especially harmful to family members who spend more time at home, such as women and children. Fuel collection may also reduce the time available for education or other income-generating activities. The burning of traditional fuels is also often very inefficient. Burning wood in an open fire may have efficiencies of less than 8%. With such low efficiencies, in some cases the poor may even pay more than others for energy services (for example, see Barnes et al. [2005]).

Although the gathering of biomass is less likely to cause deforestation, charcoal is typically made from felled trees and may contribute to deforestation. Making charcoal is a process that heats wood under anaerobic conditions to extract the carbon from the oxygen, hydrogen, and other materials. The quantity of charcoal obtained depends on how dry the wood is, the size of the operation, and the type of wood. Average numbers encountered suggest the yield is 60% by volume and 25% by weight (Nix 2013). Wood is about 50% carbon, whereas charcoal is about 90% carbon (Keita 1987). Thus, if wood for fuel is to be transported very far, it is typically converted into charcoal to lower transport cost.

Although it takes energy to create charcoal from wood, charcoal has certain advantages besides being a more concentrated fuel source. Compared to traditional biomass, it burns more steadily and hotter and emits fewer pollutants, generally with higher efficiencies. Barnes et al. (2005) cite average efficiencies of 22% for charcoal and 15% for wood. These efficiencies will vary by the type of stove used and can certainly be improved for both forms of fuel.

Now let's consider the countries where biomass, including charcoal, dominates. World Bank (n.d.) shows BioE&W share of total primary energy consumptions for 132 countries in 2011. Of these, 25 countries averaged more than one-half of their energy from BioE&W from 2001–2011, as shown in table 20–3. Two-thirds of these are in sub-Saharan Africa, four are in Asia, and the rest are in South and Central America and the Caribbean. The table also contains variables that might influence the use of bioenergy.

Table 20-3. Socioeconomic characteristics of high biofuel users

Country	BioE&W % 01–11	GDP/capita 2012	Pop×10⁶ 2012	BioE&W Kg/capita 2011	Forest Share % 2011	Urb % 2012	Charc kg/capita 2012	Access Elec % 2011	HDI 2012	Gini 01–11	% Below Pov. 01–11	TIC 2012	% GDP Fossil Fuel Rent 2012	% GDP Forest Rent 2012
Angola	65.1	7,346	20.8	391	46.81	59.9	15.1	37.8	0.51	38.6	36.6	23	42.6	0.3
Benin	61.7	1,687	10.1	216	40.01	45.6	25.3	28.2	0.44	40.0	35.5	36	0.0	4.9
Cambodia	72.7	2,789	14.9	259	56.46	20.2	2.4	34	0.54	39.7	34.8	20	0.0	4.6
Cameroon	72.3	2,551	21.7	215	41.67	52.7	20.6	53.7	0.50	44.4	40.1	25	8.5	2.9
Congo, D. R.	93.9	451	65.7	357	67.85	34.8	33.0	9	0.53	47.3	71.3	22	4.2	25.6
Congo, Rep.	58.3	5,631	4.3	184	65.59	64.1	1.0	37.8	0.30	44.9	48.6	22	71.1	2.2
Cote d'Ivoire	73.2	2,747	19.8	450	32.71	52.0	24.6	59.3	0.43	NA	41.5	27	5.1	1.8
Eritrea	73.5	1,180	6.1	101	15.12	21.8	31.2	31.9	0.35	31.7	NA	20	0.0	2.0
Ethiopia	93.5	1,218	91.7	354	12.16	17.3	42.7	23.3	0.40	41.5	34.3	33	0.0	13.1
Gabon	59.3	17,997	1.6	722	85.38	86.5	12.8	60	0.68	42.8	32.7	34	42.9	2.3
Ghana	64.1	3,638	25.4	242	21.2	52.5	67.5	72	0.56	56.4	28.5	46	5.6	5.2
Guatemala	54.4	6,990	15.1	430	33.6	50.2	1.6	81.9	0.58	59.2	52.4	29	0.7	2.4
Haiti	73.4	1,575	10.2	249	3.636	54.6	3.2	27.9	0.46	47.7	77.0	19	0.0	1.9
Kenya	76.3	2,109	43.2	348	6.072	24.4	23.3	19.2	0.52	46.4	45.9	27	0.0	3.3
Mozambique	82.0	971	25.2	329	49.35	31.5	33.7	20.2	0.33	NA	54.4	30	5.0	7.0
Myanmar	71.1	NA	52.8	203	48.16	33.2	3.2	48.8	0.50	38.3	NA	21	0.0	0.0
Nepal	86.8	2,131	27.5	322	25.36	17.3	2.9	76.3	0.46	45.9	25.2	31	0.0	5.0
Nigeria	81.1	5,440	168.8	592	9.477	50.2	24.3	48	0.47	54.0	47.2	25	28.7	1.6
Paraguay	52.1	7,215	6.7	339	43.8	62.4	108.4	98.2	0.67	39.2	38.2	24	0.0	4.1
Sri Lanka	51.2	8,862	20.3	237	29.43	15.2	0.1	85.4	0.72	35.3	15.6	37	0.0	0.7
Sudan	72.5	3,370	37.2	306	23.18	33.4	15.8	29	0.41	37.6	46.5	11	5.2	1.4
Tanzania	89.5	1,654	47.8	395	37.28	27.2	34.7	15	0.48	36.9	NA	33	0.4	4.8
Togo	83.5	1,286	6.6	351	4.913	38.5	36.0	26.5	0.46	NA	60.2	29	0.0	6.2

Zambia	80.9	2,990	14.1	498	66.32	39.6	74.0	22	0.45	51.2	NA	38	0.0	2.9
Zimbabwe	64.2	1,337	13.7	448	39.54	39.1	0.8	37.2	0.39	NA	60.5	21	1.8	4.6
World	**9.8**	**13,538**	**7043.9**	**NA**	**30.88**	**52.5**	**7.3**	**78.18**	**0.70**	**NA**	**72.3**	**38**	**3.9**	**0.3**

Sources: World Bank (n.d.); UNFAO (n.d.); UN (n.d.b); and Transparency International (2013) .
Notes: D.R. = Democratic Republic; BioE&W % = the share of total primary energy that is bioenergy and waste; Pop x 10^6 = population in millions; Urb % = the percent of the population living in urban areas; Forest Share % = the percent of the land covered by forests; Gini = the Gini Index; HDI = the UN human development indicator; Access Elec % = the percent of the population with access to electricity; BioE&W kg/capita = the per capita consumption of combustible bioenergy and waste in kilograms; Charc kg/capita = the per capita production of charcoal in kilograms; % Below Pov. = the percent of the population below the national poverty level; TIC = the Transparency Internation Corruption Perception Index; 01–11 indicates the available data is averaged for 2001 to 2011; Sudan includes the Republic of South Sudan, which separated from Sudan in 2011. *GDP per capita is reported in 2011 international dollars converted with PPP. These values would be 3.6% higher in 2013 US dollars. For more information on using PPP indices, see dahl.mines.edu/st10/st10.pdf, question 34.

BioE&W share generally trended down over the last two decades for most of these countries, as would be expected. Only five countries had a higher share in 2011 than in 1991 (the Democratic Republic of the Congo, Côte d'Ivoire, Nigeria, Zambia, and Zimbabwe). Three of the five had generally falling per capita real income over this period. Sadly, Zambia and Nigeria, with generally rising per capita income since 2000, have not seen a share reduction for BioE&W.

Clearly, income influences whether a household can afford more advanced fuels. All countries in the table have GDP per capita of less than $9,000, except for Gabon. This small oil-rich country gets almost 60% of its energy from BioE&W and has the highest per capita consumption in the group (730 kg), while having an income of $16,186 per capita. The only other country in Africa with a somewhat similar per capita income is Botswana, which consumes only about 22% of its energy from BioE&W.

Another thing that might influence use of biomass is accessibility. The average forest cover worldwide is about 31%, but it is about 85% in Gabon. This is the highest share for any country with recorded biofuels and waste consumption, suggesting a relative abundance of wood. It is also one of two countries in table 20–3 (Côte d'Ivoire is the other) with no loss of forest cover since 1990 (UNFAO, n.d.). Gabon also gets 2.3% of its GDP from forestry products, which may also provide waste products for consumption.

The majority of these countries have forest cover above the world average (31%), but again there are some notable exceptions. The countries of Haiti (0.86), Kenya (0.93), and Togo (0.39) all had less than 7% forest cover in 2011. Their 2011 forest cover divided by 1990 forest cover, shown in parentheses after each country, illustrates the decrease since 1990, suggesting stress on biomass supplies. Nigeria is another surprise. In this oil-rich country, forest cover has fallen by one-half since 1990 to 9.5% (World Bank, n.d.). It gets more than 25% of its GDP from fossil-fuel rents, but its BioE&W share is more than 80%.

Since urban households tend to consume a lower share of traditional fuels, urbanization can have an impact. More than 70% of these countries have urbanization rates below the world average of 52.5%. Again, Gabon is an exception, with more than an 85% urbanization rate. Resource-rich Angola and Republic of the Congo also have higher rates of urbanization but have accompanying high rates of biomass use. In these latter two countries, civil wars may have slowed the transition to higher quality fuels. A higher percentage of biofuels and waste typically tends to be charcoal in urban areas, but I do not see any clear-cut pattern in consumption per capita in this sample of countries. Paraguay, with more than 50% urbanization and 52% biofuels and waste, is among the higher per capita charcoal consumers. Paraguay also has a very high rate of electrification, with more than 95% of the population having access to electricity. Sadly, two-thirds of the countries in the table have electrification rates of less than 50%. Even the oil-rich countries of Angola, Republic of the Congo, and Gabon have electrification rates of less than 50%.

GDP is not the only measure we could use for economic well being. The UN Human Development Index (HDI) is based on education, life expectancy, and GDP per capita (UN, n.d.b). They normalize this index to values between 0 and 1. In 2013, the United Nations ranked Denmark first, with an HDI of 0.91. Their designation of Very High Human Development encompasses countries with indices greater than 0.793. High Human Development countries have indices from 0.698 to 0.783, Medium Human Development countries have indices from 0.522 to 0.698, and Low Human Development countries have indices less than 0.52. About two-thirds of the countries in table 20–3 are in the Low Human Development classification, and most of the rest are in the Medium Human Development category. Only Sri Lanka ranks in the high human development category.

Per capita measures only tell us an average. In a society of 10 people with a GDP of 100,000, the per capita is 10,000. However, numerous income distributions could result in the same average. For example, a situation where the sultan has 91,000 and each of the other nine members has 1,000 gives the same GDP per capita as a situation where each member has an income of 10,000. In the former case, we would have many low-income people who might still consume biomass, while in the latter case, most may have transitioned out of traditional fuels. There are two popular available measures of income inequality in table 20–3: the Gini coefficient or index and the share of the population below the poverty level.

For the Gini coefficient, the population is listed from lowest to highest income and a Lorenz curve is then created as follows. Suppose we have a population of n with an ordered sequence of incomes $\{y_1, \ldots, y_n\}$ with $y_{i+1} \geq y_i$. Then the proportion of the population with incomes less than or equal to y_i is i/n. The Lorenz curve represents the proportion of income accruing to each i/n percent of the population and is measured by the following:

$$S_i = \left. \sum_{j=1}^{i} y_j \middle/ \sum_{j=1}^{n} y_j \right.$$

The domain values of the Lorenz curve (i/n) range from 0 to 1. Figure 20–3 shows three sample Lorenz curves. In panel (a), income is equally distributed and the bottom 20% of the population gets 20% of the income, the bottom 40% gets 40% of the income, and so on. Panel (b) contains the opposite extreme, with only one person receiving all the income. Panel (c) represents an in-between case that approximates the income distribution before taxes and transfers for the United States in 2000.

The Gini coefficient is a measure of inequality that compares an actual curve, panel (c), to the perfect equality curve in panel (a), demonstrated in figure 20–4. The distance from perfect equality is the area A, perfect equality is measured as the area A + B. How far a country is from perfect equality is equal to A/(A+B). For panel (a) in figure 20–3, A = 0, B = 0.5, and the Gini coefficient is A/(A + B) = 0/

(0 + 0.5) = 0. For panel (b), A = 0.5, B = 0, and the Gini coefficient is 0.5/(0.5 + 0) = 1. Thus, the coefficient is between 0 and 1, or if reported as percentages, from 0 to 100 as in table 20–3. Smaller coefficients represent more income equality, and higher coefficients represent more income inequality. For the US case in panel (c), the Gini coefficient in percent is around 46.

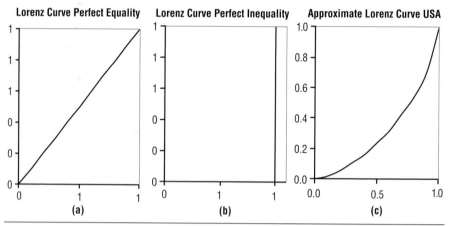

Fig. 20–3. Sample Lorenz curves

Note: Horizontal axes are the cumulative share of population. Vertical axes are the cumulative share of income.

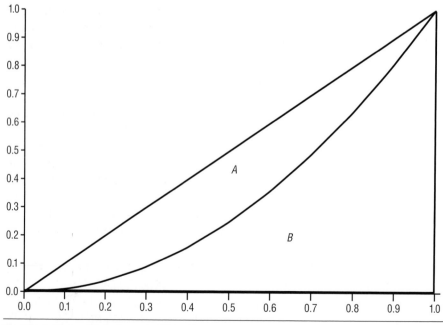

Fig. 20–4. Gini coefficient equals $A/(A + B)$

Using average data from 2001–2011, if available, or else the most recent data available before that, more than 85% of the 34 current OECD countries have Gini coefficients after taxes and transfers between 24 and 38.5. (A Gini coefficient of 24 is the equivalent of the top 10% having 27% of the income, with the other nine deciles sharing the remainder of the income equally. A Gini coefficient of 38.5 is the equivalent of the top 10% having about 43% of the income, with the other nine deciles sharing the remainder of the income equally.) The lowest Gini coefficients are in egalitarian Scandinavian countries. All OECD Western European countries, except Turkey, fall below the 38.5 cutoff. The only OECD countries higher than the cutoff in order of increasing inequality are Israel (39.2), the United States (40.8), Turkey (41), Mexico (48.9), and Chile (53.5) (World Bank, n.d.).

As income inequality often decreases with increasing income, it is not surprising that table 20–3 shows that almost three-fourths of the countries with available statistics have Gini coefficients above the 38.5% cutoff. Five countries have Gini indices less than the 38.5% cutoff (Eritrea, Myanmar, Sri Lanka, Sudan, and Tanzania). Three of these countries are quite poor with income per capita below $1700 or not reported—Eritrea, Myanmar, and Tanzania. Sudan has a bit higher GDP per capita but more than 45% of its population is below the poverty level.

Of the four fossil-rich countries with more than 25% of their GDP from fossil fuels (Angola, Republic of the Congo, Gabon, and Nigeria), all have Gini coefficients above most European countries. Of these, Angola has the best Gini coefficient at 38.6 but still has more than one-third of its population below the poverty line. The richest of the five, Gabon, is slightly worse with a Gini coefficient at 42.5, but even there, almost one-third of the population is below their national poverty line, suggesting that the oil wealth may be trickling up, not down. The situation appears even worse in the other two countries with more than 45% of their population below the poverty line.

Even before the Magna Carta in 1215, there were challenges to ruling authorities and demands for better government. Over the centuries, democratic governments evolved, and in more developed countries, there is rule of law to prevent corruption and to protect basic rights. However, such institutions are often poorly developed in low-income countries. Corruption and other abuses may be a part of everyday life in many of these countries and may prevent wealth from trickling down to the poor and weaker members of their societies (Bardhan 1997). Transparency International defines corruption as "the abuse of entrusted power for private gain" and provides an index of corruption called the Corruption Perceptions Index. I refer to their index as TIC to avoid confusing it with economists' abbreviation for the consumer price index.

Corruption can be in the public sector, such as government officials demanding bribes or siphoning off rents from public resources, as well as in private organizations, such as corporations. Since corruption is often illegal, it can be hard to quantify. However, Transparency International collects information from surveys and expert opinions to devise an index from 0 to 100 by country. The closer to 100, the less corrupt the country; the closer to 0, the more corrupt. The latest indices for 177 countries, as well as historical values, along with their 13 data sources, are available online (Transparency International 2013).

Although no country is squeaky clean, some come closer than others. Denmark and New Zealand come in at the top, with TIC = 91. The top 20 least-corrupt countries have indices above 72, and three-fourths of them are OECD countries. Northern European countries, including Scandinavia, tend to dominate in the top 20 and be less corrupt than Southern and Eastern Europe. Australia, Japan, and OECD countries in North and South America, except for Mexico, are also in the top 20. Non-OECD countries in the top 20 are Singapore, Hong Kong, Barbados, and Uruguay. Most OECD countries rank above the median index of 38, except for Mexico, which makes the worst showing at 34. Considering the high biomass countries in table 20–3, most are at or below the median TIC except for Ghana. Of the resource-rich countries, Gabon does the best at TIC = 34, with a rank of 106, with the rest considerably lower.

Thus, these high biomass countries have a host of problems, including poverty, declining forest cover, income inequality, poor human development, and weak governments that contribute to their low scores. Even the fossil-rich countries do not make a particularly good showing, and we will come back to such countries and their problems again in the next chapter.

Since this book is about wise management of energy resources, in the coming section we will consider models that relate to the use of biomass, both noncommercial and commercial.

Collecting Wood from the Commons

For landless households with no animals in developing countries, biomass is often collected from areas with public access. These areas may have become publically accessible through tradition or through government ownership with access allowed. Let's contrast such areas, often called *commons*, with privately owned property. Privately owned property has four features: *exclusivity* (others can be restricted from using it); *use privilege* (the owner receives all the income or benefit from it and is responsible for all the costs incurred); *controllability* (the owner can manage, use, and improve the property); and *transferability* (the owner can transfer the property to others for an agreed-upon consideration). Recall from chapter 12 the concept of a pure public good and associated market failure. Such goods, which are nonexcludable and nonrivalrous, do not satisfy this definition of private property.

Common goods share the trait of nonexcludability with public goods, but they are rivalrous. If I chop a tree down and remove it, the tree is no longer available for you to chop down. If my cattle eat the grass, you cannot harvest it to make biofuel. Now let's compare behavior regarding wood-growing land that is private with similar land that is a common area with no restrictions on access. First, take a privately owned plot where wood is growing. Suppose you have a fixed plot of land from which you will be gathering biomass to sell in a competitive market. The price of biomass harvested on your land is P, and the production function for biomass is $f(K,L)$. K represents your holding of land that is fixed, and L is the amount of labor applied to the land. P_L is the price of labor, or if you are self-employed, it is the opportunity cost of labor.

The profits or net benefits (π) from the plot of land are the following:

$$\pi = PF(K,L) - P_L L$$

Your challenge is to decide how much labor to apply to the land. Optimizing with respect to L yields the following:

$$\pi_L = PF_L - P_L = 0 \rightarrow PF_L = P_L \tag{20-1}$$

The expression PF_L is the price of an additional unit of biomass (P) times the marginal product of labor, F_L. Thus, it is the amount of biomass collected by unit L. The amount of biomass collected multiplied by the price of biomass results in the amount of cash earned for unit L. You will recall from chapter 16 that economists call this expression the marginal revenue product. The math tells you to apply labor up to the point the marginal revenue product equals the labor price.

Next let's consider figure 20–5 to illustrate what is going on. Note P, the price of output, is fixed in the market. The marginal revenue product (PF_L) in this example increases up to L_2 and then decreases. From 0 to L_3, the marginal revenue product is above the average revenue product (PF/L), pulling it up. After L_3, the marginal revenue product is below the average, pulling it down. Our first-order conditions above indicate you will optimize where $PF_L = P_L$.

Notice there are two places where PF_L crosses P_L, at L_1 and L_2. So let's check our second-order conditions at each candidate as follows:

$$\pi_{LL} = PF_{LL} < 0 \rightarrow PF_{LL} < 0 \rightarrow < 0$$

The second-order condition requires that the marginal revenue product slopes downward. Since P is positive, it also requires that marginal product slopes downward, or we have diminishing marginal product. At L_1, we have increasing marginal product, so it is not a maximum. Before L_1, marginal revenue product is below the price of labor, and we lose money on each hour of work. After L_1, however, the marginal revenue product is above the price of labor, and we start making money as we add labor. Our other candidate for a maximum is L_{Pv}. At this point, marginal revenue product is sloping downward. It's easy to see why L_{Pv} is a maximum. Before L_{Pv}, PF_L is above the wage, and adding labor adds to profits. After L_{Pv}, PF_L is below the wage, and adding labor reduces profits.

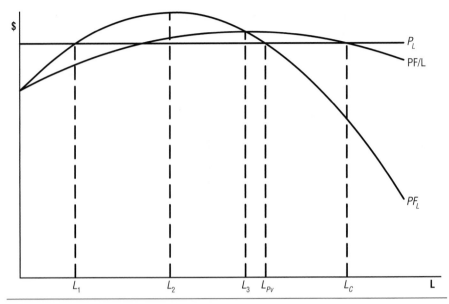

Fig. 20–5. Allocating labor on private (L_{Pv}) and common property (L_C)

Now let's compare the private property case to the commons case. In a commons, workers can freely enter, and each additional worker gets the average product. Their effect on other worker's output is an externality, and if each worker optimizes independently, these externalities will be ignored. Thus, each worker will receive the average product, and their optimum will be at L_C. This higher level suggests a commons will be overexploited compared to the private property solution. As long as the wage does not cross where average and marginal products cross, the private and common solutions will differ.

So let's compare the two solutions from society's point of view. First, take the case in figure 20–5 as reproduced in figure 20–6 panel (a). After L_{Pv}, the cost of each additional hour of labor is higher than the marginal revenue product and reduces society's welfare. Thus, the shaded gray area is the social loss from the commons compared to private property.

So does common property always get overexploited? Take the case in panel (b) where marginal and average costs are the same no matter how many labor hours are spent gathering, and L_{max} is the maximum amount of labor available. In this case, an additional worker does not affect the output of the other workers. The biomass is so abundant relative to the population that scarcity is no longer an issue.

The static solution in panel (a) may also understate the long-term problem. If the overexploitation of a biological resource exceeds its regenerative capacity, the resource may degrade, lowering its productive capacity over time and adding even larger losses in the future. For example, cutting down open access forests to make charcoal for nearby city dwellers may lead to deforestation, erosion, and even desertification.

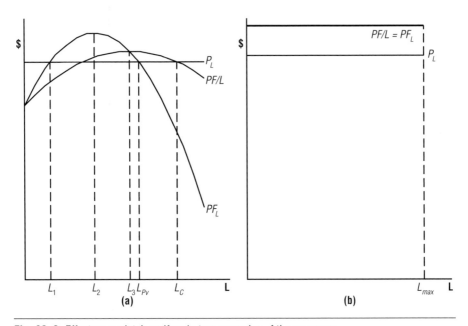

Fig. 20–6. Effect on society's welfare in two examples of the commons

In the case of scarce common resources, the solutions, of course, are some sort of restricted access. Possibilities include private property or governmental control of the resource that somehow restricts access. For example, permits could be required to use the resource. Governments could auction off the rights to develop a commonly owned resource as in oil or timber leasing on public land. Where governments are well functioning, this solution often works reasonably well. However, where governments are dysfunctional, this may be a recipe for failure. Government officials may require bribes for access and then appropriate rents for themselves. Even if access is theoretically allocated in an optimal way, a dysfunctional government may not be able to enforce the access restrictions.

A third alternative is to give rights to a group that manages their use and restricts others. Such a group may be a local community that lives near a forest, with the responsibility to develop allocation rules that maximize the long-term social value of the resource. Such a group may not only place an economic value on exploiting the resource but may also consider multiuse value. Additional value may occur from aesthetic and recreational opportunities and biodiversity, bringing tourism and other economic opportunities.

Elinor Ostrom, the first woman to receive the Nobel prize in economics, and others have developed norms that improve the chances of success for such group decision making and management of common pool resources (CPR). These norms include the following:

- Access and exclusion are clearly defined.
- The rules are adapted to local culture and conditions.
- The rules allow a role in decision making to most users with access to the resource.
- There is efficient monitoring either by or on behalf of those with access rights.
- There are punishments for rule violations that increase with the seriousness and frequency of violations.
- Communication and conflict resolution are facilitated.
- Higher authorities recognize and sanction the group.
- There is an appropriate nesting of authority for larger common pool resources, but groups are kept small at the local level, if possible.
- There is trust and reciprocity within the group.

See Poteete and Ostrom (2010) for more discussion and references to the interesting ideas and case studies used to develop these norms.

Other examples of commons include the early law of capture in the United States, where those with the property rights at the surface could keep as much oil as they extracted, despite the fact that the oil was a common pool extending under the property of many surface holders. States stepped in with regulations on well spacing and other rules to reduce those withdrawals. Water, considered in the next section, is also often a common property.

Energy and Water

Water, both at the surface and underground, is usually not privately owned with restricted access, which may have implications for future energy scenarios. Water and energy are inextricably intertwined. Energy is needed to collect, move, and remove impurities from water that might conflict with end-use needs. Such impurities include biological organisms, organic runoff from human activities and agriculture, chemical wastes from human activity, and naturally occurring minerals, including salts.

Energy production also requires water. Biomass production, including that used for biofuels, is thought to consume more fresh water than any other activity on earth. Biofuels are also water intensive if produced on irrigated land. Such water is considered consumed if it is not returned to surface collections (rivers and lakes) or aquifers.

Currently about 15% of water withdrawals are for energy production. Cooling for nuclear and fossil-based power plants are the largest users. However, such water is considered nonconsumptive, as it is typically returned to its source. However, its release is at a higher temperature, which can change local ecology. More water is also needed for unconventional shale gas and oil than for their conventional cousins. For more on the water/energy interdependence, see IEA (2012b, ch. 17).

Hydropower, high up on the energy ladder, is a sustainable option. Worldwide, hydropower accounted for about 16% of electricity generation in 2011. Of the regions still moving up the development ladder, non-OECD Americas generate the highest share of their power from hydroelectricity (around two-thirds). Brazil leads the way with about 80% of its power from hydroelectricity and more than 45% of region's kilowatt hours. Paraguay, in our high biomass user group, produces essentially all its electricity from hydropower. Other countries within the region that generate more than one-half of their electricity from hydropower include Colombia, Costa Rica, Panama, Peru, Uruguay, and Venezuela (IEA, n.d.d).

Although South Africa is a heavy coal user for electricity generation and generated more than 35% of Africa's power in 2011, if we omit South Africa, the rest of the continent generates about one-fourth of its power from hydroelectricity. Mozambique is the largest producer followed by Egypt. Together they account for about one-quarter of African hydro production. The countries in Africa that get more than one-half of their power from hydroelectricity include Angola, Cameroon, Democratic Republic of the Congo, Republic of the Congo, Ethiopia, Ghana, Mozambique, Namibia, Togo, Zambia, and Zimbabwe. All of these countries but Namibia also get more than one-half of their energy from biofuels (IEA, n.d.d; World Bank, n.d.).

Just as Brazil dominates the hydropower scene in South America, China dominates in developing Asia by producing more than 80% of the region's total hydropower in 2011. Still, hydropower is only about 15% of Chinese generation, with the remainder largely coal. Three countries, Nepal, Myanmar, and North Korea, get more than one-half of their power from hydroelectricity. The first two of these are also high biomass users (IEA, n.d.d).

None of the countries in table 20–3 generates any power from nuclear, and the above brief survey suggests that high biomass consumption share is often accompanied by low fossil-fuel power generation.

Hydropower has advantages, disadvantages, and potential for expansion. It has no fuel costs and can easily be turned on and off in a system with intermittent wind or solar power. It also has potential for expansion. WEC (2013) indicates we could likely add more than 2.5 times our current capacity. Although the large sites in the OECD have been developed, there is still potential for small-scale sites. The potential for large-scale projects still exists in Latin America, Africa, and China.

Although it is renewable, hydropower is not without its problems. It does not consume water, putting it back into the system, but it does change the timing and location of the water. This reallocation can change the ecosystem. New dams may displace humans and reduce plant and animal habitat, and thus biodiversity. The filling of new dams also inundates biomass, producing methane, a potent greenhouse gas. Sediments that previously improved soil fertility downstream may be trapped or may contain harmful components that become concentrated. A failure of a dam could release these toxins, as well as imposing human casualties and collateral property damage. Dams can fragment river transportation systems and interfere with fish migration.

Renewable Energy Policies

We know that at some point we will need to transition from fossil fuels. It could be sooner if fears of climate change finally cause governments to respond in a responsible way. It could be later if we wait for depletion to force an economic solution upon us. Right now it appears that renewable sources, which once dominated our energy systems, will be making a comeback and governments will play a role. Johansson et al. (2012, 86) summarize a number of policies related to renewable energies under seven categories:

- *Regulatory.* Renewable targets.
- *Access-related policies.* Net metering, priority access to a network, and priority dispatch.
- *Quota-driven policies.* Obligations, mandates, and renewable portfolio standards (RPS), tendering/bidding renewable quotas, and tradable renewable certificates.
- *Price-driven policies.* Feed-in tariffs (FITs) and premium payments.
- *Quality-driven policies.* Green energy purchasing and green labeling.
- *Fiscal policies.* Accelerated depreciation, investments, grants, subsidies, rebates, renewable energy conversion payments, and investment tax credits.
- *Other public policies.* Research and development (R&D), public procurement, information dissemination, and capacity building.

Policy analysis typically requires analytical support, and we will consider some models in the next section relating to wood production.

Optimal Timber Rotation

In areas where biofuels have become scarcer and prices have risen, there has often been a market response. Consumers have responded by consuming less, and markets have responded by producing more. Nevertheless, most of the lower income countries in table 20–3 have lost forest cover since 1990. However, with proper care and management, forests can be sustainable. For example, Finland and Sweden both have significant forestry industries. Sweden gets about 0.5% of its GDP from forestry rents, while Finland gets slightly more (World Bank, n.d.). Both countries have increased their production of pulpwood, round wood, and fuel wood, while at the same time increasing forest cover (UNFAO, n.d.).

To develop such sustainable operations requires that we understand the growth rate of the species under cultivation and do not harvest more than can be replaced or recultivated. In this section, we will consider such sustainable consumption in a privately managed forest. Suppose the volume of biomass yield from our forest is $V(t)$; how much the yield changes in year t is $V'(t)$, the time-derivative $(\frac{\partial V}{\partial t})$.

A typical S-shaped yield curve for a long-life tree is shown by the solid black curved line in figure 20–7. When young, the tree volume increases at an increasing rate, ($V'(t) > 0$ and $V''(t) > 0$); as it gets more mature, its volume still increases, but at a decreasing rate ($V'(t) > 0$ and $V''(t) < 0$). Eventually the tree reaches its maximum volume, and with increasing vulnerability to disease and pests, it then starts to rot and decay ($V'(t) < 0$).

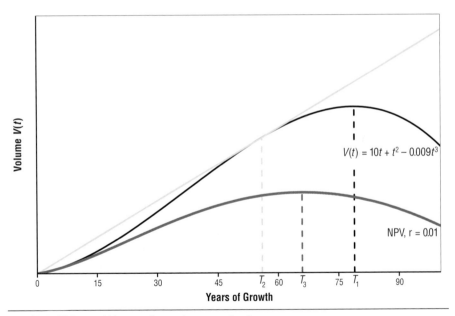

Fig. 20–7. Volume of biomass from a long-growing tree

Our challenge as forest owners is to pick the optimum time to harvest our trees, referred to as the *optimal rotation period*. Assume all trees in our forest are identical, and we will clear cut and harvest them at the same time (T). As we hate wood ticks and are allergic to pine needles, we get no other satisfaction from our forest than to sell the timber. Our choice of T initially depends solely on biological criteria relating to the volume of wood grown. If we wanted to maximize the volume of wood received from our forest, we would harvest at T_1 in figure 20–7. However, since the forest grows faster when it is young, the sooner we replant, the faster the forest will grow on average. Thus, we might prefer to maximize the average amount of volume we get every year. This average volume or yield, $V(t)/t$, is often called the *mean average increment* (MAI) or *average sustainable yield* (ASY). Maximizing this value is a bit more complicated, but calculus comes to our rescue:

$$\frac{\partial\left(\frac{V(t)}{t}\right)}{\partial t} = \frac{tV'(t) - V(t)}{t^2} = 0 \ \rightarrow \ tV'(t) - V(t) = 0 \ \rightarrow \ V'(t) = \frac{V(t)}{t} \quad \textbf{(20–2)}$$

The last expression in equation (20–2) indicates that we should choose the rotation period T to be when the average volume of our forest equals the marginal change in volume. This amount T_2, is where the slope of a line from the origin (the slope of line V/t) is equal to the slope of the volume curve ($V'(t)$) as shown in figure 20–7. The volume at this point is called the *maximum sustainable yield* (MSY). Since this maximizes the average volume of biomass produced, it also means it would sequester the maximum average amount of carbon removed every year by our standing forest.

However, suppose as the owners of the forest with all those ticks and pine needles, all we really care about is the money earned from the forest. Thus, we care about the price of the timber when it is sold (P), the discount rate (r), and any costs incurred along the way for planting (c_1), thinning (c_2), and harvesting (c_3). Now let's consider when we should harvest, if our economic criteria is to maximize the discounted present value of the timber for one rotation of T years. Assume there are no planting and thinning costs. You will sell the timber for P per cubic meter (m^3) of biomass at time T, and it costs c_3 to harvest per cubic meter. Since timber grows continuously, we will want to use continuous discounting and T to maximize as follows:

$$NPV = (P - c_3)V(t)e^{-rT}$$

The solid gray curved line in figure 20–7 shows this NPV for $P = 1$, $r = 0.01$, and $c_3 = 0$. The maximum NPV is at T_3. Notice that when we discount the future, even by a small amount, the optimal rotation becomes shorter. Now let's develop the general principal to apply to arrive at this maximum.

Our first-order conditions tell us the following:

$$\frac{\partial NPV}{\partial t} = \frac{\partial[(P - c_3)V(t)e^{-rt}]}{\partial t} = (P - c_3)V'(t)e^{-rt} + (P - c_3)V(t)(-r)e^{-rt} = 0$$

We can rearrange the above expression as follows:

$$(P - c_3)e^{-rt}\,[V'(t) - V(t)r] = 0$$

$P - c_3$ must be positive, since we must more than cover our harvesting costs to be in business. So we need to concentrate on the interpretation of the expression in the square brackets, which must be zero for the first-order condition to hold. The last expression requires that $\frac{V'(t)}{V}$, which is the change in volume over the volume or the growth rate of volume, should be equal to the interest rate (r). The conditions suggest that if our trees grow slower than the interest rate, we should harvest and put the money in the bank at r percent. If our trees grow faster, they are yielding a better return and we should wait to cut them. Notice for the single rotation case, the selling price and harvest cost, which do not change over time, do not enter into our decision, provided the price is higher than the harvest cost.

The second-order conditions are the following:

$$\frac{\partial^2 NPV}{\partial t^2} = (P - c_3)\frac{\partial[V'(t)e^{-rt} - V(t)re^{-rt}]}{\partial t} < 0$$

Taking the derivative and rearranging yields the following:

$$= (P - c_3)[V''(t)e^{-rt} + V'(t)(-r)e^{-rt} - V'(t)re^{-rt} - V(t)r(-r)e^{-rt}] < 0$$
$$= (P - c_3)e^{-rt}[V''(t) - V'(t)r - V'(t)r + V(t)r(r)^t] < 0$$
$$= (P - c_3)e^{-rt}[V''(t) - V'(t)r - r\{V'(t) - V(t)r\}] < 0$$

From the first-order condition, the expression in curly brackets is zero, and we are left with the following:

$$\frac{\partial^2 NPV}{\partial t^2} = (P - c_3)e^{-rt}[V''(t) - V'(t)r] < 0$$

Since $(P - c_3)$ and the discount factor e^{-rt} are positive, the second-order condition requires that the change in the annual increment of timber volume be smaller than the incremental timber volume at time t multiplied by the rate of interest earned if the timber were harvested. Thus, timber cut is more valuable than timber standing.

Since our forest is renewable, we could replant with each harvest. If this went on forever, (the infinite rotation problem), our net present value would then be as follows:

$$NPV = (P - c_3)V(T)e^{-rT} + (P - c_3)V(T)e^{-2rT} + (P - c_3)V(T)e^{-3rT}... = \sum_{i=1}^{\infty}(P - c_3)V(T)e^{-irT}$$

Although the above infinite series looks quite daunting, some clever mathematicians have worked out a simpler rendition of the formula:

$$NPV = \frac{(P - c_3)V(T)}{e^{rT} - 1}$$

We can pick the optimal rotation T as follows:

$$\frac{\partial NPV}{\partial T} = \frac{\partial\left(\frac{(P-c_3)V(T)}{e^{rT}-1}\right)}{\partial t} = \frac{(e^{rT}-1)(P-c_3)V'(T)-(P-c_3)V(T)re^{rT}}{(e^{rT}-1)^2} = 0 \quad \text{(20–3)}$$

$$(e^{-rt}-1)(P-c_3)V'(T)-(P-c_3)V(T)re^{rT} = (P-c_3)[(e^{rT}-1)V'(T)-V(T)re^{rT}] = 0 \quad \text{(20–4)}$$

$$(P-c_3)V'(T) = \frac{(P-c_3)V(T)re^{rT}}{(e^{rT}-1)} \quad \text{(20–5)}$$

The expression on the left side of (20–5) is the value of the gain in board feet of delaying harvest by one period at T. The right-hand side is the entire future value of interest gained on the board feet if harvested at T. Thus, the right-hand side is the financial opportunity cost of waiting one period, which is weighed against the gain in value if we wait one period to harvest.

Alternatively, we can rearrange equation (20–3) as follows:

$$(P-c_3)[(e^{rT}-1)V'(T)-V(T)re^{rT}] = 0 \;\rightarrow\; \frac{V'(T)}{V(T)} = r\frac{e^{rT}}{[e^{rT}-1]}$$

Now instead of setting the growth rate of trees equal to the interest rate, we set it to a number larger than the interest rate. Since $V(T)$ is positive and $V'(T)$ is positive but increasing at a decreasing rate ($V'' < 0$), this indicates that we harvest our timber sooner than in the single rotation case. The sooner we harvest, the sooner we get to the new stands, which grow faster.

The optimal rotation is dependent upon the yield curve. Hard woods typically take longer to grow than soft woods. Understanding the yield curve of various species as well as developing faster growing pest-resistant species that require less water can help move us closer to sustainable biomass use.

Summary

As humans reach higher economic levels, they are likely to move up the development ladder from biomass to more efficient and convenient sources of energy, such as fossil fuels. They are also likely to gain access to electricity. This climb may not be smooth, and consumers may energy stack, consuming multiple fuels along the way. Past transitions into fossil fuels have taken many decades. The transition out may also take time as technologies need to develop and infrastructure needs to be built or rebuilt. However, the fast transition out of biomass in China illustrates how swiftly the transition might occur when alternative fuels are available.

Traditional biomass is currently consumed by more than 2 billion people worldwide for cooking as well as heat and light. Biomass usage largely occurs in Latin America, developing Asia, and especially Africa. The countries with the heaviest use by share are those in the low-income bracket, with some heavy users in the lower middle-income brackets. In such places, the residential sector is typically the largest user, there is limited access to electricity, and fossil fuels are too expensive for electricity generation. Such countries also typically have high income inequality as measured by the Gini coefficient. They also have a large part of the population below the poverty line, a larger share of the population in the rural areas, a relatively larger share of forest cover, and higher levels of corruption.

Biomass use may be sustainable in many places. However, the drudgery of collecting the fuel and stoking the fire, and the harmful effects of breathing the emissions, suggest we would rather not sustain such practices. Further, in many cases these burdens fall disproportionately on women and children, who are mired in poverty. Although collection of biomass causes less biostress than once thought, it still has an impact. This impact is noticeable where wood is used for charcoal near urban areas, biomass growth is particularly slow relative to the population, or drought conditions prevail. It may cause deforestation, climate change, erosion, and reduction in biodiversity.

Modern biofuels in the form of ethanol and biodiesel, along with industrial and municipal waste and biogas, are typically policy-driven luxuries of the wealthier nations. However, China has had an active biogas program for many years. Current biofuels for transport are largely based on food crops. Current concerns about raising food prices or diverting land areas rich in biodiversity into monoculture crop areas have led to some backtracking on biofuels policy in Europe. There has been much research to develop competitive biofuels from nonfood sources. Models have been designed that could help determine the optimal harvesting rate for continuous cropping products, as well as optimal crop rotation to sustain soil fertility.

Models can help us to understand issues related to biomass collection and to growing long-life biomass crops. Traditional biomass in many cases is from noncommercial fuel sources. It is not bought and sold in the market but is collected from animal refuse, agricultural waste, fallen tree limbs in the forest, and shrubby growth. These wastes may be collected from areas that by tradition are commons open to the public. In such a case, property rights are not well-defined. The common area is nonexcludable, but it is likely to be rivalrous unless it has such a high abundance of biomass that all residents can collect their fill, with biomass to spare. Models suggest that in such a situation, collectors will collect up to the point that the average revenue from collecting is equal to the wage or opportunity cost of labor. Thus, collectors do not consider their effects on the collection of others. However, the socially optimal collection rate is the lower rate at which the marginal revenue of collecting is equal to the wage rate. We could expect such a rate to prevail if the collection area were privately owned and excludable.

Solutions for suboptimal use of scarce common resources include excludability by making the area private or having government control of the area, with restricted access. Another possibility is some sort of group management that can design restrictions on use and develop methods for monitoring, punishment for violations, and mechanisms for conflict resolution. Local accountability and decision making, cultural compatibility, good communication, and trust and reciprocity, along with the support of higher authorities, improve the chances that such groups will manage common pool resources efficiently.

Water is an additional common pool resource that needs energy to be produced. Energy production is also a heavy user of water. Biomass uses water to grow; power plants use it for cooling; biofuels need water in their manufacture. Water is used for fracing and pressure maintenance, and it is also produced with oil and requires disposal. Our hydropower is reliant on the hydrological cycle for continuous replenishment. For all these needs, water must be managed.

Hydropower is renewable and has potential for expansion, especially in the developing world. It can be a means of helping people to the top rung of the energy ladder. Once established, it does not produce greenhouse gases; it is flexible and fits well with other intermittent renewable sources. Best of all, the fuel is free. Stochastic models can be designed to analyze optimal power systems using hydropower as backup. The water, however, depends on the hydrological cycle. Thus, droughts can cause shortages and run-ups in power prices, as happened in Brazil in 2001 and as is currently threatening. If water is a scarce resource, dynamic models can also be designed to optimally manage reservoirs. If I produce electricity with water today, and the reservoir does not fill fast enough, I may not have the water tomorrow.

Problems with hydropower that need to be managed include changing ecology, displacing human population, flora, and fauna, and methane emissions from inundated rotting biomass. In addition, there are sediment issues and the potential for loss of life and property if the dam fails.

We know the transition out of fossil fuels is inevitable. The question is not if, but when. Fossil fuels are still relatively abundant, and without intervention, are likely to be with us for decades. Peak oil, peak unconventional oil, peak gas, and peak coal, among others, are yet to come. It will not be easy to replace these relatively cheap and concentrated sources of energy. However, policies in a carbon-constrained world may bring those peaks closer to the present. Policies to accomplish the increasing use of renewable energy resources include renewable targets, feed-in tariffs, labeling, subsidies, tax incentives, technology development and dissemination, and information programs.

If biomass is to continue to play a strong and sustainable role in our energy future, it is likely that biomass will need to be grown commercially in well-managed energy systems. For long-growing species, such as trees, models can help us to design optimal rotation systems. Typically, the growth pattern of a tree follows an S-shape, growing faster when young, slower as it ages, and eventually dying. Biologists might target maximizing sustainable yield. With an S-shaped growth curve, this requires that harvest occurs when average yield and marginal yield are equal. Economists would want to maximize the net present value of the timber.

With a single rotation, we harvest when the growth rate of the tree equals the interest rate. Thus, the economist takes into account the opportunity cost of tying up money in the tree and harvests the tree sooner. With an infinite rotation, the economist notes that the sooner we cut the tree, the sooner we replant faster growing young trees, and the sooner we will harvest. With a higher discount rate, we discount the future more and would want to harvest sooner. If prices increase in the future, the trees are more valuable later, and we would harvest later. If harvest costs go up later, the value of the tree would be lower later, and we would harvest sooner.

The timber rotation model could be complicated in numerous ways. We could add investment or research into new species. Other costs and benefits could be introduced. If having a larger standing forest had recreational or carbon sequestration value, the optimal rotation would be longer. If we value biodiversity, we might want to avoid clear cutting, with some sort of staggered plan to our harvest. However, often such forest benefits are externalities, and we might need a way to reimburse a tick-hating owner to consider those benefits in choosing the timing of their harvests.

21

Sustainable Wealth in Fossil Fuel–Rich Developing Countries

It took us 125 years to use the first trillion barrels of oil. We'll use the next trillion in 30.

—Dave O'Reilly, former CEO, Chevron

Introduction

Some developing countries have been especially blessed with fossil fuels and are highly dependent on them. For them, sustainability is likely to entail how to maintain their income flow after it is no longer profitable to produce their resources. Or at some point, they may need to retain the resources for local consumption. In this chapter, I consider those countries, shown in table 21–1, which average more than 14% of their GDP from fossil fuels from 2008–2012. I go down to 14% to include Norway. It is the only OECD country with this level of dependence on natural resource rent and is a role model of how resource rents can be managed. I will hereafter refer to these 31 fossil fuel–rich countries as *FR countries*.

I will consider specific issues facing these 31 FR countries and some models to help us think about how to sustain the rents from their resources for future generations. One commonality is that all of these countries receive some rents from oil. It is the dominant source of fossil-fuel income for 26 countries in the table. All of the OPEC countries are included in the table. Natural gas dominates in four other countries: Bolivia, Qatar, Trinidad and Tobago, and Uzbekistan.

Coal is dominant only in Mongolia. Although I find recorded production for coal in Mongolia as far back as 1957, sustainable exports of coal only took off in 2005. Such exports increased an astonishing tenfold from 2005 to 2011 (US EIA, n.d.d), with coal rent share of GDP increasing from 4% to more than 20% (World Bank, n.d.). However, this GDP share backed off to 14% in 2012 with the reduction in world coal prices.

Table 21–1. Countries averaging more than 14% of GDP from fossil-fuel rents from 2008–2012

Country	GDP per capita	Pop in 10^6	Fossil % GDP	Coal % GDP	Oil % GDP	Ngas % GDP	Dom Prod First	Export % Coal	Export % Oil	Export % Ngas
Year	2012	2012	08–12	08–12	08–12	08–12	year	2012	2012	2012
Algeria	13,470	38.5	27.9	0	18.1	9.8	1944	–	52.8	56.7
Angola	7,743	20.8	47.8	0.0	47.7	0.1	1958	–	93.6	–
Azerbaijan	16,748	9.3	47.2	0.0	42.1	5.1	<1900	–	87.7*	41.8*
Bahrain	42,858	1.3	25.8	0.0	19.0	6.8	1932	–	–	–
Bolivia	5,956	10.5	18.9	0.0	5.7	13.1	1955	–	–	79.7
Brunei	74,927	0.4	48.2	0.0	30.1	18.0	1913	–	92.5*	74.9
Chad	2,112	12.4	30.7	0.0	30.7	0.0	2003	–	100.0	–
Congo, Rep.	5,936	4.3	64.4	0.0	64.3	0.1	1973	–	41.1	–
Ecuador	10,436	15.5	20.0	0.0	19.9	0.1	1917	–	71.1	–
Egypt	11,263	80.7	14.7	0.0	8.8	5.9	1886		6.4*	12.1
Eq. Guinea	39,507	0.7	52.0	0.0	52.0	0.0	1997	–	100.0	68.6
Gabon	18,971	1.6	43.3	0.0	43.1	0.2	1958	–	42.1	–
Iran	16,298	76.4	32.7	0.0	25.8	6.9	1913	–	62.1	2.4
Iraq	15,313	32.6	46.8	0.0	46.2	0.6	1927	–	81.2	–
Kazakhstan	22,669	16.8	34.6	3.4	27.5	3.7	1963	24.6	84.4*	7.1
Kuwait	88,743	3.3	54.8	0.0	52.9	1.9	1946	–	78.6	–
Libya	19,242	6.2	22.0	0.0	20.0	2.0	1961	–	70.4	53.1
Mongolia	8,737	2.8	16.2	14.0	2.2	0.0	1957	65.6	100.0	–
Nigeria	5,734	168.8	29.7	0.0	27.1	2.6	1958	–	94.0	79.5
Norway	68,522	5.0	14.7	0.0	10.7	3.9	1971	3.3	79.3	93.9
Oman	46,905	3.3	44.8	0.0	37.0	7.8	1966	–	83.6	30.8
Qatar	142,578	2.1	31.4	0.0	15.0	16.4	1963	–	37.9	77.3
Russia	24,439	143.5	20.9	1.0	14.5	5.4	<1900	29.7	58.7	28.8
Saudi Arabia	53,539	28.3	49.8	0.0	46.6	3.2	1936	–	76.9	–
Sudan	3,552	37.2	17.9	0.0	17.9	0.0	1992	–	80.0*	–
Trinidad & Tobago	30,660	1.3	42.5	0.0	11.6	30.9	1955	–	–	44.9
Turkmenistan	13,134	5.2	37.6	0.0	22.2	15.4	1958	–	19.8*	65.2
UAE	60,131	9.2	24.1	0.0	21.0	3.1	1962	–	94.8	–
Uzbekistan	4,959	29.8	31.5	0.1	5.6	25.7	1962	–	–	16.2
Venezuela	18,597	30.0	26.5	0.1	24.8	1.6	1917	90.4	75.0	–
Yemen	4,213	23.9	23.2	0.0	20.8	2.4	1986	–	62.6*	87.3
World	14,300	7,043.9	4.1	0.5	3.0	0.7	<1850	–	–	–

Sources: API (1971); US EIA (n.d.d); IEA (n.d.f); Mitchell (2007a); Mitchell (2007b), Mitchell (2008); OPEC (2013), World Bank (n.d.).

Note: GDP per capita is measured in real 2013 international dollars converted using purchasing power parity. For more information on using PPP indices, see dahl.mines.edu/st10/st10.pdf, question 34. % GDP indicates the percent of GDP that rents from the particular fuel provided. Rents are defined as the value of the fuel at world prices minus costs of production. Ngas = natural gas; Dom Prod First indicates the year of recorded sustainable production for the dominant fuel. Gray highlighted cells indicate the country's dominant fuel; – indicates country does not have recorded net exports of this fuel; * indicates data is for 2010; 08–12 indicates data are averaged from 2008 to 2012.

Although all these countries share in being fossil-fuel dependent, they vary widely in many ways. The Republic of Congo, Equatorial Guinea, and Kuwait are highly subject to the vagaries of the oil markets, with more than one-half of their recent GDPs coming from oil. Five other countries get 40%–50% of their GDP from fossil-fuel rents. The most diversified economies, with less than 20% dependence on their fossil fuel rents, are Bolivia, Egypt, Mongolia, Norway, and Sudan.

These countries vary in population size from tiny Brunei, with less than 1 million, to Nigeria and Russia, with more 100 million people each. Although median income for these FR countries is about $16,200 per capita (2013$ PPP), countries vary from the abject poverty of Chad (near $2,100 per capita) to the super wealth of Qatar (>$100,000 per capita). However, even Chad is not in the World Bank's low-income category (≤$1,062 per capita). Twelve of the countries are in the World Bank's middle-income category ($1,063 to $12,924 per capita) with the remaining 18 countries in the high-income category (>$12,924 per capita in 2013$). Recent Libyan income is not reported, but data as of 2009, before the civil war, suggest it was then in the high income category.

However, only seven countries have per capita incomes in 2012 greater than the median income of OECD countries of about $36,000 per capita. They are Bahrain, Brunei, Equatorial Guinea, Kuwait, Qatar, Saudi Arabia and the United Arab Emirates. All except Bahrain have large reserves relative to their population. All but Saudi Arabia have populations of less than 6 million each (see table 21–2, column 3.)

Some of the countries have recorded sustainable production of their dominant product a century or more ago, such as Azerbaijan, Brunei, Egypt, Iran, and Russia. Slightly more than one-half have first recorded sustainable production of their dominant fuel from 1955 to 1986. Only three countries, Chad, Equatorial Guinea, and Sudan, have first recorded sustainable production since 1990. All of these newcomers are in Africa. As with most of the rest of African producers, Chad and Sudan have income per capita well below the median of the FR countries. However, Equatorial Guinea is more like relatively rich Gabon, with a population of less than 2 million and relatively high oil reserves per capita.

The last three columns of the table indicate how dependent the economy is on the export market for their rents by showing the share of each fossil fuel that is exported. A dash indicates the country is not a net exporter. Most countries export close to or more than one-half of their dominant fossil fuel. For Chad and Equatorial Guinea, with no operating refining capacity in 2012, all of their crude was exported, while they imported all their products. Bahrain exports no primary fossil fuels and is a net importer of oil. Its oil and gas rents arise from oil and gas production, but their income also benefits from the production and exports of oil products and petrochemicals. A number of oil exporters are natural gas importers, including Iran, Kuwait, the United Arab Emirates, and Venezuela. Trinidad and Tobago export more than 40% of their natural gas production (all as LNG), but import around one-fourth of their oil. Mongolia exports more than one-half its coal on a tonnage basis. However, since it exports most of its higher quality coking and bituminous coal and burns its lignite at home, its exports are closer to 75% based on energy content.

Table 21–2. Reserves, energy consumption, and GDP value added by sector

Country	R/P (dom)	R/Pop (dom) (toe)	E (mtoe)	E/Pop (toe)	Subsidy	Value Added by Sector % Gdp Ag	% GDP Mfg	% GDP Serv
Year	2012	2012	2011	2011	2012	2012	2012	2012
Algeria	22	46	49	1.4	G,D	9.3	48.5	42.2
Angola	15	67	6	0.4	G,D	10.0	59.7	30.3
Azerbaijan	21	110	14	1.5	G,D	5.5	63.1	31.5
Bahrain	8	14	14	11.5	G,D	NA	NA	NA
Bolivia	15	25	6	0.6	NA	13.0	38.7	48.3
Brunei	21	390	4	9.3	G,D	0.7	71.1	28.2
Chad	39	18	0	0	No	55.8	12.7	31.5
Congo, Rep.	16	54	2	0.5	No	3.4*	76.6*	20.0*
Ecuador	39	68	14	0.9	G,D	9.9	36.9	53.3
Egypt	22	8	90	1.1	G,D	14.5	39.2	46.3
Eq. Guinea	10	218	2	3	G,D	NA	NA	NA
Gabon	23	179	1	0.6	No	3.9	62.2	33.9
Iran	122	289	235	3.0	G,D	NA	NA	NA
Iraq	131	642	40	1.3	NA	NA	NA	NA
Kazakhstan	54	261	65	3.8	D	4.7	39.5	55.8
Kuwait	108	4676	37	14.6	G,D	NA	NA	NA
Libya	94	1,119	12	2.1	G,D	NA	NA	NA
Mongolia	79*	465	2	0.8	No	17.1	32.9	50.0
Nigeria	40	32	18	0.1	G	22.4	26.7	50.9
Norway	9	155	45	9.8	No	1.2	42.0	56.8
Oman	16	243	22	7.4	G,D	NA	NA	NA
Qatar	161	11,430	35	19.1	G,D	NA	NA	NA
Russia	17	61	746	5.3	No	3.9	36.0	60.1
Saudi Arabia	74	1380	213	8.2	G,D	2.2	62.6	35.2
Sudan	122	20	8	0.2	G,D	27.7	31.2	41.1
Trinidad & Tobago	9	265	23	19.1	NA	0.6	57.4	42.0
Turkmenistan	8	17	23	4.7	G,D	14.5	48.4	37.0
UAE	96	1553	92	18.0	G,D	0.7	60.5	38.8
Uzbekistan	29	57	55	2.0	No	18.9	32.4	48.7
Venezuela	252	1,030	82	3.0	G,D	5.8#	52.2#	42.1#
World	55	32	12,993	1.9	–	3.1*	26.7*	70.2*
Yemen	49	18	7	0.3	G,D	7.7#	29.4#	62.9#

Sources: US EIA (n.d.d), World Bank (n.d.).
Note: R/P (dom) = reserves divided by annual production for the dominant fuel; R/Pop (dom) = reserves divided by population for the dominant fuel. All cells in column two (R/P) and three (R/Pop) are for oil except for highlighted cells. Highlighted cells are all for natural gas except for Mongolia, which is for coal. The reserves for the dominant fuel have been converted to toe assumming 5,800,000 Btu/barrel of oil and country heat rates from US EIA (n.d.d) for natural gas and coal. E (Mtoe) = total primary energy consumption in million tonnes of oil equivalent. E/Pop (toe) = energy consumption per capita in tonnes of oil equivalent. One toe = 39,683,000 Btu = 41,868,000 kJ. Ag = agriculture; Mfg = manufacturing industry; Serv = the service industry; NA = number is not available; Subsidy = the country subsidizes gasoline (G), diesel fuel (D), or neither (No).

NGLs are not included in the fossil rents, but about two-thirds of these countries produce some NGLs. Theses liquids may be stripped out of natural gas at natural gas plants, or they may be produced in refineries. Worldwide about 60% of NGLs come from gas plants and the rest from refineries (Keller 2012). Although both crude oil and NGLs vary depending on their mix of hydrocarbon contents, NGLs tend to have around 60% to 70% of the energy content of crude oil. If we value NGLs at 60% of the value of crude oil, it accounted for 2% or more of GDP for Algeria, Bahrain, Brunei, Equatorial Guinea, Libya, Qatar, Saudi Arabia, and Trinidad and Tobago in 2010. The majority of these countries export a sizable chunk of their NGLs. The most significant of these is Saudi Arabia, with around 19% of world production. It is the second largest producer of NGLs in the world after the United States. LNG might contribute up to 6% of Saudi GDP. It exports little directly but separates NGLs into naphtha and LPGs, and exports a big chunk of them, with some used domestically for petrochemicals and in the residential sector. Algeria and Qatar each produce between 3% and 4% of the world's NGLs. They export between one-third and two-thirds directly, but also separate out naphtha and LPGs and export them as well (US EIA, n.d.d; IEA, n.d.m).

As we have seen above, these countries have many differences, but all would like to sustain their fossil rents. In the next section, we will consider how abundant their fossil reserves are and how these countries use their fossil fuels domestically.

Fossil Future for FR Countries

R/P is one indication of how long their reserves might last. As noted in chapter 14, we do not know the accuracy of reserves estimates in most of these FR countries. With that caution, let's consider what the reserves estimates suggest.
In table 21–2, we see a wide variation in the R/P ratio. Private companies typically like an inventory of developed reserves of 10 to 15 years. With big discoveries, this number may go higher, but they are unlikely to search hard for new reserves that will not be sold for many decades. If companies are publicly listed on stock exchanges in industrial countries, the countries typically have specific regulations on what can be booked as proven reserves. In table 21–2, Angola, Bahrain, Bolivia, Republic of the Congo, Equatorial Guinea, Norway, Russia, Turkmenistan, and Trinidad and Tobago all have an R/P for their dominant fuel of less than 20 years. The dominant fuel in all these but Trinidad and Tobago is oil. All but Bahrain have a presence of privately owned national and multinational oil companies.

The choice of including private participation is influenced by a country's culture. For example, Norway has a strong commitment to capitalism. The choice also may be driven by necessity. Some countries, like Chad and Sudan, do not have the expertise or financial resources to find and develop oil reserves. Indeed, the industries in many of these FR countries were built up by foreign investment and expertise. However, countries with huge resources often have the incentive to eventually develop their own domestic expertise.

All OPEC countries except Angola have an R/P for their dominant fuel greater than 20 years. The places expected to be rich in oil the longest as indicated by R/P are the Gulf countries of Iran, Iraq, Kuwait, Saudi Arabia, and the United Arab Emirates in oil and Qatar in natural gas. Venezuela also has a high R/P but if you take out its heavy oil deposits in the Orinoco basin it falls to 65 years. Libya also has a strong R/P, which is a bit inflated because of lower production in 2012 as a result of civil war. For countries that are relatively new producers or relatively new to the export markets, such as Chad, Mongolia, and Uzbekistan, I would expect the R/P to fall as infrastructure is put into place to produce and move more fossil fuels into the export market. Although Turkmenistan earned more from oil in 2012, natural gas also makes a strong contribution to its economy and its long term prospects are much stronger with an R/P of more than a century in 2011 (World Bank, n.d.).

Reserves per capita give us some indication of how much the fossil fuels could contribute to the average person's wealth. The reserves for the dominant fuel have all been converted to tonnes of oil equivalent (toe). The median value per capita for the FR countries is about 155 toe. The reserves per capita tell us a somewhat similar story. Qatar is rich by any measure and is expected to be so for a long time. Other small OPEC Gulf countries, such as Kuwait and the United Arab Emirates, are also relatively rich. There are a few surprises. Saudi Arabia has one-half or less of the reserves per capita of Kuwait, but more than twice the reserves per capita of Iran and Iraq. Turkmenistan (natural gas) and Mongolia (coal) have more reserves per capita than Iran and Iraq as well. However, since both coal and natural gas are more expensive to transport, and both Turkmenistan and Mongolia are landlocked, their reserves will not give them proportionately more economic value. This is also reflected in the higher current economic value Turkmenistan received from oil. Iran, Oman, and Trinidad and Tobago have similar reserves per capita for their dominant fuel. Nigeria, with its population topping 160 million, fares rather poorly and is in the bottom quintile for reserves per capita.

For most of these countries, substantial revenues are generated in the export markets. How much is available for export is influenced by how much is used in the domestic market. The fourth column in table 21–2 shows how much total energy is consumed in each country. In absolute magnitude, Russia is the largest energy consumer in the table and the third largest energy consumer in the world after China and the United States. It accounted for between 5% and 6% of global energy consumption in 2011, while China and the United States accounted for more than 19%–20% each. Russian consumption is about 40% more than consumption in Japan and about 25% more than consumption in India. Iran and Saudi Arabia each accounted for 1% to 2% of global energy consumption. None of the other countries accounted for more than 1%, and the combined total for the FR countries is about 15% of global energy consumption. In 2012, these countries also accounted for around 65% of global crude oil and lease condensate production and 50% of dry natural gas production, but less than 10% of coal production (US EIA, n.d.d).

Domestic consumption per capita varies widely across the countries (see table 21–2, column 5). The global average annual energy consumption is 1.9 toe per year, and more than half of the FR countries exceed this value. Although I have not found detailed information on all energy subsidies, comparing gasoline and diesel prices in World Bank (n.d.) suggests that more than 80% of these FR countries subsidized gasoline and diesel, as indicated in table 21–2, column 6. They likely subsidize other fossil fuels as well, and we might expect them to be more energy intensive even without subsidies.

There is some anecdotal evidence that these countries are more energy intensive. Norway consumes 9.8 toe per capita, whereas the European average is about 3.4. The FSU average (3.9) is a little larger than the European average, but Russia and Turkmenistan consume more than 20% more energy per capita than other FSU countries. (World Bank, n.d.)

The three north African FR countries beat the African average of 0.4 toe per capita by a considerable margin as does Equatorial Guinea, but the other Sub-Saharan FR countries do not meet or exceed the average by very much. However, if we take away South Africa, the African average falls to a mere 0.1 toe, and then all but Chad, a newcomer on the block, and Nigeria beat the average.

Some countries that have developed large export refining, petrochemical, or LNG exports have per capita consumption even larger than Canada (8.9 toe), which is the largest for any OECD country outside of Norway. Comparing these FR countries to countries with similar GDPs in their region also suggests these countries are more energy intensive (e.g. Bahrain, Brunei, Kuwait, Qatar, Trinidad and Tobago, and the United Arab Emirates).

Some of the exceptions—Kazakhstan, Mongolia, Nigeria, and Yemen—are relatively new exporters or exporters with per capita incomes below $6,000. They tend to have low levels of industrialization and a larger share of GDP from agriculture and services (see table 21–2, columns 7–10). Bolivia and Ecuador are exceptions as well. Although they have resource industries of longer standing, they too are less industrialized, with a large share of GDP coming from agriculture and services. The remaining exception is Azerbaijan. It is fairly industrialized, with slightly more than 60% of the value added from the industrial sector. However, industry only takes about 10% of final energy demand. It is a bit unusual in other ways as well, because almost all of its electricity is produced in combined heat and power plants, which should make this sector more efficient, while a relatively high share of energy (45%) is used in the residential sector.

Energy consumption varies depending on per capita income, as well as other variables. Figure 21–1 shows a plot of real GDP per capita and energy per capita.

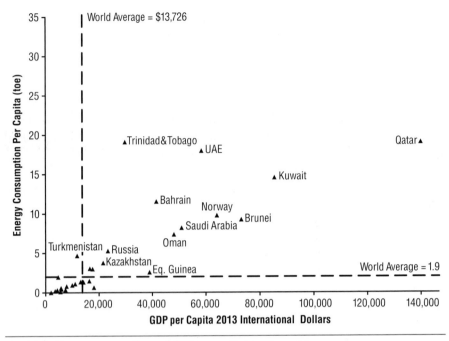

Fig. 21–1. Energy consumption and GDP per capita for FR countries
Sources: World Bank (n.d.); US EIA (n.d.b).

The dotted lines show the global averages. It is quite easy to see the positive relationship between energy consumption per capita and income per capita. A simple regression suggests an income elasticity of about 1.3. Four countries seem to be further off of the beaten track than others. Bahrain, Qatar, Trinidad and Tobago, and the United Arab Emirates. All have small populations and large exports of LNG, petrochemical, or refinery products, and are relatively more energy intensive. Equatorial Guinea, a relatively new exporter, seems less intensive.

Thus as these countries get richer, they are likely to be using more fossil products domestically, sometimes for home consumption and sometimes to produce goods for the export market. This growth in consumption will also be an important determinant in how long domestic reserves last. With this in mind, let's consider the composition of energy consumption and how energy consumption and its composition have been changing in each of these countries (see table 21–3).

From table 21–3 we can see the dominant fuel for domestic energy consumption in 2011. For each of the five countries where oil is not the dominant revenue earner, their dominant revenue earner is also their dominant domestic fuel source: coal in Mongolia, and natural gas in Bolivia, Qatar, Trinidad and Tobago, and Uzbekistan. For 26 countries, oil was the dominant revenue source, but 17 of these countries are diversified out of their dominant revenue earner in the domestic market. Hence, they use a larger share of nonliquid fuels that are more expensive to transport. Most often this other source is natural gas (9 countries), followed by traditional bioenergy (5 African countries), coal (Kazakhstan), and primary electricity (Norway).

Table 21–3. Fossil-rich countries' energy consumption shares and growth rate by source

Country	% Energy Consumption by Source					Average Annual Growth Rate 2001–11 (%)						
	Coal 2011	Oil 2011	Ngas 2011	Elec 2011	BioE&W 2011	Coal	Oil	Ngas	Elec	BioE&W	Tot E	GDP
Algeria	0.6	32.9	66.3	0.2	0.0	−5.8	4.3	4.8	16.5	−12.0	4.5	3.7
Angola	0.0	32.2	4.8	6.9	56.0	–	7.0	3.2	13.1	3.1	4.7	10.7
Azerbaijan	0.0	31.3	63.5	4.5	0.7	–	−4.0	0.5	6.1	19.3	−0.9	13.0
Bahrain	0.0	18.2	81.8	0.0	0.0	–	7.4	3.2	–	–	3.9	5.5
Bolivia	0.0	37.6	29.8	7.9	24.6	–	2.2	11.0	0.8	9.7	5.7	4.1
Brunei	0.0	22.2	77.8	0.0	0.0	–	3.5	8.3	–	–	7.0	1.3
Chad	0.0	100.0	0.0	0.0	0.0	–	2.7	–	–	–	2.7	9.1
Congo, Rep.	0.0	25.0	38.8	7.3	28.9	–	8.2	31.9*	2.6	2.5	9.4	4.5
Ecuador	0.0	73.3	1.9	20.1	4.7	–	4.4	4.1	5.0	−0.3	4.2	4.3
Egypt	1.1	43.3	50.1	3.8	1.7	2.8	3.3	7.0	−1.2	–	4.7	4.6
Eq. Guinea	0.0	11.9	88.0	0.1	0.0	–	12.1	39.9	12.0	–	29.6	9.4
Gabon	0.0	34.4	2.9	9.2	53.5	–	1.1	−1.7	−1.4	1.9	1.2	2.4
Iran	0.6	37.4	60.7	1.2	0.1	1.7	2.8	7.8	8.0	7.7	5.6	5.0
Iraq	0.0	95.1	2.0	2.9	0.1	–	3.9	−11.7	19.8	0.0	3.3	3.9
Kazakhstan	60.9	16.9	19.1	3.0	0.1	5.8	0.1	−0.7	−1.1	14.1	2.9	7.4
Kuwait	0.0	59.1	40.9	0.0	0.0	–	4.4	6.7	–	–	5.3	5.1
Libya	0.0	57.6	41.0	0.0	1.4	–	−5.3	−0.2	–	1.9	−3.4	NA
Mongolia	57.2	37.2	0.0	0.0	5.6	2.9	6.7	–	–	−0.3	3.9	7.6
Nigeria	0.0	10.5	4.2	1.2	84.1	23.3	−2.5	−1.7	−1.0	2.5	1.6	8.5
Norway	1.9	23.5	9.2	61.6	3.8	−1.1	0.3	−0.2	−1.0	3.8	−0.5	1.4
Oman	0.0	28.6	71.4	0.0	0.0	–	8.1	9.9	–	–	9.4	4.1
Qatar	0.0	23.1	76.9	0.0	0.0	–	12.0	9.8	–	–	10.2	12.5
Russia	15.0	20.8	52.2	11.1	0.9	0.7	1.6	1.9	0.6	0.3	1.5	4.6
Saudi Arabia	0.0	60.9	39.1	0.0	0.0	–	5.0	5.2	–	0.0	5.0	6.0
Sudan	0.0	34.5	0.0	8.7	56.9	–	8.3	–	13.1	−0.4	2.7	5.1
Trinidad & Tobago	0.0	67.6	29.2	3.2	0.0	0.0	2.2	4.2	−1.0	2.3	2.6	4.9
Turkmenistan	0.0	8.8	91.1	0.0	0.1	–	3.4	7.1	−4.3	−10.6	6.7	8.4
UAE	0.0	23.2	76.8	0.0	0.0	–	3.5	7.0	−0.2	–	6.1	4.1
Uzbekistan	1.4	35.7	62.8	0.0	0.1	42.1	6.5	6.6	–	11.8	6.7	7.1
Venezuela	2.6	9.2	83.7	4.5	0.0	0.9	−4.2	1.2	4.7	−0.9	0.7	3.2
Yemen	0.0	87.3	11.2	0.0	1.4	–	2.5	–	–	3.1	3.7	2.7
World	26.4	31.4	21.4	12.1	8.7	4.4	1.2	2.9	2.2	2.4	2.6	3.9

Sources: US EIA (n.d.d); World Bank (n.d.).
Note: Ngas = natural gas; Elec = electricity; BioE&W= biocombustible energy and waste; Tot E = total energy. GDP growth is measured using real local currency except for the world, where real PPP international dollars are used. * indicates the growth rate is since 2005.

The amounts and composition of energy are continually changing in all of these countries. From 2001 to 2011, global energy consumption annual growth averaged around 2.6%, while energy consumption growth for the combined total for these countries averaged more than 3%.

For more than three-fourths of the FR countries, energy growth was higher than the world average. For nine of these high-growth countries, energy growth has been faster than GDP growth, suggesting their economies are becoming more energy intensive. Seven countries had energy consumption growth increases less than the world average. For all of these latter countries, energy also grew slower than GDP, making their economies less energy intensive in 2011 than in 2001. Of the five transitional countries, Azerbaijan and Russia are in this latter category. They had recovered from the transition, and their income was significantly higher by 2011. However, with the restructuring of their economies and replacement of old Soviet capital, their economies became more energy efficient. GDP per unit of energy increased for them more than 50% during the period (World Bank, n.d.). In the case of Azerbaijan, efficiency increased enough to reduce energy consumption over the period.

Income in Venezuela with Chavez at the helm increased a bit more than the world average but well below most other FR countries. Its percent energy increase was well below others. Surprisingly, Nigeria, with higher-than-average world income growth, saw reductions in all commercial energy except for small amounts of coal, but biomass consumption increased over the decade.

As a result of civil war, Libya consumed less energy in 2011 than 2001. Norway also consumed less energy in 2011 than in 2001. Its energy consumption from 2000 to 2007 was relatively flat, even as income increased, as the result of rising energy prices and policies to decrease fossil fuel consumption and promote renewables. The high prices and global recession took a further toll, with Norwegian energy consumption falling by about 1.2% from 2007 to 2009 as its real GDP fell about 1.7%. Although income rebounded in 2010 and 2011, energy consumption continued to fall in 2010 and 2011 (US EIA, n.d.d; World Bank, n.d.)

One way to stretch a dominant fossil reserve is to substitute another fossil fuel, or better yet, a nonfossil energy source for the dominant fuel. Consider whether countries have become more or less diversified in the last decade from the growth rates of various energy sources in table 21–3. In the seven countries where oil provides the largest source of fossil rent and is the dominant domestic fuel, oil has been the fastest growing fuel only for Chad, suggesting diversification either toward natural gas (Kuwait and Saudi Arabia), primary electricity (Ecuador and Iraq), or bioenergy (Libya and Yemen).

In the thirteen countries where oil provides the largest source of rent and natural gas is the dominant domestic fuel, natural gas has grown faster than oil for all but one of them. By moving more deeply into gas, they can free up oil for the export market. The exception is Bahrain, which is exporting more than four times as much oil products as they consume domestically. Thus, much of the Bahraini growth in oil consumption is to service the export market. Some of these thirteen countries are also diversifying into other energy sources, and the fastest growing of which are biofuels in Azerbaijan and Turkmenistan, and primary electricity in Algeria, Iran, and Venezuela.

Of the four African countries where oil provides the largest source of rent and traditional biomass is the dominant domestic fuel, only Gabon and Nigeria have moved more deeply into traditional biofuels. Sudan and Angola have been modernizing with a rapid expansion of oil consumption and new hydro power.

Kazakhstan is the only country where oil is the largest source of rent, but coal is the dominant domestic fuel. Surprisingly, its traditional biomass has seen the largest growth over the decade but from a very small base. Thus, coal is still its domestic mainstay and the second fastest growing fuel.

Coal use is not as prevalent as the other two fossil fuels in these FR countries. It is only consumed in nine of the countries. It only exceeds 3% of domestic energy use in three countries: Kazakhstan (60.9%), Mongolia (57.2%), and Russia (15%). Because it is a solid and more expensive to transport, it is often consumed closer to home. Coal rents dominate only in Mongolia, where it is also the dominant domestic fuel. However, oil products have seen faster growth. With no refineries, Mongolia exports the little oil it produces and imports products, the majority of which are used in the transport sector.

What is notable in the above discussion is the wide differences in fossil fuel use, the degree of fossil fuel diversification, and growth patterns in domestic fossil fuel use. However, for most of these countries, fossil fuels provide more than 75% of their domestic fuel. Only for four African countries, where traditional combustible biofuels dominate, and Norway, where hydropower dominates, do fossil fuels provide less than one-half of their energy. In the next section, I will consider more closely diversification into electricity and modern biofuels.

Primary Electricity and Modern Biofuels

As we move forward into a carbon-constrained world, we expect that there will be some shift toward primary electricity and modern biofuels. Although the bulk of these countries have made great strides toward higher standards of living and use of fossil fuels, most have not yet started the shift to the next generation of energy sources. More than a dozen of these countries produce no primary electricity. Table 21–3 shows the world average share of primary electricity in total primary energy is about 12%, with the division by source shown in figure 21–2. When these FR countries do produce primary electricity, it is most often the most traditional and ubiquitous source of renewable electricity, hydropower.

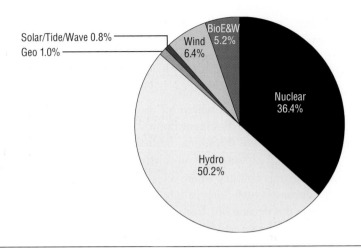

Fig. 21–2. World primary electricity production by source, 2011
Source: US EIA (n.d.c).

Only four countries have a higher share of primary electricity than the global average. Norway gets more than one-half of its energy from hydropower, Venezuela gets slightly more than one-fourth from hydropower, and Ecuador gets about 17%. Russia gets about 12.5% of its energy from primary electricity, and about one-half of that is hydropower.

The use and potential for hydropower varies considerably. Fourteen countries produce little or no hydropower. They are all in the Middle East and North Africa (MENA), plus Equitorial Guinea, Trinidad and Tobago, Brunei, and Mongolia. Most of these countries tend to be very dry or very flat, or both. Of these, WEC (2010) indicates that only Mongolia has hydropower resources that could be economical to develop. This considerable resource is estimated to be more than twice current Mongolian electricity production. Of the remaining countries, those that could potentially replace all of their current fossil electricity generation with hydropower are Bolivia, Ecuador, Norway, Iraq, Gabon, Nigeria, and Sudan.

Russia is the only one of the FR countries that has any significant nuclear power, and Iran produced a small amount in 2011. Russian nuclear production is about equal to its hydropower production. It currently has 33 plants, with 10 under construction as of 2014. However, there is also more nuclear power in the works. The United Arab Emirates started construction of its first and second out of four planned nuclear power plants in 2012 and 2013 (World Nuclear Association 2014a). Saudi Arabia has announced targets to build 16 nuclear power plants for 20% of their power generation by 2030 (Fukuyama 2013).

There is also the ongoing struggle with Iran over its enrichment program. Many doubt Iran's declarations that enrichment capacity is only to create fuel for their fledgling nuclear power industry (one operating nuclear plant as of 2013), as such capacity can be diverted to creating nuclear weapons. This originally concealed enrichment program first gained world attention and concern in 2002. Since then there has been escalating pressure on Iran to cease and desist on enrichment (World Nuclear Association 2014b). The United Nations issued six increasingly stronger resolutions with sanctions from 2006–2010. However, it was the sanctions from the United States, European Union, and other countries culminating in 2012–2013 on importing crude oil from Iran along with exclusion from the global financial network and insurance, that finally had an effect. Iran's GDP fell an estimated 1.9% in 2012 and another estimated 1.3% in 2013. Oil exports fell from 2.2 Mbbl/d to 0.7 Mbbl/d by May of 2013. In November of 2013, Iran made an agreement with the international community to curtail its enrichment activities and not to fuel or start up its Arak heavy water reactor. In exchange, some of the sanctions have been lifted (BBC News 2014).

Saudi Arabia has announced targets to build 16 nuclear power plants for 20% of its power generation by 2030 (Fukuyama 2013). You can follow these and other nuclear developments worldwide at IAEA (n.d).

Norway, of course, is a global exception, with more than one-half of its primary energy from primary electricity, more than 90% of which comes in the form of hydropower. Norway has begun forays into the more trendy renewable electricity alternatives, the nonhydro renewable electricity (NHydEl) sources, such as wind, solar, and bioenergy and waste. Globally, wind provides the largest share for these newer options, providing between 6% and 7% of primary electricity in 2011, followed by bioenergy and waste, geothermal, and then solar, tides, and waves, which is mostly solar. OECD countries, plus China and Brazil, produced about 90% of nonhydro renewable electricity worldwide. The FR countries altogether produce less than 1% of such electricity.

Norway produces some of each of these options except for geothermal, but in very small quantities. Less than 1.5% of its electrical power in 2011 came from nonhydro renewable sources, with a majority from wind followed by biomass and waste. Norway is one of thirteen FR countries to produce renewable nonhydro electricity. It is second after Russia, which is the largest producer, with a majority from biocombustible energy and waste, followed by some geothermal. Egypt is the only other FR country that produces more than a billion kWh from other renewables, all from the wind.

Bolivia, Ecuador, Trinidad and Tobago, Gabon, and Sudan produce small amounts of power from bioenergy and waste, and Bahrain, Ecuador, Kazakhstan, Iran and Mongolia produce a little bit of wind power. Bolivia and the United Arab Emirates produce a bit of solar power.

Although solar power is still rather expensive compared to the other renewable nonhydro sources of electrical power production, it is a huge but diffuse resource. Muller (2009) puts this source in perspective for us as follows. About 1 kilowatt (kW) falls per square meter (1.19 square yards) when the sun is shining. If the sun shines for an hour, we have the equivalent of 1 kWh. In 2011, a US household consumed an average 1.4 kWh/hour. If a solar photovoltaic cell has an efficiency of 15% (usable watt per watt of solar radiation or insolation reaching the earth's surface), and usable sunshine averages 5.5 hours per day, the number of square meters of PV required to supply such a home would be the following:

$$\text{Meters}^2 = \frac{\dfrac{\text{kWh}}{\text{hour}}\dfrac{\text{hours}}{\text{day}}}{\dfrac{\text{UseableW}}{\text{W}}\dfrac{\text{kW}}{\text{meters}^2}\dfrac{\text{hoursSun}}{\text{day}}} = \frac{1.4 \times 24}{0.15 \times 1 \times 5.5} = 40.727$$

So the average household would need about 41 square meters of PV. This might be doable for single-family dwellings but would be more problematic for multiunit dwellings, which currently comprise about 40% of US residences (U.S. Census 2011).

With more efficient appliances and PVs, we could, of course, reduce the required PV surface area. Currently, the most efficient collectors have efficiencies greater than 40%, but they tend to be quite expensive. Also note this computation assumes we are able to store power with no losses and have it accessible whenever needed.

Muller (2009) gives us another perspective for solar transportation. A kilowatt is roughly equal to 1 horsepower (hp). Our collector above could produce about 0.15 kilowatt per hour or about 0.15 hp (about equivalent to the energy used to pedal a bicycle). It would take quite a collector to power a 1,200 hp Lamborghini.

Some of the desert countries in our FR countries have very good solar resources. They average more than six hours of sunshine a day. Although they must solve the problem of cleaning sand off the collectors, three countries in particular are embarking on more solar in their future. Saudi Arabia has established King Abdullah City for Atomic and Renewable Energy (K.A.CARE) and is targeting solar to replace oil and gas for one-third of its electricity by 2032. In addition, K.A.CARE is recommending that Saudi Arabia set a target of 50% of domestic energy coming from renewables and nuclear by 2032 (K.A.CARE 2013). Solar and other renewables are receiving increasing interest in the United Arab Emirates. The International Renewable Energy Agency (IRENA), established in 2009, is headquartered in the United Arab Emirates. Abu Dhabi, the largest and oiliest of the emirates, started up a 100 MW solar collecting power plant in March 2013. The plant, one of the largest of its type in the world, signals an increasing commitment to renewable energy, and Abu Dhabi is targeting 7% of its electricity to come from solar in 2020 (Todorova 2011). Saudi Arabia and the United Arab Emirates, along with Norway, are the only FR countries ranked in the top 40 countries in Ernst & Young's Renewable Energy Country Attractiveness Indices (CAI) (*Peninsula* 2012).

Qatar will also have a completed solar power plant that will raise its share of electricity generation from renewables from 0% to 16%. The electricity is slated to replace gas-fired power for desalinization plants. Qatar has a target of 20% of its energy from renewable resources by 2020 (Pamuk 2012).

Desertec is a proposal to generate solar power in the Sahara Desert for local consumption in North Africa and the Middle East, as well as for export to Europe via high-voltage, direct-current (HVDC) cables. In theory, an area the size of Austria in the Sahara could supply the world's current generation of electricity. However, given the large distances and high costs, this project has not moved forward yet. Indeed, cost is not the only issue. Europe is unlikely to view electricity from such a politically volatile part of the world as very secure.

In addition to modern renewable sources of electricity, renewable liquid from biofuels is starting to make some small global inroads (0.5% of primary energy consumption in 2011). However, in the FR countries, their contribution is even more minuscule. Only four of the FR countries—Ecuador, Trinidad and Tobago, Norway, and Sudan—produce any biofuels. Norway consumes some of each fuel, but the other countries only consume ethanol. Biofuels contribute less than a half a percent of primary energy in each of these four countries.

Another way to extend oil resources is to increase efficiency in extraction. With primary recovery, oil rises to the surface naturally or with the help of lift from pumps. Typical primary recovery rates are 10%. With secondary recovery, water or natural gas is injected into the field to increase the pressure, which may bring recovery rates up to 40%. With more advanced tertiary recovery, also called *enhanced oil recovery* (EOR), the viscosity of the oil is decreased to allow it to move to the wellbore more easily. EOR includes injection of steam, chemicals including polymers and detergents, CO_2, or even microbes. EOR can raise recovery rates to 60% or even better. Norway's Statoil has a target of raising average recovery rates from 50% to 60%, while Saudi Aramco has an even more impressive target of raising such rates from 50% to 70%.

A tax or permit requirement on carbon dioxide could raise costs of fossil fuels, resulting in more left in the ground. Carbon capture and storage (CCS) could be an option, allowing more to be recovered, and much research is being done to investigate its cost and feasibility. Figure 21–3 shows commercial projects in North America where CO_2 is captured and used for EOR.

CO_2 EOR was first patented in 1952. The first large-scale project started up in 1972 in West Texas. Currently, more than 100 projects exist worldwide, with the majority in the United States and Canada. Great Plains Synfuels Plant in North Dakota converts lignite to natural gas and partners in the world's largest CCS project. It has captured and sold CO_2 commercially since 2000 for EOR in the Weyburn and Midale oil fields in Saskatchewan, Canada (Dakota Gasification Company 2013).

CCS can also be used for CO_2 disposal without EOR, such as injecting captured CO_2 into abandoned oil and gas fields and saline formations. For CCS, it is important to make sure that the CO_2 leakage back into the atmosphere is minimal. The Weyburn project is also being monitored as the world's largest demonstration CCS project.

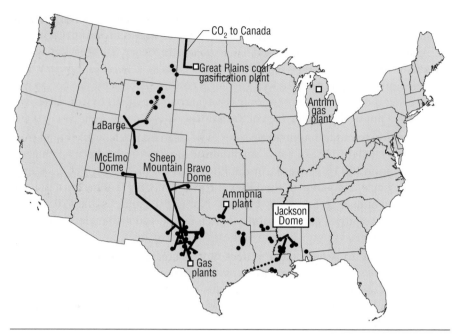

Fig. 21–3. CO_2 sources and pipelines
Source: Sweatman et al. (2009).

MIT's CCS database includes projects in three FR countries (Carbon Capture & Sequestration Technologies 2013). Since 2004, a demonstration project in Algeria has removed CO_2 from export natural gas and injected it into a saline formation below the gas field. Norway has three projects. They have been removing CO_2 from Sleipner natural gas before export and injecting it into a saline formation since 1996. They have been capturing CO_2 from an LNG plant in Snovit and injecting it into a depleted gas field since 2007, and the CO_2 Technology Centre in Mongstad has demonstration projects capturing CO_2 from a refinery and a gas-fired power plant and injecting it into saline formations. Abu Dhabi has a CCS project that will capture CO_2 from a steel plant and use it for EOR. The project has been delayed but is expected onstream in 2015 or 2016.

Economic Issues in Fossil Fuel–Rich Countries

Although fossil fuels should be a blessing giving increased productive capacity and a better life for a country's inhabitants, this does not always seem to be the case. Nankani (1979) found that mineral-exporting developing countries did not seem to outperform developing countries without mineral exports. Since this unexpected result, there has been much research looking into development issues in FR countries. Mahon (1992) is credited with first coining the term *resource curse*.

Davis and Tilton (2005) provide a nice summary of the conventional neoclassical view and the curse view on natural resources. Neoclassical economists represent the technical productive capacity of an economy with a production function. If you plug in resources, capital and labor, you get an output. Capital can include not only the plant and equipment, but also natural resources, human capital such as education, and institutional capital such as legal systems and government institutions. The more capital, the more output. However, for fossil resources, which are a gift of mother nature, the capital must be exploited to increase economic output.

In the curse view, as suggested by Davis and Tilton (2005), Ross (1999), Stevens (2003), and others, resource wealth and exploitation do not automatically generate economic growth and development. The reasons relate to the following: (a) terms of trade, (b) volatility in the resource sector, (c) Dutch disease, (d) crowding out, (e) weak linkages between the resource and other sectors, (f) the cost of resource development and exploitation, but not the benefits, may be born locally, (g) an increasing role for the government, and (h) poor use of resource rent. Evidence for these various hypotheses was surveyed by van der Ploeg (2011). I will briefly discuss them, along with some of his conclusions and some measures of performance for the FR countries.

Terms of trade

Argument (a), that terms of trade disadvantage resource producers, was popular in the 1950s. This argument suggested that over time, resource prices trend down and the prices of manufactured goods trend up. However, since 1973 we have seen no such universal tendency; we have seen both resource price spikes as well as crashes. Thus, terms of trade may be disadvantageous at times but advantageous at other times.

Volatility in the resource sector

Price fluctuations from the boom-or-bust nature of fossil-fuel markets lead to (b), the argument that volatility hurts resource exporting countries. Volatile resource prices and revenues make the macroeconomy more volatile, increase exchange rate fluctuations, enhance economic risk, raise transaction costs, and may reduce investment, innovation, and economic growth. It hurts the welfare of risk-averse households, and downturns may be especially harmful for the more vulnerable members of the country. Such volatility is likely to be less detrimental for more diversified economies.

Dutch disease and crowding out

Arguments (c) and (d), Dutch disease and crowding out, relate to how the resource sector, which trades internationally, interacts with nonresource sectors. Parties trade because each trading partner benefits from the transaction. Sellers sell things that they are better at producing and buy things others are better at producing. This same logic can hold for countries.

Let's explore this idea a bit further with a simple example at the country level. Suppose there are two goods produced and consumed: resources (R) and a composite nonresource good (N). Also assume that there are two countries, an FR country ("Sandy Desert" or "Sandy"), and a country that has fewer fossil-fuel resources ("Diversified Land" or "Dland") but is better at producing the other good. Each country has 100 workers. In Sandy, 1 tonne of R requires 1 worker-day, and 1 tonne of N requires 2 worker-days. In Dland, 1 tonne of R requires 2 worker-days, and 1 tonne of N requires 1 worker-day. In this example, Sandy requires less resources to produce a tonne of R (1 day instead of 2) and has an absolute advantage in R. Dland requires less resources to produce a tonne of N (1 day instead of 2). We can summarize the amount of labor required for each unit of R and N in each country as follows:

	Sandy	Dland
Worker-day per tonne Resource (R)	1	2
Worker-day per tonne Nonresource (N)	2	1

Assume constant returns to scale in each country, so we scale production up or down in each sector and country at a constant rate. Thus, one more Sandy worker-day producing R gives us 1 tonne; one less Sandy worker-day producing N gives us 0.5 less tonne of N. If I move one worker day from N to R, we get 1 more tonne of R and 0.5 less tonne of N.

Now let's measure the unit gains from specialization. Have Sandy produce 1 less N and move those two workers to produce 2 units of R. Likewise, have Dland produce 1 less unit of R and move those two workers to produce two units of N. The global effect of moving one worker in Sandy to sector R and one in Dland to N is 1 additional unit of R and 1 additional unit of N, as shown in table 21–4.

Table 21–4. Unit gains of trade from specialization (absolute advantage)

Product	Sandy	Dland	World
Resource (R)	2	−1	1
Nonresource (N)	−1	2	1

We can get some further insight by representing each country's production possibility frontier shown by the solid lines in figure 21–4. Panel (a) is for Sandy. If all 100 workers produce R, we could have no N and 100 R. If all workers produced N, we could have 50 tonnes of N and no R. If we moved 1 worker from N to R, we would get 1 more R and 0.5 less N. Similarly, we can develop the production possibility frontier for Dland as in panel (b).

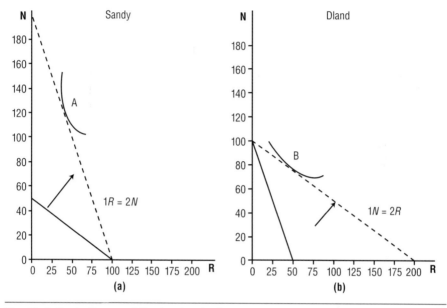

Fig. 21–4. Production possibility frontiers for Sandy and Dland at their own and each other's terms of trade

Market prices in the two countries should reflect these differences in relative costs. In Sandy, we can trade 1 R for 0.5 N, so N should be twice as expensive as R, or 1 N = 2 R. In Dland we can trade 1 R for 2 N, so N should be 0.5 as expensive, or $1N = (0.5)R$. If Sandy produced only R and traded at Dland's relative costs, Sandy would get 2 N for every R traded, changing their production possibility curve to the dotted line, as in figure 21–4, panel (a). If Dland produced only N and traded at Sandy's relative costs, Dland would get 2 R for every N, changing their production possibility curve to the dotted line, as in figure 21–4, panel (b). We can see that each country could be better off with trade unless Sandy only wanted to consume R and Dland only wanted to consume N.

Recall from chapter 16 that consumers should pick their most desired point on their budget constraint. For these aggregate economies, the production possibility frontiers represent the budget constraints. Recall also that we used indifference curves to represent preferences. Let the curved lines in figure 21–4 represent aggregate preferences. Then with specialization, Sandy would want to be at A, producing R and trading for N, and Dland would want to be at B, producing N and trading for R.

However, it is little more complicated than this. With an opening up of trade, arbitrage would cause relative prices to equalize up to transport cost across the two countries. Ignoring transport costs, we would expect terms of trade to end up somewhere between $1\,R = 0.5\,N$ and $1\,R = 2\,N$. For markets to balance, the terms of trade would also have to be a value where the desired exports of one country would have to equal the desired imports of the other for each product.

Now let's take the following more interesting case. Baltica and Pacifica each have 100 workers with days of labor required per tonne for R and N as follows:

	Baltica	Pacifica
Worker-days per tonne of Resource (R)	2	9
Worker-days per tonne of Nonresource (N)	4	6

Baltica uses fewer resources to produce both R and N and has an absolute advantage in both goods. However, notice the relative prices in the two countries differ. In Baltica $R = 0.5N$, or multiply by 2 to get $N = 2\,R$. In Pacifica, $R = 1.5\,N$ or $N = 0.667R$. R is relatively cheaper in Baltica and N is relatively cheaper in Pacifica. Now let's reallocate resources a little. Have Baltica produces 1 less N and move the labor to R and have Pacifica produce 1 less unit of R and move the labor to N. The global gains from this reallocation are shown in table 21–5.

Table 21–5. Specialization gains Baltica (R) and Pacifica (N) (comparative advantage)

Product	Baltica	Pacifica	World
Resource (R)	2	−1	1
Nonresource (N)	−1	1.5	0.5

Again, we can also see this potential for gain by specialization in figure 21–5, unless Baltica only consumes R and Pacifica only consumes N. With trade, arbitrage would change the terms of trade until they were equal in each country up to transport costs somewhere between $1\,R = 1.5\,N$ and $1\,R = 0.5\,N$.

Taking advantage of comparative advantage also works in one's personal life. An anecdotal description often used to illustrate comparative advantage on a more personal level is the lawyer and the assistant. Suppose the lawyer is better at lawyering and at assisting. It still pays for the lawyer to hire an assistant and spend her time at the activity where she has a comparative advantage.

In the above example, we have abstracted from the transaction details. However, in international trade we will be faced with foreign exchange markets. So again let's digress a bit and take the US dollar market and the case of floating or flexible exchange rates. Assume one trading partner, China, to keep the problem two dimensional. Currency markets have similarities to other markets. Buyers of the currency are demanders and sellers of the currency are suppliers. In this framework, the Chinese are demanders and want to buy US currency to buy US goods, invest in the United States, pay interest on US investments in China, and become tourists in the United States. Any time someone in China wants to spend money in the United States, they are demanding dollars. (They are also supplying Chinese yuan to the yuan market.) Likewise, any time an American wants to spend money in China, they are supplying dollars and demanding yuan.

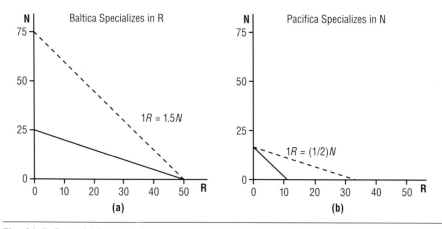

Fig. 21–5. Potential for gains from trade with comparative advantage

One difference in currency markets is that you are buying one currency with another. Thus, when an American supplies dollars to China, they are doing so to buy Chinese yuan. An additional difference is that the price paid for the other currency is the exchange rate, which tells us how much of one currency another will buy. For example, the exchange rate of Chinese yuan (¥) currently is $1 = ¥6.20, or $0.16 = ¥1.

So now let's set up the dollar market. As in other markets, quantity (dollars) is on the horizontal axis and price (exchange rate) is on the vertical axis. In the dollar market, it is most convenient to write the exchange rate as foreign currency per dollar (¥/$). Then as we go up the vertical axis, we are getting more yuan per dollar, and the dollar is appreciating or rising. As we move down the axis, the dollar is depreciating or falling.

We expect demand to slope down and supply to slope up, as in any market, for the following reasons. As the dollar depreciates, it takes less yuan to buy a dollar. Thus, US goods and services cost less for the Chinese. The Chinese would be inclined to buy more US goods, invest more in the US economy, and will need to buy more dollars. Thus, the demand for dollars is likely to slope down, as shown in figure 21–6.

On the supply side, as the dollar appreciates, the dollar buys more yuan, and Chinese goods and services become cheaper. Americans would want to buy more Chinese goods and supply more dollars to the Chinese. Thus, the supply of dollars is likely to slope up. Equilibrium is, as usual, where supply equals demand, as shown in figure 21–6. At an exchange rate above P_e, the quantity of dollars supplied exceeds the quantity demanded, and the price of dollars should fall. Similarly, at a price below P_e, the excess quantity demanded should cause the dollar to rise.

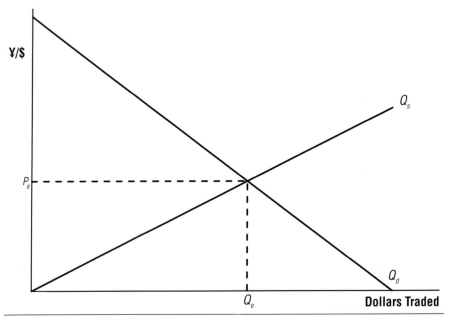

Fig. 21–6. Dollar market

Now we are ready to consider the Dutch disease effect. Suppose with the US shale revolution, the Chinese start to import more oil and LNG from the United States. These imports shift the demand for dollars out, as in figure 21–7, and the value of the dollar rises. This appreciation of the dollar makes manufactured exports more expensive and may reduce the manufacturing sector. Such an effect is labeled *Dutch disease*, as it was first observed in the Netherlands, with its large discoveries and exports of natural gas.

Now our comparative advantage examples suggest benefits from this decline in the manufacturing sector, which allows movement into the more prosperous resource sector. However, a number of situations have been put forward where such a shift could have some negative effects. If productivity increases faster in manufacturing than in resource extraction and export, manufacturing countries may experience faster productivity gains and economic growth. Furthermore, if technical progress and human capital is enhanced from learning by doing, once the country runs out of the resource, it may have a smaller technology and human capital base to fall back upon. Some have also argued that resource extraction may require less education and skill, causing FR countries to invest less in education. Column 2 in table 21–6 shows the share of gross national income spent on education.

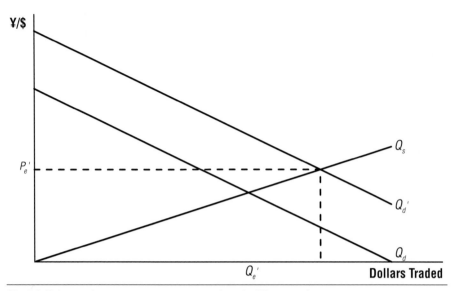

Fig. 21–7. Increasing resource exports appreciate the FR country's currency

Worldwide, WDI (World Bank, n.d.) low- and middle-income countries both on average spent 3.2% of GNI on education in 2012. Eleven FR countries were in this middle-income range in 2010; four of them invest less than these averages, and seven invest more. Uzbekistan gets the prize at 9.4%. All of the low spenders are in Africa. Nineteen FR countries are in the WDI (World Bank, n.d.) high-income range. However, only three of them, Norway (6.2%), Saudi Arabia (7.2%), and Venezuela (5.8%) invest more than the world average for high-income countries (4.7%).

If resource production is less labor intensive than manufacturing, and labor markets are not flexible and well developed or adjust very slowly, the resource market may not soak up all the labor released from manufacturing. This could result in an increase in unemployment, at least for a time. FR countries without a manufacturing sector may find that the resource industries are not labor intensive enough to soak up all new entrants into the labor market, while manufacturing does not develop as it cannot compete with imports and again unemployment may be higher.

Taking a cursory look at employment, column 3 in table 21–6 shows unemployment for the FR countries. Worldwide, World Bank (n.d.) shows middle-income countries have an average unemployment rate of 5.6%. Eight of the eleven middle-income FR countries exceed the global average for their income group. Six of them are in Africa. However, except for Egypt and Sudan, these rates are not so high by African standards. Yemen (11.3%) and Uzbekistan (17.6%), however, are considerably higher than their regional cohorts and neither has seen any improvement in unemployment rates in the last two decades (World Bank, n.d.)

Table 21–6. Development indicators for FR countries

Country	% GNI Educ	% U	% Mig	HDI	Gini	% <Pov	% Urb Pop	% Acces Elec	% Imprv H₂O	Life Exp	% GDP Health	TIC	Peace Idx
Year	2012	2012	2010	2011	01-11	2011	2012	2011	2012	2012	2012	2013	2013
Algeria	4.5	9.8	0.7	0.71	35	NA	73.7	99.4	95.2	70.9	5.2	36	61.3
Angola	3.6	7.5	0.3	0.51	43	36.6	59.9	37.8	60.1	51.5	3.5	23	58.7
Azerbaijan	2.8	5.4	2.9	0.73	35	6.0	53.9	NA	82.0	70.6	5.4	28	64.8
Bahrain	3.0	7.4	25.2	0.80	36*	NA	88.8	99.4	99.2	76.5	3.9	48	61.0
Bolivia	6.5	3.2	1.4	0.68	58	51.3	67.2	86.8	46.4	66.9	5.8	34	53.9
Brunei	2.0	3.8	37.0	0.86	35*	NA	76.3	99.7	NA	78.4	2.3	60	NA
Chad	2.0	7.8	3.3	0.34	40	46.7	21.9	NA	11.9	50.7	2.8	19	70.1
Congo, Rep.	2.5	7.1	3.5	0.53	47	46.5	64.1	37.8	14.6	58.3	3.2	35	60.6
Ecuador	4.1	4.5	2.6	0.72	52	27.3	68.0	95.5	83.1	76.2	6.4	32	55.9
Egypt	4.4	11.9	0.3	0.66	31	25.2	43.7	99.6	95.9	70.9	5.0	19	70.4
Eq. Guinea	1.0	7.6	1.1	0.55	65*	76.8	39.7	NA	NA	52.6	4.7	34	57.0
Gabon	3.1	20.3	18.3	0.68	41	32.7	86.5	60.0	41.4	63.1	3.5	25	53.3
Iran	4.1	13.1	2.9	0.74	38	NA	69.2	98.3	89.4	73.8	6.7	16	66.8
Iraq	NA	15.1	0.3	0.59	31	19.8	66.5	98.0	84.7	69.2	3.6	26	92.5
Kazakhstan	4.4	5.3	18.9	0.75	33	3.8	53.5	NA	97.5	69.6	4.2	43	58.9
Kuwait	3.2	1.5	70.1	0.79	30*	NA	98.3	100.0	100.0	74.4	2.5	15	46.0
Libya	NA	8.9	11.3	0.77	36*	NA	77.9	99.8	96.6	75.2	3.9	38	67.2
Mongolia	5.0	5.2	0.4	0.68	35	27.4	69.3	88.2	56.2	67.3	6.3	25	48.7
Nigeria	0.9	7.5	0.7	0.47	46	47.2	50.2	48.0	27.8	52.1	6.1	86	74.2
Norway	6.2	3.2	9.9	0.96	26	NA	79.6	NA	100.0	81.5	9.0	47	37.6
Oman	4.2	8.1	29.5	0.73	32*	NA	73.7	98.0	96.6	76.6	2.6	68	51.8
Qatar	1.8	0.6	74.6	0.83	41	NA	98.9	99.6	100.0	78.5	2.2	28	40.8
Russia	3.5	5.5	8.6	0.79	39	11.0	74.0	NA	70.5	70.5	6.3	46	83.3

Saudi Arabia	7.2	5.6	26.7	0.78	32*	NA	82.5	99.0	100.0	75.5	3.2	11	54.9
Sudan	0.9	14.8	1.7	0.41	35	46.5	33.4	29.0	23.6	61.9	7.2	38	92.1
Trinidad & Tobago	2.9	5.8	2.6	0.76	40*	NA	14.0	99.0	92.1	69.8	5.4	17	56.6
Turkmenistan	NA	11.3	4.1	0.70	41	NA	49.1	NA	99.1	65.3	2.0	69	57.3
UAE	NA	3.8	39.0	0.82	31*	NA	84.6	100	97.5	77.0	2.8	17	47.9
Uzbekistan	9.4	11.3	4.1	0.65	36	16.0	36.3	NA	100.0	68.1	5.9	20	59.7
Venezuela	5.8	7.8	3.5	0.75	48	25.4	93.7	99.6	NA	74.5	4.6	18	66.0
World	4.3	5.9	3.1	0.69	NA	NA	52.5	78.2	63.6	70.8	10.2	NA	NA
Yemen	4.1	17.6	2.3	0.46	38	34.8	32.9	39.9	53.3	62.9	5.5	23.0	72.0

Sources: World Bank (n.d.); UN (n.d.b); Transparency International (2014); Vision of Humanity (2014).

Note: % GNI Educ = the percent of gross national income spent on education, % U = percent of unemployment, % Mig = percent of population that are international migrants, HDI = the UN human development indicator, Gini = the Gini Index, 01-11 indicates numbers were averaged from 2001–2011 or if not available the next earlier value was used. * missing value taken from Avakov (2012). % Urb Pop = the percent of the population living in urban areas, % <Pov = the percent of the population below the national poverty level, % Acces Elec = the percent of the population with access to electricity, % Imprv H_2O = percent of population with access to improved water supply, Life Exp = life expectancy at birth in years, %GDP Health = percent of GDP spent on health care, TIC = the Transparency International Corruption Perception Index, Peace Idx = a peace index with higher numbers indicating more violence, and NA = not available.

Worldwide, World Bank (n.d.) shows high income countries have an average unemployment rate of 7.8%. Only 8 of the 20 high income FR countries exceed this rate. Since the richest of these countries tend to have high reserves relative to their populations, let's consider an alternative hypothesis. Resource-rich countries with large reserves relative to their population may have a scarcity of labor with low unemployment rates and a large percent of migratory or guest workers to fill in the labor gaps. For these countries, unemployment may be quite low.

Column 4 in table 21–6 shows percent of the work force that is migratory. The world average is near 3%. Most of the median income FR countries have low rates of migratory workers although a few exceed the world average. However, the majority of the high income countries have larger than average shares of migratory workers. The majority of these also have average lower rates of unemployment. The most extreme cases are Kuwait and Qatar, with more than 70% of their population migratory with less than 2% of population unemployed.

A corollary to the labor shortage is crowding out (d). It suggests that the resource sector, if large relative to the rest of the economy, takes so many resources, particularly capital, that other sectors have a hard time developing.

Weak linkages between the resource and other sectors

Just as there are arguments that the resource sector may experience slower technological change, according to argument (e), it also may have weaker linkages to other sectors of the economy. A *backward link* occurs when the sector buys its inputs locally and stimulates local industry. Thus, if a manufacturer buys local steel to produce its product, it has a stronger backward link than if it imports its steel from abroad. A *forward link* would be when local industries use the resource as an input. Thus, a forward link would be a petrochemical or refining industry that used local gas and oil products as inputs.

Some countries have implemented local content policies to try to enhance backward linkages with varying degrees of success. For example, Norway has had such policies and has been very successful at developing companies to provide petroleum industry services domestically and for export. Indonesia and Angola have also had local content policies but with much less success than Norway. Other countries such as Saudi Arabia, Kuwait, and Qatar do not have such policies (Anouti 2013).

Costs, not benefits, borne locally

Just as different sectors in the economy may receive or lose resources, so too may different factions or regions bear the costs and the benefits. Usually fossil resources are owned and controlled by the government, which may receive much or even all of the net revenues. However, as pointed out in argument (f), much of the external cost of resource development and exploitation, but not much of the benefits, may be borne locally. To see how these countries fare on human development and how the resource wealth is shared, look at some of the same development indicators considered in chapter 20.

Column 5 in table 21–6 shows the UN Human Development Index (HDI) going from 0 to the highest of 1. Norway has the highest HDI, not only for the FR countries, but for all countries ranked. It gives us optimism that resources do not have to be a curse. The median HDI for OECD countries is 0.89, and the lowest HDI in the OECD is Turkey's 0.72. The median HDI for all 187 countries ranked is 0.69. Of the middle-income FR countries, only Ecuador ranks at about the global median. Most of the high-income FR countries are above the global median of 0.69 with the exception of Equatorial Guinea, Gabon, and Iraq. Only Norway is above the OECD median. Of the high-income FR countries, only Gabon and Equatorial Guinea have HDIs below 0.70. Since GDP per capita has a prominent weight in the HDI, it varies considerably, just as GDP per capita varies.

The Gini coefficient, described in chapter 20 and shown in column 6 in table 21–6 represents income inequality and gives us a measure of how well wealth is distributed within these economies. These values are the Gini coefficients after taxes and transfers. With much missing data, the available World Bank (n.d.) numbers are averaged from 2001–2011 or the most recent available data before 2001 is used. Even so, the data are spotty and only available for 19 of the 31 countries. Missing data were filled in from Avakov (2012).

Again, Norway fares the best, with the most equality of the FR countries, and it is also near the most equal among OECD countries at 26. The Gini coefficient varies rather widely for OECD countries from 24 to 53, with a median of 33. Only two of the FR countries have a Gini coefficient outside this range: Bolivia (58.5) and Equatorial Guinea (65). However, most of FR countries have Gini coefficients above the OECD median of 33. The exceptions outside of Norway are Egypt, and a number of the Gulf States—Iraq, Kuwait, Oman, Saudi Arabia, and the United Arab Emirates. The highest Gini values tend to be in South America and Africa.

Percent of population below the national poverty level is another indicator of inequality, and World Bank's WDI (n.d.) has data for 18 of the 31 FR countries. Norway has not reported a value, but with its income equality, it may have little if any of its population below the poverty level. For four out of eight of the high-income countries with reported values, more than 19% of the population is below the poverty level in each. With the percent below the local poverty level in parenthesis, they are Equatorial Guinea (76.8%), Gabon (32.7%), Iraq (19.8%), and Venezuela (25.4%). For the 11 middle-income countries with reported values, almost all have more than one-quarter of their population below their national poverty level.

Since amenities tend to be higher in urban areas, lets see how urbanized the FR countries are in column 7 of table 21–6. Urbanization is fairly high in the FR countries with almost three-fourths of them higher than the global average of 52.5%. The majority of FR countries with low urbanization are in Africa. Seven of the countries have an urbanization rate higher than the 80% for the OECD. All are high-income countries with the majority of them in the Gulf Cooperation Council (GCC).

Next consider some more direct measures of wellbeing. They are access to electricity, access to improved water supply, life expectancy, and GDP spent on health (columns 9–12 in table 21–6). World Bank (n.d.) reports electrification rates for 23 of the 31 FR countries in 2011. Worldwide, high-income non-OECD countries have an electrification rate of more than 99%. Although the transition economies do not report electrification rates, given the emphasis on electrification in the USSR, I assume they fall in this category. Presumably electrification is almost universal in Norway as well. For high-income FR countries, the population with reported access to electricity is greater than 98% for all but Gabon, where about 40% of the population does not have access to electricity. Although not reported, Equatorial Guinea most surely has substandard access to electricity as well.

Worldwide, middle-income countries have an electrification access of 85%. Five of the eleven middle-income FR countries beat this rate—Bolivia, Ecuador, Egypt, Mongolia, and Uzbekistan. For the other six, all in Africa, the rates of electrification are typically less than 50%. The richest of these countries, Angola, sadly has an electricity access rate of less than 40%. Decades of civil war have not been conducive to infrastructure investment. Other non-FR countries with incomes per capita between $7,000 and $8,000 (2013 international dollars) have access rates double those of Angola. Similarly, the other five have electrification rates below those for non-FR countries with similar incomes.

The access to improved water tells a somewhat similar story to that of electricity. Globally, high-income countries average 99.5% access. For the seventeen high-income FR countries with statistics, six exceed 99%, while most of the others range from a few to many percentage points lower. Again, Gabon trails at 41%.

Worldwide, middle-income countries have a 90% access rate to improved water. Of the eleven middle-income FR countries, only Egypt and Uzbekistan beat this average. Most often, the other middle-income FR countries have lower access rates than non-FR countries with fairly similar incomes. Angola once again makes an exceptionally poor showing compared to other countries of similar income.

Two other statistics relate to health. Life expectancy for high-income OECD countries worldwide is 81 years. Of the high-income FR countries, only Norway beats this average. For all other FR countries with per capita GDP above $20,000 (2013 international dollars), their life expectancy is lower than non-FR countries with somewhat similar incomes. FR countries with GDP per capita less than $20,000 have life expectancies in the range of non-FR countries with somewhat similar incomes but are nearer the bottom range.

Countries can influence the health of their residents through investments in health care. Let's see how the FR countries fare in this category. Health expenditures as a percent of GDP for high-income OECD countries were 12.9% in 2012. No FR country beats this rate but Norway comes the closest at 9%. However, it is within the range of the other Scandinavian countries. All other high-income FR countries are below 7% and most are below other non-FR countries with similar incomes per capita. The exceptions are Russia and Iran with 6.3% and 6.7% going to health expenditures, respectively. Most middle-income FR countries are below other non-FR countries with similar per capita incomes. However, there is wide variation in both share of expenditures and per capita expenditures in both FR and non-FR countries. Also, the majority of FR countries have more rapidly growing younger populations, which likely require less health care.

From the above descriptive summary, we cannot infer causality, but there seems to be a tendency for the FR-countries to have poorer infrastructure and health than their non-FR counterparts. Also there are enough exceptions to indicate that this tendency is by no means irrefutable.

Increasing government role and poor use of resource rent

The last two issues, arguments (g) and (h), relate to quality of government, along with government management of resources and resource rent. Since the government owns the resource and must manage it, there is typically an increasing role for the government in FR countries. In places where governments are competent, this may not present problems. However, the resource may be poorly managed in places where governments are weak, the rule of law does not prevail, and accountability is low. Problems encountered in chapter 6, such as X-inefficiency, may also prevail.

Temptations for rent-seeking behavior and corruption may increase with resource windfalls. Column (13) in table 21–6, contains Transparency International's Corruption Perception Index (TIC in this table) for the FR countries. Recall from chapter 20 that the corruption index goes from 0 to 100, with a higher number indicating less corruption. The median index for the OECD is 71. Again, only Norway ranks above this median at 86, but its score is lower than for the other Scandinavian countries, except Iceland. About two-thirds of the FR countries have corruption indexes below the range for non-FR countries with similar incomes, and most of the rest have indexes at the lower end of those countries with similar incomes.

Resource riches may aid corruption and reduce institutional quality by making it easier for governments to buy off dissenters, enrich their cronies, and avoid being accountable. With poorly defined property rights, problems of the commons may be prevalent, or armed conflict may arise. The last column in table 21–6 is a measure of peacefulness in a country. The Institute for Economics and Peace has been ranking countries since 2007 based on 23 indicators, including crime statistics, military expenditures, and weapons trade. The 2014 ranking included 159 countries, with smaller numbers indicating less violent activity and more peaceful countries. I have normalized the numbers to lie between 0 and 100. For

FR countries, Norway is the most peaceful with an index of 37.6, followed by rich little Qatar at 40.8. However, two thirds of FR countries are more violent and less peaceful than the world median of 55. The majority of FR countries are also less peaceful than non-FR countries at similar income levels.

Countries may make poor use of resource rent. Entrepreneurial talent may focus on maximizing rent creation and capture instead of developing productive resources within the economy. Welfare reducing subsidies, poor industrial policy that protects industries that should be abandoned, and government projects that turn out to be "white elephants" are also examples of poor use of resource rents.

Although FR countries should have advantages in developing their economies, from the above descriptive discussion, this hardly seems to be a hard-and-fast rule. Nor does is seem to be a hard-and-fast rule that resources are also a curse, but often FR countries do not seem to perform particularly well. In his comprehensive survey, van der Ploeg (2011) asserts that FR countries often have bad growth records and high inequality, especially in countries with weak institutions, high corruption, and absence of rule of law. Countries with good institutions, openness to trade, and high investment in exploration technology are apt to do better. See van der Ploeg (2011) for the discussion of the empirical literature that supports his conjecture.

Investing Fossil Rents for a Sustainable Future

Fossil resources are one form of capital, but there are others, including producible capital (e.g., machinery or infrastructure), human capital in the form of skills and education, and even institutional capital. Fossil resources may be extended by technology, but in the end, economically profitable reserves will eventually run out. However, it is possible to extend the revenues from natural resources by investing the rents in foreign assets or other forms of capital. Thus, the fossil reserves can provide the funds to build up other forms of capital and diversify their economies. This section contains some simple models to help us think about how to invest rents to provide a sustainable future.

Numerous governments have created sovereign wealth funds to sustain some of the resource rents for future generations. The Sovereign Wealth Fund Institute has a ranking for these funds by size (Sovereign Wealth Fund 2013a). Table 21–7 contains the funds for the FR countries for which asset values are available. The Linaburg-Maduell Transparency Index is based on 10 points. These points include history of fund, purpose, contact information, independent audits, and fund investments (Sovereign Wealth Fund 2013b); 10 is the most transparent, and 1 is the least.

Of the 31 FR countries, 19 have sovereign wealth funds from fossil fuels. Most often, the source of funds is oil, and a few are from oil and gas. Norway has the largest fund dating from 1990, which is also the most transparent. Azerbaijan has a younger and smaller fund that also receives the highest transparency rating, but it provides far fewer assets per capita than Norway's fund. Three of the emirates have one or more sovereign wealth funds, with varying degrees of transparency from 3 to 10. Abu Dhabi has the most assets per capita at more than $0.4 million per person in its combined funds. Most countries have less than $100,000 worth of assets per capita.

Table 21–7. Sovereign wealth funds from fossil fuels in fossil-rich countries

Country	Assets (billion dollars)	Assets (per capita)	Year of Creation	Source of Funds	Linaburg Maduell Transparency Index
Norway	$715.90	$145,864	1990	Oil	10
UAE Abu Dhabi	$627.00	$404,867	1976	Oil	5
Saudi Arabia	$532.80	$19,330	NA	Oil	4
Kuwait	$342.00	$95,477	1953	Oil	6
Russia*	$175.50	$1,228	2008	Oil	5
Qatar	$115.00	$67,647	2005	Oil	5
UAE Dubai	$70.00	$39,536	2006	Oil	4
UAE Abu Dhabi	$65.30	$42,166	1984	Oil	9
Libya	$65.00	$9,927	2006	Oil	1
Kazakhstan	$61.80	$3,760	2000	Oil	8
Algeria	$56.70	$1,601	2000	O&G	1
UAE Abu Dhabi	$53.10	$34,288	2002	Oil	10
Iran	$42.00	$562	2011	O&G	5
Azerbaijan	$32.70	$3,614	1999	Oil	10
Brunei	$30.00	$72,464	1983	Oil	1
Oman	$8.20	$2,751	1980	O&G	1
Saudi Arabia	$5.30	$192	2008	Oil	4
Angola	$5.00	$262	2012	Oil	NA
Trinidad & Tobago	$2.90	$2,200	2000	Oil	4
UAE Ras Al Khaimah	$1.20	$6,981	2005	Oil	3
Nigeria	$1.00	$6	2011	Oil	NA
Venezuela	$0.80	$27	1998	Oil	1
Gabon	$0.40	$267	1998	Oil	NA
Equatorial Guinea	$0.08	$61	2002	Oil	NA
Oman	NA	NA	2006	Oil	NA
Kazakhstan	NA	NA	2012	Oil	NA

Source: Sovereign Wealth Fund (2013a).
Note: Table last updated by Sovereign Wealth Fund Institute (SWFI) in 2013. NA = not available. *Assets estimated by SWFI. O&G = oil and gas.

Kuwait has the oldest fund, dating from 1953. It has the fourth largest fund with the third largest accumulation per capita. The majority of funds were created after 2000. Nigeria, after more than five decades of oil production, only set up a fund in 2011 and sadly has only accumulated $6 per capita.

Now let's consider a simple model of resource fund investment from Hannesson (1998) that would create a sustainable level of income from now on. Assume a stock of S reserves that last T years. You produce an equal share (S/T) at the beginning of each year. The share you save is α, so each year you save $\alpha(S/T)$; the interest rate paid on your fund is r. When reserves are gone, you have saved the following:

$$\alpha(S/T)(1 + r)^{T-1} + (1 + r)^{T-2}\,\alpha(S/T) + \ldots + (1 + r)^0\,\alpha(S/T) = \sum_{i=0}^{T-1} \alpha(S/T)(1 + r)^t \quad (21\text{--}1)$$

Your consumption each year, which you want to continue indefinitely, is $(1-\alpha)$ (S/T). Using equation (21–1), the interest on your fund accumulation after your resource is depleted should equal your sustainable consumption:

$$r\sum_{i=0}^{T-1}(1 + r)^t\,\alpha\,\frac{S}{T} = (1 - \alpha)\frac{S}{T} \quad (21\text{--}2)$$

Solving for α in the above equation yields the savings rate for sustainable production. First get all αs on the same side:

$$r\sum_{t=0}^{T-1}\alpha(1 + r)^t = (1 - \alpha) \rightarrow \alpha + r\sum_{t=0}^{T-1}(1 + r)^t\alpha = 1$$

Then:

$$\alpha = \frac{1}{1 + r\sum_{t=0}^{T-1}(1 + r)^t}$$

Further simplify the above summation as follows. Let S equal the following:

$$S = \sum_{t=0}^{T-1}(1 + r)^t$$

Note the following:

$$(1 + r)S - S = (1 + r)^T - 1$$

Multiply through in the first expression and solve for S:

$$S + rS - S = rS = (1 + r)^T - 1 \rightarrow S = \frac{(1 + r)^T - 1}{r}$$

Substitute for S in the equation for α:

$$\alpha = \frac{1}{1 + rS} = \frac{1}{1 + r\frac{(1 + r)^T - 1}{r}} = \frac{1}{(1 + r)^T}$$

If you have reserves of 60, T is 20, and the interest rate is 7% (or in decimal form = 7/100 = 0.07), how much should you save?

$$\alpha = \frac{1}{(1 + r)^T} = \frac{1}{(1 + 0.07)^{20}} = 0.26$$

Annual production is 60/20 = 3. You save 0.26 × 3=0.78 and consume 0.78 × 3 = 2.34.

Alternatively, the country could take some of its resource rents to invest in other renewable sources of capital, such as physical, human, technological, and institutional resources. To get some insights into growth, see some classic growth models that subsequently had nonrenewable resources explicitly added to them at http://dahl.mines.edu/b2101.pdf. Although such growth models are highly stylized, they focus our attention on saving and investment to provide sustainable development. Two alternative objectives have been considered in such models: how much capital is needed to maximize total utility for all generations and intergenerational equality. To share resource rents with coming generations, some or all of the rents can be invested in other forms of capital, such as physical, human, technical, institutional, and financial. These other forms of capital can provide a flow of income in the future. As with all economic choices, efficient allocation requires knowledge of the trade-off between all the available assets.

In a seminal article, Hartwick (1977) shows that if you invest all the resource rents, you will maintain constant consumption. Hamilton et al. (2006) estimated the percentage change in reproducible capital for 70 countries that would have accumulated if they had followed the Hartwick rule from 1970 to 2000. Their natural capital includes fossil fuels as well as nine other minerals. Nine of our resource-rich countries were included in their sample. The number in parenthesis shows their estimate of the percentage difference in the countries' producible capital stock (including investment in education) if the country had followed the Hartwick rule compared to actual capital stock in 2000. For eight of the countries— Nigeria (358.9%), Venezuela (272.1%), Republic of Congo (57.0%), Gabon (80.3%), Trinidad and Tobago (182.1%), Algeria (50.6%), Bolivia (116.1%), and Ecuador (95.3%)—capital stock would be significantly higher if all rents were invested. Only for Norway (−14.3%) would the producible capital stock have been less under the Hartwick rule.

Other statistical information on whether rentier economies are living off and drawing down rents or using rents to invest in human and reproducible capital comes from World Bank's WDI (n.d.). From their database, you can compute a value for net investment in reproducible capital plus expenditure on education minus depletion for energy, minerals, and forests. This is a measure of investment including the drawdown of natural capital. The data, available for 22 of the 31 resource-rich countries, is presented as a percent of gross national income in table 21–8.

Table 21–8. Net investment in producible capital plus educational expenditure minus natural capital depletion (average 2000–2010 as percent of GNI)

Algeria	22.7	Gabon	−0.7	Saudi Arabia	1.3
Angola	−32.4	Iran	−2.3	Sudan	−6.5
Azerbaijan	−10.5	Kazakhstan	−12.8	Syria	−4.8
Bahrain	2.6	Kuwait	10.1	Trinidad and Tobago	−10.7
Bolivia	0.5	Mongolia	8.9	Venezuela	3.7
Brunei	−1.4	Norway	14.9	Yemen	−14.9
Congo, Rep.	−49.8	Oman	−11.5	China	31.7
Ecuador	0.7	Russia	2.7	LowY/MedY/HighY	12.4/12.7/8.5

Source: World Bank (n.d.).
Note: LowY/MedY/HighY = the WDI Low-, Medium-, and High-Income Country categories.

The majority of the FR countries are depleting more natural capital than they are investing in producible capital and education, hence the negative total net investment. For comparison purposes, I include the averages for the World Bank's (n.d.) low-, medium-, and high-income category countries. Only Algeria and Norway are investing a larger percent of their GDP than the average for their income group. China is investing almost one-third of its GNI and by comparison is investing well above the middle-income average of 12.7%.

Summary

Energy reserves can be a blessing and a curse. They can bring in government finances, foreign exchange reserves, funds for investment, technology transfer, and create employment. However, they may also bring corruption, inflation, political crisis from rising expectations, increasing income inequality, and rapid urbanization. Governments may squander public revenue by making poor use of resource rents. In FR countries, a high-valued currency may damage domestic industry and agriculture by making them noncompetitive in world markets.

To enjoy the blessing while exorcising the curse, energy-rich countries must design policies and institutions to use these resources wisely in the present to sustain the income flows for coming generations. Although Norway illustrates the possibility of this task, the other 30 fossil-rich countries considered in this chapter suggest the task is fraught with difficulties.

The countries considered receive more than 14% of their GDP from fossil-fuel rents. Most get the bulk of their fossil-fuel rents from oil, a few from natural gas, and only Mongolia gets the bulk from coal rents. These countries vary in many ways, including population, fossil dependence, income per capita, reserves per capita, the length of time they have had a high fossil-fuel dependence, how much of their fuel is exported, reserves relative to production, fuel diversification, and potential energy sources. Publically available data is increasingly easy to access and

I have used it here to compare and contrast the situation in these countries. I use this data to provide a snapshot of the situation in these countries as well as some historical trends, with a caution about inaccuracies always available in such data sets. These inaccuracies are likely larger the poorer and less developed the country and the further back in time we go.

There is some evidence that these countries are more energy intensive, with faster energy growth often spurred by subsidies. The transitional economies are exceptions, with slower energy consumption growth and falling energy intensity. Meanwhile, energy consumption actually fell in Norway from 2000 to 2010.

Often these countries do not have a highly diversified energy supply, but many have been moving to diversify out of their dominant fuel. Most do not produce much primary electricity. Norway is an exception with its huge hydropower production. Most primary electricity in these countries is traditional hydropower. Only Russia has nuclear power. Both these countries, as well as a few others, have begun forays into the production of wind and solar. Traditional biofuel is still prominent in Sub-Saharan Africa, but modern combustible waste and biofuels for electricity and transportation are limited in most of the countries considered. Using CO_2 for enhanced oil recovery could extend the life of oil and gas fields, and experimental as well as commercial projects are gaining interest, as is the interest in modern renewables.

Conventional neoclassical economic theory suggests that more inputs should produce more outputs and generate more income. Absolute and comparative advantage should allow trade to enhance these gains. Much empirical work has catalogued the performance of resource-rich countries and compared them to countries not considered rich in resources. Often the resource-rich countries come up short in these comparisons, including access to electricity and improved water, investment in health and education, income equality, life expectancy, corruption, and violence.

Since correlation is not necessarily causation, much work has been done to explain these failures. Theories considered include weakening of terms of trade for natural resources relative to other goods and macro volatility induced by volatile resource markets discouraging other investment. Another theory is that exports from the resource sector raise the value of the currency and hurt other sectors (Dutch disease). Other theories consider the role of resources in using productive inputs and raising input prices, making inputs scarcer to other sectors (crowding out). Weaker linkages between the resource and other sectors have also been considered, which could leave less forward and backward linkages to promote development. Other theories have considered that externalities and theft may separate those who bear the costs of resource development and exploitation from those who receive the benefits. In addition, weak institutions, including the government, may make poor use of the rent, contributing to inequality, rent seeking, low investment rates, and corruption. Although all these issues may provide challenges to the resource-rich, the exceptions suggest that these challenges can be overcome, making resources the blessing that they should be.

Development and improving standards of living are highly dependent on increasing the capital stock. Fossil-rich countries have been blessed with an abundance of natural capital. Drawing down this abundance yields income. To make this income sustainable requires that they invest it in other forms of capital.

They can invest in international financial capital through a natural resource fund, and the majority of their countries have set up one or more such funds. If they have a stock of reserves (S) that they will produce over T years, r is the interest rate that their fund pays, and they want to provide a constant stream of income in perpetuity, the share they should save is $\alpha = \dfrac{1}{1 + (1 + r)^T}$.

They could also take some of natural resource rents and invest it in other forms of capital, such as producible, human, and social. Many of the resource-rich countries considered in this chapter have negative capital accumulations and are drawing down resources faster than they are accumulating producible capital. For those that are accumulating capital, most are accumulating at a slower rate than the average for other countries in a similar income category.

22

Managing in the Multicultural World of Energy

We see things not as they are but as we are.

—Talmud

Introduction

Multinational oil companies are big and getting bigger. National oil companies are reorganizing and going global. Many of these large energy companies participate internationally through exports and imports, licensing and franchising of technologies and products, direct foreign investment, and joint ventures. Information and technology are pervasive and bringing us closer together. In this 21st century environment, multicultural communication and interaction are increasing, whether the culture is technical, corporate, or national. Business success requires the ability to operate effectively across diverse cultures, and such effectiveness requires an understanding of, and appreciation for, numerous aspects of culture.

In this chapter, let's focus first on this human element of management and consider the various aspects of intercultural differences across nations, how these differences can cause problems, and how these differences can be turned to competitive advantage. Superb management requires that these intercultural and human issues are combined with the economic principles and analytical tools developed throughout the book, which are summed up later in this chapter.

Managing in global industries, such as energy, requires a wide skill set. A good manager must plan, organize, and lead by maintaining financial control, building enthusiasm, developing innovative marketing, training personnel, measuring personnel performance, and overseeing product quality. Enhanced corporate performance may require closure of failing operations or modifications to reach new or more promising markets.

Often these more promising areas involve an international component. This has been made easier as privatization and deregulation of energy markets have caused major flows of international capital. This process is matched by increasing activity of state energy companies outside their national borders, either starting their own operations or setting up joint ventures with foreign partners.

Increased globalization, increasing attention to ethical social responsibility, changing demographic and skill requirements, and consideration of employee needs are important elements in such management. The manager must not only manage work, organization, production, operations, and technology, but the human dimension, including employees and customers. This latter dimension requires knowledge and social skills to succeed across a variety of cultures.

Culture

A culture is the shared values, attitudes, and behaviors of a group, or their customary ways of perceiving and doing things to satisfy needs. Maslow (1970) organizes needs into a five-category hierarchy, beginning from the most basic:

- physiological needs, like food and shelter
- safety needs
- love and belonging needs
- esteem needs for self and others
- need for self-actualization

Cultures satisfy these needs in different ways depending on a number of factors. These include religion, language, economic and political philosophy, educational system, physical environment, settlement pattern, social organization, family structure, literature, civil organization, government organization, and law. Cultures evolve, and their tenets are typically passed on through the family, education, workplace, religion, and even through the media.

Hooker (1998) likens culture to a group ecosystem. The group may be a nation, in which case its culture includes language, ethics, religion, and customs. The group may also be a profession, such as engineering, geology, or economics, with its own jargon, common methodologies, and analytical frameworks. It may be a corporation, with a culture that includes decision-making processes that could be bureaucratic, centralized, or entrepreneurial. It may be a part of an organization, such as a foreign division, a world headquarters, a refinery, or a research and development division. In each of these contexts, there is a prevailing culture. Understanding its norms and how the parts work together can lead to greater success in the given environment and a stronger chance of survival. Though the focus in this chapter is on national cultures, many of the concepts have wider application to the professional and corporate world.

Culture is learned, and national culture is currently accepted to be relative, rather than right or wrong when compared to some global absolute. However, various cultures have dominated across history (e.g., the ancient Romans, the 19th century British, and the 20th century Americans). Within national cultures, we find a wide variation in individual values and behavior. For example, suppose the cultural trait is the value placed on individualism. Let this trait be measured by an index that goes from 1 to 20, with higher values indicating a greater preference for individualism. Suppose in figure 22–1, the left-hand probability distribution with a mean of 7 represents Japan, and the right-hand distribution with a mean of 15 represents the United States.

From this figure, we see that the United States puts a higher value on individualism, on average, than the more group-oriented Japanese. Knowing this can be useful when trying to decide how to motivate personnel, market and sell products, or organize work assignments across individuals and teams in various cultures. It allows managers to avoid misunderstanding and to use the differences to their competitive advantage. Nevertheless, care must also be taken to not stereotype individuals, since wide differences exist within as well as across cultures. For example, in the figure, the Japanese individual represented by *J* values individualism more than the American represented by *U*. Country rankings for individualism and numerous other cultural preferences are shown in table 22–1.

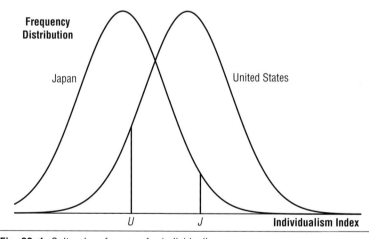

Fig. 22–1. Cultural preferences for individualism

The two most oft-cited authors classifying cultural differences, particularly relating to the corporate world, are Hofstede (1991) and Trompenaars (1993). Hofstede notes four cultural elements in work-related activities:

- individualism/collectivism (IND)
- masculinity/femininity (MAS)
- power distance (PD)
- uncertainty avoidance (UNC)

Table 22–1. Cultural differences

Country or Region	IDV 100%=high	PD 100%=high	UA 100%=high	Mas/Fem 100%=More Mas	UNI/PAR 100%=More UNI	Neu/Aff 100%=More Neu	Spe/Dif 100%=More Spe	Ach/Asc 100%=More Ach	In/Out 100%=More In	ConD 100%=More L Term
Arab C.	42	77	61	56						
Australia	99	35	46	64	84		98	90	96	26
Austria	60	0	63	83	74		79	47	54	55
Brazil	42	66	68	52	83			67	74	
Canada	88	38	43	55	91		87	95	73	19
Czechoslovakia					90		26	0	39	
China					54		25	68	46	100
France	78	65	77	45	76	41	89	73	67	
Greece	38	58	100	60	64		77	74	63	
Guatemala	0	91	90	39						
India	53	74	36	59	66		67	90	65	52
Indonesia	15	75	43	48	50	90	44	51	87	
Iran	45	56	53	45						
Ireland	77	27	31	72	100		96	88	70	
Israel	59	13	72	49						
Japan	51	52	82	100	82	100	70	41	39	68
Malaysia	29	100	32	53	52		77		43	
Mexico	33	78	73	73	67		77	74	43	
Nigeria					70		62	71	62	
Norway	76	30	45	8	86	73	88	100	69	14
Pakistan	15	53	63	53	70		77	88	100	0
Philippines	35	90	39	67	69		79	81		16
Russia			0		43		24	57	50	
Singapore	22	71		51	60	51	70	550	46	41

South Africa	71	47	44	66			100	85	28
Sweden	78	30	26	0	77		86	57	55
Turkey	41	63	76	47	72		96	97	58
USA	100	38	41	66	90	48			76
Yugoslavia	30	73	79	22	0		0	0	25

Sources: Hoecklin (1995): PD (table 2.1, 30), UA (table 2.5, 36), MAS (table 2.7, 38), IDV (table 2.3, 32), UNI/PAR (figure 4.1, 37; figure 4.2, 40; figure 4.3, 42), Neu/Aff (figure 6.1, 70), SPE/DIF (figure 7.3, 87; figure 7.6, 93), ACH/ASC (figure 8.1, 103; figure 8.2, 104), IN/OUT (figure 10.1, 139; figure 10.2, 140); Trompenaars (1993): UNI/PAR (figure 4.1, 37; figure 4.2, 40; figure 4.3, 42).

Note: IDV = individualism; PD = power distance; UA = uncertainty avoidance; MAS = masculinity; UNI = universalism; PAR = particularism; NEU = neutral; AFF = affect; SPE = specific; DIF = diffuse; ACH = achievement; ASC = ascription; OUT = outer; IN = inner; L Term = a more long-term view; Arab C. = Arab countries. All scores were normalized to go from 0 to 100. Blank cells indicate missing data. 100% is always more of the first attribute in the label at the top. The same attributes for a number of other countries can be seen at http://dahl.mines.edu/st22/st22.pdf, question 2.

Individualism/collectivism

In the most individualistic cultures (higher IDV scores) in Hofstede's survey, such as the United States, Australia, and the United Kingdom, individual initiative and leadership are most valued. People are permitted (and expected) to have their own opinions, and promotion is more likely to be based on merit and individual accomplishment. In collective societies, which are the majority worldwide, the group is more highly valued, and the individual receives value from being in the group. In return, the group is responsible for taking care of its individual members. Promotion is from within the group and tends to be based on seniority. Socialist countries in the past were very group-oriented, as are many East Asian and Latin American countries in Hofstede's sample. Thus, a brash individualistic American management style may fall flat in Asia or tribal Africa, where the group defines the individual and consensus is important.

Adler (1997) suggests that groups tend to be better at establishing objectives and evaluating and choosing alternatives to meet those objectives, whereas the individual tends to be better at coming up with objectives. The effectiveness of the orientation also depends upon the individual's cultural background. For example, Earley (1989) found that Chinese workers performed administrative tasks better while working anonymously in a group, while Americans performed the same administrative tasks better when working separately, with personal attribution of the tasks.

Power distance

Power distance represents the degree of equality in a group. Cultures vary by how authority is distributed within groups. The more hierarchical and centralized the management of the group, the larger the power distance and the higher the score in the table. Power distance is higher if the boss's decision is accepted, right or wrong.

In high power-distance contexts, the manager is viewed as an expert; in low power-distance contexts, the manager is viewed as a problem-solver in conjunction with the group. Egalitarian managers in high power-distance contexts may be viewed as weak and incompetent, or employees may interpret the manager's help as a signal that employees are doing poorly. Authoritarian managers in a low power-distance context may be viewed as dictatorial.

Thus, managers with a more egalitarian approach may not work as well in Latin America, Arab countries, and Indonesia, which tend to maintain more power distance than in the more egalitarian Northern Europe, United States, and Canada. Even within Europe we see differences. In a BP finance office, Germans tended to be more hierarchical; the Dutch, Scandinavians, and British were more likely to challenge authority, and the French accepted management authority more or less as a right and obligation (Hoecklin 1995).

Uncertainty avoidance

Uncertainty avoidance represents attitudes toward risk. Regions with high uncertainty avoidance (and larger scores in the table) include Japan, Southern European countries that have traditionally been more predominantly Roman Catholic or Orthodox, and South America. They are more uncomfortable with ambiguity, dislike conflict in organizations, and may prefer formal rules. They may be fairly conformist, with originality neither rewarded nor valued. Those with low uncertainty avoidance (Singapore, Scandinavia, Canada, the United States, and the United Kingdom) deal better with ambiguity and change and are more likely to take risks for commensurate rewards. Long-term job security tends to be important in high uncertainty-avoidance cultures. The concept of jobs-for-life in the Japanese context reflects high risk-avoidance preferences. Managers are more likely to be chosen by seniority. Rules should not be broken, even for good reasons. In low uncertainty-avoidance cultures, job mobility is higher. Managers are more likely chosen by merit. There is more flexibility and judgment in interpreting and breaking rules.

Cultures have four ways of dealing with uncertainty:

- technology
- rules and law
- religion and ritual
- relationships

Technology (often favored in the West) may be used to deal with uncertainty resulting from nature. It can help provide food, improve material comfort, and conquer disease. Rules and law (or relationships) may be used to deal with uncertainty resulting from other people. We will see this aspect of culture discussed below. Religion and ritual may often help us cope with uncertainties beyond our understanding and provide a basic value system.

Masculinity/femininity

Masculinity/femininity relates to how important stereotypically masculine values such as assertiveness and success are relative to stereotypically feminine values such as relationships and physical environment. Higher values in the table reflect more masculine cultures and lower values reflect more feminine cultures. Cultures with an aging population also tend to shift somewhat to more feminine values.

These traits also indicate how important gender is in the business world. More masculine societies tend to have tighter specifications of gender-specific activities, more industrial conflict, and higher stress levels. Business women in masculine OPEC countries face special sets of problems not as prevalent in the more feminine cultures of Scandinavia. Gustavsson (1995) notes that feminine orientation toward human values is one of the important aspects of the Scandinavian management style. This style includes not only humanitarian values but also consensus decision-making and a commitment to equality. This commitment is across gender as well as socio-economic classes and discourages braggadocio and self-promotion. Equality also requires a less hierarchical structure, with management more by vision and values and less by giving orders. For example, the Swedes were early leaders in moving from assembly line production to work teams at Volvo.

Hines (1989), in a somewhat similar vein, uses an Eastern framework featuring yin (creation) and yang (completion). Each is thought to have components of the other within it, with the dominance of the two forces changing and cycling through time. Yin values are sharing, relatedness, and kinship; cultures with a more yin orientation are more relaxed, responsive, open, flexible, and passive. Aboriginal cultures may have a yin framework toward land and work. Yang values are quantification, objectivity, efficiency, productivity, reason, and logic; cultures with more of a yang orientation are more excitable, commanding, outgoing, decisive, and active. Western nations' accounting and accountability rules reflect a yang orientation (Greer and Patel 2000).

Although criticized for stereotyping by some, the best-selling *Men are from Mars and Women are from Venus* by John Gray (2012) is a fun take on personality differences between the genders that can lead to misunderstanding. The book is aimed more at personal relationships but provides some food for thought at the professional level as well. The book offers a valuable reminder that others may be interpreting the situation from a different frame of reference.

Confucian Dynamism

In addition to the previous list of cultural indicators, Hofstede and Bond (1988) add an indicator called *Confucian Dynamism*, which is particularly important in understanding and functioning in Asian cultures and relates to a culture's orientation across time. Confucian values place a high importance on long-term commitments, respect for tradition, and a work ethic that favors thrift and persistence, delaying current gratification for longer-term gain. Business takes longer to develop in this type of society and change generally takes place more slowly than cultures with low long-term time orientation.

A longer term focus also suggests that the individual may be more likely to submit to the group and its hierarchy and have a sense of shame. Shame in this context is extrinsic and relates to group approval; you bring shame if you make the group look bad, but you are only shamed if the group knows about it. Related to the concept of shame is the notion of face—*having face* means enjoying a high standing with one's peers. Being shamed causes one to lose face. One *loses face* through criticism, insults, or the failure of others to show proper respect.

Group-oriented cultures that are shame-based often do not say "no" directly, for instance; they do not want to cause others embarrassment or loss of face.

Westerners must take care to understand that something as seemingly simple as an affirmative response of "yes" may not really mean the individual is agreeing to something. In Japan it means, "I heard you." In China, answers can be evasive; something that is "inconvenient," is "under consideration," or "would be difficult" typically means "no." A "yes" said with air sucked in through the teeth also means "no." In India, too, evasive answers typically mean "no."

In more individualistic cultures with a shorter term orientation, social control and incentives may be more guilt-based. Here, the obligation is to the self, and self-approval is important. In a guilt-based system of control, pressure is internalized: Your failures do not bring shame to the whole group, and guilt is to be felt whether others know about it or not.

Time

Trompenaars suggests another way that cultures view time, pointing out that events may be considered sequential or synchronous. In *sequential* cultures, things are done one at a time in sequence; appointments and plans are closely adhered to. Someone working out of sequence may cause a whole plan to fail. In *synchronous* cultures, many things may be done at once. Appointments and plans change, and relationships are important. Turbulent cultures may be synchronous, since many paths to the same end may increase the probability of ultimate success. However, a sequential person from the United Kingdom may be a bit disoriented in a meeting with a synchronous Arab or Latin American, who will stop the meeting for many interruptions.

Four out of five of Trompenaars' concepts deal with relationships, and his scores for these attributes are also shown in table 22–1.

Universalism and Particularism

Universalists (UNI) believe that there are norms, values, and behavior patterns that are valid everywhere; particularists believe that circumstances and relationships determine ideas and practices. Universalist cultures score higher numbers in table 22–1, while particularists score lower numbers.

In universal cultures (United States, United Kingdom, Australia, and Germany), more focus is placed on rules and formal procedures, such as detailed contracts. Hooker (1998) argues that Western rule-based cultures (which he calls "rude") are influenced by the notions of justice rooted in Judaism, Islam, and the Greek notion of rationality. People trust laws based on justice and reason to create social harmony and resolve conflict.

In particularist cultures (China, Japan, Korea, Indonesia, the former Soviet Union, and Venezuela), relationships are more important, and authority, group solidarity, and sensitivity to others are used to create harmony. Hooker calls these "polite" cultures, since more attention is often paid to the feelings of others.

In parts of Asia, a strong influence of Confucius and respect for authority within families and within organizations prevails. In China, relationships between families and organizations (*guānxi*) are developed through mutual obligations based on gifts and favors. Over time, trust is built and forms the basis for their interactions. Koreans are somewhat similar, but hierarchy is stronger than in China, with gifts and favors larger and more important. Although the Japanese also yield to authority, they place more emphasis on the group than the other two countries. Group harmony and social ritual are extremely important to the Japanese culture.

Although Latin America evolved out of the Western tradition, Hooker (1998) argues that its repressive and turbulent past has contributed to its being particularist rather than universalist and rule-based. In a violent society with erratic and repressive rulers, family and friends were a source of safety and survival; only they could be trusted, and great care was taken to not offend others.

Particularist culture small talk and socializing are part of the familiarization and trust-building process. In a let's-get-down-to-business universalist culture, such activities might be considered a waste of time. Contracts can obviate the need for trust to a universalist, whereas the detailed contracts of a universalist might signal a lack of trust to a particularist, who expects that contracts and relationships will be modified over time.

McWorld/Jihad

Along somewhat similar lines, Barber (1996) looks at a social confrontation he calls "McWorld versus Jihad." The McWorld point of view sees the world as one large market connected by information networks moving toward automation and homogenization. Centralized multinational companies with a headquarters office (or country) and more polycentric, transnational corporations use large amounts of natural resources to serve a global market. The headquarters office takes a leading role, coordinating activities of the various subsidiaries, while each production center specializes where it adds the most value, and knowledge plays an important part. McWorld is associated with occidental (particularly American) culture.

Jihad represents the point of view that opposes modern capitalism and clings to religious beliefs, ethnic traditions, and local and national communities. Jihad elements in a culture increase the risk for capitalists doing business in them; 24% of the world's oil reserves are in areas where Jihad beliefs are prevalent.

The McWorld point of view values and promotes economic well-being but not necessarily social and political well-being, while Jihad promotes community but is often intolerant. Barber suggests taking the best from McWorld while maintaining a cultural identity and sense of community from Jihad. He believes that Japan and China have been reasonably successful at doing just that. Alternately, McWorld managers in Jihad cultures need to pay special attention to indigenous groups and cultures.

Neutral/affective

The neutral/affective trait indicates whether emotions are expressed or not. In neutral cultures (high numbers in table 22–1) such as Japan, the United Kingdom, Singapore, and Indonesia, expressing emotions, particularly intense emotions, is viewed with disfavor and is considered unprofessional. More affective cultures (lower numbers in the table) such as Mexico, the Netherlands, China, and Russia are much more comfortable expressing emotions in public and may view people from neutral cultures as cold or deceitful.

A further division within these categories is whether emotions should be exhibited and whether emotions should be separated from reasoning processes for business decisions. Scandinavians tend to not exhibit their emotions and feel that emotions should be separated from business decisions. Southern Europeans tend to exhibit their emotions and not separate them from rationality in decision-making; Americans tend to exhibit but separate.

Specific/diffuse

Specific/diffuse relates to how a culture views private and public relationships. An individual presents a public space to everyone and keeps a private space that is shared with selected individuals. In the specific cultures of Australia, the United Kingdom, and the United States, an individual has a small private space, which is compartmentalized from the public space. The public space is easily entered but is very restricted and compartmentalized.

In a diffuse culture such as China, the private space is larger and less compartmentalized, so it is harder to enter someone's public space in a diffuse culture because it allows easier entrance into the private space. Diffuse cultures may at first seem cold to those from a warmer culture; specific cultures may be viewed as shallow and superficial by their more diffuse neighbors. People from specific cultures may find it very time-consuming to work in diffuse cultures. For people from specific cultures, the requisite social activities may seem not only time-consuming but perhaps even an invasion of privacy.

Achievement/ascription

Hofstede explores how power and authority vary across a group. In Trompenaars' category achievement/ascription (ACH/ASC), he explores how power and status are attributed to members of the group. In achievement cultures such as Australia, the United States, Switzerland, and the United Kingdom, one's status is determined by how well one performs desirable functions for the group; the emphasis is on the task. Status and power in an ascriptive culture is more on who you are, than what you are. Status and power are conferred by things often ascribed at birth, such as gender, family, and social connections.

Ascriptive cultures include Venezuela, Indonesia, and China, where the emphasis is more on relationships than achievements. However, ascriptive status and power may enable leaders to get things done just as well as in an achievement culture that is likely to be more rule based.

Our relationship to our environment may also vary by the degree of control we feel over our destinies. Trompenaars designates cultures whose members feel that they are in control of their fates as inner-directed (IN). Those who feel they are merely "pawns in the game" controlled by fate are outer-directed (OUT). Higher numbers in table 22–1 indicate more inner-directedness, while lower numbers indicate more outer-directedness. North Americans and Europeans tend to be more inner-directed. In contrast, in Arab nations someone may say, "Insha'Allah," which means "God willing" and is often spoken after statements of coming events, suggesting a more outer-directed view of the world. Native Americans would also fall more in the category of believing in fate.

Cognitive Styles

Cognitive style relates to how our culture organizes and processes sensory information, including what we absorb and what we filter out. Coyne (2001) refers to two cognitive styles: open-minded and closed-minded. A more open-minded culture examines external information and considers all points of view; a more closed-minded culture filters out a wider range of external information, particularly if it does not agree with preconceived notions.

The United States is an example of a more closed-minded culture. Theocratic societies, ruled by religion, also tend to be closed-minded. Indeed, Coyne contends that most cultures are closed-minded if things are going well. Japan began to modernize only when threatened by foreign encroachment in the 1860s. US complacency about American management techniques from the 1950s and 1960s began to break down in the 1980s and early 1990s as the Japanese took center stage. With this threat to national pride, American corporations began to adopt or consider many of the tenets of Japanese management. These tenets include total quality management, lean manufacturing, just-in-time inventories, and an incremental approach to reducing waste and inefficiency by empowering people at all company levels. They also include consensus decision-making and a more paternal attitude toward labor to improve long-term loyalty.

After data is filtered, it is processed and assimilated in various ways. An associative individual tries to associate new data with personal experience. An abstract individual analyzes the components and tries to develop theories about processes and relationships. Whether someone is associative or abstract is heavily influenced by the educational system. Teaching by rote develops associative thinkers, while teaching by problem solving produces more abstract thinkers, who are better able to deal with things that are completely new.

Life Values

Social cultures value people; economic cultures value acquiring wealth and satisfying physical needs. Theoretical systematic cultures value the acquisition of knowledge with little deference to beauty or usefulness. Power cultures value the acquisition of power and control over others, while religious cultures value unity with the cosmos. Truth is typically established by faith, fact, or feelings, with feelings being the most common.

Business Protocols

Coyne (2001) indicates a number of items of business protocol that vary considerably across cultures and should be mastered before doing business in another culture.

They include personal space, greetings, expectations about punctuality, necessity for making appointments, pace of business, and value of personal contacts. Delivery performance, business cards, when business discussions can take place, forms of greeting, use of titles, eye contact, gift expectations, and appropriate business dress are also important.

Personal space

Hall and Hall (1990) note that personal space and territory vary across cultures. The Japanese stand farther apart from one another than do North Americans, who in turn stand farther apart from one another than Middle Easterners and Latin Americans. Latin Americans touch more frequently than either North Americans or the Japanese. Greetings vary, as well. Respecting personal space and understanding culturally appropriate greetings will pay dividends in business dealings.

Ritual

Ritual varies across cultures. Asia tends to have high ritual cultures, where behavior tends to be more structured and follows set rules. For example, in Japan, rules govern gift giving, including the gift, the manner of presentation, the manner of acceptance, and how the gift receiver reciprocates. Another example is that of business cards in Asia. The card is presented with ceremony and is not to be shoved in the pocket after a glance (Morrison and Conaway 2006).

Human nature

Adler (1997) notes various cultural conceptions of human nature. Cultures that view people as basically good tend to trust people until they are proven untrustworthy. Cultures that view people as basically untrustworthy tend to use safeguards to protect themselves from people until they are proven trustworthy. Some cultures may be neutral or believe that each individual varies in his or her moral character. Some cultures believe that character is changeable, while others think it is not. If humans are changeable, as the Chinese believe, they will spend more time and effort on training and encouraging personal improvement. If personalities and qualities are more immutable, more resources will be spent on selection and screening, as is done in the United States.

Humans and nature

Cultures vary in how they interact with the world and nature. They may feel they dominate nature, are in harmony with nature, or are dominated by nature. Some cultures may view the world as stable and predictable, while others view it as random and turbulent. Western cultures are more likely to feel they dominate nature, whereas Eastern cultures typically want to be in harmony with nature.

For example, the Chinese practice of feng shui (derived from Taoism) illustrates the belief that there are natural laws and cycles whose energies can be harnessed as it flows through all things to be in harmony with nature. Form, shape, and particularly spatial alignment are used to bring the environment into alignment with natural energy flows. Thus, in a Far Eastern environment, office furniture alignment and location are important considerations for a smoothly flowing office and should not be left to chance.

Relation to work

Another aspect of the human relationship to nature, according to Kluckhohn and Strodtbeck (1961) is the orientation toward activity or the purpose of work. They establish three culture types: doing, being, or becoming. Doing cultures, such as the United States, focus on outward accomplishments for tangible rewards. Being cultures, such as Latin America, enjoy the here and now and tend to be more spontaneous. They are more likely to accept circumstances and try to make the best of them, rather than working to change circumstances. Becoming cultures focus more on the inner rewards of personal growth and self-actualization often associated with meditation and spiritual growth, as featured in Buddhism and Hinduism.

Understanding a culture's relationship to nature and work often helps in motivating employees. Two management theories are associated with these concepts.

Theory X suggests that people dislike work but are motivated by basic needs of safety and security. In this doing context, a manager directs, controls, and coerces employees to get the job done. Theory Y maintains that people are motivated by achievement and self-actualization. In this becoming context, employees will work toward things to which they have a commitment. Managers should seek to motivate and then allow employees to grow and develop as they move toward their goals.

Adler (1997) notes advantages and disadvantages of the more decentralized Theory Y. Decentralization encourages decision-making and problem-solving skills, and improves creativity and job satisfaction. It can, however, require more expensive training, higher quality employees, increased information flows, and a need to develop accountability measures.

Communication

Communication is another area where misunderstandings and problems can arise across cultures. There are a number of aspects to communication. At the verbal level there are three components:

1. What you say
2. What you mean
3. What the listener understands

What you say may be interpreted differently in two cultures because of differences in meanings across cultures.

Cultures have their own icons in the form of symbols, heroes, and rituals that represent underlying values. Idioms, similes, and metaphors that represent these icons may convey meanings and emotions that do not translate across boundaries. Cowboy images may not be meaningful to a Japanese person; Samurai images may not translate from East to West. My Egyptian student was very puzzled when I said, "Don't throw the baby out with the bath water."

Words may have different meanings in different contexts. For example, the statement "Bill Clinton was born in Hope and grew up in Hot Springs," translated into Italian (and back into English) by Altavista's Machine Translation service, reads, "The invoice Clinton has been taken in the hope and it has been developed in warm motivating forces." The Chevrolet Nova did not sell well in Mexico, perhaps because *no va* in Spanish means "doesn't go." One does not expect that the Iranian laundry soap Barf would sell well in the United States.

Low/high context

Hall and Hall (1990) refer to low-context and high-context situations and cultures. In a low-context situation, both parties know little about the context and nothing can be taken for granted. Everything must be spelled out. For example, the following sentence would not make sense in a low-context situation: "This book describes step-by-step procedures for setting up a DHCP server, securing your intranet with a firewall, running on an Alpha system, and configuring your kernel." However, an advanced Linux operator would know exactly what is meant.

In a high-context situation, two parties understand the context and very little needs to be spelled out. Cultures that are more homogenous and well connected (such as the Japanese, Arabs, and Mediterraneans) are typically high-context cultures. Cultures that are more individualistic and have more compartmentalized lives (North Americans and Northern Europeans) are typically lower context.

Explaining too much in a high-context culture may be taken as condescension; explaining too little in a low-context culture may lead to a lack of understanding.

Body language

Adler (1997) suggests that words communicate 7% of meaning, tone of voice adds another 39%, and the rest is conveyed through nonverbal means such as gesture, posture, and facial expression. Nonverbal communication may reinforce, contradict, or help clarify the verbal portion. If nonverbal actions contradict the verbal, the nonverbal is more likely to be the true signal. But nonverbal actions will only help clarify matters if the nonverbal signal means the same thing in other cultures. In some cases, nonverbal signals are identical across cultures. For instance, a smile is a greeting, and a frown is a signal of displeasure. But some nonverbal signals vary across cultures.

Nodding one's head up and down means "no" to a Bulgarian, "yes" to an American, and "I'm listening" to a Japanese person. A North American may feel that someone who will not maintain eye contact is shifty but may find the prolonged eye contact common to Arabian cultures aggressive. A Chinese or Japanese person, however, feels that direct eye contact is rude and aggressive. When nonverbal signals are contradictory, the most believable (in order of reliability) are: autonomic (involuntary) signals such as weeping; leg, foot, trunk, and hand gestures; facial expressions; and then verbal communication. Often, someone who is lying may often decrease the number of hand movements, increase the number of shoulder shrugs, increase the number of body shifts, or increase the frequency of hand-to-face contacts, particularly nose-touching, a sign of tension, and mouth covering. Slow-motion cameras can often pick up on slight facial changes that the individual attempted to suppress, even if they are typically difficult to detect with the naked eye.

Nonverbal communication typically takes the form of some sort of action that may be inherited, discovered, absorbed, or learned. Morris (2002) considers the whole gamut of gestures in a business setting. *Baton signals* are used while speaking and tend to give rhythm and emphasis to spoken thoughts. Often they are hand movements but can include the head and body or even the feet. Hands can be palms up (supplication), palms forward (a barrier), palms down (calming or suppressing effect), palms wide (welcoming), fingers together to a point (emphasizing detail), hands pounding (a point), or a fist raised (warning). Such gestures vary across individuals, cultures, and occasions. Italians gesticulate a lot; the Japanese, very little. The less educated tend to gesticulate more than the more educated, and the inarticulate gesticulate more than the articulate.

Pointing with a finger (or where this is rude or taboo, a whole hand or even the head) may indicate a direction. Beckoning may take the form of the whole hand waving with the palm down (as in parts of Asia, Africa, Italy, Spain, and South America) or the palm up for most of rest of the world. Using just one finger is riskier as it may be teasing, sexual, sarcastic, or rude; a slight head jerk may also have a sexual connotation.

Salutation displays for greetings, good-byes, and other life transformations are universal but take various forms in various cultures. The amount of body and eye contact vary by how well the parties know each other, how long they have been separated, the privacy of the meeting, and local cultural display rules. Handshakes and bows are more formal; embraces and kissing, less so. Ritual talk shows concern and pleasure.

Cultures universally have ways of establishing and showing the dominance-submissive ordering. This varies from culture to culture but may be indicated by the depth of the bow, the seating at the table, who dominates the conversation, who is introduced first, and so forth. It is a good idea to learn these cues before dealing with a new culture. When friends of equal status are interacting, often they echo each other's gestures and posture and may synchronize their movements as they talk. You may want to try this to put a subordinate or equal at ease; however, it may be considered an aggressive act with a dominant person.

Even clothing acts as a display, with each article transmitting a signal. Knowing the appropriate dress code for each culture you deal with is useful. Inappropriate or overly revealing clothing may violate local taboos and put you at a severe commercial disadvantage, as can touching areas of the body that are not be touched. Each culture has body areas that are public zones and others that are private or taboo zones that vary by relationship and gender. These zones may not exist for babies and those who are most intimate with one another, but for most other relationships they do. For example, one should never touch the head of a Buddhist, and males should be careful to avoid body contact with females in devout Muslim cultures.

Most cultures eat a few large meals over the course of the day, despite the fact that smaller meals are better physiologically. Morris (2002) argues that this custom evolves from our hunting past, where you feasted and shared food after the kill. This tendency is reinforced by human sociability. However, table manners and food customs vary considerably. A belch that may be a sign of pleasure in one culture is rude and boorish in another. Many cultures use the fork in the left hand and the knife in the right, except the United States, or where fingers or chopsticks are used. Food and drink taboos, often of religious origin, include pork for Jews and Muslims, alcohol for Muslims, and beef for Hindus. Toasting rituals vary; when and if business is conducted at a meal varies as well.

Paying attention to verbal and nonverbal nuances and shared cultural traits can be especially important to a successful advertising campaign. Observing advertising from another culture can also provide useful information on that culture's values.

Political culture

Democratic, market-based industrial economies typically function under rules of law. There is a generally accepted notion that if everyone acts within the law, society will perform reasonably well.

Centrally planned command economies tackle the complex task of trying to produce and allocate goods and services to millions of people. Strict central planning and adherence to the plan is their accepted norm. Economic incentives are not built into these systems, leading to weak motivation for work, along with shoddy products, shortages, and queues. Controlling the economy becomes harder as products become more complex and consumers more sophisticated. In such settings, those who sidestep the legal channels may make an impossible system possible. Getting around the system, rather than working within it, becomes an accepted activity.

With the fall of the USSR, some Western economists naively thought that privatization, liberalized prices, and market-based economies would fix the problems of the planned economies. Instead, powerful elites took control of the government and economic resources. The mafia's corruption became pervasive in the economy. Western laws were transplanted without the institutions or political will to enforce them. In the absence of the checks and balances developed over centuries in the West, cronyism, rather than liberal capitalism, evolved, much like the age of the robber barons in the United States. Interpersonal relationships and connections become especially important in dealing with these and other corrupt cultures.

Company culture

Companies have cultures, not unlike nations. Numerous mergers, privatizations, and the movement of national companies into international business have forced many energy companies to adapt to new business cultures. Trompenaars (1993) lists three important aspects of corporate cultures:

- the relationship between employees and their organization, or how employees view the organization and their place in it

- the relationship between superiors and subordinates
- whether the organization is person- or task-oriented

These relationships affect how the organization thinks, learns, and changes, as well as how it motivates, rewards, and resolves conflicts. Trompenaars considers four different types of corporate structure that vary by their vertical hierarchy and by whether they are task- or person-oriented (fig. 22–2):

- family
- Eiffel Tower
- guided missile
- incubator

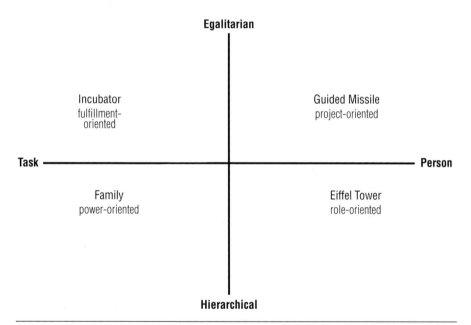

Fig. 22–2. Vertical structure and orientation for four corporate structures

The family culture, prevalent in Japan, China, Venezuela, and India, is people-oriented but very hierarchical. It maintains control through the threat of loss of affection or place in the family and motivates with praise and appreciation. It is more prevalent in high-context, diffuse cultures that are more comfortable with intuitive than with rational knowledge. Development of people is a high priority and may be accomplished through training, mentoring, coaching, and apprenticeships. Change and conflict resolution come from evolving personal relationships.

The Eiffel Tower culture, prevalent in Western industrial countries, is also very hierarchical but its orientation is toward the task rather than the person. There is a bureaucratic division of labor, a preference for order and stability, and conflict is considered irrational. Each level in the hierarchy has a clear function. Relationships are specific, status is ascribed by role, and professional qualifications are important. The organization has a purpose that is outside and separate from personal needs for power or affection. Salary is influenced by difficulty, complexity, and responsibility of the assigned job role. Learning occurs in order to acquire the skills required for advancement. Organizational change is accomplished by changing the rules (which can be a slow process), while conflict is resolved by applying the rules. It is important to follow the rules, and the means are as important as the end.

The guided missile culture is also impersonal and task-focused but is more egalitarian. In this culture, the end is what is important, and companies that follow this culture are often peopled by professionals who are cross-disciplinary. This culture is expensive because professionals tend to be expensive. Groups are formed for projects and dissolved when the project is completed. It is a neutral rather than an affective culture. Status derives from a member's contribution to problem solving for the task at hand. Members are evaluated by their peers, but it is harder to quantify individual contribution than in the Eiffel Tower model. Members tend to be practical rather than theoretical, and the group is organized around a problem rather than a discipline. Change is accomplished as projects come and go, along with the people involved in them. People are more loyal to their profession or projects than to the company. Motivation is intrinsic, with management-by-objective and pay-for-performance guidelines for operation.

Incubator cultures, or Silicon Valley–type firms, are egalitarian and people-oriented, and often are founded by a creative team or entrepreneur. They may be motivated by hope and idealism, with the organization existing for self-expression and fulfillment. The goal in such companies is to minimize routine tasks to free time for creative pursuits. Management confirms, criticizes, and provides resources for the creative process. Authority comes from the individual's ideas or inspiring vision. Individuals are motivated intrinsically and are committed more to work than to people. Change in such companies can be fast and spontaneous. Selfish power plays do not move the group forward and are anathema to it. Working for such companies can be intense and all absorbing but can lead to burnout. Conflict is handled by breaking apart and going separate ways and by experimenting to find the best course of action.

Knowledge of human management and the intercultural skills discussed in this chapter are complementary to the economic principles and analytical tools learned throughout this book. These are highlighted in the coming sections.

Human Dimensions of Managing Technology

Schumpeter (1942) argues that technology destroys the old to make way for the new in a process he calls *creative destruction*. Long-wave theory suggests that periods of accelerated technical change or creative destruction and groups of inventions spur economic growth and create cycles. For example, textiles provided change in the late 1700s and early 1800s, as did coal and steel in the late 1800s, electricity in the early 1900s, and petroleum and petrochemicals in the 1950s. The information revolution of the 1990s was spurred by development of telecommunication protocols, electronic improvement of network automation, the transition from analog to digital, and growth in bandwidth.

Moore's Law suggests that computer power doubles every 18 months or so; Metcalfe's Law suggests that the value of a network increases with the square of the number of users. Coase's theory on organization suggests that market governance will be that which minimizes transactions costs (Coase 1937). Users must consider physical, human, and organizational costs, along with the benefits of information, in deciding how much to use. Benefits include lowering costs by changing asset specificity, frequency of transactions, speed of transactions, uncertainty, and taking advantage of economies of scale and scope.

We can think about technology as another factor of production. Thus, economic theory tells us to invest up to the point where the marginal product of technology divided by its unit cost is equal to that of all the other factors. Where the market produces or transfers too little, we can consider government policies to promote these activities.

There is also a human dimension to technical change. Technological determinism suggests that such groups of inventions influence many aspects of daily life, including social change, income distribution, individual and social rights, employment, migration, privacy, sense (or lack) of community, and appropriate management styles.

In his book Understanding Media: The Extensions of Man, McLuhan (1964) argues that the "medium is the message." Thus, whether the medium is storytelling or machine-produced print, audio, or video, the medium shapes our cognition of the message. Indeed, Carr (2011) cites more recent work that the medium even changes the physical characteristics of the brain. He complains that the calm, linear, deep analytical thought encouraged by print is giving way to a fragmented hyperlinked feast of print, audio, video, and interactive communication. We can find and disseminate more information faster than ever before in an intoxicating and addicting movement from idea to idea and media to media. The cost of all this interactivity, however, is a more fragmented, cursory, less attentive reading of ideas. Our learning and thinking become shallower as information does not get processed from working memory into permanent memory. We can lose the rich assortment of permanent ideas that lead to creativity.

Carr (2011) also notes a spate of books that show concern about the way the Internet is influencing many aspects of our lives as follows. Lanier (2010) argues the Internet hurts creativity and may even make us less human. Powers (2010) notes the constant buzz from our digital gizmos makes us less able to enjoy the once quieter and simpler pleasures. Turkle (2011) argues that communication technologies may push us apart rather than pull us together.

Thus, we live in exciting but complicated times. Efficient management of our business as well as our personal lives requires us to choose technologies and the energy that runs them wisely. It requires us to consider not only such input and technology choices but also human-to-human interactions, human-to-technology interactions, and technology-to-technology interactions. Economists want not only a smart grid, but all our network and other activities to be smart, as well.

Think Like an Economist

Basic physical and emotional human needs are universal. Likewise, economists believe that the basic economic principles covered throughout this book can be universally applied to better meet personal and professional objectives. The first step in making any economic decision is to determine the objective. You don't want to climb the ladder of success only to find at the top that your ladder has been leaning on the wrong wall.

Next you need to consider the costs and benefits of activities that move you toward your objective, as well as any constraints you face. With increasing marginal costs and decreasing marginal benefits of such activity, you will typically optimize where marginal benefit equals marginal cost. Exceptions will occur if a constraint prevents you from getting to such an optimum. Such an interior solution should also be compared to a corner solution, such as undertaking no activity at all.

Some of the constraints considered throughout the book that need to be considered include those on resources, time, ability to understand, laws of science, technology, and laws passed by humans, such as taxes, subsidies, price controls, and quotas.

Managing on the Margin

We have seen the preceding marginal principal applied in many contexts throughout the book. In a perfectly competitive market, we can think of demand as marginal benefit and supply as marginal cost. In such an idealized world, no player has any market power, and we do not need to take into account the strategy of other players in the market. Marginal revenue is the market price, which is set by the marginal players. That is the marginal cost of the most expensive unit produced, and the marginal benefit of the least valued unit consumed. Although no market is perfect, this model is useful in many contexts, where markets are reasonably competitive and contestable (Baumol et al. 1982). This model provided some insights into the global coal market in chapter 3 and provides the basic building blocks for many of the models considered throughout the book.

When a market player is able to influence price, economists say that player has some market power. We considered three cases where only one player has market power. In the extreme case of one seller (a monopoly) facing competitive buyers, the seller has market power. Under normal circumstances, the monopolist will try to raise the market price above the competitive level, unless the monopolist can price discriminate. Then the marginal benefit is not the price but the marginal revenue. We considered one such extreme case in chapter 5 with electricity generation as a natural monopoly.

Another extreme is the case of one buyer (a monopsony) facing competitive sellers. In such a case, the buyer will typically want to push the price below that of a competitive market. Further, since the buyer can influence the price, marginal cost is no longer the market price (unless the monopsonist can price discriminate), but rather the marginal factor cost. Such a case was considered in the context of the Asia-Pacific LNG market.

The third case we considered with market power for only one player was the dominant firm using OPEC as an example in chapter 7. There, OPEC faces the world demand minus the supply from a competitive fringe. Such a model typically yields a kinked demand curve. Again, market price is not marginal revenue, but marginal revenue can be derived from the kinked demand curve to find where marginal cost crosses marginal revenue.

Although fixed sunk costs should be ignored in the short run, all other costs, including opportunity and transaction costs, should be included in the analysis. Indeed, transaction costs may also influence the kind of governance that evolves, as was shown in chapter 8 in the context of the US natural gas industry. For example, spot markets, contracts, and vertical integration are all possibilities, and all have prevailed at times in the US natural gas markets as regulation and market conditions have changed.

Once we add more players with market power, the situation becomes more complicated, and the strategy of one's opponent needs to be considered. For example, if a monopolist faces a monopsonist (a bilateral monopoly), bargaining will come into play, as considered in chapter 9.

With two buyers or two sellers facing competitive markets and taking the opponent's quantity as given, each player should consider the effect that quantity has on the price. Again, marginal revenue will typically not equal the price. However, the situation changes if players take their opponent's price as given, or one or both players try to become a dominant player. Such models have been considered in chapter 10 in the context of the European natural gas market.

In such small number decision making, as in chapter 10, the strategy of any opponent with market power becomes relevant. Facing one opponent in the market or sharing the market with two or three is very different from the multi-participant markets of perfect competition. With few players, it is important to know who these "big kids on the block" are and whether they are friends or foes. (For an online course that covers numerous game theory models allowing various strategies, see Polak [2007].)

Since energy products are more typically commodities and reasonably homogeneous, we have not considered strategies such as advertising and product differentiation, as would be much more important in industries such as automobiles and fashion. However, at one time, a fair amount of advertising dollars were spent in US markets to convince drivers that a fairly homogeneous product, gasoline, was really not homogeneous. For example, you could "Put a Tiger in Your Tank" or benefit from platformate. Giving coupons redeemable for catalogue items and attaching convenience stores to gas stations (still a prevalent marketing strategy) were also used to garner customer loyalty.

When our objective function and constraints are linear, our marginal benefit/ marginal cost rule breaks down, and optimums are typically at corners of the constraint set. Linear programming can then be employed to solve for the optimum. This methodology was demonstrated in chapter 17, where it was applied to refinery blending problems and energy transportation. We did not cover corner solutions for the nonlinear case, but they can typically be solved using Kuhn-Tucker conditions. For examples, see Chiang (1984, ch. 21).

In economic situations, what gets measured tends to be what gets done. Therefore, you should measure what you treasure. That leads us to two other important cost and benefit concepts: the average and the total. Marginal benefits and costs help us choose the optimal amount of activity. Averages and totals are better measures of how we are doing overall.

Managing across Time

When our decisions today influence our objective function in another time period, we need to consider the marginal cost and marginal benefits today and in the future. Such time dependence is especially true for nonrenewable resources. Typically, such requirements call for discounting to put future costs and benefits on par with those today. The dynamic models in chapter 14 help us think about how to make such decisions under various circumstances. Levelized costs in chapters 12 and 15 help us distribute our costs across our production profile.

Clearly we cannot know the future, which means we will be making dynamic decisions under uncertainty. This leads to attempts to forecast demand, as considered in chapters 2 and 16, as well manage price risk through futures and options markets, as discussed in chapters 18 and 19.

Energy is intimately bound up with the economy and its evolution across time. Energy enters the supply side as a factor of production and the demand side to provide energy services to consumers. We considered input-output models for such economy/energy interactions in a static world by sector in chapter 2. We considered the interaction of energy and development in chapter 20. The world's lowest income populations still typically satisfy a large portion of their energy needs with traditional biomass sources. When the financial situation improves for consumers, they tend to climb the energy ladder to cleaner and more convenient fossil fuels and also gain access to electricity. As fossil fuels become more expensive, either from scarcity or policy, presumably we will move into modern renewable energy sources. Models and markets, along with government policy, can help us determine the optimal production profile of forest and other biomass products and stimulate technologies to produce gas and liquids from waste and other biomass. They will also aid in transitioning our electricity supply out of dependence on fossil fuels and into biomass, nuclear, solar, wind, hydro, and geothermal.

Fossil-rich countries will also need to transition out of their income dependence. Dynamic models in chapter 21 can help them consider this transition and the need to invest a portion of their fossil rents into financial, physical, human, technological, and institutional capital.

When Markets Fail

Most economists believe there is a role for government in providing basic protection for people and property as well as providing and enforcing ground rules for markets and commerce, among other roles. However, economists most often favor a competitive market for the production and distribution of goods and services. Under ideal conditions, competitive markets with their free choice will maximize social welfare, defined as consumer plus producer welfare, as developed in chapter 4. We saw how taxes, subsidies, and price controls in energy markets would move us away from this market optimum.

However, we have considered a number of cases where markets may fail to provide the socially optimal level of output, and governments may be asked to step in. Market power may develop, giving pricing power to some market players as in our monopoly, monopsony, dominant firm, and other oligopoly models. With a natural monopoly or monopsony, it may make economic sense to not encourage entry of more players but rather to regulate behavior to try to simulate a competitive outcome. Various regulatory pricing schemes were considered in chapters 5 and 6 for the electricity sector. Since governments, too, can fail and often do, there have been efforts to restructure the electricity sector away from vertically integrated government regulated or owned utilities. Instead, restructuring leans toward only controlling the natural monopoly segments of transportation and distribution but promoting competition in generation, retailing, and wholesale markets. Prominent examples of such restructuring were included in chapter 6.

Where there is no natural monopoly, economists urge policies to make the exercise of market power illegal, monitor whether market power is being exercised, and even go so far as to break up firms if necessary.

Another area where markets are likely to fail occurs with poorly defined property rights. In such cases, some of the costs or benefits accrue to others than those directly involved in an economic activity. For example, pollution and accidents discussed in chapters 11 through 13 have negative externalities, and governments may use taxes or regulations to move us toward a more optimal level of activity. Common pools and common property may be overexploited, as modeled in chapter 20, and governments may act to restrict access. The market is likely to underproduce public goods, which are nonexcludable. Thus CO_2 abatement, investment in more efficient new technologies, and the dissemination of information may be underfunded by the market, leaving room for government policy.

We also need to recall that governments fail as well. In democratic countries, governments are largely composed of lawyers, and the primary motivation of such governments tends to be political power rather than economic efficiency. Thus, we may get regulations and policies that work, but at a very high cost. Or we may even get regulations that do not work at a very high cost. Policies that defy the laws of science and economics never fare well.

Governments are composed of individuals who may value their own well-being above the common good. For example, the revolving door may lead regulators to take the point of view of the regulated. Where huge natural resource rents are available, governments may have their hands in the till, and they also may make poor public choices for the use of the rents. Poor management of rents is not a universal law. Norway is a role model for careful use of resource rents. However, several factors increase the likelihood that rents will be poorly used, including less transparency in government actions and less democracy in a government's structure. Other factors that increase the likelihood of misuse occurring are having a lower income population that is less educated and more disenfranchised, and having a weak rule of law. See Johnson (1991) for additional discussion of many of the issues related to government decision making and public choice.

Summary

Economic models are simplifications of reality that help us understand reality and make better economic decisions. Optimization decisions typically have several requirements. We must understand our objective, measure costs and benefits, make activity decisions based on marginal costs and benefits, and measure overall performance based on average and total costs and benefits. All costs except fixed sunk costs should be considered, including opportunity, nonpecuniary, and transactions costs. When costs and benefits are related across time, we typically need to discount the future and deal with future uncertainty.

Perfect competition is the benchmark that maximizes social welfare under ideal conditions. Adding in a single player with market power into an otherwise competitive situation usually yields predictable results under profit maximizing behavior. When more than one player has market power, the outcomes will typically depend on bargaining and strategies, and the human elements in the equation become important.

When markets fail, governments may step in with policy. They may try to measure and offset market power by targeting competitive outcomes with fines, jail, regulations, merger prevention, or the breakup of large companies into smaller ones. They may try to measure and internalize externalities by using taxes, subsidies, and regulations. Governments may provide public goods or restrict access to common goods.

Whether you are a corporation making and implementing business decisions or a government bureaucrat making and implementing policy decisions, understanding the economic analysis is not enough. All implementation will be done in a cultural context. Understanding this cultural context also may be vital to successful implementations. Culture is an ecosystem of shared values and customary ways of thinking, doing things, and reacting to each other and to the environment. It is passed on at home, at school, at work, and at play, and it determines how basic human needs for sustenance, safety, belonging, esteem, and self-fulfillment are satisfied. Despite individual variations within a culture, there is often a great deal of similarity, as well. Understanding and using these similarities, while allowing for the differences, can help one flourish in any culture, whether national, professional, or corporate.

In any intercultural situation, it is important to determine people's attitudes toward authority, risk, and uncertainty. Dichotomies to consider are the relationship of the individual to the group, guilt or shame-based methods of behavior control, and whether emotions are displayed in public or not. Another consideration is whether the culture values assertiveness and success more or less than relationships and the physical environment. A culture with a longer time horizon puts a premium on long-term commitments and tradition; those with a shorter horizon value immediate results. Cultures may view events as sequential or synchronous; they may view values and norms as particular or universal. Public and private faces and personal space vary across cultures, as well.

In achievement cultures, you earn power and status by what you do or by your contribution to the tasks at hand and group goals. In ascription cultures, you earn status and power by who you are, which may be associated with family, gender, or social connections.

We organize and process information in different ways. A more open-minded culture considers more information from a wider point of view; a more closed-minded culture considers less information from a narrower point of view. Cultures at threat are more likely to be open-minded than those for whom things are going well. An associative culture (more likely taught by rote) tries to associate new data with old experience and categories. An abstract culture (more likely taught by problem solving) tries to develop theories about processes and relationships that will allow one to understand and better cope with completely new things.

Values differ, truth is affirmed by various means, and perceptions of people are not universal. Aesthetic cultures value beauty and life experiences; social cultures value people and relationships. Economic cultures value wealth and physical comfort. Theoretical, systematic cultures value knowledge for knowledge's sake. Power cultures value power and control over others. Religious cultures value unity with the cosmos. Truth is verified by faith, fact, or feelings. Some cultures view people as trustworthy; others view them as untrustworthy. Some view personalities as immutable; others view them as mutable. Some view people as working to satisfy physical needs; others view them as working for self-fulfillment.

Business protocols address such concerns as personal space, greetings, the value of punctuality, the need to make appointments, and the pace of business. Other concerns include the value of personal contacts, delivery performance, and business cards. Protocol issues may arise concerning when business discussions can take place, forms of greeting, use of titles, eye contact, gift expectations, taboo zones, table manners, food taboos, and appropriate business dress.

Verbal communication includes what you say, what you mean, and what the listener understands. Communication can go awry at each level. The emotional character of idioms and symbols may not translate across time and space. Low-context cultures require one to be more explicit than high-context cultures, which are more homogeneous, with many shared values and experiences.

More than one-half of communication may be nonverbal, in the form of gestures, posture, and facial expressions, which may be inherited, discovered, absorbed, or learned. Baton signals give rhythm and emphasis to spoken words. Guide signs indicate direction. Salutation displays are for hellos, good-byes, and life transformations. Signs of dominance and submission signal the social hierarchy. Echo postures are often observed among friends and can be used to put strangers and subordinates at ease.

How corporate cultures think, learn, change, motivate, reward, and resolve conflicts is determined by the relationship between employees and their organization, the relationship between superiors and subordinates, and whether the organization is person- or task-oriented.

Four corporate structures are the family, the Eiffel Tower, the guided missile, and the incubator. The family culture is hierarchical and people-oriented; the Eiffel Tower is hierarchical and task-oriented; the guided missile is egalitarian and task-oriented; and the incubator is egalitarian and people-oriented.

Cultural and corporate differences influence who makes decisions, how fast decisions are made, an acceptable level of risk, and how problems are viewed and solved. A Westerner may view life as a series of problems to be solved using scientific and analytical thought. For example, I would likely take a sequential approach and make incremental decisions based on economic models and analysis.

An important implication of how decisions are implemented depends on the ethical, institutional, and legal framework in the operating country. Environmental standards vary across countries. A gift that may be viewed as a bribe in the United States may be a normal part of business in Korea. Labor unions may negotiate national contracts in some countries, but not in others. Cartels may be illegal in some places but encouraged in others.

Cultural differences also impact negotiations. The style may vary with the underlying values and assumptions of the culture and might be based on fact and logic, emotion, or ideals. Ritual may influence the opening offer, the amount of conflict, the size and timing of concessions, and the response to concessions. The autonomy and number of negotiators is often related to the power structure and individualist tendencies of the culture. Cross-cultural joint ventures, mergers, and teams must also learn how to set goals, adapt, and move forward together.

New technologies can also affect the way we think, process information, and interact. As a heavy user of information technologies, I have found the Internet to be a valuable source. However, I must be careful to extract the nuggets of truth from the ore, because false and misleading information is also available. As with Carr, I also find too much time on the Web makes it harder to concentrate. Writing this book, however, has required concentrated and critical thinking, and my hope is that your reading of it will instill depth and discernment in your thinking processes.

So you have come to the end of this particular road and can now move on to bigger and better things. As a member of the global village, you will want to use your analytical skills to understand pricing, policy, and profits and make optimal use of energy resources subject to the constraints you face. Having traveled to more than 120 countries, I have found that dealing with the human and cultural constraints have also added to the challenge and fun of working in these exciting industries.

A

Energy Conversions

International energy comparisons are fraught with difficulties. Not only are there the difficulties of currency conversions, but fossil fuels come in a variety of metric and nonmetric units, including barrels, gallons, liters, metric tons, short tons, long tons, cubic feet, cubic meters, and energy units. Energy units include kilocalories, British thermal units, kilowatt-hours, kilojoules, and tons of oil or coal equivalent. In addition, fossil fuels are not homogeneous but rather vary by energy content and impurities. Throughout the book, I have deliberately used a variety of units, along with conversions, to familiarize the reader with common ways energy is measured. This brief appendix summarizes some of the more common conversions. There are numerous resources on the Internet for the metric to nonmetric conversions. For links to some of them, and other conversion information, see http://dahl.mines.edu/EConv.pdf.

Metric Prefixes/Suffixes

kilo = 10^3 (k)

mega = 10^6 (M)

giga = 10^9 (G)

tera = 10^{12} (T)

peta = 10^{15} (P)

deci = 10^{-1} (d)

centi = 10^{-2} (c)

milli = 10^{-3} (m)

micro = 10^{-6} (μ)

nano = 10^{-9} (n)

Metric Length

1 meter = 100 centimeters = 1,000 millimeters = 0.001 kilometers.

It is easy to convert between these equivalences. For example, if you want to know what 1 centimeter equals in terms of the other units, divide each value by 100. That is,

0.01 meter = 1 centimeter = 10 millimeters = 0.00001 kilometers.

Nonmetric Length

1 mile = 1,760 yards = 5,280 feet = 0.869 nautical mile.

1 foot = 12 inches = 1/3 yard.

Length Conversion: Metric ⟷ Nonmetric

To convert lengths from nonmetric to metric, multiply by the value in the table below.

To convert lengths from metric to nonmetric, divide by the value in the table below.

	Metric →			
	Millimeter	**Centimeter**	**Meter**	**Kilometer**
Inch	25.4	2.54	25.4×10^{-3}	25.4×10^{-6}
Foot	304.8	30.48	0.3048	304.8×10^{-6}
Yard	914.4	91.44	0.9144	914.4×10^{-6}
Mile	1.6093×10^{6}	1.6093×10^{5}	1.6093×10^{3}	1.6093
Nautical mile	$1,852 \times 10^{3}$	$1,852 \times 10^{2}$	1,852	1.852

(Nonmetric →)

Metric Area

1 square kilometer = 1×10^{6} square meters = 100 hectares.

Nonmetric Area

1 square mile = 27.8784×10^{6} square feet = 3.0976×10^{6} square yards = 640 acres.

Area Conversion: Metric ⟷ Nonmetric

To convert from nonmetric to metric, multiply by the value in the table below.
To convert from metric to nonmetric, divide by the value in the table below.

	Metric→		
	Sq. meter	Sq. kilometer	Hectare
Sq foot	92.9×10^{-3}	92.9×10^{-9}	9.29×10^{-6}
Sq yard	0.8361	836.1×10^{-9}	83.61×10^{-6}
Sq mile	2.59×10^{6}	2.59	259
Acre	4,047	4.047×10^{-3}	0.4047

Metric Volume

1 cubic meter = 1,000 liters = 1 kiloliter = 10^6 cubic centimeters.

Nonmetric Volume

1 US gallon = 4 quarts = 8 pints = 231 cubic inches = 0.833 Imperial gallons.

1 US barrel = 42 US gallons = 5.615 cubic feet.

Volume Conversion: Metric ⟷ Nonmetric

To convert from nonmetric to metric, multiply by the value in the table below.
To convert from metric to nonmetric, divide by the value in the table below.

	Metric→	
	Liter	Cubic meter = kiloliter
Quart	0.946	0.946×10^{-3}
Cubic foot	28.317	28.317×10^{-3}
US gallon	3.785	3.785×10^{-3}
Barrel	159	0.159

Metric Weight

1 metric ton = 1,000 kilograms = 1,000,000 grams.

Nonmetric Weight

1 long ton = 2,240 pounds = 1.12 short ton.

1 pound = 16 ounces.

Weight Conversion: Metric $\longleftrightarrow$ Nonmetric

To convert from nonmetric to metric, multiply by the value in the table below.
To convert from metric to nonmetric, divide by the value in the table below.

| | Metric $\rightarrow$ | | |
	Gram	Kilogram	Metric tonne
Ounce	28.35	28.35×10^{-3}	28.35×10^{-6}
Pound	453.6	453.6×10^{-3}	453.6×10^{-6}
Short ton	907.2×10^3	907.2	0.9072
Long ton	1.016×10^3	1,016	1.016

(Nonmetric $\rightarrow$)

Metric Temperature

Degrees Celsius (Centigrade) (°C) = 273.5 + Kelvin (K).

Nonmetric Temperature

Degrees Fahrenheit (°F)

Temperature Conversion: Metric ⟷ Nonmetric

$$°C = (9/5)(°F + 32).$$

$$°F = (5/9)(°C - 32).$$

$$°F = (5/9)(K + 273.15 - 32).$$

Metric Energy/Heat Values

1,000 calories = 1 kilocalorie (which is our food calorie [*Calorie*]) = 4.187 kilojoules.

1 kilowatt hour (kWh) = 859.84 kilocalories.

TOE is one tonne of oil equivalent, often considered to be 10 million kilocalories. TCE is one tonne of coal equivalent, about two-thirds of a ton of oil equivalent.

Nonmetric Energy/Heat Values

Horsepower-hour = 2,544.4 Btu.

1 quad = 10^{15} Btu.

1 therm = 100,000 Btu.

Conversion Energy/Heat Values:
Metric ⟷ Nonmetric

To convert from nonmetric to metric, multiply by the value in the table below.
To convert from metric to nonmetric, divide by the value in the following table.

Nonmetric ↓	Metric→		
	Calorie = kilocalorie	kilojoule	kWh
1 Btu	0.252	1.055	293.07×10^{-6}

Approximate Heat Values and Barrels per Metric Tonne for US Energy Products

	Btu/unit	Unit	Barrels/tonne
Coal	20,200,000	short ton	NA
Oil	5,800,000	barrel	7.33
Dry natural gas	1,022	cubic foot	NA
Natural gas liquids	3,700,000	barrel	10.40
Gasoline	5,253,000	barrel	8.53
Jet fuel/kerosene	5,670,000	barrel	7.93
Naphtha jet fuel	5,355,000	barrel	8.27
LNG	3,820,000	barrel	13.98
Distillate	5,825,000	barrel	7.46
Residual	6,287,000	barrel	6.66
Lubricant	6,065,000	barrel	7.00

Sources: US EIA (2011c); US EIA (n.d.d.); IGU (2012).

LNG

1 metric ton of LNG = 2.222 cubic meters LNG = 1.22 tons of oil equivalent
= 1,360 cubic meters of natural gas (BP 2014; IEA 2005).

B

Bibliography

ACER (Agency for the Cooperation of Energy Regulators). n.d. "Home Page." http://www.acer.europa.eu/Pages/ACER.aspx.

Adelman, M. A. 1972. *World Petroleum Market*. Washington, DC: Resources for the Future.

———. 1990. "Mineral Depletion with Special Reference to Petroleum." *Review of Economics and Statistics* 72 (1): 1–10.

———. 1992. "Oil Resource Wealth of the Middle East." *Energy Studies Review* 4 (1): 7–22.

———. 1993. *The Economics of Petroleum Supply*. Cambridge, MA: Massachusetts Institute of Technology.

Adelman, Morris, and Manoj Shahi. 1989. "Oil Development-Operating Cost Estimates 1955–1985." *Energy Economics* 11 (January): 2–10.

Adler, N. J. 1997. *International Dimensions of Organizational Behavior*. Cincinnati, OH: South-Western.

Ajanovic, Amela, Carol A. Dahl, and Lee Schipper. 2012. "Modeling Transport (Energy) Demand and Policies—An Introduction." *Energy Policy, Special Section: Modeling Transport (Energy) Demand and Policies* 41 (February): iii–xiv.

Alaska Department of Revenue Tax Division. 2013. "Petroleum Production Tax." December 3. http://www.tax.alaska.gov/programs/programs/forms/index.aspx?606.

Aleksandrova, Olga. 2009. "The Druzhba Oil Trunkline." *Oil of Russia*: 4.

Allen, Steven. 2012. *Financial Risk Management: A Practitioner's Guide to Managing Market and Credit Risk*. 2nd ed. Hoboken, NJ: Wiley.

Alley, William M., and Rosemarie Alley. 2013. *Too Hot to Touch: The Problem of High-Level Nuclear Waste*. Cambridge: Cambridge University Press.

Allnutt, Liz, and Sarah Lilly. 2012. "MRRT: New Tax on Australian Mining Projects." May 13. Australia: Norton Rose Fulbright. http://www.mondaq.com/australia/x/177150/Income+Tax/MRRT+New+tax+on+Australian+mining+projects.

Alyeska Pipeline Service Company. n.d. "Prudhoe to Valdez. 800 Miles." http://www.alyeska-pipe.com/TAPS/PipelineOperatons/Throughput.

An, Feng, Robert Earley, and Lucia Green-Weiskel. 2011. *Global Overview on Fuel Efficiency and Motor Vehicle Emission Standards: Policy Options and Perspectives for International Cooperation.* New York: UN Department of Economic and Social Affairs, Commission on Sustainable Development. Background Paper No. 3, CSD19/2011/BP3. May. http://www.un.org/esa/dsd/resources/res_pdfs/csd-19/Background-paper3-transport.pdf.

Anouti, Yahya Faisal. 2013. "Local Content Policies in the Oil and Gas Sector: Justification, Tradeoffs and First Best Policy." February 3. Draft, Colorado School of Mines.

API (American Petroleum Institute). 1959. *Petroleum Facts and Figures.* Washington, DC: American Petroleum Institute.

———. 1971. *Petroleum Facts and Figures.* Washington, DC: American Petroleum Institute.

———. 2012. *Basic Petroleum Databook.* Washington, DC: American Petroleum Institute.

Arnold, J. E. M., G. Köhlin, and R. Persson. 2006. "Woodfuels, Livelihoods, and Policy Interventions: Changing Perspectives." *World Development* 34 (3): 596–611.

Association of American Railroads. 2013. "Class I Railroad Statistics." July 15. https://www.aar.org/StatisticsAndPublications/Documents/AAR-Stats-2013-07-09.pdf.

Audubon. 2009. "The Real Deal on Compact Fluorescent Light Bulbs." http://web4.audubon.org/GlobalWarming/files/CFLs.pdf.

Automation and Power Technologies. 2013a. "The Evolution of Power Cable Systems: Power Cable Systems." http://www.abb.com/cawp/db0003db002698/b0dd84c5a65b9027c1257a1300319830.aspx.

———. 2013b. "The History of HVDC Transmission." http://www04.abb.com/global/seitp/seitp202.nsf/0/a521beb28ac88e75c12572250046e16a/$file/HVDC+history.pdf.

Bakhsh, Nidaa. 2013. "Fracking Comeback in U.K. as Browne Seeks Shale Bonanza: Energy." January 28. http://www.bloomberg.com/news/2013-01-28/fracking-comeback-in-u-k-as-browne-seeks-shale-bonanza-energy.html.

Bank for International Settlements. 2013. *BIS Quarterly Review* (March). http://www.bis.org/statistics/otcder/dt1920a.pdf.

Barber, Benjamin R. 1996. *Jihad vs. McWorld.* New York: Ballantine Books.

Bardhan, Pranab. 1997. "Corruption and Development: A Review of Issues." *Journal of Economic Literature* 35(3) (September): 1320–1346.

Barnes, D., K. Krutilla, and W. Hyde. 2005. *The Urban Household Energy Transition: Social and Environmental Impacts in the Developing World.* Washington, DC: Resources for the Future.

Barnes, Douglas F., Priti Kumar, and Keith Openshaw. 2012. *Cleaner Hearths, Better Homes: New Stoves for India and the Developing World.* New Delhi, India: Oxford University Press. http://www.esmap.org/sites/esmap.org/files/Cleaner%20Hearths,%20Better%20Homes_Book_Small.pdf.

Barth, James R., and Donald McCarthy. 2012. "Trading Losses: A Little Perspective on a Large Problem." October. http://www.milkeninstitute.org/pdf/Trading_Losses.pdf.

Baumol, W., J. Panzar, and R. Willig. 1982. *Contestable Markets and the Theory of Industry Structure.* San Diego, CA: Harcourt Brace Jovanovich.

Baumol, William J., and Wallace E. Oates. 1988. *Theory of Environmental Policy.* Cambridge, UK: Cambridge University Press.

BBC Europe. 2014. "Russia's President Putin Moves towards Annexing Crimea." March 18. http://www.bbc.com/news/world-europe-26624789.

BBC News. 2014. "Q&A: Iran sanctions." January 20. http://www.bbc.com/news/world-middle-east-15983302.

Becker, Gary S. 1968. "Crime and Punishment: An Economic Approach." *Journal of Political Economy* 76 (2): 169–217.

———. 1983. "A Theory of Competition Among Pressure Groups for Political Influence." *Quarterly Journal of Economics* 98 (3): 371–400. http://www2.bren.ucsb.edu/~glibecap/BeckerQJE1983.pdf.

Belyaev, Lev S. 2011. *Electricity Market Reforms: Economics and Policy Challenges.* New York: Springer.

Berger, Bill D., and Kenneth E. Anderson. 1992. *Modern Petroleum: A Basic Primer of the Industry.* 3rd ed. Tulsa, OK: PennWell.

Beuerlein, Jim. n.d. "Bushels, Test Weights and Calculations." Department of Horticulture and Crop Science. Ohio State University FactSheet. http://ohioline.osu.edu/agf-fact/0503.html.

Bichpuriya, Yogesh K., and S. A. Soman. 2010. "Electric Power Exchanges: A Review." 16th National Power Systems Conference. December 15–17, 115–120.

Bindemann, Kirsten. 1999. *Production-Sharing Agreements: An Economic Analysis.* WPM 25. October. Oxford, UK: Oxford Institute for Energy Studies.

Bjerkholt, Olav, Eystein Gjelsvik, and Øystein Olsen. 1990. "The Western European Gas Market: Deregulation and Supply Competition." *Recent Modeling Approaches in Applied Energy Economics.* London: Chapman and Hall. 3–28.

Blumsack, Seth, and Dmitri Perekhodtsev. 2009. "Electricity Retail Competition: An International Review." In *International Handbook on the Economics of Energy*, edited by Joanne Evans and Lester C. Hunt: 663–684.

Bohi, Douglas, and Michael Toman. 1996. *The Economics of Energy Security*. Washington, DC: Resources for the Future.

Box, George E. P., and Gwilym M. Jenkins. 1970. *Time Series Analysis: Forecasting and Control*. San Francisco: Holden-Day.

BP (British Petroleum). 2012. *Statistical Review of World Energy*. The most current version of this annual publication can be found at http://www.bp.com/en/global/corporate/about-bp.html.

———. 2013. *Statistical Review of World Energy*. The most current version of this annual publication can be found at http://www.bp.com/en/global/corporate/about-bp.html.

———. 2014. *Statistical Review of World Energy*. The most current version of this annual publication can be found at http://www.bp.com/en/global/corporate/about-bp.html.

Braziel, Rusty. 2012. "Fifty Shades of Condensates—Which One Did You Mean?" RBN Energy. October 22. http://www.rbnenergy.com/fifty-shades-of-condensates%E2%80%93which-one-did-you-mean.

Breisford, Robert, Warren R. True, and Leena Koottungal. 2011. "Western Europe Leads Global Refining Contraction." *Oil & Gas Journal* (December 2): 33–48.

Büyüksahin, Bahattin, and Michel A. Robe. 2012. "Does It Matter Who Trades Energy Derivatives?" *Review of Environment, Energy and Economics*. Fondazione ENI Enrico Mattei (FEEM). March 1. http://dx.doi.org/10.7711/feemre3.2012.03.001; http://re3.feem.it.

CAISO (California ISO). 2013. "Market Issues and Performance Reports." http://www.caiso.com/market/Pages/MarketMonitoring/MarketIssuesPerfomanceReports/Default.aspx.

California Public Utilities Commission. 2013. "California Renewables Portfolio Standard (RPS)." http://www.cpuc.ca.gov/PUC/energy/Renewables/index.htm.

Canadian Nuclear Safety Commission. 2012. "CNSC Action Plan: Lessons Learned from Fukushima Nuclear Accident." May 3. http://www.oecd-nea.org/nsd/fukushima/documents/CanadaMay-3-2012-Staff-Presentation-CNSC-Action-Plan_e.pdf.

Capen, E. C., R. V. Clapp, and W. M. Campbell. 1971. "Competitive Bidding in High-Risk Situations." *Journal of Petroleum Technology* 23 (6): 641–653.

CARB (California Air Resources Board). 2012. "Key Events in the History of Air Quality in California." February 6. http://www.arb.ca.gov/html/brochure/history.htm.

Carbon Capture & Sequestration Technologies. 2013. "Power Plant Carbon Dioxide Capture and Storage Projects." November 22. http://sequestration.mit.edu/tools/projects/index_capture.html.

Carr, Nicholas. 2011. *The Shallows: What the Internet Is Doing to Our Brains*. New York: W. W. Norton.

Carroll, Chris. 2011. "Wyoming's Coal Resources." Wyoming State Geological Survey. http://www.wsgs.uwyo.edu/public-info/onlinepubs/Coal-Resources.aspx.

CEER (Council of European Energy Regulators). n.d. "Home Page." http://www.energy-regulators.eu/portal/page/portal/EER_HOME.

Center for Energy Economics. circa 2004. *Natural Gas and Electric Power Marketization in North America*. University of Texas, Austin. http://www.beg.utexas.edu/energyecon/new-era/case_studies/Natural_Gas_Marketization_in_North_America.pdf.

Central Electricity Regulatory Commission (India). n.d. "Home Page." http://www.cercind.gov.in/.

Chambwera, Muyeye. 2004. "Economic Analysis of Urban Fuelwood Demand: The Case of Harare in Zimbabwe." PhD thesis: Wageningen University.

Chanal, Margaux. 2012. *How Is 100% Renewable Energy Possible in South Korea by 2020?* Global Energy Network Institute. August. http://www.geni.org/globalenergy/research/100-percent-renewable-for-south-korea/100-percent-Renewable-for-South-Korea.pdf.

Chance, Clifford. 2010. *The Dodd-Frank Act and the Proposed EU Regulation on OTC Derivatives: Impact on Asian Institutions*. December. http://www.isda.org/uploadfiles/_docs/Impact%20on%20Asian%20Institutions.pdf.

Charemza, W. W., and D. F. Deadman. 1997. *New Directions in Econometric Practice: General to Specific Modelling, Cointegration and Vector Autoregression*. 2nd ed. Cheltenham, UK: Edward Elgar Pub.

Chiang, Alpha C. 1984. *Fundamental Methods of Mathematical Economics*. New York: McGraw-Hill.

Chiang, Alpha C., and Kevin Wainwright. 2006. *Fundamental Methods of Mathematical Economics*. New York: McGraw-Hill Higher Education.

China Bystander. 2013. "China Forces More Coal Industry Consolidation." May 23. http://chinabystander.wordpress.com/2012/03/23/china-forces-more-coal-industry-consolidation/.

China Coal Report. 2011. "Chinese Coal Company Stock Prices." *China Coal Report* 0203: 2. January 6. http://www.coalportal.com/temp_files/CCR-203.China%20Coal%20Report.6-Jan-2011.pdf.

Choi, Charles. 2012. "Humans Used Fire 1 Million Years Ago." *Discovery News.* April 2. http://news.discovery.com/history/archaeology/human-ancestor-fire-120402.htm.

Christensen, Laurits R., and William H. Greene. 1976. "Economies of Scale in U.S. Electric Power Generation." *Journal of Political Economy* Part I 84 (4): 655–676.

CIGB (Commission Internationale Des Grands Barrages). circa 2012. "Classification by Installed Capacity with Energy." http://www.icold-cigb.org/gb/world_register/general_synthesis.asp?IDA=213.

CME Group. 2012a. *Introduction to Crack Spreads.* http://www.cmegroup.com/trading/energy/ files/EN-211_CrackSpreadHandbook_SR.PDF.

———. 2012b. "Twenty Years of CME Globex." June 21. http://www.cmegroup.com/education/files/globex-retrospective-2012-06-12.pdf.

———. 2013. "Henry Hub Natural Gas Options." http://www.cmegroup.com/trading/energy /natural-gas/natural-gas_quotes_settlements_options.html#prodType=AME.

———. n.d. "All Products—Codes and Slate." Input into dropdown boxes: Energy, All Venues, All Exchanges, All Cleared as Futures, Options. http://www.cmegroup.com/trading/products/#pageNumber=1&sortField=oi&sortAsc=false&group=7&page=1.

Coase, Ronald. 1937. "The Nature of the Firm." *Economica* 4 (16): 386–405.

———. 1960. "The Problem of Social Cost." *Journal of Law and Economics* 3 (1): 1–44.

Consensus Economics. 2013. "Energy & Metals Consensus Forecasts." http://www.consensuseconomics.com/download/Energy_and_Metals_Price_Forecasts.htm.

Considine, Jennifer I., and William A. Kerr. 2002. *The Russian Oil Economy.* Northhampton, MA: Edward Elgar Publishing, Inc.

Consumer Reports. 2009. "By the Numbers: How Long Will Your Appliances Last? It Depends." March 21. http://www.consumerreports.org/cro/news/2009/03/by-the-numbers-how-long-will-your-appliances-last-it-depends/index.htm.

Cookenboo, Leslie. 1980. "Production Functions and Cost Functions: A Case Study." In *Managerial Economics and Operations Research: Techniques, Applications, Cases,* edited by Edwin Mansfield. New York: W. W. Norton and Company. 52–75.

Cornell Center for Materials Research. 1999. "Earth's Dynamism Creates Heat." December 1. http://www.ccmr.cornell.edu/education/ask/index.html?quid=215.

Coyne, Sr., Edward J. 2001. "Syllabus and Course Material for BUSA 470, Business in the International Environment." Samford University. Spring. http://faculty.samford.edu/~ejcoyne/busa470.htm.

Cramer, Jacqueline. 2007. *Testing Framework for Sustainable Biomass.* Final report of the Sustainable Production of Biomass Project Group. Commissioned by the Energy Transition's Interdepartmental Programme Management (IPM). March. http://www.lowcvp.org.uk/assets/reports/070427-cramer-finalreport_en.pdf.

Cropper, Maureen L., Alexander Limonov, Kabir Malik, and Anoop Singh. 2012. "Estimating the Impact of Restructuring on Electricity Generation Efficiency: The Case of the Indian Thermal Power Sector." September. University of Maryland. http://www.econ.umd.edu/research/papers/613.

Crude Oil Peak. 2012. "Historical Oil Disruptions." http://crudeoilpeak.info/historical-oil-disruptions.

CSIS (Center for Strategic and International Studies). 2013/14. "The Ukraine Crisis Timeline." http://csis.org/ukraine/index.htm. November 21, 2013–June 7, 2014.

Cupcic, François. 2003. "Extra Heavy Oil and Bitumen—Impact of Technologies on the Recovery Factor: The Challenges of Enhanced Recovery." PowerPoint presentation given at the ASPO Annual Meeting, May 26–27. http://www.peakoil.net/iwood2003/ppt/CupcicPresentation.pdf.

Curtis, Glenn E. 1992. *Czechoslovakia: A Country Study.* "Appendix B: The Council for Mutual Economic Assistance—Soviet Union." Washington, DC: Federal Research Division of the Library of Congress. http://lcweb2.loc.gov/frd/cs/soviet_union/su_appnb.html.

Dahl, Carol A. 1993. "A Survey of Energy Demand Elasticities in Support of the Development of the NEMS." Prepared for the US Department of Energy. Golden, CO: Department of Mineral Economics, Colorado School of Mines. October 19.

———. 2004. *International Energy Markets: Understanding Pricing, Policies, and Profit.* March. Tulsa, OK: PennWell.

———. 2007. "Oil and Oil Product Demand." In *Encyclopedia of Hydrocarbons.* (English and Italian). Rome, Italy: Istituto della Enciclopedia Italiana Treccani, 49–74.

———. 2010. "Survey of Econometric Energy Demand Elasticities." Draft. Mineral and Energy Economics Program. Colorado School of Mines.

———. 2012. "Measuring Global Gasoline and Diesel Price and Income Elasticities." *Energy Policy, Special Issue on Transport Modeling* 41 (February): 2–13.

———. 2013. "Self Tests to Accompany *International Energy Markets: Understanding Pricing, Policies, and Profit,* 2nd Edition." September. Posted and maintained by Carol A. Dahl, Colorado School of Mines. http:/dahl.mines.edu.

———. 2014. "Supplemental Material to Accompany *International Energy Markets: Understanding Pricing, Policies, and Profit,* 2nd Edition." Posted and maintained by Carol A. Dahl, Colorado School of Mines. http:/dahl.mines.edu.

Dahl, Carol A., and Thomas K. Matson. 1998. "Evolution of the U.S. Natural Gas Industry in Response to Changes in Transaction Costs." *Land Economics* 74 (3): 390–408.

Dakota Gasification Company. 2013. "CO_2 Capture and Storage: The Greatest CO_2 Story Ever Told." http://www.dakotagas. com/CO2_Capture_and_Storage/.

Davis, Graham A., and John E. Tilton. 2005. "The Resource Curse." *Natural Resources Forum* 29 (3): 233–242.

Davis, Jerome D. 1984. *Blue Gold: The Political Economy of Natural Gas.* London: George Allen Unwin.

Deaton, Angus, and John Muellbauer. 1980. *Economics and Consumer Behavior.* Cambridge, UK: Cambridge University Press.

Deaves, Richard, and Itzhak Krinsky. 1992. "Risk Premiums and Efficiency in the Market for Crude Oil Futures." *Energy Journal* 13 (2): 93–118.

Deepwater Horizon Study Group. 2011. "Investigation of the Macondo Well Blowout Disaster." March 1. http://ccrm.berkeley.edu/pdfs_papers/bea_pdfs/dhsgfinalreport-march2011-tag.pdf.

Delphi. 2012/2013. *Worldwide Emissions Standards: Passenger Cars and Light Duty Vehicles.* http://www.ttiinc.com/object/Delphi-emissions-standards-cars-vehicles.html.

DieselNet. 2013. "European Union Emission Standards: Cars and Light Trucks." July. http://www.dieselnet.com/standards/eu/ld.php.

Dowling, Edward T. 1992. *Introduction to Mathematical Economics.* Schaum's Outline Series. New York: McGraw-Hill.

Drèze, Jacques H. 1964. "Some Postwar Contributions of French Economists to Theory and Public Policy: With Special Emphasis on Problems of Resource Allocation." *American Economic Review* 54 (4), Supplement, Surveys of Foreign Postwar Developments in Economics Thought: 1–64.

Dunn, Stephanie, and James Holloway. 2012. "The Pricing of Crude Oil." *Reserve Bank of Australia Bulletin*, September Quarter. http://www.rba.gov.au/publications/bulletin/2012/sep/8.html.

Durham, Louise S. 2010. "Advancements Push 'Salt' Plays." *AAPG Explorer* (February). http://www.aapg.org/explorer/2010/02feb/subsalt.cfm.

Duval, Claude, Honoré Le Leuch, André Pertuzio, and Jacqueline Lang Weaver. 2009a. "International Petroleum Agreements 1: Politics, Oil Prices Steer Evolution of Deal Forms." *Oil & Gas Journal* (September 7). http://www.ogj.com/articles/print/volume-107/issue-33/general-interest/international-petroleum.html.

———. 2009b. "International Petroleum Agreements 2: Embargo of '73 Launched Era of Mutual Adjustment." *Oil & Gas Journal.* (September 14). http://www.ogj.com/_search?q=International+Petroleum+Agreements-2&x=35&y=16.

Earley, P. C. 1989. "Social Loafing and Collectivism: A Comparison of United States and the People's Republic of China." *Administrative Science Quarterly* 34: 565–581.

Economic Commission for Africa. 2009. *Assessment of Power-Pooling Arrangements in Africa.* Addis Ababa, Ethiopia: Economic Commission of Africa.http://www.uneca.org/sites/default/files/publications/powerpooling_assessmentreport.pdf.

Economic Consulting Associates, Ltd. 2010. "The Potential of Regional Power Sector Integration: Literature Review." Central American Electric Interconnection System (SIEPAC) Transmission & Trading Case Study. April. http://www.esmap.org/sites/esmap.org/files/BN004-10_REISP-CD_The%20Potential%20of%20Regional%20Power%20Sector%20Integration-Literature%20Review.pdf.

———. 2012a. "Solar Tariffs: Sunspots; American Tariffs on Chinese Solar Panels Are Dangerous and Pointless." May 24. http://www.economist.com/node/21555958.

———. 2012b. "Natural Gas: Shale of the Century." May 31. http://www.economist.com/node/21556242.

———. 2012c. "Focus: Natural Gas Reserves." June 5. http://www.economist.com/blogs/graphicdetail/2012/06/focus.

———. 2013. "Solar Power Sunset for Suntech : The Troubling Bankruptcy in a Troubled Business." March 27. http://www.economist.com/news/business/21574534-troubling-bankruptcy-troubled-business-sunset-suntech.

EDGAR (Emission Database for Global Atmospheric Research). n.d. "Emission Database for Global Atmospheric Research." Joint project of the European Commission Joint Research Centre and the Netherlands Environmental Assessment Agency. The latest update of this database is at http://edgar.jrc.ec.europa.eu/index.php#.

EEGA (East European Gas Analysis). 2013. "Gazprom Pipelines of West Siberia." April 20. http://www.eegas.com/fsu.htm.

EIO-LCA. n.d. "Economic Input-Output Life Cycle Assessment." Pittsburg, PA: Carnegie Mellon University. http://www.eiolca.net/.

Electricity Authority. 2013. "About the Market." February. http://www.ea.govt.nz/industry/market/about-the-market.

Elert, Glenn. 2013. "Viscosity." In *The Physics Hypertextbook*. http://physics.info/viscosity/.

Ellerman, Denny, and Paul L. Joskow. 2008. *The European Union's Emissions Trading System in Perspective.* Prepared for the Pew Center on Global Climate Change. Massachusetts Institute of Technology. May. http://www.c2es.org/publications/european-union-emissions-trading-system.

Ellsworth, William L. 2013. "Injection-Induced Earthquakes." *Science* July 12. 341 (6142). doi: 10.1126/science.1225942; http://www.sciencemag.org/content/341/6142/1225942.

EMF (Energy Modeling Forum). n.d. "Home page." Stanford University: Palo Alto, CA. http://emf.stanford.edu/.

Encyclopedia of Earth. 2012. "Climate Change Timeline." July 7. http://www.eoearth.org/article/Climate_Change_Timeline.

Energy Library. 2009. "North American Electricity Grid." http://www.theenergylibrary.com/node/647.

Energy Markets Group. 2012. *Chronology of New Zealand Electricity Reform.* Energy & Communications Branch, Ministry of Business, Innovation & Employment. June. http://www.med.govt.nz/sectors-industries/energy/electricity/industry/chronology-of-new-zealand-electricity-reform/.

Energypedia. 2012. "Biogas Technology in China." November 22. http://energypedia.info/wiki/Biogas_Technology_in_China.

ENTSOE (European Network of Transmission System Operators for Electricity). n.d. "Information upon the Lengths of Circuits on December 31st (in km)." https://www.entsoe.eu/db-query/miscellaneous/lengths-of-circuits/.

ENTSOG (European Network of Transmission Operators for Gas). n.d. "Home Page." http://www.entsog.eu.

Ernst and Young. 2012. *2012 Global Oil and Gas Tax Guide.* http://www.ey.com/GL/en/Industries/Oil---Gas/2012-global-oil-and-gas-tax-guide.

Etherington, John, Torbjorn Pollen, and Luca Zuccolo. 2005. *Comparison of Selected Reserves and Resource Classifications and Associated Definitions.* SPE Oil and Gas Reserves Committee (OGRC). Mapping Subcommittee Final Report. December. http://www.spe.org/industry/docs/OGR_Mapping.pdf.

EU (European Union). 1998. "Natural Gas Directive 98/30/EC." *Official Journal* L 204, 21/07/1998 P. 0001 – 0012. http://eur-ex.europa.eu/LexUriServ/LexUriServ.do?uri=CELEX:31998L0030:EN:HTML.\

———. 2003. "Natural Gas Directive 2003/55/EC." *Official Journal* (July 15). http://eur-lex.europa.eu/LexUriServ/LexUriServ.do?uri=OJ:L:2003:176:0057:0057:EN:PDF.

———. 2009a. "Natural Gas Directive 2009/73/EC." *Official Journal* (August 14). http://eur-lex.europa.eu/LexUriServ/LexUriServ.do?uri=OJ:L:2009:211:0094:0136:en:PDF.

———. 2009b. "Natural Gas Transmission Regulation (EC) No 715/2009." *Official Journal* L 211 (August 14): 36–54. http://eur-lex.europa.eu/LexUriServ/LexUriServ.do?uri=OJ:L:2009:211:0036:0054:en:PDF.

——— 2009c. "Renewable Energy." Official Journal L 140 (June 5): 16–62.. http://eur-lex.europa.eu/legal-content/EN/ALL/?uri=CELEX:32009L0028.

———. 2013. "Reduction of Pollutant Emissions from Light Vehicles." http://europa.eu/legislation_summaries/environment/air_pollution/l28186_en.htm.

———. n.d.a."Home Page." http://europa.eu/index_en.htm.

———. n.d.b. "Countries." http://europa.eu/about-eu/countries/index_en.htm.

———. n.d.c. "The history of the European Union." http://europa.eu/about-eu/eu-history/index_en.htm.

Eurostat. 2013. "European Transport Statistics Database." April 12. http://epp.eurostat.ec.europa.eu/portal/page/portal/transport/data/main_tables.

Evans, Joanne, and Lester C. Hunt, eds. 2009. *International Handbook on the Economics of Energy*. Northampton, MA: Edward Elgar Publishing.

Federal Reserve Bank of St. Louis. n.d. "Federal Reserve Economic Data (FRED)." http://research.stlouisfed.org/fred2/.

Feibelman, Alan, and Michael Britt. 2012. "Are Utility Economies of Scale Real, or a Mirage?" May. Oliver Wyman, Marsh & McLennan Companies. http://www.oliverwyman.com/media/LON-UTL91501-001_.pdf.

Feiveson, Harold, Zia Mian, M. V. Ramana, and Frank von Hippel. 2011. "Managing Nuclear Spent Fuel: Policy Lessons from a 10-Country Study." *Bulletin of the Atomic Scientists* (June 27).

Felsenthal, David, Gareth Old, and David Yeres. 2011. "An Introduction to the US Cleared Swap Infrastructure." The Harvard Law School Forum on Corporate Governance and Financial Regulation. July 17. http://blogs.law.harvard.edu/corpgov/2011/07/17/an-introduction-to-the-us-cleared-swap-infrastructure/.

Fielden, Sandy. 2012. "Uptown-top Ranking." RBN Energy. http://www.rbnenergy.com/uptown-top-ranking.

Flach, Bob, Sabine Lieberz, Karin Bendz, and Bettina Dahlbacka. 2011. *EU-27 Annual Biofuels Report*. June 24. http://gain.fas.usda.gov/Recent%20GAIN%20Publications/Biofuels%20Annual_The%20Hague_EU-27_6-22-2011.pdf.

Fletcher, Sam. 2002. "LNG Costs Falling as Demand Climbs, Says Phillips Petroleum Executive." *Oil & Gas Journal*. (February 7). Online. http://www.ogj.com/articles/2002/02/lng-costs-falling-as-demand-climbs-says-phillips-petroleum-executive.html.

Food and Agricultural Policy Research Institute. 2006. "Biofuel Conversion Factors (Approximate, from Various Sources)." http://www.fapri.missouri.edu/outreach/publications/2006/biofuelconversions.pdf.

Frank, Robert H. 2010. *Microeconomics and Behavior*. 8th ed. New York: McGraw-Hill Higher Education.

Frankel, Paul. 1969. *Essentials of Petroleum: A Key to Oil Economics*. New York: A. M. Kelley.

Freed, Daniel. 1997. "An Analysis of the Deregulation of the Electricity Industries of New Zealand, Norway, Sweden and the United Kingdom." Draft. Colorado School of Mines.

Fukuyama, Takashi. 2013. "Japan Offers to Aid Saudi Arabia in Nuclear Power Development." *Asahi Shimbun* (February 11). http://ajw.asahi.com/article/economy/business/AJ201302110107.

Fusaro, Peter, and Tom James. 2005. "Energy Hedging in Asia: Market Structure and Trading Opportunities." Basingstoke, UK: Palgrave Macmillan.

FXCM. 2013. "X-Rates Currency Calculator.: US Dollar to Euro." http://www.x-rates.com/calculator.html.

Gas Infrastructure Europe. 2013. *Storage Map*. http://www.gie.eu.com/index.php/maps-data/gse-storage-map.

Gazprom. 2013a. "Companies with Gazprom's Participation and Other Affiliated Entities." http://www.gazprom.com/about/subsidiaries/list-items/.

———. 2013b. "How Is Gas Transported in Russia? What is the Unified Gas Supply System of Russia?" http://eng.gazpromquestions.ru/?id=6#c310.

———. 2014. "Transmission: Unified Gas Supply System of Russia." http://www.gazprom.com/about/production/transportation/.

Gazprom Export. 2013. "Transportation." http://www.gazpromexport.ru/en/projects/transportation.

Geological Society of America. 2012. "GSA Geologic Time Scale." November. http://www.geosociety.org/science/timescale/.

Geography of Transport Systems. 2013. "Tanker Size." http://people.hofstra.edu/geotrans/eng/ch5en/appl5en/tankers.html.

Georgescu-Roegen, Nicholas. 1979. "Energy Analysis and Economic Valuation." *Southern Economic Journal* 45 (4): 1023–1058.

Giddy, Ian H. n.d. *Global Financial Markets Update (Part 2): Foreign-Exchange Prediction and Hedging Tools*. http://www.stern.nyu.edu/~igiddy/gfmup2.htm.

Gillingham, Kenneth, Richard G. Newell, and Karen Palmer. 2009. "Energy Efficiency Economics and Policy." *Annual Review of Resource Economics* 1 (1): 597–620.

Gipe, Paul. 2012. "Tables of Feed-in Tariffs Worldwide." Wind-Works.org. http://www.wind-works.org/cms/index.php?id=92.

———. 2013. "A Summary of Fatal Accidents in Wind Energy." Wind-Works.org (May 1). http://www.wind-works.org/cms/index.php?id=128&tx_ttnews%5Btt_news%5D=414&cHash=5a7a0eb3236dd3283a3b6d8cf4cc508b.

GIZ (Deutsche Gesellschaft für Internationale Zusammenarbeit GmbH). 2013. "International Fuel Prices 2012/2013." http://www.giz.de/expertise/downloads/Fachexpertise/giz2013-en-ifp2013.pdf.

Global Alliance for Clean Cookstoves. 2013. "Resources." http://www.cleancookstoves.org/resources/.

Global Subsidies Initiative of the International Institute for Sustainable Development. 2008. *Biofuels—At What Cost? Government Support for Ethanol and Biodiesel in China.* November. http://www.iisd.org/gsi/sites/default/files/china_biofuels_subsidies.pdf.

Global Transmission Report. 2009. "Focus on UHV AC: China Shows the Way by Energising 1,000 kV Line." March 2. http://www.globaltransmission.info/archive.php?id=1434.

Glomsrød, Solveig, Taoyuan Wei, and Knut H. Alfsen. 2012. "Pledges for Climate Mitigation: The Effects of the Copenhagen Accord on CO_2 Emissions and Mitigation Costs." *Mitigation and Adaptation of Strategies for Global Change* (April 3). doi:10.1007/s11027-012-9378-2. http://link.springer.com/article/10.1007/s11027-012-9378-2/fulltext.html.

Goodstein, Eban S. 1999. *Economics and the Environment.* 2nd ed. Upper Saddle River, NJ: Prentice Hall.

Gordon, Richard L. 1987. *World Coal: Economics, Policies and Prospects.* Cambridge, UK: Cambridge University Press.

Gosden, Emily. 2013. "BP Profits Triple to £10.7bn on Russian Deal as New Production Helps Beat Expectations." April 30. http://www.telegraph.co.uk/finance/newsbysector/energy/oilandgas/10026908/BP-profits-triple-to-10.7bn-on-Russian-deal-as-new-production-helps-beat-expectations.html.

Grace, John D. 2005. *Russian Oil Supply: Performance and Prospects.* Oxford, UK: Oxford University Press.

Greene, William H. 2012. *Econometric Analysis.* 7th ed. Upper Saddle River, NJ: Prentice Hall.

Greer, Susan, and Chris Patel. 2000. "The Issue of Australian Indigenous World-Views and Accounting." *Accounting, Auditing and Accountability Journal* 13 (3): 307–329.

Gregory, Regina. 2010. "The EcoTipping Points Project, Models for Success in a Time of Crisis: China—Biogas." November. http://www.ecotippingpoints.org/our-stories/indepth/china-biogas.html.

Griffin, James, and Henry Steele. 1986. *Energy Economics and Policy.* 2nd ed. Orlando, FL: Academic Press.

Grübler, Arnulf. 1998. *Technology and Global Change.* Cambridge, UK: Cambridge University Press.

———. n.d. "Technology and Global Change: Data Appendix." http://user.iiasa. ac.at/~gruebler/Data/TechnologyAndGlobalChange.

Gulf Oil and Gas. 2011. "Salini Builds the Largest Dam in Africa." April 18. http://www.gulfoilandgas.com/webpro1/main/mainnews.asp?id=15024.

Gustavsson, Bengt. 1995. "The Human Values of Swedish Management." *Journal of Human Values* 1 (2): 153–172.

Gustafson, Thane. 2012. *Wheel of Fortune: The Battle for Oil and Power in Russia.* Cambridge, MA: The Belknap Press of Harvard University Press.

Hall, Edward T., and Mildred Reed Hall. 1990. *Understanding Cultural Differences: Germans, French and Americans.* Yarmouth, ME: Intercultural Press.

———. 2011. "Historical Oil Shocks." NBER Working Paper No. 16790. February. http://www.nber.org/papers/w16790.pdf.

Hamilton, Kirk, Giovanni Ruta, and Liaila Tajibaeva. 2006. "Capital Accumulation and Resource Depletion: A Hartwick Rule Counterfactual." *Environmental and Resource Economics* 34: 517–533.

Hannesson, Rögnvaldur. 1998. *Petroleum Economics: Issues and Strategies of Oil and Natural Gas Production.* Westport, CT: Quorum Books.

Harris, Nancy L., Sandra Brown, Stephen C. Hagen, Sassan S. Saatchi, Silvia Petrova, William Salas, Matthew C. Hansen, Peter V. Potapov, and Alexander Lotsch. 2012. "Baseline Map of Carbon Emissions from Deforestation in Tropical Regions." *Science* 336 (6088) (June 22): 1573–1576. doi:10.1126/science.1217962.

Hartwick, John M. 1977. "Intergenerational Equity and the Investing of Rents from Exhaustible Resources." *American Economic Review* 67 (5): 972–74.

Hassmann, Heinrich. 1953. *Oil in the Soviet Union.* Translated by Alfred M. Leeston. Princeton, NJ: Princeton University Press.

Heather, Patrick. 2012. *Continental European Gas Hubs: Are They Fit for Purpose?* Oxford: Oxford Institute for Energy Studies. NG 63. June. http://www. oxfordenergy.org/wpcms/wp-content/uploads/2012/06/NG-63.pdf.

Henisz, Witold, and Bennet A. Zelner. circa 2011. "The Cycling of Power between Private and Public Sectors: Electricity Generation in Argentina, Brazil and Chile." http://www-management.wharton.upenn.edu/henisz/finaly_henisz_zelner.pdf.

Herbert, John H. 1993. "The Relations of Monthly Spot to Futures Prices for Natural Gas." *Energy* 18 (11): 1119–1124.

Herron, James, and Jared A. Favole. 2011. "IEA to Release 60 Million Barrels of Oil from Emergency Stocks." *Wall Street Journal* (June 23). http://online.wsj.com/article/SB10001424052702303339904576403480626606792.html.

Hines, R. D. 1989. "The Sociopolitical Paradigm in Financial Accounting Research." *Accounting, Auditing & Accountability* 2 (1): 52–76.

Hinrichs, Roger A. 1996. *Energy: Its Use and the Environment.* 2nd ed. Ft. Worth, TX: Saunders College Publishing.

Historical Archive. 2007. "The History of Electricity—A Timeline." February 13. http://www.thehistoricalarchive.com/happenings/57/the-history-of-electricity-a-timeline/.

Hoecklin, Lisa. 1995. *Managing Cultural Differences: Strategies for Competitive Advantage.* New York: Addison-Wesley.

Hoel, Michael, Bjart Holtsmark, and Jon Vislie. 1990. "The European Gas Market as a Bargaining Game." In *Recent Modeling Approaches in Applied Energy Economics,* edited by Olav Bjerkholdt, Øystein Olsen, and Jon Vislie, 49–65. London: Chapman and Hall.

Hofstede, Geert. 1991. *Cultures and Organizations: Software of the Mind.* New York: McGraw-Hill.

Hofstede, Geert, and M. H. Bond. 1988. "Confucius & Economic Growth: New Trends in Culture's Consequences." *Organizational Dynamics* 16 (4): 4–21.

Högselius, Per, Arne Kaijser, and Anna Åberg. 2010. "Natural Gas in Cold War Europe: The Making of a Critical Transnational Infrastructure." Draft chapter for the EUROCRIT edited volume (version of 5 May 2010). http://perhogselius.files.wordpress.com/2010/09/natural-gas-in-cold-war-europe-5-may-2010.pdf.

Hooker, J. N. 1998. "The Polite and the Rude." Shorter version published as "The Polite and the Rude: Etiquette Abroad." *Monash Mt. Eliza Business Review* 1 (3): 40–49.

Hotelling, Harold. 1931. "The Economics of Exhaustible Resources." *Journal of Political Economy* 39 (2) (April): 137–175.

Hubbert, M. K. 1962. "Energy from Fossil Fuels, Energy Resources." National Academy of Sciences, National Research Council Publication 1000-D: 22–94.

Hull, John C. 2000. *Options, Futures, and Other Derivative Securities.* 4th ed. Upper Saddle River, NJ: Prentice Hall.

———. 2011. *Options, Futures, and Other Derivatives and DerivaGem CD Package.* 8th ed. Upper Saddle River, NJ: Prentice Hall.

Hunt, S., and G. Shuttleworth. 1996. *Competition and Choice in Electricity.* New York: John Wiley & Sons.

IAEA (International Atomic Energy Agency). 2006. *Chernobyl's Legacy: Health, Environmental and Socio-Economic Impacts and Recommendations to the Governments of Belarus, the Russian Federation and Ukraine.* 2nd rev. ed. April. https://www.iaea.org/sites/default/files/chernobyl.pdf

———. 2014. *Managing Spent Nuclear Fuel: Global Overview* (Issue Brief). November 17. http://www.iaea.org/newscenter/focus/radwaste/nuclfueloverview.html.

———. n.d. "Home Page." http://www.iaea.org.

ICE (Intercontinental Exchange). 2012a. "Existing ICE OTC Rulebook." October 12. https://www.theice.com/publicdocs/otc/advisory_notices/ICE_Advisory_10_12_003.pdf.

———. 2012b. *Natural Gas.* https://www.theice.com/publicdocs/ICE_NatGas_Brochure.pdf.

———. 2013a. "Home Page." http://www.theice.com.

———. 2013b. "WebICE." https://www.theice.com/webice.jhtml.

IEA (International Energy Agency). 2005. *Energy Statistics Manual.* http://www.iea.org/publications/freepublications/publication/statistics_manual.pdf.

———. 2011a. *25 Energy Efficiency Policy Recommendations: 2011 Update.* Paris: OECD/IEA. http://www.iea.org/publications/freepublications/publication/25recom_2011.pdf.

———. 2011b. *World Energy Outlook.* Paris: OECD/IEA. http://www.worldenergyoutlook.org/.

———. 2012a. *2012 Annual Questionnaires.* Paris: OECD/IEA. http://www.iea.org/statistics/resources/questionnaires/annual/#twelve.

———. 2012b. *Key Energy Statistics.* Paris: OECD/IEA. http://www.iea.org/publications/freepublications/publication/name,31287,en.htm.

———. 2012c. *World Energy Outlook.* Paris: OECD/IEA. http://www.worldenergyoutlook.org/.

———. 2013a. *Key Energy Statistics.* Paris: OECD/IEA.

———. 2013b. *World Energy Balances: Documentation for Beyond 2020 Files.* Paris: OECD/IEA. http://wds.iea.org/wds/pdf/documentation_worldbal.pdf.

———. 2013c. *World Energy Statistics: Documentation for Beyond 2020 Files.* Paris: OECD/IEA. http://wds.iea.org/wds/pdf/documentation_worldbes.pdf.

———. 2014. *PVPS Report: Snapshot of Global PV-1992-2013.* Paris: OECD/IEA. http://www.iea-pvps.org/index.php?id=266.

———. n.d.a. *Closing Oil Stock Levels in Days of Net Imports.* http://www.iea.org/netimports.asp.

———. n.d.b. *Coal and Peat Database.* Paris: OECD/IEA. http://www.iea.org/statistics/statisticssearch/.

———. n.d.c. *Coal Information Database.* Paris: OECD/IEA. http://data.iea.org/ieastore/statslisting.asp?.

———. n.d.d. *Electricity and Heat Database.* Paris: OECD/IEA. http://www.iea.org/statistics/statisticssearch/.

———. n.d.e. *Electricity Information Database.* Paris: OECD/IEA. http://data.iea.org/ieastore/statslisting.asp?.

———. n.d.f. *Energy Balances Database.* Paris: OECD/IEA. http://www.iea.org/statistics/statisticssearch/.

———. n.d.g. *Energy Prices and Taxes Database.* Paris: OECD/IEA. http://data.iea.org/ieastore/statslisting.asp?.

———. n.d.h. *Extended World Energy Balances Database.* Paris: OECD. http://www.oecd-ilibrary.org/energy/data/iea-world-energy-statistics-and-balances_enestats-data-en.

———. n.d.i. *Gas Trade Flows in Europe.* Paris: OECD/IEA. http://www.iea.org/gtf/index.asp.

———. n.d.j. *Glossary of Statistical Terms.* Paris: OECD/IEA. http://stats.oecd.org/glossary/index.htm.

———. n.d.k. *Indicators.* Paris: OECD/IEA. http://www.iea.org/statistics/statisticssearch/.

———. n.d.l. *Natural Gas Database.* Paris: OECD/IEA. http://www.iea.org/statistics/statisticssearch/.

———. n.d.m. *Oil Database.* Paris: OECD/IEA. http://www.iea.org/statistics/statisticssearch/.

———. n.d.n. *Oil Information Database.* Paris: OECD/IEA. http://data.iea.org/ieastore/statslisting.asp?.

———. n.d.o. *RD&D Statistics Database.* Paris: OECD/IEA. http://www.iea.org/statistics/RDDonlinedataservice/.

———. n.d.p. *Renewable Database.* Paris: OECD/IEA http://www.iea.org/statistics/statisticssearch/.

———. n.d.q. *World Balance Graphical Database.* Paris: OECD/IEA. http://www.iea.org/Sankey/index.html.

IEA-ETSAP and IRENA (International Energy Agency—Energy Technology Systems Analysis Programme and International Renewable Energy Agency). 2013. *Concentrating Solar Power: Technology Brief.* E10. January. http://www.irena.org/DocumentDownloads/Publications/IRENA-ETSAP%20Tech%20Brief%20E10%20Concentrating%20Solar%20Power.pdf.

IEW (International Energy Workshop). 2005. "IEW Aims and History." http://webarchive.iiasa.ac.at/Research/ECS/IEW/history.html.

IEX (Indian Energy Exchange). n.d. "Home Page." http://www.iexindia.com.

IGU (International Gas Union). 2011. *World LNG Report 2011*. http://www.igu.org/gas-knowhow/publications/igu-publications/LNG%20Report%202011.pdf.

_____. 2012. *Natural Gas Conversion Pocketbook*. http://www3.scribd.com/doc/175710893/Natural-Gas-Conversion-Pocketbook.

_____. 2013. *World LNG Report 2013*. http://members.igu.org/news/igu-world-lng-report-2013.pdf.

_____. 2014. *World LNG Report 2014*. http://www.igu.org/sites/default/files/node-page-field_file/IGU%20-%20World%20LNG%20Report%20-%202014%20Edition.pdf.

IIASA (International Institute for Applied Systems Analysis). n.d. "Home Page." Laxenburg, Austria. http://www.iiasa.ac.at/.

IMF (International Monetary Fund). 2015. *World Economic Outlook (WEO), Uneven Growth: Short- and Long-Term Factors*. April. http://www.imf.org/external/pubs/ft/weo/2015/01/.

———. n.d. *World Economic Outlook Database*. http://www.imf.org/external/pubs/ft/weo/2012/01/weodata/index.aspx. The most recent update of this database is at http://www.imf.org/external/ns/cs.aspx?id=28.

Infoplease. 2013. "Oil Spills and Disasters." http://www.infoplease.com/ipa/A0001451.html.

International Swaps and Derivatives Association. 2013. "ISDA Members." http://www2.isda.org/membership/members-list/.

Institute for Economics and Peace. 2013. *Global Peace Index 2013*. http://www.visionofhumanity.org/sites/default/files/2013_Global_Peace_Index_Report_0.pdf.

INTERTANKO (International Association of Independent Tanker Owners) n.d. "Home Page." http://www.intertanko.com/.

Invensys. 2010. *LNG Regasification: Real-Time Operations Management*. http://iom.invensys.com/EN/pdfLibrary/IndustrySolution_Invensys_RealTimeOperationsMgmtUpstreamLNG_04-10.pdf.

IOGCC (Interstate Oil and Gas Compact Commission). n.d. "Home Page." http://www.iogcc.state.ok.us/.

Intergovernmental Panel on Climate Change (IPCC). 2007. *Climate Change 2007: IPCC Fourth Assessment Report*. http://www.ipcc.ch/publications_and_data/ar4/syr/en/contents.html.

———. 2012. *Renewable Energy Sources and Climate Change Mitigation: Special Report of the Intergovernmental Panel on Climate Change.* Cambridge, UK: Cambridge University Press.

———. 2013. *Climate Change 2013: The Physical Science Basis. Contribution of Working Group I to the Fifth Assessment Report of the Intergovernmental Panel on Climate Change.* (T. F. Stocker, D. Qin, G. K. Plattner, M. Tignor, S. K.Allen, J. Boschung, A. Nauels, Y. Xia, V. Bex, and P. M. Midgley (eds.)]. Cambridge, UK: Cambridge University Press.

———. n.d. "Reports." https://www.ipcc.ch/publications_and_data/publications_and_data_reports.shtml.

IRENA (International Renewable Energy Agency). 2012a. *Renewable Energy Technologies: Cost Analysis Series, Volume 1: Power Sector Issue 2/5, Concentrating Solar Power.* June. http://www.irena.org/DocumentDownloads/Publications/RE_Technologies_Cost_Analysis-CSP.pdf.

———. 2012b. *Renewable Energy Technologies: Cost Analysis Series, Volume 1: Power Sector Issue 4/5, Solar Photovoltaics.* June. http://www.irena.org/DocumentDownloads/Publications/RE_Technologies_Cost_Analysis-SOLAR_PV.pdf.

Itoh, Shoichi. 2010. *Russia Looks East: Energy Markets and Geopolitics in Northeast Asia.* A Report of the CSIS, Russia and Eurasia Program, July. http://csis.org/files/publication/110721_Itoh_RussiaLooksEast_Web.pdf.

Jalali-Naini, Ahmad R., and Maryam Kazemi Manesh. 2006. "Price Volatility, Hedging and Variable Risk Premium in the Crude Oil Market." *OPEC Energy Review* 30 (2): 55–70.

Jaramillo, Paulina, W. Michael Griffin, and H. Scott Matthews. 2008. "Comparative Analysis of the Production Cost and Life-Cycle GHG Emissions of FT Liquid Fuels from Coal and Natural Gas." *Environmental Science and Technology* 42 (20): 7559–7565.

Jenkins, Gilbert. 1989. *Oil Economists' Handbook.* New York: Elsevier Applied Science.

Jensen, James T. 2011. "Natural Gas Pricing: Current Patterns and Future Trends." Presentation to the Beijing Energy Club, Shanghai, February 18.

Jevons, William Stanley. 1965. *The Coal Question: An Inquiry Concerning the Progress of the Nation, and the Probable Exhaustion of Our Coal Mines.* Reprints of Economic Classics Series. Originally published in 1865. New York: August M. Kelly.

JIT (Joint Investigation Team), US Bureau of Ocean Energy Management, Regulation and Enforcement (BOEMRE)/US Coast Guard. 2011. "*Deepwater Horizon* Joint Investigation Team Releases Final Report." September 14. http://www.boem.gov/BOEM-Newsroom/Press-Releases/2011/press0914.aspx.

JODI (Joint Organizations Data Initiative). n.d. "Data Downloads." The most recent update of this database is at http://www.jodidata.org/database/data-downloads.aspx.

Johnson, David B. 1991. *Public Choice: An Introduction to the New Political Economy*. Mountain View, CA: Mayfield Publishing Company.

Johnston, Daniel. 1994. *International Petroleum Fiscal Systems and Production Sharing Contracts*. Tulsa, OK: PennWell.

———. 2003. *International Exploration: Economics, Risk, and Contract Analysis*. Tulsa, OK: PennWell.

———. 2008. "Changing Fiscal Landscape." *Journal of Energy, Law, and Business* 1 (1): 31–54.

Jorion, Philippe. 2006. *Value at Risk: The New Benchmark for Controlling Market Risk*. New York: McGraw-Hill.

———. 2009. *Financial Risk Manager Handbook*. 5th ed. Hoboken, NJ: John Wiley & Sons, Inc.

Joskow, P. 2010. "Market Imperfections versus Regulatory Imperfections." *CESifo DICE Report* 8 (3): 3–7.

Jull, Oliver K. 2012. "Canada Withdraws from Kyoto Protocol to Avoid Non-Compliance Penalties." Canadian Bar Association Bulletin. http://www.cba.org/cba/newsletters-sections/pdf/2012-04-international_jull.pdf.

K.A.CARE (King Abdullah City for Atomic and Renewable Energy). 2013. "Home Page." http://www.energy.gov.sa/en.

Kagel, John H., and Alvin E. Roth, eds. 1995. *The Handbook of Experimental Economics*. Princeton, NJ: Princeton University Press.

Kahneman, Daniel. 2011. *Thinking, Fast and Slow*. New York: Farrar, Straus and Giroux.

Kalt, Joseph P. 1981. *The Economics and Politics of Oil Price Regulation: Federal Policy in the Post-Embargo Era*. Cambridge, MA: MIT Press.

Kaplan, Seymour. 1983. *Energy Economics: Quantitative Methods for Energy and Environmental Decisions*. New York: McGraw-Hill.

Kar, D. P. n.d. Personal conversation with the author.

Kar, Durga Prasad. 2010. "The Lowest Cost Electricity for a Poor Rural Village in India: Rural Grid or Off-Grid SPV?" PhD thesis, Mineral and Energy Economics Program. Golden, CO: Colorado School of Mines.

Keita, J. D. 1987. "Wood or charcoal—which is better?" Unasylva No. 157–158 – *Small-Scale Forest Enterprises* 39 (3, 4). FAO Corporate Document Repository. http://www.fao.org/docrep/s4550e/s4550e09.htm.

Keller, Anne B. 2012. "NGL 101—The Basics." Midstream Energy Group. EIA—Virtual Workshop on Natural Gas Liquids. June 6. Http://www.midstreamenergygroup.com/fyi//NGL_workshop-Anne-Keller.pdf.

Kemp, Alex. 2011. *The Official History of North Sea Oil and Gas, Vol. 1: The Growing Dominance of the State* (Government Official History Series). London: Routledge.

Kennes, Walter, Jyoti K. Parikh, and Herman Stolwijk. 1984. "Energy from Biomass by Socio-economic Groups—A Case Study of Bangladesh." *Biomass* 4 (3): 209–234.

Kent, C., and E. Eastham. 2011. "Consumer and Producer Surplus Show Deadweight Loss from a Tax, Taxation of Coal: A Comparative Analysis." http://www.marshall.edu/cber/research/Coal%20State%20Compare.pdf.

King, Gilbert. 2011. "Edison vs. Westinghouse: A Shocking Rivalry." Smithsonian.com. October 11. http://blogs.smithsonianmag.com/history/2011/10/edison-vs-westinghouse-a-shocking-rivalry/.

Kluckhohn, Florence R., and Fred L. Strodtbeck. 1961. *Variations in Value Orientations*. Evanston, IL: Row, Peterson.

Kolb, Robert. 1995. *Options: An Introduction*. 2nd ed. New York: John Wiley & Sons.

Kolb, Robert W., and James A. Overdahl. 2007. *Futures, Options, and Swaps*. 5th ed. Malden, MA: Blackwell.

Korchemkin, Mikhail B. 1993. "Russia's Huge Gazprom Struggles to Adjust to New Realities." *Oil & Gas Journal*. 91 (42) (October 18): 39–44.

Kramer, Andrea, and Peter Fusaro, eds. 2010. *Energy and Environmental Project Finance Law and Taxation: New Investment Techniques*. Oxford, UK: Oxford University Press.

Krylov, N. A., A. A. Bokserman, and E. R. Stavrovsky (eds.). 1998. *The Oil Industry of the Former Soviet Union: Reserves, Extraction and Transportation*. Amsterdam: Gordon and Breach Science Publishers.

Ku, Anne. 1997. "Power Pools." April. http://www.analyticalq.com/energy/powerpools/default.htm.

Kumbhakar, Subal C., and Lennart Hjalmarsson. 1998. "Relative Performance of Public and Private Ownership Under Yardstick Competition: Electricity Retail Distribution." *European Economic Review* 42 (1): 97–122.

Labys, W. C., S. Paik, and A. M. Liebenthal. 1979. "An Econometric Simulation Model of the US Market for Steam Coal." *Energy Economics* 1 (1): 19–26.

Lane, Jim. 2013. "Biofuels Mandates Around the World: 2014." *Biofuels Digest*. December 31. http://www.biofuelsdigest.com/bdigest/2013/12/31/biofuels-mandates-around-the-world-2014/.

Lanier, Jaron. 2010. *You Are Not a Gadget*. New York: Alfred A. Knopf.

Lantau Group. 2014. "TLG on China: UHV." May. http://www.lantaugroup.com/files/tlgchina_uhv_may14.pdf.

Lawler, Alex. 2011. "UPDATE 2-Platts to Change Brent Oil Assessment in 2012." Reuters. September 16. http://uk.reuters.com/article/2011/09/16/platts-brent-idUKL3E7KG2B120110916.

Lee, Jeannette. 2013. "Financing Strategies for LNG Export Projects." Alaska Natural Gas Transportation Projects, Office of the Federal Coordinator. April 3. http://www.arcticgas.gov/financing-strategies-lng-export-projects.

Leffler, William L. 2000. *Petroleum Refining in Nontechnical Language*. 3rd ed. Tulsa, OK: PennWell.

Leiserowitz, A., E. Maibach, C. Roser-Renouf, G. Feinberg, and P. Howe. 2013. Climate Change in the American Mind: Americans' Global Warming Beliefs and Attitudes in April, 2013. Yale University and George Mason University. New Haven, CT: Yale Project on Climate Change Communication. http://environment.yale.edu/climate-communication/article/Climate-Beliefs-April-2013.

Leising, Matthew. 2013. "Energy Swaps Migrating to Futures on Dodd-Frank Rules." January 25. http://Bloomberg.com/news/2013-01-25/energy-swaps-migrating-to-futures-as-dodd-frank-rules-take-hold.html.

Lighting Science. 2012. "Global Legislation to Phase-Out Incandescent Lamps." http://www.lsgc.com/why-leds/global-ban-on-incandescent/.

Lin, Yi-min, and Tian Zhu. 2001. "Ownership Restructuring in Chinese State Industry: An Analysis of Evidence on Initial Organizational Changes." *Chinese Quarterly* (June) 166: 301–341.

LOOP (Louisiana Offshore Oil Port). 2013. "Home Page." http://www.loopllc.com.

Lovejoy, Wallace F., and Paul T. Homan. 1967. *Aspects of Oil Conservation Regulation*. Washington, DC: Resources for the Future.

Machmud, Tengku Nathan. 2000. *The Indonesian Production Sharing Contract: An Investor's Perspective*. Cambridge, MA: Kluwer Law International.

Maddison, Angus. 2010. "Maddison Project Database." http://www.ggdc.net/maddison/maddison-project/home.htm.

Mahon, J., Jr. 1992. "Was Latin America Too Rich to Prosper? Structural and Political Obstacles to Export-Led Industrial Growth." *Journal of Development Studies* 28 (January): 241–63.

Makholm, Jeff D. 2012. *The Political Economy of Pipelines*. Chicago: University of Chicago Press.

Maritime Connector. n.d. "Oil Tanker–Mont (Knock Nevis, Jahre Viking, Happy Giant, Seawise Giant)." http://maritime-connector.com/worlds-largest-ships/.

Martin, Richard, and Laverne Murach. 2012. "Large Build-Outs of HVDC Transmission Systems Will Help Integrate Renewable Energy on the Power Grid." *Daily Finance.* October 29. http://www.dailyfinance.com/2012/10/29/large-build-outs-of-hvdc-transmission-systems-will/.

Mascarelli, Amanda Leigh. 2009. "A Sleeping Giant?" *Nature* (March). doi:10.1038/climate.2009.24 http://www.nature.com/climate/2009/0904/full/climate.2009.24.html.

Maslow, Abraham H. 1970. *Motivation and Personality.* New York: Harper & Row.

MBendi. 2013. *Electrical Power in Africa.* December 4. http://www.mbendi.com/indy/powr/af/p0005.htm.

McFadden, Daniel. 1999. "Rationality for Economists?" *Journal of Risk and Uncertainty* 19 (1–3): 73–105.

McKinsey & Company. 2009. "Unlocking Energy Efficiency in the US Economy." Electric Power & Natural Gas. July. http://www.mckinsey.com/clientservice/electricpowernaturalgas/downloads/US_energy_efficiency_full_report.pdf.

McLuhan, Marshall. 1964. *Understanding Media: The Extensions of Man.* New York: Mentor.

Mead, Walter. 1994. "Toward an Optimal Oil and Gas Leasing System." *Conference Proceedings of the 17th Annual International Association for Energy Economics Annual Meeting.* Stavanger, Norway, May 25–27.

Meadows, Dennis L., Donella H. Meadows, and Jorgen Randers. 1992. *Beyond the Limits: Confronting Global Collapse, Envisioning a Sustainable Future.* Post Mills, VT: Chelsea Green Publishing.

Meadows, Donella H., Dennis L. Meadows, Jorgen Randers, and William W. Behrens. 1972. *The Limits to Growth.* New York: Universe Books.

Meadows, Donella H., Jorgen Randers, and Dennis L. Meadows. 2004. *Limits to Growth: The 30-Year Update.* White River Junction, VT: Chelsea Green Publishing.

Melling, Anthony J. 2010. *Natural Gas Pricing and Its Future: Europe as the Battleground.* Washington, DC: Carnegie Endowment for International Peace. http://carnegieendowment.org/files/gas_pricing_europe.pdf.

Metalaugmentor. 2010. "Crazy Market Thoughts: Crude Tankers and Contango." January 6. http://metalaugmentors.com/analysis/crazy-market-thoughts-crude-tankers-and-contango.html.

Miesner, Thomas O., and William L. Leffler. 2006. *Oil & Gas Pipelines in Nontechnical Language.* Tulsa, OK: PennWell.

Mitchell, Brian R. 2007a. *International Historical Statistics: Europe, 1750–2005.* 6th ed. London: Palgrave Macmillan.

———. 2007b. *International Historical Statistics: The Americas, 1750–2005.* 6th ed. London: Palgrave Macmillan.

———. 2008. *International Historical Statistics: Africa, Asia, and Oceania, 1750–2005.* 5th ed. London: Palgrave Macmillan.

Morris, Desmond. 2002. *Peoplewatching: The Desmond Morris Guide to Body Language.* Revised ed. New York: Vintage.

Morrison, Terri, and Wayne A. Conaway. 2006. *Kiss, Bow, or Shake Hands: The Best Selling Guide to Doing Business in More Than 60 Countries.* Revised ed. Avon, MA: Adams Media Corporation.

Mouawad, Jad. 2012. "Delta Buys Refinery to Get Control of Fuel Costs." April 30. http://www.nytimes.com/2012/05/01/business/delta-air-lines-to-buy-refinery. html?_r=0.

Muller, Richard A. 2009. *Physics for Future Presidents.* New York: W. W. Norton & Company.

Museum of Romanian Oil Industry. n.d. "Romanian Oil Industry: History." http://www.roconsulboston.com/Pages/InfoPages/Commentary/OilRoHistory.html.

Mushrush, George W., and James G. Speight. 1995. *Petroleum Products: Instability and Incompatibility.* Bristol, PA: Taylor & Francis.

Mutikana, Lucia. 2012. "Labor's Share of Income Falls, Widening Wealth Gap: Study." Reuters. http://www.reuters.com/article/2012/09/25/us-usa-economy-inequality-idUSBRE88O17T20120925.

NAESB (North American Energy Standards Board). 2010. "Home Page." http://www.naesb.org.

Nakićenović, N. 1984. *Growth to Limits: Long Waves and the Dynamics of Technology.* Laxenburg, Austria: International Institute for Applied Systems Analysis.

Nankani, Gobind. 1979. *Developmental Problems of Mineral Exporting Countries, 1979.* World Bank Background Paper No. 354 for the 1979 World Development Report.

NASA (National Aeronautics and Space Administration). 2013. "The Big Bang." http://science.nasa.gov/astrophysics/focus-areas/what-powered-the-big-bang.

National Energy Technology Laboratory. 2013. *Modern Shale Gas Development in the United States: Un Update.* September. http://www.netl.doe.gov/File%20Library/Research/Oil-Gas/shale-gas-primer-update-2013.pdf.

National Petroleum Council. 1972. *U.S. Energy Outlook.* Washington, DC: National Petroleum Council.

National Renewable Energy Laboratory. 2001. "Concentrating Solar Power: Energy from Mirrors." March. http://www.nrel.gov/docs/fy01osti/28751.pdf.

———. 2013a. "Concentrating Solar Power Research." November 18. http://www.nrel.gov/csp/.

———. 2013b. "Concentrating Solar Power Projects." November 25. http://www. nrel.gov/csp/solarpaces/.

National Research Council. 2010. *Adapting to the Impacts of Climate Change.* Washington, DC: The National Academies Press. http://www.nap.edu/catalog. php?record_id=12783.

Natural Gas Supply Association. 2011. "The History of Regulation." http://naturalgas.org/regulation/history/.

Nei, Hisanori. 2000. "The Transformation of Japanese Energy Policy." *The Journal of Energy and Development* 25 (2): 173–185.

NERC (North American Electric Reliability Corporation). 2013. "Home Page." http://www.nerc.com/.

Newbery, David M. 2005. "Electricity Liberalization in Britain: The Quest for a Satisfactory Wholesale Market Design." *The Energy Journal* 26 (Special Issue European Electricity Liberalization): 43–70.

New Mexico Taxation and Revenue Department. 2013. "All Taxes." http://www.tax.newmexico.gov/All-Taxes/Pages/Home.aspx.

New Zealand Ministry of Economic Development. 2001. *Chronology of New Zealand Electricity Reform.* Energy Markets Policy Group. http://www.med.govt.nz/ers/electric/chronology/.

NGI (Natural Gas Intelligence). 2013. "Anemic Prices Flatten Marketers' 4Q2011 Results." http://intelligencepress.com/features/rankings/ gas/gas_marketer_ rankings_2011.html.

Nix, Steve. 2013. "Making Lump and Briquette Charcoal" About.com, Forestry. http://forestry.about.com/od/alternativeforest/p/The–History–And–Business– Of–Making–Lump–Charcoal.htm.

Nixon, Niki. 2008. "Timeline: The History of Wind Power." *The Guardian* (October 17). http://www.theguardian.com/environment/2008/oct/17/wind-power- renewable-energy.

NOAA (National Oceanic and Atmospheric Administration). n.d. *Earth System Research Laboratory (ESRL)—Global Monitoring Division (GMD) CO_2 Database.* Mauna Loa Observatory. The latest update of this database is at ftp://ftp.cmdl. noaa.gov/ccg/co2/trends/co2_mm_mlo.txt.

Nordhaus, William D. 1973. "World Dynamics: Measurement Without Data." *Economic Journal* 83 (332): 1156–1183.

Nordhaus, William D., Robert N. Stavins, and Martin L. Weitzman. 1992. "Lethal Model 2: The Limits to Growth Revisited." *Brookings Papers on Economic Activity* (2): 1–59.

Nord Pool Spot. 2011. *Nordic Electricity Exchange and the Nordic Model for the Liberalized Electricity Market.* http://www.nordpoolspot.com/globalassets/ Download-Center/Rules-and-regulations/The-Nordic-Electricity-Exchange-and-the-Nordic-model-for-a-liberalized-electricity-market.pdf.

NVE (Norwegian Water Resources and Energy Directorate). 2012. *Annual Report 2011.* June. Oslo, Norway: Norwegian Water Resources and Energy Directorate. http://webby.nve.no/publikasjoner/rapport/2012/rapport2012_19.pdf.

NY Times. 1993. "14-Year Cleanup at Three Mile Island Concludes." August 15. http://www.nytimes.com/1993/08/15/us/14-year-cleanup-at-three-mile-island-concludes.html.

OECD (Organisation for Economic Co-operation and Development). 1992. *The Polluter-Pays Principle: OECD Analyses and Recommendations.* Environment Directorate. Paris: OECD. http://www.tradeenvironment.eu/uploads/OCDE_ GD_92_81.pdf.

———. n.d. "Input-Output Tables." The most recent updates of this database can be found at http://www.oecd.org/trade/input-outputtables.htm.

OGJ (Oil & Gas Journal) Editors. 1999. "The Mexican Gas Tariff." *Oil & Gas Journal* 97 (24): 18–19.

———. 2013a. "Chevron Confirms First Cargo from Angola LNG." June 17. http:// www.ogj.com/articles/2013/06/chevron-confirms-first-cargo-from-angola-lng. html.

———. 2014. "Gazprom, CNPC Sign 30-Year Natural Gas Supply Contract." May 21. http://www.ogj.com/articles/2014/05/gazprom-cnpc-sign-30-year-natural-gas-supply-contract.html.

———. 2001. "Gas Pipelines." Special Issue: Pipeline Economics. (September 3): 84–85.

———. 2004. "Saudi Supergiant Shaybah Oil Field Still Growing." (April 5). http://www.ogj.com/articles/print/volume-102/issue-13/general-interest/ saudi-supergiant-shaybah-oil-field-still-growing.html.

———. 2010a. "Refined Product Prices." (June 25): 53.

———. 2010b. "Worldwide Look at Reserves and Production." Special Report: Worldwide Issue. (December 6): 48–49.

———. 2011. "Survey of Top 200 US Oil & Gas Companies." http:/www.ogj.com/ ogj-survey-downloads.html.

———. 2012a. "Worldwide Construction Update." (May). http://www.ogj.com/ ogj-survey-downloads.html.

———. 2012b. "OGJ150." (September 3): 38–41.

———. 2012c. "OGJ100: Leading Oil and Gas Companies Outside the US." (September 3): 46–49.

———. 2012d. "Worldwide Look at Reserves and Production." Special Report: Worldwide Issue. (December 3): 31–32.

———. 2013a. "OGJ Survey Downloads." http:/www.ogj.com/ogj-survey-downloads.html.

——— 2013b. "OGJ150 Earnings down as US Production Climbs." (September 2): 34-51.

———. 2013c. "Freedonia: Global Demand for Fuel Additives to Surge by 2016." (January 10). http://www.ogj.com/articles/2013/01/freedonia-global-demand-for-fuel-additives-to-surge-by-2016.html.

———. 2014a. "US Crude Prices." (May 5): 110.

———. 2014b. "US Crude Prices." (June 2): 126.

OGJ Databook. 2002. "Crude Oil Assays." Tulsa, OK: PennWell. 273–349.

OGJ Newsletter. 2014. "Gazprom to Discontinue Gas Price Discount for Ukraine. March 10. http://www.ogj.com/articles/print/volume-112/issue-3a/regular-features/ogj-newsletter.html.

Olaya Morales, Yris. 2006. "LNG World Trade Model." PhD thesis, Mineral Economics Program. Golden, CO: Colorado School of Mines.

OPEC (Organization for Petroleum Exporting Countries). 2009. *Annual Statistical Bulletin*. http://www.opec.org/opec_web/en/publications /202.htm

———. 2013. *Annual Statistical Bulletin*. http://www.opec.org/opec_web/en/ publications /202.htm

———. n.d. *OPEC Basket Price*. The latest updates and archives to this database can be found at http://www.opec.org/opec_web/en/data_graphs/40.htm.

Ophardt, Charles E. 2003. "Boiling Points and Structures of Hydrocarbons." http://www.elmhurst.edu/~chm/vchembook/501hcboilingpts.html.

ORNL (Oak Ridge National Laboratory). 2011. *Biomass Energy Databook*. 4th ed. http://cta.ornl.gov/bedb/index.shtml.

———. 2013. *Transportation Energy Databook*. July 31. http://cta.ornl.gov/data/index.shtml.

Owen, Edgar Wesley. 1975. "The U.S.S.R." In chapter 21 in *Trek of the Oil Finders: A History of Exploration for Petroleum*, 1353–1415. Tulsa, OK: American Association of Petroleum Geologists (AAPG) Special Volumes. http://archives. datapages.com/data/specpubs/methodo2/data/a073/a073/0001/1350/1353.htm.

Paganie, David. 2012. "New Technology May Revolutionize Deepwater Drilling." *Offshore* 72 (7) (July 1): 10. http://www.offshore-mag.com/articles/print/volume-72/issue-7/departments/comment/new-technology-may-revolutionize-deepwater-drilling.html.

Pamuk, Humeyra. 2012. "Gas-Rich Qatar to Invest Up to $20 Billion in Solar Energy Plant." Reuters. (December 2). http://www.reuters.com/article/2012/12/02/us-climate-qatar-solar-idUSBRE8B10AC20121202.

Pardo, Robert. 1985. "Gann Lines and Angles." *Stocks and Commodities* 3 (5): 177. http://gann.su/book/eng/Pardo,%20Robert%20-%20Gann%20Lines%20and%20Angles.pdf.

Parfomak, Paul W., Neelesh Nerurkar, Linda Luther, and Adam Vann. 2012. *Keystone XL Pipeline Project: Key Issues.* Congressional Research Service, 7-5700. www.crs.gov, R41668, May 9.

Paris, Jerome A. 2006. "A Primer on Caspian Oil." November 18. http://www.theoildrum.com/story/2006/11/18/102426/08

Patel, Tara. 2014. "Natural Gas Loses Decades-Old Tie to Oil in Landmark Deal." Bloomberg (April 11). http://www.bloomberg.com/news/2014-04-10/natural-gas-loses-decades-old-tie-to-oil-in-landmark-deal.html.

Pathak, Pramod, and Govind Swaroop Pathak. 2011. "Behavioural Issues in Accidents: A Study." *Management Insight* 7 (1): 81–87.

Peninsula. 2012. "Qatar to Be Key Player in Renewable Energy." December 18. http://thepeninsulaqatar.com/qatar-business/218486-qatar-to-be-key-player-in-renewable-energy.html.

Petzet, Alan. 2012. "World Class Play Seen in Argentina's Vaca Muerta Shale." *Oil & Gas Journal* (March 27). http://www.ogj.com/articles/2012/03/world-class-play-seen-in-argentinas-vaca-muerta-shale.html.

Piccolo [pseud.]. 2008. "The Bakken Formation: How Much Will It Help?" April 26. http://www.theoildrum.com/node/3868.

Pindyck, Robert S. 1978. "The Optimal Exploration and Production of Nonrenewable Resources." *Journal of Political Economy* 86 (5)(October): 841–61.

Pipeline International. 2011. "Russia Expands Gas Pipeline Network." Issue 007 (March). http://pipelinesinternational.com/news/russia_expands_gas_pipeline_network/055348/.

Polak, Ben. 2007. "ECON 159: Game Theory." Open Yale Courses, Yale University. http://oyc.yale.edu/economics/econ-159#sessions.

Pollitt, Michael. 2008. "Electricity Reform in Argentina: Lessons for Developing Countries." *Energy Economics* 30 (4): 1536–1567.

Pooladi-Darvish, Mehran. 2004. "Gas Production from Hydrate Reservoirs and Its Modeling." *Journal of Petroleum Technology* 56 (6): 65–71.

Porter, Michael E. 1990. *Competitive Advantage of Nations.* New York: The Free Press.

Portland General Electric. 2013. *Annual Report.* http://investors.portlandgeneral.com/annuals.cfm.

Poteete, Janssen, and Elinor Ostrom. 2010. *Working Together: Collective Action, the Commons, and Multiple Methods in Practice*. Princeton, NJ: Princeton University Press.

Powers, William. 2010. *Hamlet's BlackBerry: Building a Good Life in the Digital Age*. New York: HarperCollins Publishers.

Ramseur, Jonathan. 2012. Deepwater Horizon *Oil Spill: Highlighted Activities*. CRS Report R42371. Washington, DC: Congressional Budget Office.

Ratnikas, Algis. circa 2013. "Oil and Gas Timeline." *Timelines of History*. http://timelines.ws/subjects/Oil.HTML.

Ray, G. F. 1979. "Energy Economics—A Random Walk in History." *Energy Economics* 1 (3): 139–143.

Ren21. 2013. "Global Status Report (GSR) Policy Table." http://www.ren21.org/ RenewablePolicy/GSRPolicyTable.aspx.

Reuters. 2012a. "Fitch Affirms LOOP LLC's IDR at 'A-'." http://www.reuters.com/ article/2012/08/10/idUSWNA321220120810.

Robinson, Colin. 1992. "The Demand for Electricity: A Critical Analysis of Producer Forecasts." In *Energy Demand: Evidence and Expectations*, edited by David Hawdon, 215–234. London: Surrey University Press.

Rosenthal, Edward C. 2011. *The Complete Idiot's Guide to Game Theory*. New York: Penguin Group.

Rosneft. 2013. "Home Page." http://www.rosneft.com.

Ross, Michael L. 1999. "The Political Economy of the Resource Curse." *World Politics* 51 (2): 297–322.

Royal Institution of Naval Architects. 2013. "Home Page." http://www.rina.org.uk/ LNG-Ships-Floating-Systems.

Rudnick, H., and J. Zolezzi. 2001. "Electricity Sector Deregulation and Restructuring in Latin America: Lessons to Be Learnt and Possible Ways Forward." *Institute of Electrical Engineers Proceedings, Generation, Transmission, and Distribution* 148 (2): 180–184. http://web.ing.puc.cl/~power/paperspdf/rudnickzolezzi.pdf.

Russia & India Report. 2012. "Russia Creates Strategic Petroleum Reserve." November 1. http://indrus.in/articles/2012/01/11/russia_creates_strategic_ petroleum_reserve_14148.html.

Sampson, Anthony. 1975. *The Seven Sisters: The Great Oil Companies and the World They Shaped*. New York: Viking Press.

Sandrea, Ivan, and Rafael Sandrea. 2007. "Global Oil Reserves—Recovery Factors Leave Vast Target for EOR Technologies." *Oil & Gas Journal* Part 1 (November 5) and Part 2 (November 12).

SBM Offshore. 2013. "Home Page." http://www.sbmoffshore.com.

Schrattenholzer, Leo. 1998. "The International Energy Workshop: Results of the 1997 Poll." *OPEC Review* (July): 147–158.

Schumpeter, Joseph A. 1942. *Capitalism, Socialism and Democracy*. Whitefish, MT: Kessinger Publishing, LLC.

Schwabe, Paul. 2011. "Drilling Down into the Cost of Geothermal Energy." National Renewable Energy Laboratory. May 2. https://financere.nrel.gov/finance/content/drilling-down-cost-geothermal-energy.

Schwartz, Peter. 1991. *The Art of the Long View*. New York: Doubleday.

Scotese, Christopher R. 2002. "Paleomap Project." http://www.scotese.com/climate.htm.

Selley, Richard. 1985. *Elements of Petroleum Geology*. New York: W. H. Freeman and Company. 392.

———. 1997. *Elements of Petroleum Geology*. New York: W. H. Freeman and Company.

Shell. n.d. "Shell Scenarios." http://www.shell.com/global/future-energy/scenarios.html.

Shiryaevskaya, Anna. 2013. "Novatek's Yamal LNG Backs Investment Even as Costs Climb." Bloomberg. http://www.bloomberg.com/news/2013-12-18/novatek-s-yamal-lng-backs-investment-even-as-costs-climb.html.

Sim, Li-Chen. 2008. *The Rise and Fall of Privatization in the Russian Oil Industry*. Basingstoke, UK: Palgrave Macmillan.

Singh, Gurdip. 2012. "EXMAR Launches FLRSU to Exploit Stranded Gas Offshore Colombia." *Offshore* (August 1). http://www.offshore-mag.com/articles/print/volume-72/issue-8/engineering-construction-installation1/exmar-launches-flrsu-to-exploit-stranded-gas-offshore-colombia.html.

Skov, Arlie. 1995. "An Analysis of Forecasts of Energy Supply, Demand, and Oil Prices." Presented at the Society of Petroleum Engineers, SPE Hydrocarbon Economics and Evaluation Symposium, Dallas, TX, March 26–28.

Slade, Margaret E., Charles D. Kolstad, and Robert J. Weiner. 1993. "Buying Energy and Nonfuel Minerals: Final, Derived, and Speculative Demand." In *Handbook of Natural Resources and Energy Economics (III)*, edited by Allen V. Kneese and James L. Sweeney, 935–1009.

Smil, Vaclav. 1994. *Energy in World History*. Boulder, CO: Westview Press.

———. 2005. "Limits to Growth Revisited: A Review Essay." *Population and Development Review* 31 (1): 157–180.

Smith, Christopher E. 2010. "Special Report: Pipeline Construction Plans Slow for 2010." *Oil & Gas Journal* (February 15). http://www.ogj.com/articles/print/volume-108/issue-6/technology/special-report-pipeline.html.

———. 2012. "TransCanada Moves Projected Keystone XL Startup to 2015." *Oil & Gas Journal* (February 15). http://www.ogj.com/articles/2012/02/transcanada-moves-projected-keystone-xl-startup-to-2015.html.

Sol, Patricio del. 2002. "Responses to Electricity Liberalization: The Regional Strategy of a Chilean Generator." *Energy Policy* 30: 437–446.

Solow, Robert M. 1956. "A Contribution to the Theory of Economic Growth." *Quarterly Journal of Economics* 70 (1)(February): 65–94.

Sorrell, Steve. 2007. *The Rebound Effect: An Assessment of the Evidence for Economy-Wide Energy Savings from Improved Energy Efficiency.* The UK Energy Research Centre (UKERC), Sussex Energy Group. October. http://www.ukerc.ac.uk/publications/the-rebound-effect-an-assessment-of-the-evidence-for-economy-wide-energy-savings-from-improved-energy-efficiency.html..

Sorrell, Steve, and John Dimitropoulos. 2008. "The Rebound Effect: Microeconomic Definitions, Limitations and Extensions." *Ecological Economics* 65 (3): 636–649.

Sovacool, Benjamin K. 2008. "The Costs of Failure: A Preliminary Assessment of Major Energy Accidents, 1907 to 2007." *Energy Policy* 36 (5): 1802–1820.

Sovereign Wealth Fund. 2013a. "Sovereign Wealth Fund Rankings." http://www.swfinstitute.org/fund-rankings/.

———. 2013b. "Linaburg-Maduell Transparency Index." http://www.swfinstitute.org/statistics-research/linaburg-maduell-transparency-index/.

SPE (Society of Petroleum Engineers). 2007. *Petroleum Reserves & Resources Definitions.* http://www.spe.org/industry/reserves.php.

Spiegel, Murray R. 1995. *College Algebra.* New York: McGraw-Hill.

Statistics Canada. 2012. *Energy Statistics Handbook.* First Quarter. http://www.statcan.gc.ca/pub/57-601-x/57-601-x2012001-eng.htm.

Statoil. 2012. "Statoil and Wintershall Sign Major Strategic Gas Supply Agreement." November 20. http://www.statoil.com/en/NewsAndMedia/News/2012/Pages/20Nov_Wintershall.aspx.

Stermole, Franklin J., and John M. Stermole. 2012. *Economic Evaluation and Investment Decision Methods.* 13th ed. Lakewood, CO: Investment Evaluations Corporation.

Stern, Jonathan. 1984. *International Gas Trade in Europe: The Policies of Exporting and Importing Countries.* London: Heinemann Educational Books.

———. 1995. *Russian Natural Gas 'Bubble': Consequences for European Gas Markets.* London: Royal Institute of International Affairs, Energy and Environmental Programme.

———. 2005. *The Future of Russian Gas and Gazprom.* Oxford: Oxford University Press.

Stevens, Paul. 2003. "Resource Impact: Curse or Blessing? A Literature Survey." *The Journal of Energy Literature*, 9 (1), 3–42. http://graduateinstitute.ch/files/live/sites/iheid/files/sites/mia/users/Rachelle_Cloutier/public/International%20Energy/Stevens%20Resource%20Curse.pdf.

Stigler, George J. 1971. "The Theory of Economic Regulation." *The Bell Journal of Economics and Management Science* 2 (1): 3–21.

Stillman, Charles. 2006. "Cellulosic Ethanol: A Greener Alternative." June. http://www.cleanhouston.org/energy/features/ethanol2.htm.

Su, Fubing. 2004. "Political Economy of Industrial Restructuring in China's Coal Industry." *In Holding China Together: Diversity and National Integration in the Post-Deng Era*, edited by Dali L. Yang and Barry Naughton, 226–252. Cambridge, UK: Cambridge University Press.

Su, Chua Liang. 2007. "Optimal Demand-Side Participation in Day-Ahead Electricity Markets." Ph.D. thesis, School of Electrical and Electronic Engineering, University of Manchester. http://www.ee.washington.edu/research/real/Library/Thesis/Chua-Liang_SU.pdf.

Swan, Lukas G., and V. Ismet Ugursal. 2009. "Modeling of End-Use Energy Consumption in the Residential Sector: A Review of Modeling Techniques." *Renewable and Sustainable Energy Reviews* 13 (8): 1819–1835. http://dx.doi.org/10.1016/j.rser.2008.09.033.

Sweatman, R. E., M. E. Parker, and S. L. Crookshank. 2009. "Special Report: Industry CO_2 EOR Experience Relevant for Carbon Capture and Storage (CCS)." Presented at the SPE International Conference on CO_2 Capture, Storage, and Utilization, San Diego, November 2–4.

Switchme. n.d. "New Zealand Power Companies." http://www.switchme.co.nz/residential/power-companies.php.

Symonds, G. H. 1955. *Linear Programming: The Solution of Refinery Problems.* New York: Esso Standard Oil Co.

TBEA Ltd. 2013. "Development and Cooperation of China Power Transmission and Transformation Technology." http://www.unescap.org/sites/default/files/2.5Ms_HuangXinnan_TBEA_0.pdf.

Teach a Man to Fish. 2010. "Making and Selling Solar Cookers." January. http://www.teachamantofish.org.uk/resources/incomegeneration/Solar-Cooker-Business-Guide.pdf.

Temte, Dean. 2010. "Wyoming Severance Taxes and Federal Mineral Royalties." http://legisweb.state.wy.us/budget/wyosevtaxes.pdf.

Tenenbaum, B., R. Lock, and J. Barker. 1992. "Electricity Privatization: Structural, Competitive and Regulatory Options." *Energy Policy* (December): 1134–1160.

Thakar, Nidhi. 2008. "The Urge to Merge: A Look at the Repeal of the Public Utility Holding Company Act of 1935." *Lewis and Clark Law Review* 12 (3): 903–942.

Thomas, Steve. 2001. *The Wholesale Electricity Market in Britain—1990–2001.* August. Public Services International Research Unit (PSIRU), School of Computing and Mathematical Sciences, University of Greenwich. London: PSIRU. http://www.docstoc.com/docs/28221807/ The-Wholesale-Electricity-Market-in-Britain---1990–2001.

Thompson, Eric V. n.d. *A Brief History of Major Oil Companies in the Gulf Region with Corporate Contact Information.* Institute for Global Policy Research, University of Virginia: http//www.virginia.edu/igpr/apagoilhistory.html.

Tietenberg, Tom, and Lynne Lewis. 2011. *Environmental & Natural Resource Economics.* 9th ed. New York: Prentice Hall.

Tippee, Bob. 2012. "IHS Indexes Show Near-Record High Upstream Costs." *Oil & Gas Journal* (July 2). http://www.ogj.com/articles/2012/07/ihs-indexes-show-record-high-upstream-costs.html.

Tiratsoo, Eric Neshan. 1986. *Oilfields of the World.* Houston, TX: Gulf Publishing Company.

Tirole, Jean. 1998. *The Theory of Industrial Organization.* Cambridge, MA: MIT Press.

Tissot, B. P., and D. H. Welte. 1984. *Petroleum Formation and Occurrence.* New York: Springer-Verlag.

Tobin, James. 2001. *Natural Gas Transportation—Infrastructure Issues and Operational Trends.* Natural Gas Division, US EIA. October. http://www. eia.doe.gov/pub/oil_gas/natural_gas/analysis_publications/natural_gas_ infrastructure_issue/pdf/nginfrais.pdf.

Todorova, Vesela. 2011. "Abu Dhabi's Renewable Energy Target Can Be Reached." *The National* (January 13). http://www.thenational.ae/news/uae-news/ environment/abu-dhabis-renewable-energy-target-can-be-reached.

Tolf, Robert. 1976. *The Russian Rockefellers: The Saga of the Nobel Family and the Russian Oil Industry.* Stanford, CA: Hoover Institution Press, Stanford University.

Tortora, Bob, and Magali Rheault. 2012. "In Africa, Power Reliability Similar for All Business Sectors." *Gallup World* (January 11). http://www.gallup.com/ poll/151973/africa-power-reliability-similar-business-sectors.aspx.

TransCanada. 2007. "TransCanada Reports 2006 Net Income of $1.1 Billion— Board of Directors Increases Quarterly Dividend." January 30. http://www. transcanada.com/3088.html.

———. 2011. "TransCanada Announces Completion of the Guadalajara Pipeline." June 20. http://www.transcanada.com/5776.html.

———. 2012. "TransCanada Awarded Contract to Build US$500 Million Natural Gas Pipeline Extension in Mexico." February 24. http://www.transcanada.com/5965.html.

Transparency International. 2013. "Corruption Perceptions Index 2013." http://cpi.transparency.org/cpi2013/results/.

Trompenaars, Fons. 1993. *Riding the Waves of Culture: Understanding Cultural Diversity in Business*. London: Nicholas-Brealey.

True, Warren R. and Leena Koottungal. 2011. "Global Capacity Growth Reverses; Asian, Mideast Refineries Progress." *Oil & Gas Journal*. (December 5): 30–43.

Tu, JianJun. 2010. "Industrial Organization of the Chinese Coal Industry." Working Paper #103. Palo Alto, CA: Stanford University. Draft version available at http://www.emrg.sfu.ca/media/publications/China_Coal_Value_Chain_J.Tu_Draft.pdf.

Tu, Kevin Jianjun, and Sabine Johnson-Reiser. 2012. *Understanding China's Rising Coal Imports*. Carnegie Endowment for International Peace. February 16. http://www.carnegieendowment.org/files/china_coal.pdf.

Turkle, Sherry. 2011. *Alone Together: Why We Expect More from Technology and Less From Each Other*. New York: Basic Books.

TVA (Tennessee Valley Authority) n.d. "Hydroelectric Dam." http:\www.tva.gov/power/hydroart.htm.

Tverberg, Gail. 2012. "The Long-Term Tie between Energy Supply, Population, and the Economy." *The Oil Drum* (September 5). http://www.resilience.org/stories/2012-09-05/long-term-tie-between-energy-supply-population-and-economy.

UN (United Nations). 2010. *Annual Questionnaire on Energy Statistics*. http://unstats.un.org/unsd/energy/quest.htm.

———. n.d.a. "Energy Statistics Database." The latest update of this database is at http://unstats.un.org/unsd/energy/edbase.htm.

———. n.d.b. *Human Development Index Database*. The latest update of this database is at http://hdrstats.undp.org/en/hdi/.

UN Conference on Trade and Development Secretariat. 2011. *Review of Maritime Transport 2011*. http://unctad.org/en/docs/rmt2011ch1_en.pdf.

UNFAO (UN Food and Agriculture Organization). n.d. "FaoStat Database." The latest update of this database is at http://faostat3.fao.org/faostat-gateway/go/to/home/E.

UNFCCC (UN Framework Convention on Climate Change). 2008. *Kyoto Protocol Reference Manual on Accounting of Emissions and Assigned Amount*. http://unfccc.int/resource/docs/publications/08_unfccc_kp_ref_manual.pdf.

———. 2013a. "Appendix I: Quantified Economy-Wide Emissions Targets for 2020." http://unfccc.int/meetings/copenhagen_dec_2009/items/5264.php.

———. 2013b. "Appendix II: Nationally Appropriate Mitigation Actions of Developing Country Parties." http://unfccc.int/meetings/cop_15/copenhagen_accord/items/5265.php.

———. 2013c. "Doha Amendment to the Kyoto Protocol." https://unfccc.int/files/kyoto_protocol/application/pdf/kp_doha_amendment_english.pdf.

———. 2013d. "List of Annex I Parties to the Convention." http://unfccc.int/parties_and_observers/parties/annex_i/items/2774.php.

———. 2013e "Status of Ratification of the Kyoto Protocol." http://unfccc.int/kyoto_protocol/status_of_ratification/items/2613.php.

University of California Museum of Paleontology. n.d. "Creating an Earth System II: The Electromagetic Spectrum." http://www.ucmp.berkeley.edu/education/dynamic/session5/sess5_electromagnetic.htm.

US BEA (Bureau of Economic Analysis). n.d. "U.S. Economic Accounts." The most recent version of this database is at http://www.bea.gov.

US BLS (Bureau of Labor Statistics). n.d. "Consumer Price Index." The most recent version of this database is at http://www.bls.gov/cpi/data.htm.

US BTS (Bureau of Transportation Statistics). n.d. *National Transportation Statistics.* The most recent version of this database is at http://www.rita.dot.gov/bts/sites/rita.dot.gov.bts/files/publications/national_transportation_statistics/index.html.

US CBO (Congressional Budget Office). 1982. "Accelerating Oil Acquisition for the Strategic Petroleum Reserve." Staff Working Paper. May. Washington, DC: Government Printing Office (GPO).

US CEA (Council of Economic Advisors). 1997. *Economic Report of the President.* Washington, DC: US GPO. http://www.presidency.ucsb.edu/economic_reports/1997.pdf.

US Census Bureau. 2011. "Historical Census of Housing Tables." http://www.census.gov/hhes/www/housing/census/historic/units.html.

———. 2013. *State Government Tax Collections.* http://factfinder2.census.gov/faces/tableservices/jsf/pages/productview.xhtml?src=bkmk.

US CFTC (Commodity Futures Trading Commission). 2012. "This Month in Futures Markets." November. http://www.cftc.gov/OCE/WEB/files/OCE_COT_Report_November_2012.pdf.

———. 2013. "CFTC Announces Real-Time Public Reporting of Swap Transactions and Swap Dealer Registration Began on December 31, 2012." January 2. http://www.cftc.gov/PressRoom/PressReleases/pr6489-13.

———. n.d. "Home Page." http://www.cftc.gov/.

US CIA (Central Intelligence Agency). 2010. *The World Factbook.* https://www.cia.gov/library/publications/download/download-2010/index.html.

US DOC (Department of Commerce). 1975. *Historical Statistics of United States: Colonial Times to 1970.* Part I and Part II. Washington, DC: US DOC. http://www2.census.gov/prod2/statcomp/documents/CT1970p1-01.pdf, http://www2.census.gov/prod2/statcomp/documents/CT1970p2-01.pdf.

US DOE (Department of Energy). circa 2008. *The Smart Grid: An Introduction.* http://energy.gov/sites/prod/files/oeprod/DocumentsandMedia/DOE_SG_Book_Single_Pages%281%29.pdf.

US DOL (Department of Labor). n.d. "Green Job Hazards." Washington, DC. http://www.osha.gov/dep/greenjobs/.

US EIA (Energy Information Administration). 1995. *Coal Data: A Reference.* EIA-0064. Washington, DC: US GPO. February.

———. 1997. *Effects of Title IV of the Clean Air Act Amendments of 1990 on Utilities: An Update.* EIA-0582. ftp://ftp.eia.doe.gov/pub/electricity/ef_caau1.pdf.

———. 2003a. "Photovoltaic Milestones." March 7. http://www.eia.gov/cneaf/solar.renewables/renewable.energy.annual/backgrnd/chap11i.htm.

———. 2003b. "The Global Liquefied Natural Gas Market: Status and Outlook." December. http://www.eia.gov/oiaf/analysispaper/global/lngindustry.html.

———. 2008. *Distribution of Natural Gas: The Final Step in the Transmission Process.* http://www.eia.gov/pub/oil_gas/natural_gas/feature_articles/2008/ldc2008/ldc2008.pdf.

———. 2009a. *Light-Duty Diesel Vehicles: Market Issues and Potential Energy in Emission Impacts.* January. Washington, DC: US EIA. http://www.eia.gov/analysis/requests/2009/lightduty/pdf/sroiaf(2009)02.pdf.

———. 2009b. "Natural Gas Market Centers and Hubs in Relation to Major Natural Gas Transportation Corridors, 2009." April. http://www.eia.gov/pub/oil_gas/natural_gas/analysis_publications/ngpipeline/MarketCenterHubsMap.html.

———. 2009c. *The National Energy Modeling System: An Overview 2009.* October. Washington, DC: US EIA. http://www.eia.gov/forecasts/archive/0581(2009).pdf.

———. 2010a. "Status of Electricity Restructuring by State." September. http://205.254.135.7/cneaf/electricity/page/restructuring/restructure_elect.html.

———. 2010b. "Natural Gas Residential Choice Programs by State as of December 2009." May 17. http://www.eia.gov/oil_gas/natural_gas/restructure/restructure.html.

———. 2011a. "Voluntary Reporting of Greenhouse Gases Program Fuel Emission Coefficients." January. http://www.eia.gov/oiaf/1605/coefficients.html.

———. 2011b. *Review of Emerging Resources: U.S. Shale Gas and Shale Oil Plays.* July. Washington, DC: US EIA. http://www.eia.gov/analysis/studies/usshalegas/pdf/usshaleplays.pdf.

———. 2011c. *Electric Power Annual 2010*. November. Washington, DC: US EIA. http://www.eia.gov/electricity/annual/archive/03482010.pdf.

———. 2012a. "OPEC Spare Capacity in the First Quarter of 2012 at Lowest Level since 2008." May 24. http://www.eia.gov/todayinenergy/detail.cfm?id=6410.

———. 2012b. *Annual Energy Outlook 2012: With Projections to 2035*. EIA-0383. June 25. Reference Table 4. Washington, DC: US EIA. http://www.eia.gov/forecasts/archive/aeo12/.

———. 2012c. "World Oil Transit Chokepoints." August 22. http://www.eia.gov/countries/regions-topics.cfm?fips=wotc&trk=p3.

———. 2012d. *Annual Energy Review 2011*. September. Washington, DC: US EIA. http://www.eia.gov/totalenergy/data/annual/pdf/aer.pdf.

———. 2012e. *Annual Coal Report*. November 8. Washington, DC: US EIA. http://www.eia.gov/coal/annual/.

———. 2013a. "How Much Electricity Is Used for Lighting in the United States?" January 9. http://www.eia.gov/tools/faqs/faq.cfm?id=99&t=3.

———. 2013b. *Annual Energy Outlook Retrospective Review: Evaluation of 2012 and Prior Reference Case Projections*. March. http://www.eia.gov/forecasts/aeo/retrospective/pdf/retrospective.pdf.

———. 2013c. *Updated Capital Cost Estimates for Utility Scale Electricity Generating Plants*. April. Washington, DC: US EIA. http://www.eia.gov/forecasts/capitalcost/pdf/updated_capcost.pdf.

———. 2013d. *Monthly Energy Review*. October. http://www.eia.gov/totalenergy/data/monthly/pdf/mer.pdf.

———. 2014a. *Annual Energy Outlook 2014*. May. http://www.eia.gov/forecasts/aeo.

———. 2014b. "Thermal Conversion Factor Source Data." *Monthly Energy Review*. June. http://www.eia.gov/totalenergy/data/monthly/pdf/sec13_a-doc.pdf.

———. n.d.a. "Country Analysis Briefs." The latest update of this database is at http://www.eia.gov/countries/reports.cfm.

———. n.d.b. "Data Tools and Models: Time Series." The latest update of this database is at http://www.eia.gov/tools/models/timeseries.cfm.

———. n.d.c. "Energy & Financial Markets: What Drives Crude Oil Prices?" Monthly and quarterly updates are posted at http://www.eia.gov/finance/markets/.

———. n.d.d. "International Energy Statistics." The latest update of this database is at. http://www.eia.gov/countries/data.cfm.

———. n.d.e. "Model Documentation." The latest update of model documentation is at http://www.eia.gov/reports/index.cfm?t=Model%20Documentation.

———. n.d.f. "U.S. States: State Profiles and Energy Estimates." The latest update of this database is at http://www.eia.gov/state/state-energy-profiles-notes-sources-data.cfm.

US EPA (Environmental Protection Agency). 1999. *SO$_2$, NO$_x$, Heat Input, and CO$_2$ Emissions Trend in the Electric Utility Industry.* Air and Radiation Acid Rain Division (6204J). EPA-430-R-98-020. January. Washington, DC: US EPA.

———. 2002. "Clean Energy: eGRID 2002 Archive." http://www.epa.gov/cleanenergy/energy-resources/egrid/archive.html.

———. 2007. "Summary of Current and Historical Light-Duty Vehicle Emission Standards." April. http://www.epa.gov/greenvehicles/detailedchart.pdf.

———. 2012a. "Light-Duty Vehicle and Light-Duty Truck—Tier 0, Tier 1, National Low Emission Vehicle (NLEV), and Clean Fuel Vehicle (CFV) Exhaust Emission Standards." November 14. http://www.epa.gov/oms/standards/light-duty/tiers0-1-ldstds.htm.

———. 2012b. "Light-Duty Vehicle, Light-Duty Truck, and Medium-Duty Passenger Vehicle—Tier 2 Exhaust Emission Standards." November 14. http://www.epa.gov/otaq/standards/light-duty/tier2stds.htm.

———. 2013a "Overview of Greenhouse Gases." September 9. http://epa.gov/climatechange/ghgemissions/gases/ch4.html.

———. 2013b. "Greenhouse Gas Equivalencies Calculator. " November 26. http://www.epa.gov/cleanenergy/energy-resources/calculator.html.

———. 2013c "Green Building." November 28. http://www.epa.gov/greenbuilding/.

US Federal Register. 2014. "Carbon Pollution Emission Guidelines for Existing Stationary Sources: Electric Utility Generating Units." June 18. https://www.federalregister.gov/articles/2014/06/18/2014-13726/carbon-pollution-emission-guidelines-for-existing-stationary-sources-electric-utility-generating#h-9.

US FERC (Federal Energy Regulatory Commission). 2007. *Report to Congress on Competition in Wholesale and Retail Markets for Electric Energy.* Pursuant to Section 1815 of the Energy Policy Act of 2005. http://energy.gov/oe/downloads/report-congress-competition-wholesale-and-retail-markets-electric-energy.

———. 2010. *2009 Analysis of Physical Gas Market Transactions.* December 16. http://www.ferc.gov/market-oversight/reports-analyses/special-analyses/gas-transactions2009.pdf.

———. 2012. "RTO/ISO Performance Metrics." September 18. http://www.ferc.gov/industries/electric/indus-act/rto/rto-iso-performance.asp.

———. 2013. *Regulated Entities.* http://www.ferc.gov/industries/gas/gen-info/reg-ent.asp.

USGS (US Geological Survey). 1995. *1995 National Assessment of United States Oil and Gas Resources*. US Geological Survey Circular 1118. http://greenwood. cr.usgs.gov/energy/circ1118.pdf.

———. 2012. *An Estimate of Undiscovered Conventional Oil and Gas Reserves of the World, 2012*. March. World Petroleum Resources Project. http://pubs.usgs. gov/fs/2012/3042/fs2012-3042.pdf.

US National Academies. 2009. "G8+5 Academies' Joint Statement: Climate Change and the Transformation of Energy Technologies for a Low Carbon Future." May. http://www.nationalacademies.org/includes/G8+5energy-climate09.pdf.

US NMA (National Mining Association). n.d. "Home Page." http://www.nma.org/.

US NRC (Nuclear Regulatory Commission). n.d. "Home Page." http://www.nrc.gov.

US SBA (Small Business Administration). 2013. "Borrowing Money for Your Business." http://www.sba.gov/content/borrowing-money.

van der Plas, R. 1995. "Burning Charcoal Issues." FPD Energy Note 1. April, Washington, D.C.: The World Bank Group. Quoted in Chambwera 2004.

van der Ploeg, Frederick. 2011. "Natural Resources: Curse or Blessing?" *Journal of Economic Literature* 49 (2): 366–420. doi:10.1257/jel.49.2.366.

Van Dyke, Kate. 1997. *Fundamentals of Petroleum*. 4th ed. Austin, TX: University of Texas, Petroleum Extension Service.

Vasquez Cordano, Arturo. 2012. *The Regulation of Oil Spills and Mineral Pollution: Policy Lessons for the U.S.A. and Peru from the Deep Water Horizon Blowout and Other Accidents*. Berlin: Lambert Academic Publishing.

Vatansever, Adnan. 2010. *Russia's Oil Exports: Economic Rationale Versus Strategic Gains*. Carnegie Endowment for International Peace. Energy and Climate Program, Carnegie Papers #116, December.

Viscusi, W. K., John Vernon, and J. Harrington Jr. 1995. *Economics of Regulation and Antitrust*. 2nd ed. Cambridge, MA: MIT Press.

Wälde, Thomas W., ed. 1996. *The Energy Charter Treaty: An East-West Gateway for Investment and Trade*. Boston: Kluwer Law International.

Wang, D., R. Butler, H. Liu, and S. Ahmed. 2011. "Surfactant Formulation Study for Bakken Shale Imbibition." Presented at the SPE Annual Technical Conference and Exhibition, October 30–November 2. http://www.und.edu/ instruct/dwang/Research/SPE-145510.pdf.

Wang, Zhendi, B. P. Hollebone, M. Fingas, B. Fieldhouse, L. Sigouin, M. Landriault, P. Smith, J. Noonan, G. Thouin, and James W. Weaver. 2003. *Characteristics of Spilled Oils, Fuels, and Petroleum Products: 1. Composition and Properties of Selected Oils*. US EPA. http://www.epa.gov/athens/publications/reports/ EPA-600-R03-072-OilComposition.pdf.

Watson, Joel. 2013. *Strategy: An Introduction to Game Theory*. 3rd ed. New York: W. W. Norton & Company, Inc.

WEC (World Energy Council). 2010. 2010 Survey of Energy Resources. http://www.worldenergy.org/publications/2010/survey-of-energy-resources-2010.

——— 2013. "World Energy Resources: 2013 Survey." http://www.worldenergy.org/wp-content/uploads/2013/09/Complete_WER_2013_Survey.pdf.

Weiner, Robert J. 1991. "Origins of Future Trading: The Oil Exchanges in the 19th Century." Working Paper. Department of Economics. Waltham, MA: Brandeis University.

Wilbanks, Thomas J., Paul Leiby, Robert Perlack, J. Timothy Ensminger, and Sherry B. Wright. 2007. "Toward an Integrated Analysis of Mitigation and Adaptation: Some Preliminary Findings." *Mitigation and Adaptation Strategies for Global Change* 12: 713–725. http://citeseerx.ist.psu.edu/viewdoc/download?doi=10.1.1.132.1271&rep=rep1&type=pdf.

Williamson, Oliver E. 1993. "Transaction Cost Economics Meets Posnerian Law and Economics." *Journal of Institutional and Theoretical Economics* 149 (1): 99–118.

Willrich, Mason. 2009. "Electricity Transmission Policy for America: Enabling a Smart Grid, End-to-End." MIT-IPC-0. Industrial Performance Center, Massachusetts Institute of Technology. http://web.mit.edu/ipc/research/energy/pdf/EIP_09-003.pdf.

Winston, C. 1993. "Economic Deregulation: Days of Reckoning for Microeconomists." *Journal of Economic Literature* 31 (3): 1263–1289.

Wong, Fayen. 2014. "China Targeting 2014 Launch of First Crude Oil Futures." Reuters. http://www.reuters.com/article/2014/05/28/china-crude-futures-idUSL3N0OE1LA20140528.

World Bank. n.d. *World Development Indicators*. Washington, DC: World Bank. The most recent update of this database is at http://data.worldbank.org/data-catalog/world-development-indicators.

World Coal Institute. n.d. "Coal Market & Transportation." http://www.worldcoal.org/coal/market-amp-transportation/.

World Nuclear Association. 2012. "Supply of Uranium." August. http://www.world-nuclear.org/info/inf75.html.

———. 2013a. "Nuclear Development in the United Kingdom." January. http://www.world-nuclear.org/info/Country-Profiles/Countries-T-Z/Appendices/Nuclear-Development-in-the-United-Kingdom/#.UYbGQrVlnpU.

———. 2013b. "Mixed Oxide (MOX) Fuel." May. http://www.world-nuclear.org/info/inf29.html.

———. 2013c. "World Uranium Mining Production." July. http://www.world-nuclear.org/info/inf23.html.

———. 2013d. "Conversion and Deconversion." September. http://www.world-nuclear.org/info/Nuclear-Fuel-Cycle/Conversion-Enrichment-and-Fabrication/Conversion-and-Deconversion/#.UZ7QbLVll7c.

———. 2013e. "Processing of Used Nuclear Fuel." September. http://www.world-nuclear.org/info/Nuclear-Fuel-Cycle/Fuel-Recycling/Processing-of-Used-Nuclear-Fuel/.

———. 2013f. "Uranium Enrichment." October. http://www.world-nuclear.org/info/Nuclear-Fuel-Cycle/Conversion-Enrichment-and-Fabrication/Uranium-Enrichment/#.UZ7TN7Vll7d.

———. 2013g. "World Nuclear Power Reactor & Uranium Requirements." October. http://www.world-nuclear.org/info/reactors.html.

———. 2014a. "Nuclear Development in the United Arab Emirates." April. http://www.world-nuclear.org/info/Country-Profiles/Countries-T-Z/United-Arab-Emirates/.

———. 2014b "Nuclear Power in Iran." May. http://www.world-nuclear.org/info/Country-Profiles/Countries-G-N/Iran/.

WSJ (Wall Street Journal) Market Data Group. n.d "Market Data Center." Find the latest quotes at http://online.wsj.com/mdc/public/page/2_3023–fut_metal–futures.html.

Xu, Conglin and Laura Bell. 2013. "Worldwide Reserves, Oil Production Post Modest Gain." *Oil & Gas Journal* (December 2): 30–33.

Xu, Shaofeng, and Wenying Chen. 2006. "The Reform of Electricity Power Sector in the PR of China." *Energy Policy* 34: 2455–2465.

Yergin, Daniel. 1991. *The Prize: The Epic Quest for Oil, Money, and Power.* New York: Simon and Schuster.

———. 2011. *The Quest: Energy, Security, and the Remaking of the Modern World.* London: Penguin Group.

Zhou, Moming, and Alistair Holloway. 2011. "LNG-Tanker Rates Doubling as Ship Glut Erodes: Freight Markets." February 16. http://www.bloomberg.com/news/2011-02-16/lng-tanker-rates-seen-doubling-as-ship-glut-disappears-freight-markets.html.

Zickel, Raymond E., ed. 1991. *Soviet Union: A Country Study.* Federal Research Division, US Library of Congress. http://lcweb2.loc.gov/frd/cs/sutoc.html.

Index

F

G

game theory
    Bertrand model, 264, 270, 273
    Cournot duopoly, 264–267, 272–273
    duopoly compared to competitive market, 267–269
    limit pricing model, 271
    monopoly compared to duopoly market, 269–270
    Stackelberg model, 264, 270, 273
game theory models, 39
gas
    aboveground costs for, 383–387, 394
    inground costs for, 383–387
gas and oil, aboveground costs for, 383–387, 394
gas consumers, 203
gas marketers, 201
gas oil, 414, 441, 445
gas shortage, 144–145
gas turbines, 105, 139
gasoline blending, 444–446, 447
    octane number, 444
    raising oxygen levels, 445, 463–464
    Reid vapor pressure, 444, 445, 463
gas-to-liquids (GTL) technology, in Qatar plant, 246, 255–256, 356, 403
gathering pipelines, 461
Gazprom natural gas company, 251, 259, 263–264
generating companies (gencos), 135
generation capacity, 122, 126, 131, 144, 147, 225
generation diversity, 142
generation plant outages, 144
gentailers, in New Zealand, 135
Georges Besse II nuclear power plant, 373
geothermal, solar, tide, and wave (Geo&STW), 257, 258
geothermal energy, 377, 381–383, 393
Getty, 166
Giant Troll field, 32
Gini coefficient, 523–525, 537, 567
glasnost, 32
glass and glass products energy-intensive industry, 406
global carbon policy, 297–298, 308–311
global climate change, 8, 9, 275
Global Laser Enrichment (GLE), 373
global pollutants, 292
global warming, 184, 292, 296, 297, 311, 377, 378

Global Warming Solutions Act (2006), of California, 147
globalization, 578
going long, 489
Goldfinger, Steve, 511
governance structures, 7
    bilateral governance, 192, 195
    market governance, 192
    trilateral governance, 193, 196
    unified governance, 193, 196
government
    disruptions response by, 325–327, 333–334
    energy security and, 325–327, 330, 334
    energy stockpiles, 325, 326–327, 334
    nuclear power promotion by, 330–333
    pollution policies, 280
    production sharing agreement, 360
    role, resource rent and, 569–570
    role increase, in fossil fuel-rich countries, 569–570
    safety regulations, 330, 334
    taxing and bidding decisions, 78, 357–362, 365
government controls, 75–78, 357
    in European natural gas market, 259
government failure, 307–308, 313
government ownership
    corporation, 141, 151–152
    of mineral rights, 82, 357, 364–365
government policies, 49, 69, 70
    for natural monopolies, 111
    on pollution, 280
government taxes, 78, 357–362, 365
government-held stocks, 325, 326–327
government-owned corporation, 141, 151–152
gravity, 2
Gray, John, 584
Great Leap Forward, China, 516
greenhouse gases (GHG), 291, 292, 308, 310, 311, 312, 377
grid-connected solar photovoltaic cell (PV), 380
Groningen natural gas field, 248
gross electricity generation, 94, 386
gross inland electricity consumption, 94
growth, in electricity consumption, 96
growth models, with nonrenewable resources, 573
guided missile culture, 596, 604
guilt-based culture, 585, 603
Gulf, 154, 166, 171
Gulf Intracoastal Waterway, US, 458

H

having face, 584
head, for hydropower, 375
heat content, of electricity, 95
heavy crudes and bitumen, 21–22
hedge contracts, 136
hedgers, 473, 474, 482, 489
hedging, 473–475
hedonic pricing, 288, 289
Henry Hub, 205, 206, 208, 214, 224, 475, 491, 493, 496
Herbert, John H., 482
Herfindahl-Hirschman index of concentration (HHI), 140
hertz, 4
heuristic bias, 306
Hicksian rent, 344
high absorbers, 180–181
high prices, speculation and, 484–488
high-voltage AC (HVAC) transmission line, 452–453
Hines, R. D., 584
historical trends, 28–29
Hofstede, Geert, 579, 582, 587
holding companies, 197
holdup, 195
homo economicus of economic theory, 305
homo sapiens, 305
homogenous of degree zero in income and prices, 419–420
homogenous products in competitive markets, 51
Hooker, J. N., 578, 585–586
Hotelling rent, 344, 349, 355, 357, 363, 365
household or consumer demand, 417–426, 434
Hubbert's model on total energy resources, 391–392
Human Development index (HDI), 523, 566
human nature, 590
humans and nature, 590
Hunt, S., 127, 128, 260, 263
hydrocarbons (HO), 274–275, 277
hydrocracking, 442
hydrofluorocarbons (HFCs), 309
hydropower, 95, 97, 129, 257, 324, 393, 513, 530
    in California, 143, 144
    from dam, 375–376
    head, 375
    in New Zealand, 135, 136, 139
    in Norway, 136, 140, 238
    problems from, 531–532, 538
    production, 375–377, 403–404
    Shimantan Dam failure, 317
    in Sweden, 138
hydrotreating of crude oil, 439, 463

I

ideal market, 38, 48
idiosyncratic investment, 193
IEA. *See* International Energy Agency
import-export monopoly, in Norway, 137
in the money option, 11, 494, 505, 506
incandescent lightings cost, 300–302
incentive-based policies, on pollution, 280–281
incidence of tax, 89
income consumption curve, 423
income elasticities, 63–64
income frame regulation, 137
income growth, 32
income inelasticity, 63
increasing costs, 101
incubator culture, 596, 604
independent power companies (IPCs), 130
independent power producers (IPPs), 130, 143
independent system operators (ISOs), 128, 143–144, 147, 149
index pricing, 208, 211
indifference curve, 420–422
individualism/collectivism (IND), 579
industrial capital, 335
industrial sector, in total final consumption of energy, 402
industry restructuring
    in California, 142
    of natural gas, 186–191
inferior goods, 64
inflation, 117
inflation adjustment, 113
information, as public good, 305
information asymmetries, 193, 195, 196
information disclosure rates, in New Zealand, 136
information flow, 591
information revolution, 597
information technologies, 33, 142, 305, 466, 605
inground costs, for gas and oil, 383–387
inner-directed (IN) culture, 588
input-output models, 33–36

Q

R

T

X

Y

Z